To test for DC backup or liquid flooding:
- Reboiler in manual, turn up reflux

To test for entrainment flooding:
- reflux in manual, turn up heat

Signs of Flooding:
↓
accum of
liquid

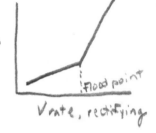

ΔP

Flood point

V rate, rectifying

- high ΔP > 0.2 psi/tray

- For top tray only, overall ΔP will ↑ by 0.5 psi + ↓ also in cycles

- reduction of bottoms flowrate, liquid not coming down

- loss of separation, temp profile

- entrainment • more reflux

Distillation
Operation

Distillation Operation

Henry Z. Kister
C. F. Braun, Inc.

Boston, Massachusetts   Burr Ridge, Illinois
Dubuque, Iowa   Madison, Wisconsin   New York, New York
San Francisco, California   St. Louis, Missouri

Library of Congress Cataloging-in-Publication Data

Kister, Henry Z.
  Distillation operation.

  1. Distillation   I. Title.
  TP156.D5K56   1989   660.2'8425   88-37746
  ISBN 0-07-034910-X

McGraw-Hill

A Division of The McGraw-Hill Companies

Printed and bound by Book-mart Press, Inc.

        13 14 15 16 17 18 19   BKM BKM   9 0 9 8 7 6 5 4 3 2 1

ISBN 0-07-034910-X

*The editors for this book were Robert Hauserman and Dennis
Gleason, the designer was Naomi Auerbach, and the production
supervisor was Richard A. Ausburn. This book was set in Cen-
tury Schoolbook. It was composed by the McGraw-Hill Publishing
Company Professional & Reference Division composition unit.*

*For more information about other McGraw-Hill materials,
call 1-800-2-MCGRAW in the United States. In other
countries, call your nearest McGraw-Hill office.*

To those who contributed so much to my work over the years—my mother, my father, and Gillian.

Contents

Preface

*"Man will occasionally stumble over the
truth, but most of the time he will pick
himself up and continue on."*
WINSTON CHURCHILL,
Commentary on Man

In an age when sophisticated computer technology has taken over the
design, and often the operation, of distillation columns, the fundamental practical aspects of distillation are rapidly becoming a forgotten
art. Rules of thumb and McCabe-Thiele diagrams, once the main tools
in the hands of the distillation practitioner, are rapidly being replaced
by powerful mathematics. Yet, when a column experiences problems
in the field, it is neither the computer nor the powerful mathematics
that is able to reinstate trouble-free operation. Instead, it is the person
in a pair of worn-out overalls with bags under his (or her) eyes from
spending 16 hours on the plant the night before who finally solves the
problem. This person rarely uses an electronic wizardry. His (or her)
main tools are a good understanding of the plant and its equipment,
a great deal of experience, good engineering judgment, and good systematic reasoning. This book is dedicated to that person—the distillation troubleshooter.

Well over a decade ago, after designing and revamping several distillation columns, I was given the duties of operating and starting up
a distillation unit. My new boss had one simple warning: nine out of
ten problems that occur in distillation equipment are not caused by
poor design or inadequate theory, but by overlooking some practical
guideline. I asked him if he knew of a text where such guidelines are
documented. He made one request: if I ever found one, he would like
to get a copy.

In the following decade, I spent a lot of time talking to people, collecting experiences and guidelines, combing the literature for published articles describing typical problems, and often calling or writing

to their authors to discuss these problems. I converted their stories into notes, molded the notes into guidelines, turned the guidelines into chapters, and the chapters into a book.

The purpose of this book is to show what has gone wrong in the past and to suggest means of preventing similar incidents in the future. Unfortunately, the process industries have short memories: people move on, and lessons are forgotten. It is my hope that the book retains these memories and makes the experience of several generations available to those who operate, start up, and design distillation and absorption equipment.

The suggestions and recommendations made in this book are given in good faith, but without any warranty. What works in one case may not work under even slightly different circumstances. In fact, you may even feel that some of the recommendations are not appropriate for your installation. Congratulations. Good chefs do not blindly follow recipes, even when they like the recipes. Likewise, you must critically examine any suggestions, recommendations, or ideas before utilizing them in a specific application.

If you have chosen a career as a designer or troubleshooter, you have selected a tough road. Many long, sleepless nights at the plant are ahead. You will spend many worrisome moments wondering whether your design is going to work. Many times your expectations will be shattered when a "fix" you could swear by does not make an ounce of difference. Your journey will be tiring and rough, yet exciting and strangely rewarding. Perhaps this book can help smooth your path.

Henry Z. Kister

Acknowledgments

The author wishes to express his gratitude to C. F. Braun, Inc., particularly to Walt Stupin, for their backing and support for this effort. Special thanks are also due to my parents for their warm encouragement and inspiration throughout, and for the German translations. Appreciation is also due to my mentors over the years, particularly Ian Doig, Dick Harris, Trevor Whalley, Reno Zack, Tom Hower, Walt Stupin, and Jack Hull, for their immense encouragement of my work; much of their teaching found its way into the following pages.

The author is grateful to the many contributors that willingly shared their experiences with him, and whose stories erected the foundations for many useful guidelines in this book. Special thanks are due to Tom Hower, Scott Golden, Derek Reay, Layton Kitterman, and Norm Lieberman, who were particularly enthusiastic about contributing their experiences to this work. Thanks are also due to Jim Litchfield for his invaluable comments on Chapter 9, and to Fred Ledeboer for his help with the section on forced-circulation reboilers. Finally, the author gratefully acknowledges the efforts on Sherlyn Severson, Karen Moody, and Sandy Shemula who tirelessly and flawlessly typed this manuscript.

The author will be pleased to hear any comments or experiences any readers may wish to share for possible inclusion in a future edition. Write to H. Z. Kister, C. F. Braun, Inc., 1000 S. Fremont Ave., Alhambra, CA 91802, USA. All communications will be answered.

Distillation Troubleshooting

A well-known sales axiom states that 80 percent of the business is brought in by 20 percent of the customers. A sales strategy tailored for this axiom concentrates the effort on the 20 percent of the customers without neglecting the others. Distillation operation and troubleshooting follow an analogous axiom. A person engaged in operating or troubleshooting distillation columns must develop a good feel and understanding for the factors that cause the vast majority of column malfunctions. For these factors, this person must be able to distinguish good from poor practices and correctly evaluate the ill effects of poor practices and their relevance to the assignment at hand. While a good knowledge and understanding of the broader field of distillation will be beneficial, the troubleshooter can often get by with a shallow knowledge of this broader field.

It is well accepted that troubleshooting is a primary job function of operating engineers and supervisors. Far too few realize that distillation troubleshooting starts at the design phase. Any designer wishing to achieve a trouble-free column design must be as familiar with troubleshooting and operation as the person running the column.

The next 19 chapters journey through the vast majority of factors that cause column malfunctions. They distinguish good from poor practices and propose guidelines for avoiding and overcoming troublesome designs and operation. Prior to embarking on this journey, it is necessary to define the problem areas and examine the tools available for uncovering malfunctions. This first chapter focuses attention on these aspects.

This chapter first reviews the common causes of column malfunctions and shows where these causes fit in the book. It then looks at the basic troubleshooting tools: the systematic strategy for troubleshooting distillation problems and the dos and don'ts for formulating and testing theories. Detailed techniques for testing distillation columns are described in Chap. 14.

1.1 Causes of Column Malfunctions

Close to 300 case histories of malfunctioning columns were extracted from the literature and abstracted in Chap. 20. Table 1.1 classifies the malfunctions described in these case histories according to their principal causes. If one assumes that these case histories make up a representative sample (note exclusions and limitations in Chap. 20), then the analysis below has statistical significance. Accordingly, Table 1.1 can provide a useful guide to the factors most likely to cause column malfunctions and can direct troubleshooters toward the most likely problem areas.

Note that the general guidelines in Table 1.1 often do not apply to a specific column or even plant. For instance, foaming is one of the least likely causes of column problems according to Table 1.1; however, in amine absorbers foaming is a common trouble spot. The author therefore warns against blindly applying the guides in Table 1.1 to any specific situation.

An analysis of Table 1.1 suggests the following:

- Instrument and control problems, startup and/or shutdown difficulties, and malfunctioning column internals are the major single

TABLE 1.1   Causes of Column Malfunctions (Based on the Analysis of Case Histories in Chap. 20)

Cause	Number of reported cases	Percent of reported cases	Reference chapter
Instrument and control problems	52	18	5, 16–19
Troublesome column internals	51	17	2–5, 7, 8
Startup and/or shutdown difficulties	48	16	11–14
Operational difficulties	38	13	13,14
Reboilers, condensers	28	9	15
Primary design: VLE, column size, packing type, etc.	21	7	Outside scope
Foaming	18	6	14
Installation mishaps	16	6	10
Tray and downcomer layout	13	4	6
Relief problems	12	4	9
	297	100	

causes of column malfunctions. Among them, they make up more than half of the reported incidents. Familiarity with these problems, therefore, constitutes the "bread and butter" of persons involved in troubleshooting and operating distillation and absorption columns. For this reason, these topics receive primary emphasis in this book; the relevant chapters dealing with them are marked in Table 1.1.

- Reboilers, condensers, and operation difficulties amount to about half of the remaining problems. Thus three out of four incidents are caused by either these or the factors previously mentioned. Familiarity with these problems, therefore, is of great importance to persons involved in distillation and absorption operation and troubleshooting. For this reason, these topics are also highlighted in this book.

- Primary design problems, foaming, installation mishaps, relief problems, and tray and downcomer layout problems make up the rest of the column malfunctions. Familiarity with these problems is useful to troubleshooters and operation personnel, but only one incident out of every four is likely to be caused by one of these factors. All these topics, except the primary design, are relatively narrow and are treated accordingly in this book.

- Primary design is an extremely wide topic, encompassing vapor-liquid equilibrium, reflux-stages relationship, stage-to-stage calculations, unique features of multicomponent distillation, tray and packing efficiencies, scale-up, column diameter determination, flow patterns, type of tray, and size and material of packing. This topic occupies the bulk of most distillation texts (e.g., 38, 193, 371, 409), and perhaps represents the bulk of our present distillation know-how.

While this topic is of prime importance for designing and optimizing distillation columns, it plays only a minor role when it comes to distillation operation and troubleshooting. Table 1.1 suggests that only one column malfunction in fourteen is caused by problems incurred at the primary design stage. The actual figure is probably higher for a first-of-a-kind separation, but lower for an established separation. Due to the bulkiness of this topic in relation to its likelihood to cause malfunctions, and due to the coverage that the topic receives in several texts (e.g., 38, 193, 371, 409), it was excluded from this book. The reader is referred to the cited sources for detailed coverage.

The above statements must not be interpreted to suggest that operation personnel and troubleshooters need not be familiar with the primary design. Quite the contrary. A good troubleshooter must have a solid understanding of primary design because it provides the foundation of our distillation know-how. However, the above statements do suggest that in general, when a troubleshooter examines the primary design for the cause of a column malfunction, he or she has less than one chance out of ten of finding it there.

1.2 Column Troubleshooting—A Case History

In Section 1.3, the systematic approach recommended for tackling distillation problems is mapped out. The recommended sequence of steps is illustrated with reference to the case history described below.*

The following story is not a myth; it really happened. One morning as I sat quietly at my desk in corporate headquarters, the boss dropped by to see me. He had some unpleasant news. One of the company's refinery managers was planning to visit our office to discuss the quality of some of the new plants that had been built in his refinery. As an example of how not to design a unit, he had chosen a new gas plant for which I had done the process design. The refinery manager had but one complaint: "The gas plant would not operate."

I was immediately dispatched to the refinery to determine which aspect of my design was at fault. If nothing else, I should learn what I did wrong so as not to repeat the error.

Upon arriving at the refinery, I met with the operating supervisors. They informed me that, while the process design was fine, the gas plant's operation was unstable because of faulty instrumentation. However, the refinery's lead instrument engineer would soon have the problem resolved.

Later, I met with unit operating personnel. They were more specific. They observed that the pumparound circulating pump (see Fig. 1.1a) was defective. Whenever they raised hot oil flow to the debutanizer reboiler, the gas plant would become destabilized. Reboiler heat-duty and reflux rates would become erratic. Most noticeably, the hot-oil circulating pump's discharge pressure would fluctuate wildly. They felt that a new pump requiring less net positive suction head was needed.

Both these contradictory reports left me cold. Anyway, the key to successful troubleshooting is personal observation. So I decided to make a field test.

When I arrived at the gas plant, both the absorber and debutanizer towers were running smoothly but not well. Figure 1.1b shows the configuration of the gas plant. The debutanizer reflux rate was so low it precluded significant fractionation. Also, the debutanizer pressure was 100 psi below design. Only a small amount of vapor, but no liquid, was being

---

*Reproduced from Norman P. Lieberman, *Troubleshooting Process Operations* (2nd ed., PennWell Books, Tulsa, 1985). This case history is a classic example of how to perform a systematic troubleshooting investigation. The permission of PennWell Books and Norman P. Lieberman for reproducing this material is gratefully acknowledged.

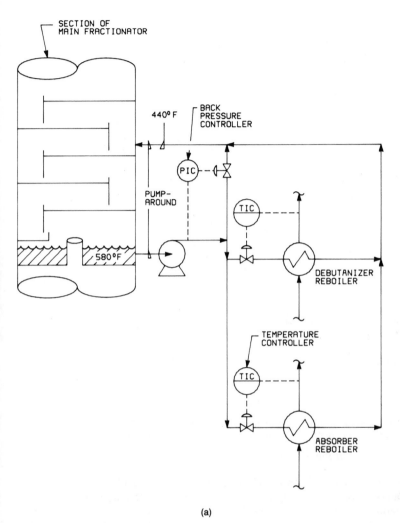

SECTION OF
MAIN FRACTIONATOR

440° F

BACK
PRESSURE
CONTROLLER

PIC

PUMP-
AROUND

TIC

580°F

DEBUTANIZER
REBOILER

TEMPERATURE
CONTROLLER

TIC

ABSORBER
REBOILER

(a)

Figure 1.1 Column troubleshooting case history. (a) Hot oil from the fractionator supplies heat to gas-plant reboilers.

produced from the reflux drum. Since the purpose of the gas plant was to recover propane and butane as a liquid, the refinery manager's statement that the gas plant would not operate was accurate.

As a first step, I introduced myself to the chief operator and explained the purpose of my visit. Having received permission to run my test, I switched all instruments on the gas-plant control panel from automatic over to local/manual. In sequence, I then increased the lean oil flow to the absorber, the debutanizer reflux rate, and the hot-oil flow to the debutanizer reboiler.

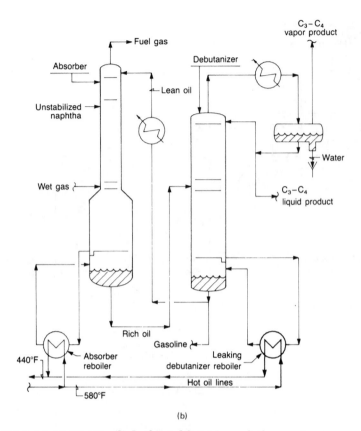

(b)

Figure 1.1 (Continued) (b) Leaking debutanizer reboiler upsets gas plant. (*Parts a and b from Norman P. Lieberman* Troubleshooting Process Operations *2nd ed., PennWell Books, Tulsa, 1985. Reprinted courtesy of PennWell Publishing Co.*)

The gas plant began to behave properly. The hot-oil circulating pump was putting out a steady flow and pressure. Still, the plant was only producing a vapor product from the debutanizer reflux drum. This was because the debutanizer operating pressure was too low to condense the C_3–C_4 product. By slowly closing the reflux drum vapor vent valve, I gradually increased the debutanizer pressure from 100 psig toward its design operating pressure of 200 psig.

Suddenly, at 130 psig the hot-oil flow to the debutanizer's reboiler began to waiver. At 135 psig, the debutanizer pressure and the hot-oil flow plummeted. This made absolutely no sense. How could the debutanizer pressure influence hot-oil flow?

To regain control of the gas plant, I cut reflux to the debutanizer and lean-oil flow to the absorber. I was now back where I started. The thought of impending of failure loomed.

I repeated this sequence twice more. On each occasion, all went well until the debutanizer pressure was increased. By this time it was 3 a.m. Was it also time to give up and go home?

Just then, I noticed a commotion at the main fractionator control panel. The operators there stated that the fractionator was flooding again—for the third time that night. The naphtha production from the fractionator had just doubled for no apparent reason.

In every troubleshooting assignment there always occurs that special moment, the moment of insight. All of the bits and pieces fall into place, and the truth is revealed in its stark simplicity.

I cut the debutanizer pressure back to 100 psig and immediately the flooding in the main fractionator subsided. The operators then closed the inlet block valve to the hot-oil side of the reboiler and opened up a drain. Naphtha poured out instead of gas oil. This showed that the debutanizer reboiler had a tube leak.

Whenever the debutanizer pressure reached 130 psig, the reboiler pressure exceeded the hot-oil pressure. The relatively low-boiling naphtha then flowed into the hot oil and flashed. This generated a large volume of vapor that then backed hot oil out of the reboiler. The naphtha vapors passed on into the main fractionator and flooded this tower. Thus, the cause of the gas plant instability was neither a process design error, instrument malfunction, nor pumping deficiency. It was a quite ordinary reboiler tube failure.

1.3   Strategy for Troubleshooting Distillation Problems

In almost any troubleshooting assignment, it is desirable to solve a problem as rapidly as possible with the least amount of expenditure. In a surprisingly large number of cases, this objective is only partially achieved. One of the major obstacles to achieving this objective is a poor (often nonexistent) strategy for tackling the problem.

When devising a troubleshooting strategy, it is useful to think in terms of a "doctor and patient" analogy. The doctor's troubleshooting strategy in treating a patient is well-established and easily understood by most people. Applying similar principles to solving distillation problems can often map out the most effective and least expensive course of action.

The sequence of steps below is often considered optimum for tackling a troubleshooting problem. It is based on the author's experience as well as the experience of others (131, 150a, 239, 415), and makes reference to the doctor and patient analogy. Actions described in Lieberman's case history (Sec. 1.2) are used to demonstrate the optimum sequence of steps. A good troubleshooting strategy always proceeds stepwise, starting with the simple and obvious.

1. Assess the safety or environmental hazard that the problem can create. If a hazard exists, an emergency action is required prior to any troubleshooting efforts. In terms of the medical analogy, measures to save the patient or prevent the patient's problem from af-

fecting others have priority over investigating the cause of the problem.

2. Implement a temporary strategy for living with the problem. Problem identification, troubleshooting, and correction take time. Meanwhile, adverse effects on safety, the environment, and plant profitability must be minimized. The strategy also needs to be as conducive as practicable for troubleshooting. The strategy, and the adverse effects that are to be temporarily tolerated (e.g., instability, lost production, off-spec product), usually set the pace of the troubleshooting investigation.

In the debutanizer case history, the short-term strategy was to run the column at a pressure low enough to eliminate instability and to tolerate an off-spec bottom product. In the medical analogy, the short-term strategy is hospitalization, or going to bed, or just "taking it easy." This strategy usually sets the urgency of treatment.

3. Obtain a clear, factual definition of the symptoms. A poor definition of symptoms is one of the most common troubleshooting pitfalls. In the debutanizer case history above, the following definitions were used by different people to describe the symptoms of a reboiler tube leak problem:

- "The gas plant would not operate."
- "The gas plant's operation is unstable because of faulty instrumentation. However, the problem will soon be resolved by the instrument engineer."
- "The oil circulating pump is defective. Whenever the oil flow to the reboiler is raised, reboiler heat duty and reflux rate would become erratic, and the pump's discharge pressure would fluctuate wildly. A new pump requiring less net positive suction head is needed."
- "The column was running smoothly but not well. Reflux rate was too low, so it precluded significant fractionation. The column pressure was 100 psi below design. Only a small amount of vapor, but no liquid, was being produced from the reflux drum, which should have produced mainly liquid. Other problems noticed by plant personnel are as described above."

The above represents a typical spectrum of problem definitions. The last definition, supplied by a troubleshooting specialist, can clearly be distinguished. The first two definitions were nonspecific and insufficiently detailed. The third described part of the story, but left out a major portion. The first three definitions also contained implied diagnoses of the problem, none of which turned out to be correct.

The doctor-patient equivalents to the first three definitions are statements such as "I feel I am going to die," "I am feeling a bit off, but I will be OK soon," and "I do have a sharp headache (without mentioning other pains and having a temperature as well)." It is apparent that these statements do not provide the doctor with the entire story.

4. Examine the column behavior yourself. This is imperative if the problem definition is poor. In the debutanizer example above, the troubleshooter would have been oblivious to a major portion of the problem definition had he based his investigation entirely on other people's observations. Some communication gap always exists between people, and it is often hard to bridge. In a similar manner, a doctor always needs to examine the patient before starting treatment.

   In some circumstances, it may be impractical or too expensive for the troubleshooter to visit the site (e.g., a column located on another continent). In this case, the troubleshooter must be in direct (i.e., phone) communication with the operating person, who should be entirely familiar with the column, its operation, and its history. The problem definition in this case must be particularly sharp.

5. Learn about the column history. The question, "what are we doing wrong now that we did right before?" is perhaps the most powerful troubleshooting tool available. If the column is new, closely examine any differences between the column and columns used for identical or at least similar services. In addition, examine any differences between the expected and the actual performance. Each difference can provide a major clue. Doctors always ask patients about their health histories, searching for similar clues. In the debutanizer example above, the troubleshooter included a comparison to design performance in the problem definition (he was working with a new column).

   Digging into the past may also reveal a recurring ("chronic") problem. If so, finding the correct link between the past and present circumstances can be very illuminating. Be cautious when identifying the link; a new problem may give the same symptoms as a past problem but be caused by an entirely different mechanism.

   A history search may also unveil a hidden flaw. In one case (150a), a column modification caused a loss in column efficiency. The loss was unnoticed, and the reduced performance became the norm. The problem was noticed several years later.

6. Search and scan events that occurred when the problem started. Carefully review operating charts, trends, computer, and operator logs. Establish event timing in order to differentiate an initial problem from its consequences. Harrison and France (150a) use a

case history with actual operating charts to demonstrate the value of analyzing event timing. Their column experienced premature flooding, resulting from the following sequence of events: a temporary loss of bottoms pump; a rise of base level above the reboiler return nozzle; a collapse of bottom trays (see Sec. 13.2); raising of reflux to meet purity and reduced trays; flooding. An analysis of event timing using the operating charts clearly mapped this se-. quence. In terms of the medical analogy, doctors always ask patients if they did something different about the time when the trouble started, and what happened first.

Include events that may appear completely unrelated, as these may be linked in an obscure manner to the problem. In the debutanizer example, it was the observation that flooding in the fractionator coincided with the debutanizer becoming unstable that gave the troubleshooter the vital clue. At first glance, the two appeared completely unrelated.

7. Listen to shift operators and supervisors. Experienced people can often spot problems, even if they cannot fully explain or define them. Listening to those people can often provide a vital clue. In the debutanizer example, some of the important observations were supplied by these people.

8. Do not restrict the investigation to the column. Often, column problems are initiated in upstream equipment. Doctors frequently look for clues by asking patients about people they have been in contact with or their family health history.

9. Study the behavior of the column by making small, inexpensive changes. These are particularly important for refining the definition of symptoms, and they may contain a vital clue. Record all observations and collect data; these may also contain a major clue, which can easily be hidden and become forgotten as the investigation continues. In the debutanizer example, the troubleshooter increased column pressure and watched its behavior. This led him to the observation that the debutanizer pressure affected oil flow—a major step in refining the problem definition. In the doctor and patient analogy, this is similar to the doctor asking the patient to take a deep breath or momentarily stop breathing during a medical examination.

10. Take out a good set of readings on the column and its auxiliaries, including laboratory analyses. Misleading information supplied by instruments, samples, and analyses is a common cause of column malfunctions. Always mistrust or suspect instrument or lab-

oratory readings, and make as many crosschecks as possible to confirm their validity. Instruments may malfunction even when the instrument technician can swear they are correct. In one example (415) an incorrect pipe design caused an erroneous reading of a reflux flow meter. Survey the column piping for any unusual features such as poor piping arrangement, leaking valves, "sticking" control valves, and valves partially shut. Compile mass, component, and energy balances; these function as a check on the consistency of instrument readings and the possibility of leakage. This step is equivalent to laboratory tests taken by a doctor on the patient. Scan the column drawings carefully for any unusual features. Check the column internals against good design practices, and determine whether any have been violated. If so, examine the consequences of such violation and its consistency with the information. Carry out a hydraulic calculation at test conditions to determine if any operating limits are approached or exceeded. If a separation problem is involved, carry out a computer simulation of the column; check against test samples, temperature readings, and exchanger heat loads.

1.4   Dos and Don'ts for Formulating and Testing Theories

Following the previous steps, a good problem definition should now be available. In some cases (e.g., the debutanizer), the cause may be identified. If not, there will be sufficient information to narrow down the possible causes and to form a theory. In general, when problems emerge, everyone will have a theory. In the next phase of the investigation, these theories are tested by experimentation or by trial and error. The following guidelines apply to this phase:

1. Logic is wonderful as long as it is consistent with the facts and the information is good.

2. When formulating a theory, attempt to visualize what is happening inside the column. One useful technique is to imagine yourself as a pocket of liquid or vapor traveling inside the column. Keep in mind that this pocket will always look for the easiest path. Another useful technique is to think of everyday analogies. The processes that occur inside the column are no different from those that occur in the kitchen, the bathroom, or in the yard. For instance, blowing air into a straw while sipping a drink will make the drink splash all over; similarly, a reboiler return nozzle sub-

merged in liquid will cause excessive entrainment and premature flooding (Sec. 13.2).

3. Do not overlook the obvious. In most cases, the simpler the theory, the more likely it is to be correct.

4. An obvious fault is not necessarily the cause of the problem. One of the most common troubleshooting pitfalls is discontinuing or retarding further troubleshooting efforts when an obvious fault is uncovered. Often, this fault fits in with most theories, and everyone is sure that the fault is the cause of the problem. The author is familiar with many situations where correcting an obvious fault neither solved the problem nor improved performance. Once an obvious fault is detected, it is best to regard it as another theory and treat it accordingly.

5. Testing theories should begin with those that are easiest to prove or disprove, almost irrespective of how likely or unlikely these theories are. If it is planned to shut the column down, and shutting it down is expensive, it is often worthwhile to cater to a number of less drastic theories even if some are longer shots.

6. Refrain from making any permanent changes until all practical tests are done.

7. Look for possibilities of simplifying the system. For instance, if it is uncertain whether an undesirable component enters the column from outside or is generated inside the column, consider operating at total reflux to check it out.

8. Do not overlook human factors. Other people's reasoning is likely to differ from yours, and they will act based on their reasoning. The more thoroughly you question their design or operating philosophy, the closer you will be able to reconstruct the sequence of events leading to the problem. In many cases, you may also discover major considerations you are not aware of.

9. Ensure that management is apprised of what is being done and is receptive to it (415). Otherwise, some important nontechnical considerations may be overlooked. Further, management is far less likely to become frustrated with a slow-moving investigation when it is convinced that the best course of action is being followed.

10. Involve the supervisors and operators in each "fix." Whenever possible, give them detailed guidelines of an attempted fix, and leave them with some freedom for making the system work. The author has experienced several cases where actions of a motivated operator made a fix work, and other cases where a correct fix was unsuccessful because of an unmotivated effort by the operators.

11. Beware of poor communication while implementing a "fix." Verbal instruction, rush, and multidiscipline personnel involvement generate an atmosphere ripe for communication problems (150a). Ensure any instructions are concise and sufficiently detailed. If leaving a shift team to implement a fix by themselves, leave written instructions. Be reachable and encourage communication should problems arise. Call in at the beginning of the shift to check if the shift team understood your instructions.

12. Recognize that modifications are hazardous. Many accidents have been caused by unforeseen side effects of even seemingly minor modifications. Ban "back of an envelope" modifications, as their side effects can be worse than the original problem. Properly document any planned modification, and have a team review it systematically with the aid of a checklist such as a "hazop" checklist (225). Before completion, inspect to ensure the modification was implemented as intended.

13. Properly document any fix which is being adopted, the reasons for it, and the results. This information may be useful for future fixes.

1.5 Using This Book: A Troubleshooting Directory

The two common troubleshooting practices are analogous to those of medicine. One common practice is to wait until the illness strikes before calling for help. A healthier practice is "preventive troubleshooting," which aims at eliminating the cause of illness before it occurs. Although preventive troubleshooting is seldom perfect, it can go a long way toward reducing the chances, severity, and pain of potential ailments.

The sequence of discussions in this book is best tailored for preventive troubleshooting. Table 1.2 caters to the practice of "wait until the illness strikes" troubleshooting. Table 1.2 describes common column ailments and links them with topics discussed in this book. The entry 1, 2, or 3 means that the discussion of the topic contains some:

1. Common causes of this type of problem

2. Possible causes of this type of problem

3. Possible, but relatively uncommon, causes of this type of problem

No entry means that it is an uncommon or unlikely cause or not a cause of this type of problem.

Table 1.2 is only intended to serve as a general directory. It proposes some potential causes of a given problem and refers the user to sections in this book where relevant information can be found. The entries are subjective and based on the author's experience. Needless to state, several may not apply to specific situations; for instance, the chimney tray entries will be irrelevant to troubleshooting a column that does not contain one.

An additional application of Table 1.2 with the "wait until the illness strikes" troubleshooting approach is for guiding preparation of a troubleshooting checklist. Once a problem is identified, the numeral marked in the relevant column can serve as a priority guide. For instance, if the problem is identified as premature flooding, items marked with the numeral 1 in the premature flooding column are worthy of being given first priority consideration. If the problem is not fully identified, it may be necessary to examine entries in several of the Table 1.2 columns. For instance, unstable control may be caused by premature flooding. Here entries both under unstable control and premature flooding need reviewing.

The author stresses that Table 1.2 is meant to serve only as a general directory and must not be blindly applied to any specific situation. The type of service and past experiences with similar columns are prime considerations that must be incorporated in the preparation of a troubleshooting checklist.

TABLE 1.2 Using This Book for Troubleshooting Operation and Designs

Chapter	Topic	Premature flooding	Excessive entrainment (premature)	Foaming	Excessive pressure drop (nonflooding)	Plugging/fouling	Poor tray efficiency	Tray maldistribution	Poor packing efficiency	Packing maldistribution	Local loss of efficiency	Tray/packing damage	Other internals damage	Localized damage	Pump damage	Catastrophic failure	Hazard to personnel	Unstable control	Inefficient control	Misleading measurement	Poor reboil action	Poor condenser action	Piping problems	Gravity piping only	Poor turndown	Leakage	Drainage	Low liquid rate problem	Troubleshooting techniques	Column tests (operation)	Precommissioning checks	Startup/shutdown	Ease of maintenance/access	Internals strength
1	Troubleshooting	2	2		3			3			1		2	1										3					1	1			3	3
2	Top & reflux inlets	1	1	3	1	2	2	2			1		3	1									1										3	3
	Intermediate feed inlets	2	2	2	1	1		1			1		3	1									3										3	3
	Distributors & multipass inlets	2	2	3	1	3			1	1	2		3			1																	3	3
3	Liquid distributors	3	2	3	3	1			1	1	2		3	3			2	2		2			3		1	2	3	1						3
	Distributor inlets	2	2		1	2	2		1	1	2	3	3	3											2	2	3							
	Redistributors	1	1		1	2			1	1	2	3	3	1				3					1		1			3						1
	Flashing distributors	1	1						1	1	2	1	1	1				3					3		2									1
	Vapor distributors	1	1		1				1	1	1	2	3	1	1								2		3									2
4	Reboiler return inlet	2	2	3	1	3		3	3	1											1		3									3		
	Spargers	2	3			3				1	2				1			2			3		1			2	1					3		
	Bottom sumps & baffles	1	1			3				1	1		2	2	1			2			3		1			1	2							
	Once-through reboiler pan	1			1	3				1	1				1						1		1			1	1	1						
	Chimney trays			3		3		3			2			1	3								1		1	1	2	1						
	Downcomer trapouts	3	1			3			2		2						2						1		3	1	2						2	
	Collectors		1						2	2							1								1								2	
	Vapor outlets									2			2	2				1																
5	Gravity lines	1		3		1					1		2	1		1	1	1		1				1						1	1	1		
	Instrument connections	1			3								1			1	1	1		1										1			1	
	Manholes		1			1	2	2				3	3	3		1	1			1	3		3										1	2

TABLE 1.2 Using This Book for Troubleshooting Operation and Designs (Continued)

Chapter	Topic	Premature flooding	Excessive entrainment (premature)	Foaming	Excessive pressure drop (nonflooding)	Plugging/fouling	Poor tray efficiency	Tray maldistribution	Poor packing efficiency	Packing maldistribution	Local loss of efficiency	Tray/packing damage	Other internals damage	Localized damage	Pump damage	Catastrophic failure	Hazard to personnel	Unstable control	Inefficient control	Misleading measurement	Poor reboil action	Poor condenser action	Piping problems	Gravity piping only	Poor turndown	Leakage	Drainage	Low liquid rate problem	Troubleshooting techniques	Column tests (operation)	Precommissioning checks	Startup/shutdown	Ease of maintenance/access	Internals strength
6	Tray spacing	3					3																										1	
	Hole diameters	3	3		1	1	3																		1	1								
	Fractional hole area	2	2		2	1	3	1																	1	3								
	Valve tray layout	3	3		3	1	2	2																	3	2								
	Calming zones	1	2		2		3	3																	1	1								
	Outlet weirs	3	1				2	2				2			1		3											1						
	Splash baffles/vapor hoods				1		3	3																		2		1						
	Internal demisters	3		1			2							1												1		2						
	Reverse-flow baffles						3	3						1											1			1						
	Stepped trays	3	1		1		3	1																	3									
	Multipass trays	1					1	1			2														1									
	Change in no. of passes					3	3	3																	3									
	Flow-induced vibrations															3																		
	Downcomer type	3			3	2	3	3					3	3																				
	Downcomer width	1	3		3	2						2																1				1		
	Antijump baffles	3					3																											
	Downcomer sealing	1	1			1	1	1				2	3															1						
	Clearance under dc	1	1		3																				3	1	1	1						
	Inlet weirs/seal pans	1	1			1	3	3																		1	1	1						

TABLE 1.2   Using This Book for Troubleshooting Operation and Designs (Continued)

Chapter	Topic	Premature flooding	Excessive entrainment (premature)	Foaming	Excessive pressure drop (nonflooding)	Plugging/fouling	Poor tray efficiency	Tray maldistribution	Poor packing efficiency	Packing maldistribution	Local loss of efficiency	Tray/packing damage	Other internals damage	Localized damage	Pump damage	Catastrophic failure	Hazard to personnel	Unstable control	Inefficient control	Misleading measurement	Poor reboil action	Poor condenser action	Piping problems	Gravity piping only	Poor turndown	Leakage	Drainage	Low liquid rate problem	Troubleshooting techniques	Column tests (operation)	Precommissioning checks	Startup/shutdown	Ease of maintenance/access	Internals strength
7	Materials of construction	3	3									1	1	1		1	1						3											1
	Thickness of parts	3							3	3		2	2	2			2											1						1
	Tray supports	3	3					2				2	2	2			2									1								1
	Fastening	3					3					2	2	2			2									2							1	1
	Manways															1	1																1	
	Thermal expansion	3	3		1		3	2	3			3	3	2											2	2		2						2
	Tray levelness						3						3								3				2	1	1	1				2		
	Tray drainage					1		2				2	2		3										2	1		1						
	Leakage											1	1													2								
	Cartridge trays						2					2	2		1										2	1								
8	Packing supports	1			1				3	3	3	1					1	3					3											1
	Support structure	3	3		3				2	1	3						1	3		3	3		3											1
	Retaining devices	3	3		3				2	2		2	2		2	1	1								2	1		1						1
	Packed column vertically								2	2		1	2	1		1	1								2	1		1						
9	Overpressure relief						2		2		2	2	2	1		1	1								2	1	1	1						
10	Preassembly dos & dont's	2	2		2		2	2	2	2	2	2	2	1	1	2	1	3		2					2								2	3
	Safety precaution inside											3	3	.3		2	1																1	
	Removing trays/packing	1	1	3	2		2	1	1	1		1	1	1		1	1						3		2	1	1	1					1	
	Tray/packing installation	1	1		2	1	1	1	1	1	1	1	1	1		2	1				2	2	3		1	1	1	1					1	2
	Prestartup inspection	1	1		2	1	1	1				1		1		1	1	2		2	3	3	3		1	1	1	1					2	3

17

TABLE 1.2 Using This Book for Troubleshooting Operation and Designs (*Continued*)

Problem classification →	Premature capacity limitation					Poor separation					Damage & Hazards						Control and instruments			Auxiliaries				Low rates				Trouble-shooting/testing				Mech-anical	
Nature of problem / Chapter – Topic	Premature flooding	Excessive entrainment (premature)	Foaming	Excessive pressure drop (nonflooding)	Plugging/fouling	Poor tray efficiency	Tray maldistribution	Poor packing efficiency	Packing maldistribution	Local loss of efficiency	Tray/packing damage	Other internals damage	Localized damage	Pump damage	Catastrophic failure	Hazard to personnel	Unstable control	Inefficient control	Misleading measurement	Poor reboil action	Poor condenser action	Piping problems	Gravity piping only	Poor turndown	Leakage	Drainage	Low liquid rate problem	Troubleshooting techniques	Column tests (operation)	Precommissioning checks	Startup/shutdown	Ease of maintenance/access	Internals strength
11 Line blowing, pressuring, depressuring, purging	3	3			1			3	3		2	2			1	1						1								1	1	1	
Blinding/unblinding															1	1						1								1	1	3	2
Leak testing	3		2		1						2	2			2	2										1				1	1	3	1
Washing			3		1						1	1	2		2	2										1				1	1	3	
Steam-water operation						3					1	1	2		2	2												1	2		1		
Steaming						3					1	1	2		1	1										1		3					
Dehydration, liquid circulation					3									2																			
Solvent testing											2	2												2				2	1	3			
12 Startup/shutdown	1	1			1						1	1	2	2	2	1				2						1					1		
Cooling/heating										2			2		2	2															3		
Excess subcool/superheat	1	1											2	2	2	2								2					2				
Total reflux				3	2						1															2			2		1		
Drainage			2			3	1																	2		1	1		2				
Sealing	1	1				3	1								1											1							
Reverse flow											2	1	2		1	2				2											1		
13 Tray dislodging/damage	1	1				3					1	1	2	2	2	1	1										1				1		
Liquid level in column	1			3		1					1	1	2	2	3	3		3		1											3		
Liquid level in accumulator						3					1	2	2	2	1	3			1		2								2		2		
Pressure surges	1	1				1					2	2	1	2	3	3	1												2		1		
Intermediate impurity	1	1	2								1	1	2		1	1													2	1	1		
Absence of component					1	1					1				1	1	2										1				3		

18

TABLE 1.2 Using This Book for Troubleshooting Operation and Designs (Continued)

Chapter	Topic	Premature flooding	Excessive entrainment (premature)	Foaming	Excessive pressure drop (nonflooding)	Plugging/fouling	Poor tray efficiency	Tray maldistribution	Poor packing efficiency	Packing maldistribution	Local loss of efficiency	Tray/packing damage	Other internals damage	Localized damage	Pump damage	Catastrophic failure	Hazard to personnel	Unstable control	Inefficient control	Misleading measurement	Poor reboil action	Poor condenser action	Piping problems	Gravity piping only	Poor turndown	Leakage	Drainage	Low liquid rate problem	Troubleshooting techniques	Column tests (operation)	Precommissioning checks	Startup/shutdown	Ease of maintenance/access	Internals strength
		Premature capacity limitation					**Poor separation**					**Damage & Hazards**						**Control and instruments**			**Auxiliaries**				**Low rates**				**Troubleshooting/testing**				**Mechanical**	
13	Cooking, hydrates, freezing, precipitation	2	2	3		1	1		1			3	3	2		1	1	3		3	3	2	3		2							1		
	Chemical reaction	2	3	2		1	1		1			2	2	2				3		3	3	3			2			3				3		
	Heat exchanger leaks	1				1															1	1										1		
	Heat integration spins	1	2															2														1		
	Instrument problems	1	1				2					1	1	1	1	1	1	1		1					3						1	1		
14	Flooding			2	2													1			2	2							1	1				
	Foaming				1		1		1		1																		1	1				
	Efficiency/performance tests			1	1		1			3	1							1											1	1				
	Other tests							3												1									1	1				
	Radioactive troubleshooting	1	1	2		2	2		2			1	1	1				2		3			1		1				1	1				
15	Reboilers	2	2	2		2	2		3			2	2	2		3				1	1		1										3	
	Condensers																			1		1	2											
16 –19	Column control	2	2			3	3		3		2		2			3		1	1	1	1	1	1		2			2						

Reflux and Intermediate Feed Inlets for Tray Columns

The main consideration for introducing reflux or intermediate feed into a tray tower is to achieve adequate hydraulics in the inlet area. Failure to achieve this may result in premature flooding, excessive entrainment, and mechanical damage. When the tower contains multipass trays, it is also important to split the feed or reflux adequately among the passes. With single-pass trays, and sometimes with two-pass trays, achieving good distribution of the feed to the tray is of secondary, but not negligible, importance.

This chapter examines common practices of introducing reflux and intermediate feed into tray columns, outlines the preferred practices, highlights the consequences of poor practices, and supplies guidelines for troubleshooting and for reviewing designs of reflux and intermediate feed inlets.

2.1 Top-Tray Feed and Reflux Inlet Arrangements

Figure 2.1 shows methods for introducing top-tray feed or reflux into a column. Table 2.1 lists the dimensions restricting each design. All of the arrangements shown in Fig. 2.1 are suitable for liquid reflux or top feed. If the feed contains some vapor, only arrangements b, d, e, and h of Fig. 2.1 are suitable. Arrangements a, b, c, e, and f of Fig. 2.1 are usually preferred for cost reasons. Arrangements d and h are usually used when there is a distinct advantage for orienting the inlet nozzle at an angle other than about 0° to the liquid flow. Arrangements a, d, h, and to a lesser extent g have the disadvantage of inducing weeping through the tray inlet rows of perforations or valves because of hydraulic jump over the inlet weir.

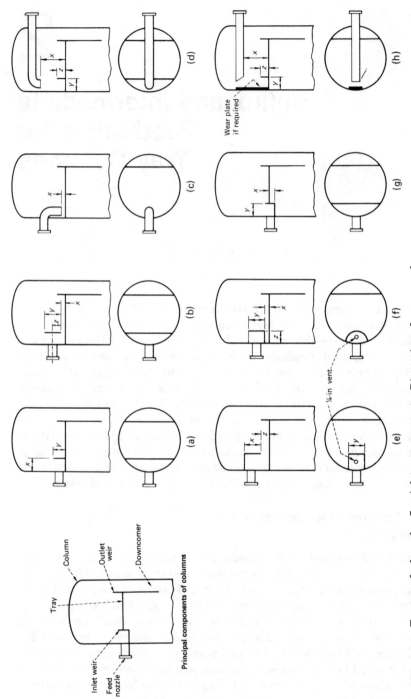

Figure 2.1 Top-tray feeds and reflux-inlet arrangements. Dimensions for x, y, and z are listed in Table 2.1. (*Henry Z. Kister, excerpted by special permission from Chemical Engineering, May 19, 1980, copyright ©1980, by McGraw-Hill, Inc., New York, NY, 10020.*)

TABLE 2.1   Dimensions for Top Feed/Reflux Inlet Arrangements

Fig.	1a	1b	1c	1d	1e	1f	1g	1h
Maximum nozzle dia., in	6	—	—	6	—	—	6	—
Note	3	3	—	—	1	1	—	2
Reference	207	354	354	138, 207	207	354	179, 207	138, 207
Pure-liquid feed								
Dimension, x, in	W_d	h_{cl}	$d_n/2$	>12	$2d_n$	$d_n/2$	4	>12
Dimension, y, in	4-6	$2d_n$	—	W_d	$2d_n$	$2d_n$	W_d	W_d
Dimension, z, in	—	d_n	—	4-6	d_n	$1.5d_n$	—	4-6
Vapor/liquid feed	NS		NS			NS	NS	
Dimension x, in		$2d_n^*$		>12	$2d_n^*$			>12
Dimension y, in		$2d_n^*$		W_d	$2d_n^*$			W_d
Dimension z, in		d_n^*		4-6	$2d_n^*$			4-6

d_n = Inlet pipe dia., in
h_{cl} = Clearance under downcomer, in
W_d = Downcomer width, in
NS = Not suitable.
*Dimensions as recommended by the author. All other dimensions recommended by the cited reference.
Note 1: Drill a ¼-in vent hole on top.
Note 2: Wear plate may be required.
Note 3: Ensure nozzle enters behind the baffle. If it does not, hydraulic jump could be a problem.
Internal inlet pipes should be removable for maintenance.
SOURCE: Reproduced with permission from *Chemical Engineering*, May 19, 1980, p. 139.

Arrangement b (often referred to as the *false downcomer*) is popular. It offers better liquid distribution than the others, does not suffer from hydraulic jump, and provides some flexibility in inlet nozzle orientation. The width of the false downcomer should be the same as the width at the bottom of a downcomer (307). For feed into center or off-center downcomers it is recommended to make the $x + y$ dimension 12 in (307). If entrainment due to liquid splashing in the false downcomer is a concern, a horizontal baffle, with dimensions of about $2d_n$ by W_d can be installed directly above the nozzle entry into the false downcomer. This baffle is installed at some clearance above the false downcomer. In most cases, there is no need for this baffle.

Arrangement e is also popular. It is one of the least expensive arrangements, does not suffer from hydraulic jump, and minimizes inlet splashing. The baffle is open both at its sides and bottom.

2.2   Intermediate Feed Inlet Arrangements

Figure 2.2 shows methods for introducing an intermediate feed to the the column. Table 2.2 summarizes the applications for which each arrangement is suitable.

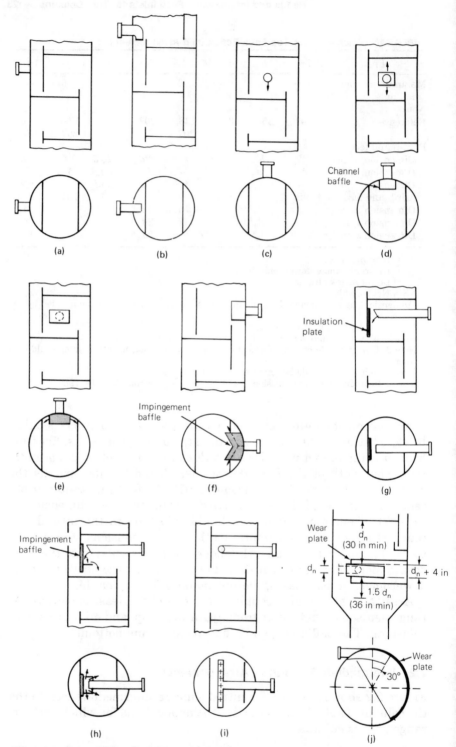

Figure 2.2 Intermediate-feed arrangements for columns (*Parts a, c, d, e, g, h Henry Z. Kister, excerpted by special permission from* Chemical Engineering, *May 19, 1980, copyright © by McGraw-Hill Inc., New York, NY 10020.*)

TABLE 2.2   Intermediate Feed Inlet Arrangements

Arrangement	a	b	c	d	e	f	g	h	i	j
Cold-liquid feed	Yes	Yes	Yes	Yes	Yes	Yes	Yes	Yes	Yes	No
Vapor-liquid feed	No	No	No	Yes	Yes	Yes	Yes	Yes	Yes	Yes
Vapor feed	No	No	No	Yes	Yes	Yes	Yes	Yes	No	Yes
Hot feed	No	No	No	No	No	No	Yes*	Yes*	Yes	Yes
High-velocity feed	No	No	No	No	Yes	Yes	No	Yes	Yes	Yes
High-pressure application	No	No	Yes	Yes	Yes	Yes	Yes	Yes	Yes	Yes
Downcomer capacity critical	No	No	Yes	Yes	Yes	Yes	Yes	Yes	Yes	Yes

*Assuming insulation plate is provided.
SOURCE: Reproduced with permission from *Chemical Engineering*, May 19, 1980, p. 140.

Arrangement a is suitable only for subcooled low-velocity liquids such as circulating reflux streams. If the liquid contains some vapor or is hotter than the downcomer liquid, flashing will occur and downcomer capacity will be reduced. Even with subcooled liquid feeds, this arrangement introduces turbulence in an area where phase separation is important. It should therefore be avoided where downcomer capacity is critical, such as in high-pressure systems and systems in which fouling tends to progressively limit downcomer capacity. One designer (237) recommends avoiding this arrangement altogether, because vapor may find its way into the feed (e.g., by tube leakage in an upstream heat exchanger).

Arrangement b is similar to arrangement a and is also only suitable for subcooled, low-velocity liquids where downcomer capacity is not critical. Compared to arrangement a, it may reduce the turbulence generated inside the downcomer; otherwise it suffers from the same disadvantages.

Arrangement c is also suitable only for low-velocity liquid feeds. If the feed contains vapor, impingement of the feed against liquid in the vapor space can cause premature entrainment. One designer (211), however, recommends this arrangement for general use for liquid as well as liquid-vapor feeds. This designer implies that significant interference with tray action can be avoided by locating the bottom of the feed nozzle 6 in above the tray floor for liquid feeds, and half a tray spacing above the tray floor for liquid-vapor feeds. Arrangement c has the advantage of being the least expensive.

Arrangement c can be troublesome with feeds whose temperature substantially exceeds the tray liquid. If the feed nozzle is positioned too close to the outlet weir, the joint liquid path may be too short to permit proper mixing between feed and tray liquids. Hot liquid will overflow the weir and induce vaporization, and, therefore, a capacity restriction, in the downcomer. In one column (152), this resulted in premature flooding and loss of efficiency.

Arrangement d is similar to arrangement c, but a channel baffle is added to avoid the impingement problem. This arrangement is suitable for vapor-containing feeds. The baffle may be straight or round and is open at its sides, top, and bottom.

Arrangement e is similar to arrangement d, but a plate is added below the nozzle so that the liquid initially flows sideways instead of downward. This reduces the feed velocity at the inlet and is particularly suitable for high-velocity feeds, whether liquid or vapor.

Arrangement f is similar to e, but the feed is introduced above the downcomer instead of above the tray to minimize interference with tray action. Minimizing this interference can be a distinct advantage with high-velocity feeds.

Arrangement g is often considered optimum (179, 192, 207) for columns whose outlet-weir length is less than 5 ft; in larger columns, a distributor is better. This arrangement has the advantage of introducing feed at the tray inlet, thus improving separation, minimizing interference with tray action, and providing ample mixing distance for hot liquid feeds. References (143 and 207) recommend locating the feed pipe centerline two-thirds of the tray spacing above the tray below, to orient the plane of the pipe outlet 30° from the vertical, and to make the minimum clearance between the pipe and the downcomer half a pipe diameter.

An insulation plate (Fig. 2.2g) is required on the outside wall of the downcomer if the feed enters at a temperature higher than the liquid in the downcomer. Failure to provide such a plate will cause flashing in the downcomer with a reduction in downcomer capacity.

Arrangement h is similar to g, but a wear plate is added on the outside wall of the downcomer and a horizontal impingement baffle is added below the nozzle to prevent entrainment. This arrangement is recommended for high-velocity feeds. Other advantages and recommended dimensions for this arrangement are similar to those for arrangement g above.

Arrangement i is a typical feed distributor arrangement. References (143 and 207) recommend it for all columns where weir length exceeds 5 ft. The recommended clearance between the distributor and the downcomer is 3 to 4 in, with the distributor openings oriented 45° from the vertical toward the downcomer (143, 207, 354). Additional guidelines are in Sec. 2.4. This arrangement provides similar advantages to those of arrangement g, and in addition, provides superior liquid distribution, which is important in large columns.

Arrangement j is unique for high-velocity feeds in which vapor is the continuous phase and liquid is present in the form of a spray. This arrangement is common when the feed makes up the bulk of the vapor traffic in the column section above the feed. It is also used when the feed flashes upon entry to a low-pressure column. Typical examples are feeds to refinery crude and vacuum columns and rich solution feeds to hot carbonate regenerators. A tangential helical baffle or vapor horn, covered at the top, open at the bottom, and spiraling downward, is used at the feed entry. This baffle forces the vapor to follow the contour of the vessel as it expands and decreases in velocity. Liquid droplets, due to their higher mass, tend to collide with the tower wall, which deflects them downward, thus reducing entrainment to the tray above. Large forces, generated by vapor flashing, are absorbed by the entire column wall rather than by a small area. A wear plate is required at the tower wall. Some recommended dimensions (143, 207) are shown in Fig. 2.2j.

It is important to ensure that the helical baffle spirals downward and is covered. The author is familiar with situations where failure to do this caused excessive entrainment to the trays above the feed.

Pilot-scale experiments (292) showed that compared to a radial vapor inlet, the tangential inlet gives better vapor distribution above the feed inlet zone. The tests also showed that with the tangential arrangement, vapor velocity around the periphery of the zone above the feed was higher than in its center. Addition of an annular deflection ring in the tower, above the feed zone, further improved vapor distribution. An additional improvement resulted when two tangential inlets were used instead of one.

2.3   Dos and Don'ts for Reflux, Top-Tray, and Intermediate Feed Inlets

Below are guidelines for avoiding operating problems with reflux, top feed, and intermediate feed inlets.

1. The inlet arrangements must be suitable for the service as described above. This should be checked not only during the initial design but also in any revamp and whenever feed conditions change. The author is familiar with one aromatics plant where equipment upstream of a column was revamped for energy savings. The revamp replaced the all-liquid column feed by a partially vaporized feed. The column feed distributor (similar to Fig.

2.2*i*) was not modified. The mixture issued at excessive velocities, and premature flooding resulted.

2. When the feed can contain vapor, the tray sections and baffles that contact the entering feed can be subjected to abnormally high forces. To avoid structural damage, these sections and baffles should be strengthened. Also, the feed pipe should be anchored to the tower shell.

3. Inlet lines containing two-phase feeds should be designed so that the flow is outside the slug-flow regime. When a horizontal pipe run precedes a vertical rise, a lift orifice or a trap is often advocated (421) in order to prevent liquid from accumulating in the horizontal run. Slugging at the column inlet can lead to severe hydraulic pounding and tray damage, as well as column instability.

4. When the feed is liquid, nozzle velocity should not exceed 3 ft/s (354). This ensures that the entering jet is broken up immediately on entering the column.

   For vapor or mixed feeds, it has been recommended (354,355) that the velocity head at the tower inlet not exceed 10 percent of the pressure drop across one tray or across the packed bed above [this criterion appears in equation form in Ref. 354. The equation has a misprint; the above statement of this rule of thumb is correct (355)]. The author and others (166) feel that this rule is good for packed columns. The author feels that the same rule is somewhat conservative for tray columns, where vapor maldistribution is seldom a concern. The author experienced many tray columns working well with feed velocity heads between 10 and 100 percent of a single tray pressure drop. Perhaps it is appropriate to follow the above conservative practice (355) in general, but to somewhat relax it if it incurs a substantial cost penalty (e.g., if it significantly increases column height). An alternative practice (299*a*) preferred by the author is to size the inlet nozzle to have a pressure drop lower than that of the tray above. Feed velocity heads largely exceeding a single tray pressure drop should be avoided.

   If a large-velocity head cannot be avoided at the inlet to a packed section, a vapor-distributing device (Sec. 3.12) should be used.

5. Tray spacing should be increased, usually by 6 to 12 in, if vapor is present at the feed, or if large-diameter internal feed pipes are used (138, 179, 192, 207, 208, 354). This is particularly important if the feed tray is heavily loaded.

6. High-velocity two-phase and vapor feeds should be avoided. Although wear plates can protect column internals from damage, there have been cases (61) when high-velocity vapor feeds cut holes through wear plates as thick as ¼ in.

7. All internal feed pipes should be removable.

8. It is preferable to locate large internal liquid feed pipes below the trusses of the next higher tray.

9. Components which are present in the column overhead vapor above their dew points may condense locally upon contacting highly subcooled reflux or internal reflux pipes. If the condensing component is corrosive, the reflux piping or tray areas contacting the condensate may experience severe local corrosion. A typical example is where column overheads contain hydrocarbons, steam (above its dew point), and chlorides. Water may condense on the cold surfaces, dissolve and hydrolyze chlorides, thus forming acid. Figure 2.3 (6) shows the end result.

10. Pipe supports should be located near the feed nozzle so that the pipe is not supported by the nozzle. Pipe guides should be used to prevent pipes from swaying in the wind.

11. Alternative feed nozzles are often provided to allow for uncertainties and add flexibility to the design (93, 268). The location of these nozzles should be carefully reviewed, particularly if column, tray, downcomer, or packing dimensions change from the section above the feed to the section below. Feed should be piped to an alternative nozzle only if the column section between the main feed nozzle and the alternative feed nozzle is suitable for processing the liquid and vapor loads which prevail both above and below the feed point. Failure to observe this can cause premature flooding, downcomer unsealing, or packing wettability problems when the feed is inserted into the alternative feed point or if the valve at the alternative feed point leaks.

Figure 2.3 Chloride compound corrosion (*American Petroleum Institute, "Guide for Inspection of Refinery Equipment,"* 4th ed., 1982. Reprinted courtesy of the American Petroleum Institute.)

2.4 Guidelines for Distributors and Multipass-Tray Inlets

Feed inlet distributors are recommended for large-diameter single-pass trays (207, 354). In multipass trays, feed and reflux distributors are essential to ensure uniform distribution. The only exceptions are

1. Pure liquid feeds into two-pass trays, which can be introduced into the central downcomer by arrangements similar to those shown in Fig. 2.1 (arrangements *a, b, d, g, h*) and Fig. 2.2 (arrangement *a*).

2. When feed to multipass trays is to be split unevenly, such as in a three-pass tray. In such cases, a feed trough (Fig. 2.4) is often preferred to a distributor (179, 307).

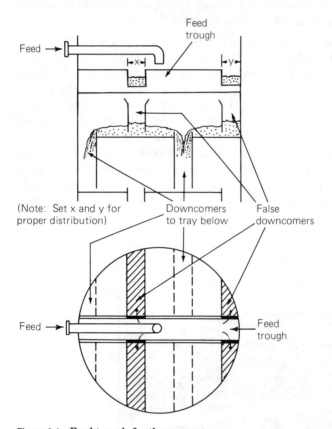

Figure 2.4 Feed trough for three-pass trays.

The guidelines listed earlier for reflux, top-tray, and intermediate feed inlets apply whether a distributor is used or not. In addition, guidelines for avoiding operating problems unique to distributors are listed below.

1. The distributor must distribute the incoming stream evenly among the passes. If a feed is not split equally, liquid maldistribution between the passes will be established that may persist throughout the trays below the feed (12, 88), with a resulting reduction in capacity and efficiency (47). Therefore, correct sizing of the distributor pipe and distributor perforations as well as proper orientation are important. This guideline is most important when more than two tray passes are used. With two-pass trays, Biddulph (35) reported that a disturbance generated by an unequal liquid split is unlikely to persist beyond the few neighboring trays below. However, the disturbance to the neighboring trays can be substantial (12).

2. When good distribution is important (Fig. 2.5a), fluid velocity through the distributor perforations should be considerably higher than through the distributor pipe. It was recommended to make the pressure drop through the distributor perforations at least 5 (237) or 10 times (319, 350) greater than the pressure drop through the distributor pipe. Alternatively, another designer (354) recommends making the hole velocity three times the distributor pipe velocity. This designer (354) also recommends a distributor pipe velocity of 5 ft/s for liquid feeds. Design procedures are discussed elsewhere (319, 350).

   This guideline need not be adhered to when the quality of inlet stream distribution is not of major importance, e.g., when distributing feed to a single-pass tray or to the center area of a two-pass tray.

   With vapor and two-phase feeds, this guideline may be difficult to adhere to. Unless very large hole velocities are acceptable, this guideline calls for low pipe velocities. This, in turn, leads to impractically large pipes. A common compromise is to use a hole area of the same order as the pipe area, at the expense of the inferior distribution profile described in item 3 below (Fig. 2.5b, c). This, in turn, may lead to a tendency of vapor to flow toward one wall (Fig. 2.6a). When vapor maldistribution can be troublesome (e.g., beneath a packed bed), flow-straightening tubes (Fig. 2.6b) can alleviate the problem. Tube length is usually two to three times the perforation diameter.

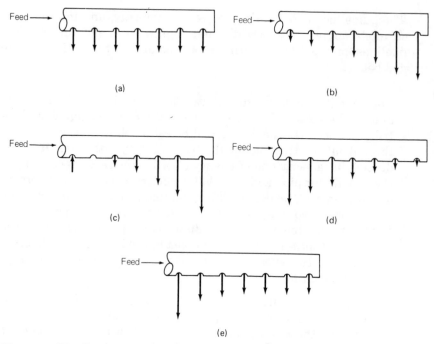

Figure 2.5 Distribution profiles of perforated pipe distributors.(*a*) Ideal distribution; (*b*) excessive fluid velocity through pipe; (c) same as for *b*, but with column vapor sucked in; (*d*) insufficient perforation pressure drop; (*e*) severe hydraulic disturbance near pipe inlet.

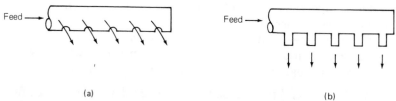

Figure 2.6 Application of flow-straightening tubes. (*a*) Feed tends to channel; (*b*) flow-straightening tubes alleviate channeling.

3. Excessive fluid velocity through the distributor pipe may cause excessive flow through perforations near its closed end. (Fig. 2.5*b*). In the extreme case, vapor from the column may even be sucked into the distributor pipe through perforations near the feed end of the pipe (Fig. 2.5c). If the feed is a subcooled liquid, sucking in vapor may cause hammering. It has been recommended (319, 350) to make the pressure drop through the distributor holes at least 10 times the kinetic energy of the inlet stream.

4. Pressure drop through the distributor perforations should range between 1 to 2 psi and 15 to 20 psi (237). A lower perforation pressure drop may cause maldistribution (Fig. 2.5d), while a higher perforation pressure drop may cause mist formation.

At turned-down conditions, perforation pressure drop is low and the pattern shown in Fig. 2.5d sets in. Perforations near the closed end of the pipe may dry, allowing vapor into the pipe. If the feed is a subcooled liquid, vapor sucked in may collapse onto the liquid, causing instability and hammering. In one case (150a), a column operating at 30 percent of its design rate was shaken by water hammer induced by this mechanism. The hammering was eliminated by orienting the perforations upward so that the pipe was kept full of liquid. A deflection bar was added above the orifices to arrest impingement into the tray above (see item 12 below). An alternative solution to the problem would have been to make distributor modifications (150 a).

5. Severe hydraulic disturbances near the distributor inlet may cause excessive (Fig. 2.5e) or insufficient flow. It is best to avoid sharp bends and high-pressure-drop fittings close to the distributor inlet.

6. Long, continuous slots that are parallel to the length of the distributor pipe are not recommended because they may partially block or partially corrode and then cause maldistribution. Circular perforations or short rectangular slots are preferred.

7. The distributor perforations are a high-velocity area and are likely to deteriorate in service, particularly if the service is corrosive or erosive. Correct material selection is therefore important.

8. Impingement of high-velocity jets issuing from distributor perforations on column walls and other internals should be minimized. Cases have been reported (215, 410) in which such impingement caused severe corrosion; in other situations, it may also cause mechanical damage and erosion.

9. Distributor design should be reviewed for simplicity. The simpler the distributor, the less expensive it is, and the less likely it is to cause trouble.

10. Feed distributors should be located at least 8 in above the tray floor for liquid feeds, and at least 12 in above the tray floor for flashing feeds (207). A distributor of the type shown in Fig. 2.2, arrangement i, is best located so that its centerline is two-thirds of a tray spacing above the tray floor (207).

11. With long distributor pipes (>10 ft), a good practice is to "tee" the feed pipe into the distributor at the center, so that the feed flows from the center toward both ends of the distributor (as shown in Fig. 2.2*i*). With shorter distributor pipes, one end of the distributor is usually connected to the feed nozzle, with feed flow from the inlet end of the distributor pipe toward the other end.

12. At times, with vapor feeds, it may appear attractive to orient the distributor (sparger) perforations upward. In one case, this was successful in overcoming a hammering problem (item 4 above). The author feels that this practice is troublesome and should only be used as a last resort. The following considerations apply to spargers with upward-oriented openings:

- A dead pocket conducive to accumulation of debris, deposits, and undesirable components forms at the sparger floor. The debris may originate in the feed or drop in through the perforations.

- Adequately located drain holes must be provided for shutdown drainage. With vapor-containing feeds, these drain holes must also prevent liquid accumulation in the sparger pipe (e.g., due to liquid pockets in the feed, or liquid weeping through the perforations, or vapor condensation at low rates). In one incident (215, 410), poor drainage of such a vapor-feed sparger caused liquid buildup, which in turn led to maldistribution and corrosion by impingement of feed on the tower wall.

- Impingement of feed on the tray above (and often also on the tower wall) must be avoided using properly placed baffles. Corrosion, maldistribution, and excessive entrainment can result from poorly baffled spargers.

- The upward orientation of sparger openings makes feed distribution patterns difficult to predict, more rate dependent, and often inferior compared to those of standard spargers.

Reflux and Intermediate Feed Distribution and Liquid Redistribution in Packed Towers

The main consideration for introducing reflux or intermediate feed into a packed tower is adequately distributing the incoming stream to the packing. Unlike most tray columns, packed towers are sensitive to distribution. Maldistribution is detrimental to packing efficiency and turndown. The main devices that set the quality of distribution in a packed column are the top (or reflux) distributor, the intermediate feed distributor, the redistributor, and sometimes the vapor distributor. Adequate hydraulics in the inlet area is also important; failure to achieve this can affect distributor performance and can also cause premature flooding.

This chapter examines common distributor and redistributor types and inlet arrangements used in packed columns, outlines the preferred practices, highlights consequences of poor practices, and supplies guidelines for troubleshooting and reviewing designs of distributors, redistributors, and feed and reflux inlets to packed towers.

3.1 Nature and Effects of Maldistribution

A detailed discussion of packed-tower maldistribution is far too bulky for inclusion here and is available elsewhere (e.g., 160, 183, 221, 222, 442, 443). Conclusions which specifically pertain to distribution equipment practices are highlighted below:

1. Packing efficiency may decrease by a factor as high as 2 to 3 due to maldistribution (221, 284, 435, 436).

2. A packed column has reasonable tolerance for a uniform or smooth variation in liquid distribution and for a variation that is totally random ("small-scale maldistribution"). However, the impact of discontinuities or zonal flow ("large-scale maldistribution") is much more severe (219, 221, 222, 442, 443).

3. The necessity for uniform liquid distribution sharply increases with the number of theoretical stages per packed bed (289, 386; see Fig. 3.2b). For less than five theoretical stages per bed, the column is relatively insensitive to the uniformity of liquid distribution, while with ten or more stages per bed, efficiency is extremely sensitive to liquid distribution. A corollary is that beds consisting of small packings or structured packings, which develop more theoretical stages per bed, are substantially more sensitive to maldistribution than equal-depth beds of larger random packings.

4. A packed bed appears to have a "natural distribution," which is an inherent and stable property of the packings (1, 160, 221, 386, 442). An initial distribution which is better than natural will rapidly degrade to it, and one that is worse will finally achieve it, but sometimes at a very slow rate. If the rate is extremely slow, recovery from a maldistributed pattern may not be observed in practice (221, 222).

5. Three factors appear to set the effect of maldistribution on efficiency (160, 193, 284, 443):

   a. Maldistribution delivers less liquid to some areas than to others. In these areas, the liquid-to-vapor ratio is relatively low, causing a composition pinch. The pinched areas contribute little to mass transfer. Vapor leaving these areas is rich with the less volatile components, which contaminate the vapor rising from the rest of the bed. Similarly, lights-rich liquid leaving these areas contaminates the liquid descending from the rest of the bed. The pinches also create nonuniform liquid and vapor composition profiles along the cross section of the column. This is referred to as the *pinching effect.*

   b. Packing particles deflect both liquid and vapor laterally. This promotes mixing of vapor and liquid and counteracts the pinching effect in (*a*) above. This is referred to as the *lateral mixing effect.*

   c. Liquid flow through the packing is uneven. Directly under the distributor, the column wall area is poorly irrigated (unless the distributor nozzles are directed toward the wall). In the bed, liquid tends to flow toward the wall. After some depth, the liquid flow in the wall region exceeds the average flow through the bed.

6. At small tower to packing diameter ratios (<10), the effect of lat-

eral mixing outweighs the pinching effect, and a greater degree of maldistribution can be tolerated without a serious efficiency loss (40, 284, 443). At high ratios of column to packing diameter (>40), the lateral mixing effect becomes too small to counteract the pinching effect (284). This implies that the effects of maldistribution on efficiency are most severe in large-diameter columns and with small-diameter packings (40, 161, 284, 443).

7. Either a shortage or an excess of liquid near the wall causes large-scale maldistribution and can substantially lower packing efficiency (160). If the wall zone is poorly irrigated at the top of the bed, it may take several feet of packing before a reasonable amount of liquid reaches the wall region. This effect is most severe with small packings, where liquid spread toward the wall is slow. On the other hand, buildup of excessive wall flow further down in the bed is most severe with larger packings, where liquid spread toward the wall is rapid.

8. In the presence of large-scale maldistribution, packing efficiency decreases as packing height increases (161, 284, 289, 386, 443). This is due to the composition nonuniformity generated by pinching and to the development of wall flow. With small packings, the above may occur even in the absence of initial maldistribution (443).

9. Liquid maldistribution tends to lower packing turndown (221, 386, 387). The "standard distributor" curve in Fig. 3.1 depicts typical variation of packing HETP (height equivalent of a theoretical

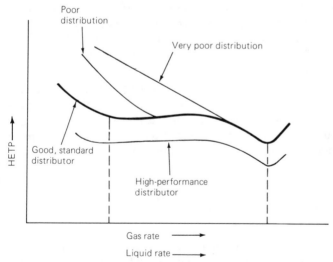

Figure 3.1 Effect of poor distribution on HETP (at constant liquid-to-vapor ratio).

plate) as a function of vapor or liquid flow rate at a constant vapor-to-liquid ratio. The two upper curves represent a progressively lower quality of initial liquid distribution. A curve similar in shape to the uppermost curve is a clear indication of poor distribution. The diagram shows that packing turndown is largely reduced with greater maldistribution.

10. Maldistribution tends to be a greater problem at low liquid flow rates than at high liquid flow rates (103, 217, 387).

11. Vapor is easier to distribute than liquid, but vapor maldistribution can also be troublesome. Vapor flow through packing tends to be uniform if the initial liquid and vapor distribution to the packing is uniform (217, 386).

A nonuniform initial vapor profile is often generated in the column vapor inlet and vapor redistribution regions (183, 237, 292, 325), especially when inlet velocities are high. Although vapor spreads radially through the packing quite rapidly (217), a nonuniform profile will persist at least for some height, causing pinching similar to that described in 5 above. In a number of 15-ft-diameter absorbers (183), vapor maldistribution persisted throughout a 50-ft bed; the resulting efficiency was about half that encountered during good vapor distribution. Vapor maldistribution is most severe in large-diameter columns (183, 289, 386), shallow beds [where the ratio of bed height to column diameter is less than 0.5 (325)], and where the packing geometry resists radial spread (385). The latter may be particularly troublesome with those structured packings that permit substantial radial spread only parallel to their sheets (385). Since the orientation of structured packing sheets usually alters every 8 to 12 in, this flow nonuniformity is unlikely to persist beyond the bottom packing element. However, the disturbance this creates to the composition profile may linger for a greater vertical distance.

Vapor maldistribution may also be induced by liquid maldistribution (217) when vapor flows are high. Areas of high liquid holdup will impede vapor rise and will channel the vapor into the lighter-loaded regions (217). Since liquid tends to accumulate near the wall, vapor will tend to channel through the center.

3.2   Quantitative Definition of Liquid Irrigation Quality

Moore and Rukovena (289) developed an index for quantifying the quality of liquid irrigation to a packed column. This index is given by

$$D_Q = 0.40(100 - A) + 0.60B - 0.33(C - 7.5) \qquad (3.1)$$

In Eq. (3.1), D_Q is the distribution quality rating index in percent. The higher D_Q, the better the irrigation quality. Typical indexes are 10 to 70 percent for most standard commercial distributors; 75 to 90 percent for intermediate-quality distributors; and over 90 percent for high-performance distributors (289). Figure 3.2a shows efficiency improvements accomplished by improving this index in various commercial columns.

In order to determine D_Q, each distributor drip point is represented by a circle. The center of the circle is located where the liquid from each drip point strikes the top of the bed. The area of each circle is proportional to the liquid flow, and the sum of all circle areas equals the tower cross-sectional area. If the liquid is evenly divided among all drip points, the area of each circle equals the tower cross-sectional area divided by the number of drip points. Terms A, B, and C in Eq. (3.1) are then evaluated as follows (Fig. 3.3a).

A is the percent of the cross-sectional area at the top of the bed which is not covered by the drip point circles. This is a direct measure of the fraction of unirrigated area at the top of the bed.

B is evaluated by selecting a continuous region at the top of the packing, occupying one-twelfth of the column cross-sectional area. This is the area in which the largest deviation from the average flow occurs. If this area is underirrigated, B is evaluated by dividing the circle area enclosed within this region by the area of the region (i.e., by one-twelfth of the column cross-sectional area). If the area is overirrigated, B is evaluated by dividing the area of the region by the circle area enclosed within the region. The lowest value of B anywhere in the column is used in Eq. (3.1). The value thus calculated is multiplied by 100 so that it is expressed as a percent. B gives an empirical measure of large-scale maldistribution.

C is the total area of overlap of adjacent drip point circles expressed as a percent of tower cross-sectional area.

Figure 3.3b to d are examples used by Moore and Rukovena (289) for illustrating the application of their technique for rating distributors. Additional examples are in their paper.

A correlation (Fig. 3.2b) proposed by Moore and Rukovena (289) can be used to determine the efficiency loss in a packed tower containing pall rings or Metal Intalox packing as a function of their distribution quality rating D_Q and the number of stages in the bed. This correlation was only recently proposed, and more experience with its predic-

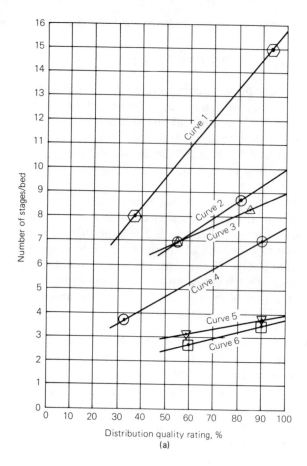

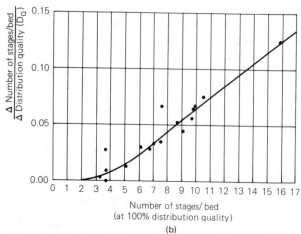

Figure 3.2 Effect of irrigation quality on packing efficiency. (a) Case histories demonstrating efficiency enhancement with higher distribution quality rating. (b) Correlation of the effect of irrigation quality on packing efficiency. (*From F. Moore and F. Rukovena, "Chemical Plant and Processing, Europe edition, August 1987. Reprinted courtesy of* Chemical Plant and Processing.)

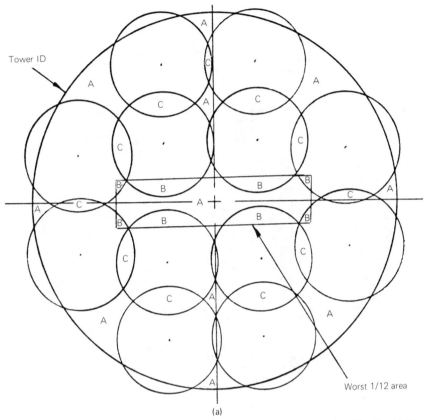

Tower ID

Worst 1/12 area

(a)

Figure 3.3 Distribution quality rating applications. (a) Areas considered in liquid distribution quality rating (A = cross-sectional tower area not covered by point circles, B = point circle area in 1/12 tower area, C = area of overlap of point circles.)

tion is required before it can be generally applied with confidence. Nevertheless, it is simple to use and may be valuable, at least as a preliminary guide.

Regardless of the validity and accuracy of the final correlation, the analysis proposed by Moore and Rukovena (289) for determining irrigation quality is a valuable tool for troubleshooting distributors and examining their performance. An inspection of Fig. 3.3b to d readily pinpoints regions of large-scale maldistribution and enables visualization of irrigation troublespots. The author strongly recommends using this or a similar analysis when evaluating distributor performance.

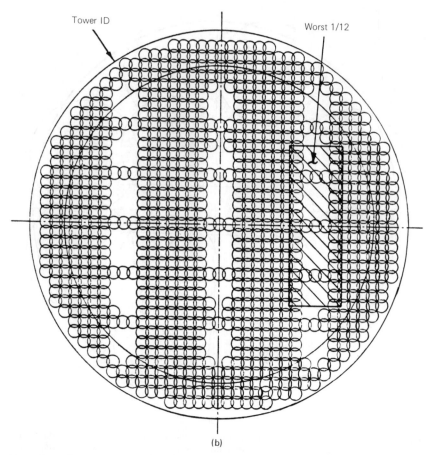

(b)

Figure 3.3 (*Continued*) (*b*) Standard quality distributor, $D_Q = 60$ percent.

3.3 Types of Liquid Distributors

Any liquid distributor gives some maldistribution, because for practical reasons, liquid can only be divided into a limited number of streams. From these point sources the liquid spreads. The main considerations in selecting a distributor for a given service are compatibility with the service and avoiding large-scale maldistribution.

Most of the common types of commercial liquid distributors are shown in Fig. 3.4 and compared in Table 3.1. Table 3.1 was compiled using information available in Refs. 74, 111, 142, 212, 224, 237, 257, 305, 319, 386, and 438, together with the author's experience.

Several modern designs, often referred to as *high-performance dis-*

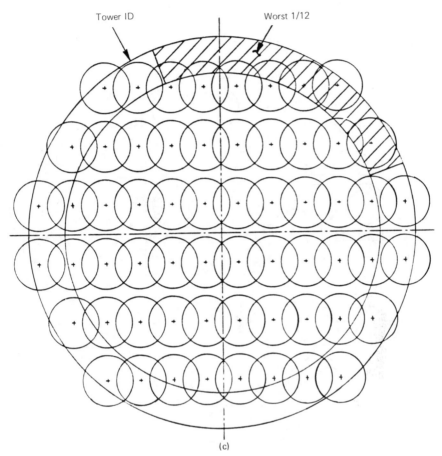

Tower ID   Worst 1/12

(c)

Figure 3.3   (*Continued*) (*c*) Standard quality distributor, $D_Q = 72$ percent.

tributors (e.g., Fig. 3.5) are sophisticated versions of the common types in Figure 3.4. They incorporate features for minimizing large-scale maldistribution and for improving distributor compatibility with the service. These high-performance distributors are usually proprietary custom-designed devices and can be expected to perform better than standard distributors (Fig. 3.1) when properly designed, fabricated, and installed. Several cases have been reported (141, 156a, 219, 289, 304) of substantial column efficiency enhancements resulting from replacement of standard distributors by their high-performance counterparts. Some of these are depicted in Fig. 3.2a. However, the nonstandard nature of high-performance distributors makes them more expensive, more complex, and more susceptible to errors. Many of the

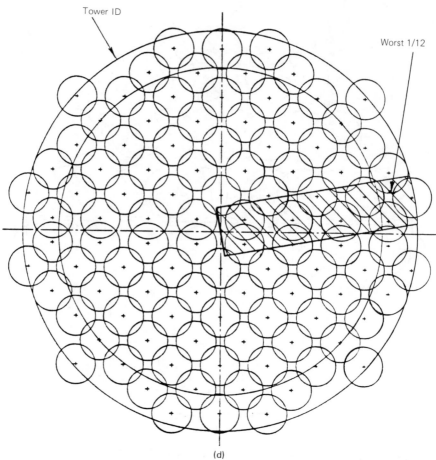

(d)

Figure 3.3 *(Continued)* (*d*) Intermediate quality distributor, $D_Q = 84$ percent. *(Parts a to d from F. Moore and F. Rukovena, "Chemical Plant and Processing, Europe edition, August 1987. Reprinted courtesy of* Chemical Plant and Processing.)

unique features incorporated in high-performance distributors will be highlighted in the following discussions.

Liquid distributors are usually classified into pressure distributors and gravity distributors (Table 3.1). In general, pressure distributors provide more open area for vapor flow and tend to be less expensive, lighter, less robust, and to require smaller lead-up piping than gravity distributors. Their disadvantages are high operating cost (because of the liquid pressure drop), susceptibility to plugging and corrosion, entrainment, and a relatively inferior quality of liquid distribution. The common pressure distributors are the perforated-pipe type and the spray type.

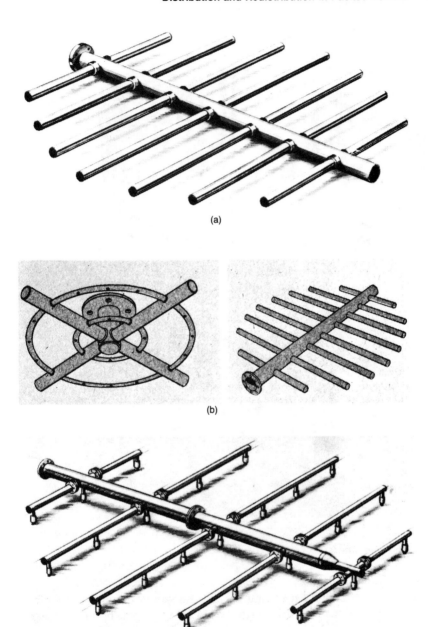

(a)

(b)

(c)

Figure 3.4   Common types of liquid distributors, (a) Ladder pipe distributor; (b) perforated ring distributor; (c) spray distributor; (d) orifice pan distributor; (e) tunnel orifice distributor; (f) notched-trough distributor; (g) weir-riser distributor. (Parts a and c to f reprinted courtesy of Norton Company; part b, reprinted courtesy of Koch Engineering Company, Inc.; part g, reprinted courtesy of Glitsch, Inc.)

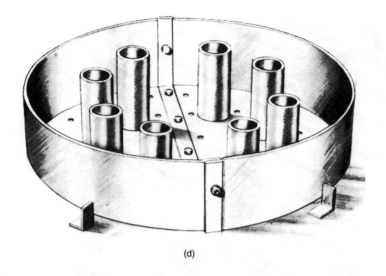

(d)

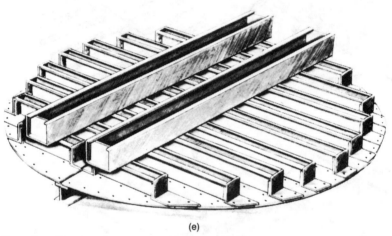

(e)

Figure 3.4   *(Continued)*

The common gravity distributors are the weir type and the orifice type. Both types can handle large liquid flow rates. The weir type is generally one of the least troublesome distributors and has an excellent turndown, but it can usually provide only a limited number of drip points and is extremely sensitive to levelness and liquid surface agitation. The orifice type may suffer from corrosion and plugging, but it can be designed with a large number of drip points to provide superior liquid distribution.

Each of the common distributor types is discussed in detail below.

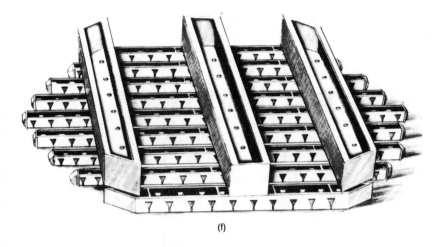

(f)

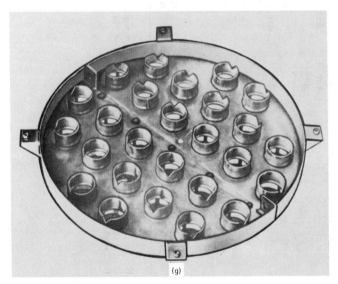

(g)

Figure 3.4   (*Continued*)

3.4   Perforated-Pipe Distributors

Perforated-pipe distributors are normally of the ladder type (Fig. 3.4*a*) or the perforated-ring type (Fig. 3.4*b*). Perforations are located on the underside of the pipes. The ladder type is usually easier to fabricate and therefore less expensive than the perforated-ring type. A high performance variation of the Fig. 3.4*a* distributor is shown in Fig. 3.5*b*.

TABLE 3.1 Liquid Distributors

	Ladder ("spider pipe")	Perforated ring	Multiple spray	Orifice pan	Tunnel orifice	Notched trough	Weir riser
Diagram	3.2a	3.2b	3.2c	3.2d	3.2e	3.2f	3.2g
Driving force	PR	PR	PR	G	G	G	G
Type	PP	PP	S	O	O	W	W
Materials available	M,P	M,P	M	M,P,C	M	M,P,C	M,P,C
Tower diameter, in	>18	>36	Any	Any usually <48	Any usually >48	Any usually >24	Any usually <48
Plugging tendency	Med	Med	L-Med	H	H	L	L
Resistance to gas flow	L	L	L	H	Med	L	H
Prone to uneven levelness	N	N	N	Mainly at low rates	Mainly at low rates	Y	Y
Affected by corrosion	Y	Y	Some	Y	Y	N	N
Prone to liquid surface agitation	N	N	N	Y	Y	Y	Y
Likely to cause entrainment	Y	Y	Y	N	N	N	N
Turndown	L	L	L	Med	L	H	Med
Approximate range of liquid rates for standard design (305), gpm/ft^2	1–10	1–10	Wide	1–30	1.5–70	1–50	1–10
Weight	L	L	L	H	Med	Med	Med
Quality of distribution	Med	Med	L-Med	H	H	Med	Med

Legend

C	= ceramic	L	= low	N	= no
G	= gravity	M	= metal	O	= orifice
H	= high	Med	= medium	P	= plastic

PP = perforated pipe   W = weir
PR = pressure   Y = yes
S = spray

48

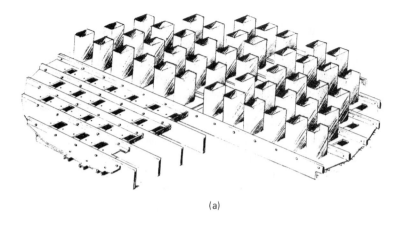

(a)

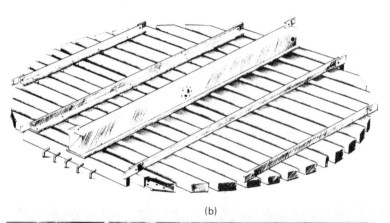

(b)

(c)

Figure 3.5 Examples of high-performance distributors. (*a*) Deck type; (*b*) lateral arm type; (*c*) tunnel-orifice type, with orifices drilled in the angled wall, just above the bottom of the troughs. Liquid leaving each orifice hits a baffle which spreads the liquid laterally. (*Parts a and b reprinted courtesy of Norton Company; part c reprinted courtesy of Glitsch Inc.*)

The quality of distribution achieved with perforated-pipe distributors is generally somewhat inferior to that achievable with orifice-type distributors (below). The higher liquid pressure drop available in perforated-pipe distributors (compared to gravity orifice-type distributors) induces a greater liquid flow per unit area; this in turn restricts the numbers of drip points. In an effort to reduce the liquid pressure drop, a gravity trough replaces the pressure header in some high-performance variations. If it is practical to provide a sufficient number of evenly spaced drip points per unit of column cross-sectional area, the perforated-pipe distributor can provide a distribution as good as orifice-type distributors. To improve irrigation evenness, some high-performance variations have some additional perforations drilled at an angle to the vertical (Fig. 3.5b).

The perforated-pipe distributor is best suited where vapor mass velocities are high and where an open area in excess of 70 percent is needed to avoid localized flooding (305). Together with the spray type, the perforated-pipe distributor offers the highest vapor flow area. However, the maximum liquid flow recommended for this type of distributor is relatively low and should not exceed 10 gpm/ft^2 of column area (305) with standard designs.

Another advantage of the perforated-pipe distributor is its low cost. Its construction is simple, it is easy to support, and it generally consumes less vertical space than most other distributors. Guidelines for selection, design, and operation of perforated-pipe distributors are given below.

1. The perforated-pipe distributor is suitable for liquid feeds only and should be avoided when vapor is present. This distributor also needs to be running full if uniform distribution is to be achieved. A method to check this is presented elsewhere (111). A case where this type of distributor performed poorly with a partially vaporized feed has been reported (76).

2. It is generally recommended that perforated-pipe distributors be located 6 to 8 in above the top of the packed bed (305) to permit vapor disengagement from the bed before passing through the distributor. At least one designer (438) feels that this distributor may be embedded directly in the packing, thus using the packing layers above it as a mist eliminator. The author would not recommend the latter practice, as it may cause premature flooding of the tower.

3. Guidelines listed in Sec. 2.4, with the exception of guidelines 10 and 12, also apply to perforated-pipe distributors. In addition, it has been recommended (111, 150a) that for perforated-pipe dis-

A pipe distributor under test. Three rows of orifices are punched in each pipe. A certain minimum irrigation rate is needed for uniform distribution

(a)

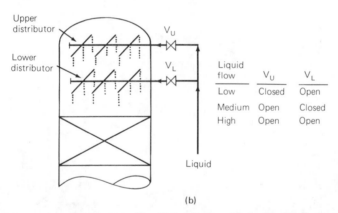

Liquid flow	V_U	V_L
Low	Closed	Open
Medium	Open	Closed
High	Open	Open

(b)

Figure 3.6   Techniques for improving liquid distribution. (*a*) A pipe distributor under test. Note that three rows of orifices are punched in each pipe to improve irrigation. (*b*) Dual liquid distributor. (*c*) An orifice pan distributor equipped with drip tubes to prevent plugging. (*d*) An orifice pan distributor, equipped with drip tubes to prevent plugging, with tubes extending all the way to the trim of the pan, and narrow criss-cross vapor risers to minimize unirrigated area and optimize vapor distribution. Note the distributor feeder arrangement; a feeder box above the troughs to minimize frothing, baffles in troughs to resist wave formation and surface unevenness, and liquid equalizing pipes between troughs. (*Part a from Chen, G. K., The Chemical Engineer, Supplement, September 1987. Reprinted courtesy of The Institution of Chemical Engineers (UK); part c, reprinted courtesy of Koch Engineering Company, Inc.; part d, reprinted courtesy of Nutter Engineering.*)

tributors, liquid velocity through the perforations should not exceed 4 to 6 ft/s.

4. The perforated-pipe distributor is best avoided in services where plugging may occur (111, 142, 305), such as when solids are present or when the liquid is close to its freezing point. A partially

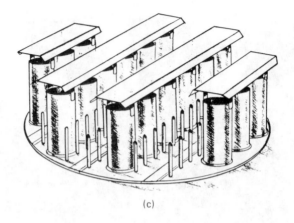

(c)

Figure 3.6 (*Continued*)

(d)

plugged distributor may perform worse than no distributor at all. If it is still desired to use this distributor with a solids-containing stream, adequate filtration (Sec. 3.8) is mandatory.

5. The perforated-pipe distributor is best avoided when the liquid may corrode, erode, or otherwise expand the orifices, because some orifices may expand more than others, resulting in maldistribution.

6. Excessive liquid pressure drop through the distributor should be avoided, because this may restrict the number of drip points. The liquid line to the distributor should contain a control valve or a restriction orifice to let down any excessive pressure. A more satisfactory alternative is to use the high-performance perforated-pipe distributor variation which lets out liquid by gravity instead of pressure.

7. The perforated-pipe distributor has a relatively low turndown ratio, roughly 2:1 to 2.5:1 (111, 305). Excessive liquid flow rates may generate fine mist, while deficient liquid flow rates may generate uneven irrigation (Fig. 2.5d). The turndown can be enhanced by using a dual liquid distributor (Fig. 3.6b; Sec. 3.8).

3.5 Spray Distributors

Spray distributors (Fig. 3.4c) are pipe headers with spray nozzles fitted on the underside of the pipes. They are most popular in heat transfer and scrubbing services and are infrequently used in fractionation (75, 386). Services where spray distributors are common include refinery crude towers (237, 298), FCC main fractionators (299a), and refinery vacuum towers (147, 297, 298, 299a, 335, 343, 386). Spray distributors are also used in very small columns (where a single spray nozzle adequately covers the entire cross-sectional area), and in applications where a large vapor-handling capacity is most important. Some designers (144) recommend avoiding spray distributors in small-diameter columns.

The quality of distribution provided by spray distributors may be inferior to any of the others because the spray cones create areas of uneven irrigation (224, 386), the spray cones are often nonhomogeneous, and because a significant amount of liquid is directed toward the wall (257). Factors such as spray angle, height of the spray nozzles above the bed, nozzle construction, and nozzle pattern set the quality of distribution. The effect of these factors on the quality of distribution is poorly understood, and good nonproprietary ground rules for spray distributor design are unavailable. In short packed beds, the sprays themselves may significantly contribute to mass and heat transfer (386). This can partly account for the favorable experiences with spray distributors in refinery vacuum towers, which typically have short beds.

Other performance characteristics of spray distributors are generally similar to perforated-pipe distributors. Like perforated-pipe distributors, spray distributors are of simple construction, are easy to support, and are inexpensive. Compared to perforated-pipe distributors, spray distributors offer an even larger open area, a greater liquid flow rate, easier replacement of corroded or eroded sections, and more extensive irrigation. On the debit side, spray distributors require higher pumping horsepower than other distributors, and an overhead mist eliminator is mandatory in order to control entrainment. Compared to perforated-pipe distributors, spray distributors also consume much more vertical space.

Guidelines for selection, design, and operation of spray distributors are listed below:

1. Typically, spray distributors use wide-angle (120°) sprays, and are located 18 to 36 in above the bed, providing irrigation area of the order of 5 to 10 ft^2 per spray nozzle. Typical pressure drops are 5 to 30 psi.

2. Some overlap of the spray footprints at the top of the packing is beneficial (1). An overlap of about 100 percent (double coverage) has been recommended for uniform irrigation (386).

3. The sprays should be as homogeneous as possible. Liquid distribution is sensitive to even a slight spray nonhomogeneity (1). The sprays should be inspected for homogeneity during distributor water tests (Sec. 3.8). If desired, a sample of the nozzles to be specified can be obtained ahead of time, hooked to a water pipe, and tested for homogeneity before the nozzles are specified.

4. Areas of either overspraying or underspraying near the column wall may severely degrade liquid distribution (1) and should be avoided.

5. Spray distributors are not suitable for vapor-containing streams.

6. Like perforated-pipe distributors, spray distributors are sensitive to plugging, corrosion, and erosion. The sensitivity to plugging is somewhat lower with spray distributors, while the sensitivity to corrosion and erosion is somewhat higher compared to perforated-pipe distributors because of the higher liquid velocity. "Nonplugging" spray nozzles are available, but their effectiveness is uncertain (297). If a spray distributor is used with solid-containing streams, adequate filtration (Sec. 3.8) is mandatory. There were cases (335) where not only the spray nozzles but also the header plugged.

7. Spray distributors should never be embedded in the packing, because this is likely to cause premature flooding (237). For good distribution, a spray distributor containing more than a single spray nozzle should be placed at least 18 in, and preferably more than 24 in, above the bed.

8. Like perforated-pipe distributors, the spray distributor turndown ratio is about 2:1 (111). Excessive liquid flow rates may create a mist problem, while low liquid flow rates reduce the cone diameter and may create poorly irrigated areas. For this reason, oversized spray nozzles must be avoided (237). As with perforated pipe distributors, a dual liquid distributor can be used to enhance turndown (Sec. 3.8).

3.6   Orifice Distributors

Orifice distributors are usually of the pan type (Fig. 3.4d) or of the tunnel type (Fig. 3.4e). The former type is best suited for small-diameter columns (<4 ft), while the latter is used in larger-diameter columns (>4 ft). High-performance variations of the Fig. 3.4e distributor are shown in Fig. 3.5a, c and 3.6a, d.

An orifice pan distributor consists of a pan equipped with circular or rectangular risers for vapor flow and perforations in the pan floor for liquid flow. The pan may rest on a support ring; alternatively, it may be supported on lugs in a manner that provides an annular space for vapor rise between the distributor and the column wall.

Orifice tunnel distributors consist of parallel troughs with perforations for liquid flow in the trough floors. Vapor rises in the space between the troughs. The troughs are often interconnected by cross channels that equalize liquid levels in different troughs (Fig. 3.6d). Level-equalizing channels are most important in columns greater than 10 ft in diameter (111, 349).

Orifice distributors can incorporate a large number of drip points and therefore have the potential for providing better liquid distribution than most other distributor types. This better liquid distribution is not always achieved, the main restricting factors being difficulty in irrigating areas beneath vapor passages and supports, a high sensitivity to uneven spread of drip points, and a high sensitivity to plugging and construction irregularities.

In some high-performance designs (e.g., Fig. 3.5c), side orifices and deflection baffles are used (75a, 224a); item 9 below has further details.

Orifice distributors are capable of handling high liquid loads, with standard orifice pan distributors and orifice tunnel distributors delivering up to 30 and 50 to 70 gpm per square foot of bed, respectively (305). The open area for vapor flow is relatively low in orifice distributors.

Orifice distributors are also generally larger, more expensive, consume more vertical space, and are more difficult to support than most other distributors. Tunnel orifice distributors provide greater open areas for vapor flow, are easier to support, and are more suitable for large-diameter columns than orifice pan distributors.

Guidelines for selection, design, and operation of orifice distributors are listed below.

1. It is frequently difficult to incorporate sufficient open area for vapor rise while maintaining proper irrigation of areas beneath vapor risers. Inadequate irrigation under vapor risers (e.g., Fig. 3.3b) may cause large-scale maldistribution, which can be detrimental to column efficiency. To avoid this problem, it is best to provide a large number of small risers (this can be expensive). The risers can be rectangular (e.g., Figs. 3.5a, 3.5c, 3.6d) or round. One designer (74) recommends a riser diameter of 4 to 6 in when round risers are used, while another designer (166) prefers 2 to 4 in. Alternative solutions are installing short tubes (e.g., Fig. 3.5b) which direct liquid to poorly irrigated areas (436), or drilling holes in the sidewalls of orifice troughs (Fig. 3.6a) or using criss-cross risers (Fig. 3.6d). Such solutions must be carefully engineered and properly tested, and prediction of their performance may be difficult.

   To incorporate sufficient open area with orifice pan distributors, a pan of smaller diameter than the column is sometimes specified, with the annular space between the pan and the wall utilized for vapor rise. This practice may leave a significant fraction of column area unirrigated. For instance, a 4-in-wide annulus in a 4-ft column will leave 30 percent of the bed cross-sectional area unirrigated.

   If it becomes impractical or too expensive to provide sufficient open area without forming poorly irrigated regions, it is best to consider an alternative distributor type.

2. Wide support beams or support rings may generate poorly irrigated areas on top of the packing, and therefore large-scale maldistribution. The distributor supports should be carefully reviewed to ensure proper irrigation underneath, especially in the wall region. Sometimes, short tubes can be used to direct liquid to unirrigated areas under distributor supports, but these need to be carefully designed and tested.

   Supporting an orifice distributor directly on the packing is not recommended, because it may be misaligned during column upsets. In addition, this practice does not permit adequate vapor disengagement from the bed and may cause maldistribution and premature flooding.

3. Orifice distributors are favored in foaming services, because liquid drip points are separated from the vapor risers (111, 349).

4. Liquid aeration in pans or troughs can be troublesome in high-pressure, high-liquid-rate, and foaming services. As the liquid enters the distributor liquid pool from above, vapor bubbles are dragged into the liquid (a "waterfall pool" effect). In low-pressure services, these bubbles readily disengage from the liquid and aeration is seldom troublesome (289). On the other hand, vapor disengagement is slow in foaming and high-pressure services, while in high-liquid-rate services, residence time for vapor disengagement may be too small (289). Frothing will then occur, causing uneven perforation flow and excessive liquid height in the distributor (also see item 7 below). The problem is often overcome by using a closed pan or trough (e.g., similar to Fig. 3.5b), or by other techniques (Sec. 3.9).

5. Methods for sizing orifice distributors were described by Chen (74) Fadel (111), and Kaiser (183a). Some highlights are described below. A most useful relationship is (74, 150a)

$$Q = 5.46Knd^2h^{0.5} \qquad (3.2)$$

Where Q = total liquid flow rate, gpm
n = number of orifices
d = orifice diameter, inches
h = liquid head loss across orifices, inches
K = orifice discharge coefficient

For punched holes, it has been recommended (74, 150a) to use $K = 0.707$ but values as low as 0.62 to 0.63 (111, 183a) are sometimes used. Note that h is equal to the liquid height in the pan or troughs minus the vapor head loss (expressed as inches of liquid) in the distributor risers.

Jets leaving the orifices may be unstable; they may move laterally or even break into spray. This can be visualized by experimenting with a household tap. Kaiser (183a) proposed a correlation in terms of the Reynolds and Weber numbers to predict jet stability. However, jet stability is also dependent on surface irregularities at the orifices, and these are difficult to predict.

To improve jet instability, some high-performance distributors have each perforation equipped with a short, flow-straightening tube. This alleviates spray formation and minimizes any lateral movement of jets. Kaiser (183a) recommends using these whenever spray formation is predicted.

6. The flow area of the vapor risers is set by the allowable pressure drop. Too large an area may promote poor irrigation; too small an area leads to excessive pressure drop. The pressure drop must be low enough to satisfy system criteria and to avoid excessive liquid backup in the pan or troughs. Excessive liquid backup will overflow into the vapor risers, leading to maldistribution and possibly premature flooding. It has been recommended (74, 166) to make riser area 15 to 45 percent of tower cross-sectional area. Typically, riser pressure drop is 0.25 in of liquid (150a). Methods for calculating riser pressure drops are detailed elsewhere (74, 111, 420).

7. Liquid depth in the distributor pan or troughs dictates the riser height. According to one designer (74), the operable liquid depth ranges from ½ in at minimum liquid flow rates to 1 in below the top of the riser at maximum flow rates. The author feels that an additional margin of ½ to 1 in or greater at each end of the range is frequently justified.

Excessive liquid depth may spill liquid into the vapor risers, causing maldistribution and possibly premature flooding. Liquid spillage is promoted when the liquid is aerated (item 4 above), agitated, or when plugging occurs. One designer (166) recommends that the normal liquid head be set at 50 to 70 percent of the riser height. Another designer (111) recommends setting the liquid level such that no spillage occurs when 10 to 15 percent of the orifices are plugged. The author is familiar with one column that achieved poor efficiency because liquid spilled into the vapor risers. The spill was caused by undersizing of the total orifice area.

Insufficient liquid depth is likely to cause maldistribution and may allow some perforations to dry, permitting vapor flow through them (111). The lower the liquid depth, the greater the sensitivity of the irrigation pattern to distributor out-of-levelness, fabrication irregularities, perforation corrosion, and liquid surface agitation. In addition, the fall height of liquid from the feed pipe to the distributor increases as liquid depth in the distributor declines, thus increasing aeration and liquid surface agitation.

Within the above limits and the turndown requirements, there is often an incentive to minimize the maximum liquid depth. The lower the maximum liquid depth, the greater is the number of drip points that can be incorporated, the smaller is the vertical space consumed by the distributor, and the lesser is the distributor cost and support requirements. Riser height is usually about 6 in for standard orifice distributors (74, 305) and 8 to 12 in for high-performance variations.

8. Turndown ratios of orifice pan distributors are relatively high, with ratios of up to 4:1 achievable with standard designs (74, 166, 305). The turndown ratio is lower with tunnel orifice distributors, with standard designs achieving ratios of about 2.5:1 (111, 305). Higher ratios can be achieved by using taller pans or troughs and taller gas risers (111).

9. Orifice distributors should be avoided in services where plugging may occur (111, 142, 305, 438), such as when solids are present or when liquid is close to its freezing point. In one reported case (75), plugging of an orifice pan distributor (with large orifices) in a fouling service caused liquid maldistribution and poor separation efficiency. Compared to perforated-pipe distributors, orifice distributors have lower liquid velocities, larger liquid residence time, and open pans (or troughs) into which solids can be carried over during upsets and which can overflow when perforations are plugged. All these factors render orifice distributors more sensitive to plugging than even perforated-pipe distributors. If it is still desired to use an orifice distributor with a solids-containing stream, adequate filtration (Sec. 3.8) is mandatory, but may be insufficient to avoid plugging. If deposits adhering to the column top head may drop into the distributor troughs (or pan), a closed-trough design (Fig. 3.5b) or trough covers should be considered.

A technique sometimes used to minimize plugging and its effect on distribution is to provide perforations or V-notches in the side of the troughs (Figs. 3.5c, 3.6a); alternatively, perforations or V-notches near the top of the troughs can be provided to ensure an even overflow.

The technique of using side openings (e.g., Fig. 3.6a) can be troublesome, because the liquid jets follow a complex trajectory motion. As a result, the points at which the jets hit the top of the packing become a complex function of the liquid head in the troughs and of the vertical distance from the openings to the top of the packing. This may generate maldistribution. To overcome this problem, deflecting baffles are often installed in front of the orifices (Fig. 3.5c), and these deflect the liquid downward to a desired location (224a). These baffles have the added advantage of breaking the liquid jets and converting them into liquid sheets (a "men's urinal wall" principle). With structured packings, liquid sheets are advantageous; they are introduced perpendicular to the crimp openings, thus ensuring even irrigation to all flow channels (224a). However, the effectiveness of the baffles in converting the point sources into liquid sheets maybe low, even nonexistent. The author had experienced one distributor water test where the baffles were effective in deflecting the liquid directly downward to the desired locations and in giving even irrigation, but completely ineffective in converting the point sources into liquid sheets.

Another technique to minimize plugging (212) is to equip each orifice with a short drip tube which rises vertically above the bottom of the pan (Figs. 3.6c, d), so that solids settling at the bottom of the pan do not enter the orifices.

The above techniques must be carefully engineered and adequately tested to ensure even irrigation.

10. Orifice distributors are best avoided in corrosive services, because some orifices may expand more than others.

11. The method of drilling the perforations is important. Fabrication irregularities on the top surface of the pan or troughs may increase the flow resistance of some perforations compared to others. On the other hand, fabrication irregularities on the bottom surface may induce liquid flow along the bottom face of the pan or trough, and therefore, uneven irrigation. For further discussion, see Sec. 3.8.

12. Orifice distributors are sensitive to out-of-levelness and to liquid surface agitation, particularly when liquid depth is either low or close to the point of overflowing the risers. Both out-of-levelness and liquid surface agitation cause uneven liquid depths, and therefore an uneven irrigation pattern to the bed below. At high liquid rates, both may cause uneven and premature liquid overflow into vapor risers. Section 3.9 discusses techniques for minimizing liquid surface agitation. Levelness tolerances of ⅛ and ¼ in have been recommended (289) for orifice distributors in towers 1.5 to 8 ft and 8 to 20 ft in diameter, respectively.

3.7 Weir Distributors

Weir distributors are usually of the weir riser type (Fig. 3.4g) or the notched-trough type (Fig. 3.4f). The former type is commonly used in small-diameter columns (<4 ft), while the latter is used in larger-diameter columns (>3 ft), but can also be used in smaller columns.

A weir riser distributor consists of a pan equipped with cylindrical risers with a V-notch cut in each riser. The V-notch allows liquid to descend countercurrently to the rising vapor. A major disadvantage which renders the weir riser distributor unpopular is the interdependence of the maximum vapor and maximum liquid flow rates. At a tower F-factor (superficial vapor velocity by square root of vapor density) of 1.0 (ft/s) $(lb/ft^3)^{0.5}$, the maximum liquid flow rate of a standard weir riser distributor is 10 gpm/ft^2; when the F-factor increases, the maximum liquid flow rate declines (305). A method for calculating the maximum liquid flow rate as a function of the vapor rate is presented elsewhere (111). The open area of this distributor is low and of similar magnitude to that of an orifice pan distributor (305).

In order to reduce the vapor-liquid interaction, standard weir riser pans are usually smaller than the column diameter and are supported on lugs, leaving an annular space for vapor rise between the distributor and the tower wall. This, however, creates an unirrigated region near the column wall, which may cause large-scale maldistribution. Other performance characteristics of this distributor are similar to those of the notched-trough distributor (below).

Notched-trough distributors consist of parallel troughs with V-notches cut in their sides for liquid flow. Vapor rises through the space between the troughs.

The quality of distribution provided by notched-trough distributors is generally somewhat inferior to that achievable by orifice-type distributors. With notched-trough distributors, it is generally difficult to incorporate more than three to four drip points per square foot of column cross-sectional area (111, 438). It may also be difficult to space these drip points evenly. If it is practical to provide a sufficient number of drip points per unit of column area, and to space them evenly, this distributor can provide a distribution as good as an orifice-type distributor.

To enhance the number of drip points per unit area, some recent designs have perforations drilled in the bottom and/or sides of the troughs. Other designs have two rows of notches; the lower row has a larger number of smaller notches, while the upper row contains fewer, larger notches. At low liquid rates, only the bottom row is active; at higher liquid rates, both rows are active, with the bottom row notches covered by liquid and acting as orifices. Designs incorporating perforations or two rows of liquid must be carefully engineered and adequately tested, and prediction of their performance may be difficult. The presence of side notches or perforations at the bottom of the troughs may also render the distributor sensitive to plugging, corrosion, or fabrication irregularities.

The notched-trough distributor is one of the most popular types of distributors (111, 305, 438). Notched-trough distributors are insensitive to plugging, corrosion, and erosion; are the least likely to give difficulty during operation; and can effectively handle a wide range of feeds at a good turndown. They are capable of handling large quantities of liquid, with standard designs delivering up to 50 gpm per square foot of column area (305). They offer a reasonably large open area for gas flow, with standard designs having up to 55 percent open area (111). In general, they also consume less vertical space, are easier to support, and are less expensive than orifice distributors. On the debit side, they are extremely sensitive to out-of-levelness, liquid surface agitation, and hydraulic gradients in the troughs.

Guidelines for selection, design, and operation of notched-trough distributors are listed below:

1. With notched-trough distributors, poor irrigation is most likely to occur beneath wide troughs, wide vapor passages, supports, or near the column wall. Inadequate irrigation in such areas can cause large-scale maldistribution which may be detrimental to column efficiency.

2. The liquid head in the trough must be between the V-notch base and apex for the entire operating range. Typically, the liquid head is about 1 and 3 in above the apex for the minimum and maximum liquid flow rates, respectively. The author would generally recommend avoiding liquid heads lower than 1 in. A lower minimum allowable liquid head permits more drip points to be incorporated, but at the expense of a higher sensitivity of the irrigation pattern to out-of-levelness, liquid surface agitation, and trough hydraulic gradients.

   Typical V-notch angles range from 30 to 60°. Smaller angles allow more drip points to be incorporated, but increase distributor cost and its sensitivity to fabrication irregularities. Often, the bottom apex of the notch is rounded into a γ shape or extended into a narrow slot to provide a smaller angle. Methods for sizing notched-trough distributors are discussed elsewhere (111, 228).

3. Most weir distributors are insensitive to plugging and corrosion. They can handle large volumes of solids as well as liquids near their freezing point.

4. Weir distributors are prone to out-of-levelness more than any other distributor because the flow rate through a triangular notch is proportional to the liquid head raised to the power of 2.5 (flow rate through an orifice is only proportional to the liquid head raised to the 0.5 power). With weir distributors, an out-of-levelness of 1 or 2 in is sufficient to cause severe maldistribution (237, 436). One case was reported (436) in which a distributor in a 40-ft tower was installed to a level tolerance of $\pm$ 1/16 in to avoid this problem. Adjustable leveling screws are often provided and should always be specified with this distributor to enable an in situ level adjustment. The problem is most severe at low liquid rates and turned-down conditions.

The same reason renders weir distributors most sensitive to liquid surface agitation and trough hydraulic gradients. Careful design of the troughs, the parting boxes, and their feeds is mandatory. Tunnels that equalize liquid level in the troughs (Fig. 3.6d) are often provided in large-diameter (> 10-ft) columns. Excessive horizontal liquid velocities in the troughs should be avoided. The trough hydraulic gradient head can be calculated from (289)

$$h = 0.187v^2 \qquad (3.3)$$

where h = hydraulic gradient head, inches of liquid
v = horizontal liquid velocity, feet per second

5. A notched-trough distributor should never be supported on the bed, as it may be misaligned during column upsets, with disastrous effects on liquid distribution.

6. Notched-trough distributors can provide high turndown ratios. Ratios of 4:1 are readily achieved with standard designs (305), and higher ratios are readily achievable with special designs. The technique of using two rows of notches (above) can also be helpful for improving the turndown of notched-trough distributors.

3.8 General Dos and Don'ts for Distributors

Guidelines for distributor design, selection, construction, and operation are presented below.

1. A liquid distributor (or redistributor) should be used in any location in a packed column where an external liquid stream is introduced.

2. It is best to have the packing manufacturer specify and supply the distributor. The user should critically examine and carefully troubleshoot the manufacturer's recommendation and design.

3. In order for manufacturers to specify or design a distributor correctly, they must be provided with concise information on the service; its plugging, corrosive, erosive, and foaming tendencies; and of any requirements which may affect distributor selection or design.

4. Drilling (or punching) holes or cutting V-notches appears a simple task, and there may be a temptation to "fabricate your own distributor" in the workshop. This practice is dangerous and may

lead to disasters, because fabrication irregularities may lead to severe maldistribution and loss of performance in the tower. It is recommended to specify that all perforations (or notches) be punched (or cut) with the smooth edge of the hole facing the liquid, and that the rough edge is ground smooth free of burrs (similar to Fig. 6.2b, but with the smooth edge facing the liquid). There have been cases where in a single distributor some troughs had the rough edge facing the liquid while others had the smooth edge facing the liquid, leading to uneven irrigation.

5. Distributor performance should always be water-tested prior to startup. A similar recommendation was made by others (318a), with emphasis on critical services and large-diameter (> 8 ft) towers. This test can be performed in situ or at the manufacturer's shop. If not performed in situ, the piping supplying liquid to the distributor should be closely duplicated at the test rig. If maldistribution is apparent, it is best to seek the manufacturer's advice. The author is familiar with experiences where severe maldistribution problems could have been detected and rectified prior to startup if a water test had been performed. One experience has been reported (349) where a water test led to the solution of an absorber separation problem which resulted from maldistribution.

The value of a water test can be appreciated by inspecting Fig. 3.6a. Experience (318a) has shown that distributors which test well with water perform well inside the column. Costs for such a test increase with column diameter, and are roughly $5000 (1988 prices) for a 10-ft-diamter distributor (318a).

6. The irrigation pattern at the top of the bed should be closely examined to identify areas of large-scale maldistribution. This should be carried out first on paper at the design stage, and then checked in the water test. A valuable technique for examining distributor performance is outlined in Sec. 3.2. The author has experienced cases where an analysis based on similar principles (i.e., drawing "drip point circles," but without calculating a distributor quality rating index) clearly identified major distributor deficiencies.

Another useful paper check originally proposed by the Norton Company (305, 318a) divides the tower cross-sectional area into three or four concentric radial zones of equal areas. The amount of liquid entering each area should ideally be equal. This check is useful for highlighting underirrigation of the wall zone.

Special attention should be paid to areas directly underneath unperforated troughs, vapor passages, support beams, support rings, and near the column wall. The manufacturer's advice should be sought if any maldistribution is detected.

7. To counteract the tendency of liquid flow toward the wall, a large percentage (> 10 percent) of the total liquid should not enter at the tower wall or within 5 to 10 percent of the tower diameter from the wall (257).

At the same time, it is important to ensure that some liquid gets to the wall (220, 349). One rule of thumb (318a) recommends maintaining a set of radially distributed pour points within one packing diameter of the wall. One experience has been reported (220) where the presence of a 3-in-wide orifice pan distributor support ring in a 4-ft column caused a major drop in column efficiency because the area under the support ring was unirrigated.

8. A minimum of four drip points per square foot of bed cross section has been recommended (103, 257, 305, 386, 387, 413). Some recent publications advocate a minimum number as high as 6 to 10 drip points per square foot of bed (74, 75a, 111, 166, 289, 318a, 404, 436) or even more (40, 75). Experience (289, 318a, 386) suggests that gains from using more than 10 drip points per square foot are marginal, if any, and that packed-bed efficiency can normally be maintained with 5 drip points per square foot. There is, however, some nonconclusive evidence (40) which may dispute this statement. A large number of drip points is often advocated for short packed beds (103), and for low liquid flow rate applications (75, 436).

In all but very clean noncorrosive services, the actual number of drip points per unit of bed area is usually dictated by the liquid flow rate and plugging tendencies, because these set the total perforation area and perforation diameter (74, 111, 386, 436). Experiences demonstrating the effect of the number of drip points on liquid distribution have been reported (40, 305, 341). Some discussion is also available in Zanetti's survey (436).

9. The drip points should be evenly spread. The author has experienced cases where a distributor provided more liquid per unit area to some central regions than to some peripheral regions, resulting in maldistribution. Such zonal maldistribution is detrimental to column efficiency (221).

10. The distributor should be located at least 6 to 12 in above the packing to permit vapor disengagement from the bed before passing through the distributor (305). Greater distances (at least 18 to 24 in) above the bed are recommended for spray distributors.

One source (289) recommends positioning high-performance distributors 4 to 6 in above the bed; another designer (166) prefers 6 to 8 in. The former source (289) is concerned that a larger space above the bed may prompt liquid streams issuing from drip points to drift away from their striking target at the top of the packing, thereby leading to uneven irrigation. This drift is less of a problem when jets leaving the orifices are stable; a photograph in Ref. 219 shows no drift with a distributor mounted 18 in above the packings. However, it may be troublesome with unstable jets (Sec. 3.6, item 5). The same source (289) also argues that the close spacing of vapor risers in high-performance distributors alleviates the vapor disengagement problem, and therefore a larger space above the bed serves little purpose. The author prefers the larger (6 to 8 in) height between the distributor and the top of the bed as a better safeguard against entrainment, frothing, and splashing.

11. The plugging potential of a service should not be underestimated. In one case (346), 1 lb of solids was sufficient to plug 80 percent of the perforations of a ladder pipe distributor in a 13.5-ft column. The author is familiar with another case where a small quantity of solids was sufficient to plug an orifice pan distributor. In a third case (150a), fungus growth caused plugging and consequent overflow of an orifice distributor in a water scrubber.

12. If the service contains solids, or the liquid is close to its freezing point, a weir-type distributor is the best choice. If it is still desired to use a perforated-pipe, spray, or orifice distributor, a filter should be installed upstream to remove particles that can block the perforations or spray nozzles (237,305). Successful applications of this technique have been reported (237,346).

   The opening of the filter elements should be considerably (preferably at least 10 times) smaller than the distributor perforations (168, 237). The filter arrangement should include a spare filter in parallel and no bypass to ensure that one filter element is always on stream. Good filter maintenance and cleaning are essential; automatically cleaning filters are sometimes used (75).

   The filters should be installed in an accessible location as close to the column as possible. Typical good locations are close to the foot of the vertical rise of liquid feed or reflux, or just upstream of the flashing control valve for flashing feeds. The line downstream of the filter should be adequately flushed or blown (Sec. 11.1) to shake free and remove loose rust particles prior to the startup. The author is familiar with one case where the metallurgy of this downstream line was upgraded from carbon to stainless steel to avoid the rust particles. One case has been reported (299a) where spray nozzles plugged due to reliance on filters that were too distant from the column with carbon steel piping downstream of the filters.

Orifice distributors with bottom perforations should be avoided in plugging services, even when filters are installed; there have been cases where small solid particles passed through the filter, agglomerated in the pan or troughs, and blocked perforations. In pressure distributors, particularly the spray type, a filter is often [but not always (297)] sufficient.

13. Perforation diameters smaller than ¼-in should be avoided in order to prevent plugging (74, 75a, 111, 166, 386); ½-in perforations are preferred (111). If the service is perfectly clean and noncorrosive, some designers (386) advocate using holes as small as ⅛-in. Corrosion, erosion, and plugging also tend to change perforation diameter (and therefore perforation flow) to a greater extent when perforation diameter is small. On the other hand, the larger the perforation diameter, the lower the number of drip points that can be incorporated in the distributor.

    Several incidents have been experienced in which small distributor perforations were plugged by scale and dirt (168). In one reported case (346), enlarging the perforation diameter and installing an upstream filter with ¹⁄₆₄-in openings eliminated a plugging problem in a ladder pipe distributor.

14. In slightly corrosive services, it may pay to use a stainless steel distributor even when carbon steel is satisfactory as the packing material. Successful applications of this practice have been reported (84, 346). Alternatively, a distributor which is insensitive to corrosion, such as the notched-trough type, can be used.

15. When a high liquid flow rate is required, notched-trough, orifice-type, or spray-type distributors are the best selections.

16. The vapor risers or channels offer resistance to vapor flow. If vapor pressure drop across the risers becomes equal to the liquid head above the distributor, the distributor will flood. It is therefore important to allow sufficient open area for vapor flow. This open area must be distributed evenly and in a manner that prevents formation of poorly irrigated regions directly beneath the vapor passages.

17. When a high rate of vapor flow is required, the orifice pan and the weir riser distributors are best avoided.

18. The area directly beneath wide troughs with no bottom perforations (e.g., in notched-trough distributors) should be closely examined to ensure absence of unirrigated regions.

19. Column turndown is commonly set by the turndown of the liquid distributor. Distributor turndown, therefore, is a most important consideration.

20. For good turndown, weir-type or some orifice-type distributors are the best selections. Alternatively, the turndown of perforated-pipe, spray, and some orifice-trough distributors can be enhanced by using a dual liquid distributor arrangement (Fig. 3.6b). This arrangement consists of two distributors, mounted one above the other. The upper distributor is designed for a higher range of liquid flow rates than the lower distributor. At low liquid flow rates, only the lower distributor is operated; at medium liquid flow rates, only the upper distributor is operated; and at high liquid flow rates, both distributors are operated.

21. Distributor levelness affects the quality of distribution, especially under turned-down conditions, when liquid head is low. Careful design and inspection are required to ensure that the distributors are level. Inspection with level gages is strongly recommended for weir-type distributors. Weir-type distributors should be specified with leveling screws to enable in situ level adjustment.

22. Leakage of liquid from the distributor or flanges on the pipes leading to the distributor may cause maldistribution. This is most severe in low liquid flow rate applications. Techniques for minimizing leakage which are also applicable to distributors are discussed in Sec. 4.10.

23. Distributor pans and troughs should be deep enough to avoid liquid overflow. The overflowing liquid is likely to give poor irrigation, which will create maldistribution in the bed below. It may also interfere with vapor rise and cause entrainment and premature flooding. Overflowing is most troublesome when liquid flow is high and when the liquid tends to foam or become aerated (e.g., high-pressure services).

3.9 Dos and Dont's for Liquid Inlets into Gravity Distributors

In most small columns (<3 to 4 ft in diameter), the inlet pipe directly feeds the distributor pan (or trough). In larger columns, liquid is fed from the inlet pipe into one or more parting boxes (Fig. 3.4e, f), which then feed liquid to the distributor pan (or troughs) via perforations or notches. Incorrect feeding of liquid into the parting boxes or into the distributor can generate uneven liquid surface in the distributor, causing maldistribution in the packed bed below. Guidelines for introducing liquid into parting boxes and distributor pans (or troughs) are presented below:

1. Feed velocities leaving the feed pipe or feed sparger should not exceed 10 ft/s (305) and preferably be less than 4 to 5 ft/s (111). High velocities may disturb the liquid surface or cause excessive aeration in the distributor or parting box.

2. Parting boxes should be sparger-fed (305). The feed points must fall between the troughs (Fig. 3.7a) and not over the perforations or weirs feeding liquid from the parting box to the troughs (111, 305). An even better practice is to feed parting boxes from a feeder box (Fig. 3.6d).

3. Sparger pipes feeding parting boxes should be oriented parallel to and centered directly above the centerline of each parting box (Fig. 3.7a). This prevents the entering liquid from missing the parting box (Fig. 3.7b).

4. Sparger pipes feeding parting boxes should be designed according to the criteria in Sec. 2.4 (except for criteria 10 and 12). Achieving uniform distribution out of the sparger perforations is not critical because the parting box equalizes liquid level.

   The sparger must be designed so that it does not induce wave formation or excessive hydraulic gradient in the parting box. Special attention is needed to item 2 in Sec. 2.4, since the horizontal velocity component depicted in Fig. 2.6a can form waves or push liquid against the narrow wall of the parting box. This, in turn, can cause poor distribution into the troughs, or induce liquid overflow at the narrow wall. In one acid tower (206b), this overflowing liquid was entrained by rising vapor, leading to an acid emission problem. Use of flow-straightening tubes (Fig. 2.6b, 3.7a) can avoid this problem.

5. When the inlet pipe directly feeds the distributor, the incoming liquid should be fed into the center of the distributor in order to ensure uniform head over all the orifices (111, 305). The feed pipe should be located about 2 to 8 in above the top edge of the distributor pan (305). It is important to ensure the feed flows into the distributor and does not enter the vapor risers. In one case (150a) poor column separation resulted from distributor feed being directly introduced into an open riser.

   It is best to "ell" the pipe down and continue it vertically downward for a short distance (Fig. 3.7c) to prevent feed from entering the vapor risers (Fig. 3.7d). The distributor should be inspected for absence of risers directly beneath the pipe outlet.

6. The liquid fall height from the inlet pipe or sparger to the parting box or distributor should be minimized to prevent excessive

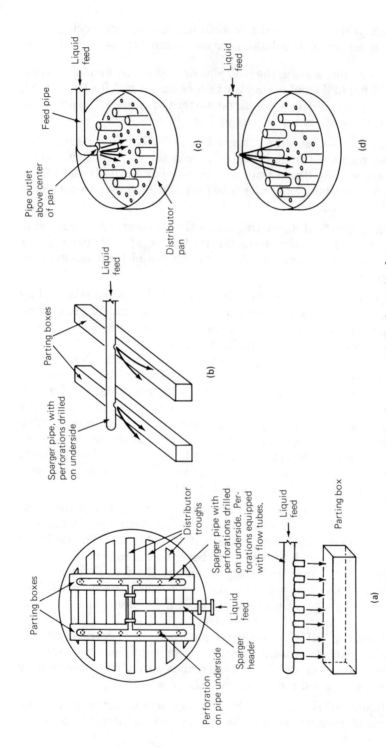

Figure 3.7 Methods of feeding parting boxes and distributor pans. (*a*) Good practice for parting box feed. (*b*) Bad practice for parting box feed. Some feed misses parting box. (*c*) Good practice for distributor feed. (*d*) Bad practice for distributor feed. Some feed enters vapor risers.

splashing, frothing, or liquid aeration, and uneven liquid surface. A feeder box (Fig. 3.6d) minimizes these problems.

7. The depth of liquid in the parting box is of major importance. If too shallow, surface disturbances in the parting box may cause maldistribution among the troughs; if too high, they may cause liquid overflow, which in turn may induce maldistribution in the packed bed below.

8. Waves on the liquid surface in the parting box must be avoided. The potential for wave formation can best be assessed in a water test. If waves are observed, either the feed sparger should be modified or baffles should be added to the parting box (e.g., Fig. 3.6d).

9. Parting boxes may contain perforations in their bottom, V-notches or rectangular notches or perforations in their sides, or both. Bottom perforations are generally preferred because they generate the least disturbance to the liquid surface in the distributor troughs. The disturbances can be further mitigated by equipping perforations with tubes submerged in the trough liquid. Liquid entering the troughs via bottom perforations possesses no horizontal momentum, thus reducing the tendency to form an uneven crest and surface waves. Bottom perforations also avoid the possibility of liquid missing the troughs and are less sensitive than V-notches to out-of-levelness. Since the number of parting box perforations is generally much smaller than distributor drip points, larger perforation diameters are used. These larger perforations are far less sensitive to plugging and corrosion than the smaller perforations used in orifice distributors.

10. Several of the guidelines in Secs. 3.8 (numbers 2 to 6, 13, 14, 21, and 23), 3.6 (numbers 4, 5, 7, 11, and 12), and 3.7 (numbers 2 to 4) can also be broadly extended to parting boxes.

11. Whenever possible, internal flanges should be avoided or minimized in the feed pipes. If one or more such flanges are necessary, they should be carefully installed and inspected to prevent leakage. Significant leakage of internal flanges in trough distributors is likely to generate large-scale maldistribution.

3.10   Liquid Redistributors

Liquid redistributors are used wherever an intermediate liquid feed is introduced into a packed column, and between packed sections, wherever liquid redistribution is required.

Apart from distributing an incoming feed, evenly irrigating the bed below, and providing a uniform vapor flow to the bed above, liquid redistribution serves four main functions. First, and most importantly, a redistributor mixes the liquid and equalizes its composition throughout the column cross section. Equalizing the liquid composition counteracts composition pinches (Sec. 3.1), thus lessening the detrimental effect of maldistribution on packing efficiency. This mixing is analogous to that produced when liquid is deflected laterally by packing particles, but in the case of a redistributor, the mixing is far more extensive. Second, a redistributor mixes the vapor and equalizes vapor composition throughout the column cross section, with consequences analogous to those of mixing the liquid. Third, a redistributor removes liquid from the wall and redirects it toward the center of the bed, thus counteracting buildup of excessive wall flow. Fourth, a redistributor improves the wetting of the packings by splitting up any "rivers" into small liquid streams.

Since the main function of a redistributor is to counteract the effects of large-scale maldistribution, less frequent redistribution is required when large-scale maldistribution is absent or minimized. For instance, less frequent redistribution is needed when the ratio of column to packing diameter is small (443), because large-scale maldistribution is well-counteracted by lateral mixing in the packing (Sec. 3.1).

Experiences have been reported where packed beds 30 to 40 ft deep between redistribution points performed well (94, 166, 386, 388, 436). Based on such experiences, some designers advocate redistribution approximately every 30 ft (103, 387, 436). Unfortunately, the presence and extent of maldistribution, and its effect on column efficiency, cannot be readily predicted in most circumstances, and many designers (74, 144, 166, 237, 257, 404, 413, 436) prefer the more conservative practice of redistributing approximately every 20 ft. The author also prefers this more conservative practice. With high-efficiency packings (HETP < 2 ft), some designers (166) advocate redistribution at intervals not exceeding 10 theoretical stages. For small towers (< 2 to 3 ft in diameter), the formation of excessive wall flow appears a greater problem, and some designers (103, 166, 257, 386, 413) advocate redistributing at intervals not exceeding 10 tower diameters. With plastic packings, some designers (144, 413, 436) recommend bed depths not exceeding 15 ft to avoid compression of packing particles.

The importance of adequate redistribution is illustrated by a field experience reported for two identical refinery debutanizers operating in parallel and in identical service (237). Both were packed with 3- and 4-in packings; the only difference between the two was the presence of redistributors. The tower containing four 19-ft beds with redistributors achieved an HETP of 39 in; the other contained two 38-ft beds and achieved an HETP of 72 in. The author also experienced a case of a poorly performing packed bed which was about 35 ft deep and did not contain a redistributor. Martin et al. (277) experimented with 10-ft-tall beds of structured packings in a pilot-scale column using an excellent top liquid distributor with no redistribution. In some experiments, the efficiency in the bottom portion of the bed was considerably less than in the top portion of the bed, while in others, the efficiency was uniform throughout the column.

Three types of redistributors are common:

1. *Orifice redistributors* (Fig. 3.8a): These are identical to orifice distributors, either the pan or the trough type, except that hats, caps, or strips are usually installed above the risers to prevent liquid from the packed bed above from entering the vapor risers, and also to promote lateral mixing of vapor.

2. *Weir redistributors:* These are identical to notched-trough distributors. Because weir redistributors cannot collect liquid from the upper section, a liquid collector such as a chimney tray or chevron collector (Chap. 4) or a collecting support plate (e.g., the upper plate in Fig. 3.8c) is usually required above the redistributor.

3. *Wall wipers (or "rosette") redistributors* (Fig. 3.8b): This is a collection ring equipped with short projections extending toward the tower center. Liquid removed from the wall is deflected into the projections ("fingers"), which transport it to a desired location in the bed. Wall wipers effectively remove liquid from the wall, but they are only partially effective for counteracting composition pinches. Their ability to counteract composition pinches diminishes as column diameter increases. Therefore, they are only suitable for small columns [< 2 to 3 ft in diameter (74, 305)], where deflection of liquid and vapor by packing particles is sufficient to counteract pinching effects, and where wall flow formation is the main problem.

Wall wipers were specifically advocated for the stripping sections of small columns (< 2 to 3 ft in diameter), producing a high-purity bottom product (386). Generally, wall wipers are spaced apart by about two theoretical stages of packed height (386).There is some uncertainty regarding the effectiveness of wall wipers. The author is familiar with many small columns (2 to 3 ft in diameter) that operate well without wall wipers. The author also knows of many designers that prefer to avoid wall wipers due to their capacity restriction potential (see item 7 below). The author even heard of a case where removing wall wipers from an existing small column improved its efficiency.

Several types of structured packing are equipped with wall wipers (168). Experiments (40) have shown that without them, effi-

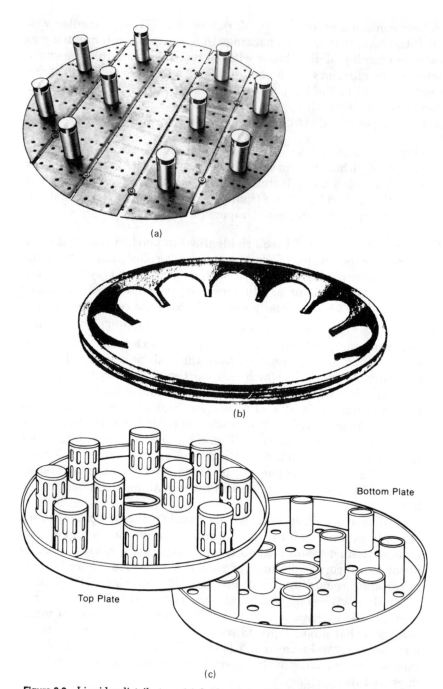

(a)

(b)

Top Plate

Bottom Plate

(c)

Figure 3.8 Liquid redistributors. (a) Orifice redistributor; (b) "rosette" wall wiper; (c) support mixer redistributor. (*Part a Gilbert K. Chen, excerpted by special permission from* Chemical Engineering, *March 5, 1984, copyright © by McGraw-Hill, Inc., New York, NY 10020; part b reprinted courtesy of Norton Company; part c reprinted courtesy of Koch Engineering Company, Inc.*)

ciency can drop by as much as 20 percent in small-diameter (<1.5-ft) columns. Unlike the rosette ring, these wall wipers are fixed to the packing layers rather than to the column wall.

One source (404) reports success with using a layer of structured packing as a redistributor. The author has limited inconclusive evidence to suggest that this technique may not always be effective. Additional experience will be required before this technique can be applied with confidence.

Generally, redistributors for large-diameter (> 3-ft) columns are of the orifice or weir type. The orifice type is more popular because it does not require the addition of a liquid collector, which consumes vertical space and increases column cost and complexity. Other pros and cons of orifice and weir redistributors, as well as application guidelines for each type of redistributor, are identical to those described earlier for orifice and weir distributors. The general dos and don'ts for distributors and for liquid inlets into distributors also extend to redistributors. Additional guidelines unique for selection, design, and operation of redistributors are presented below.

1. The redistributor must be compatible with the bed support above. It is best to have the manufacturer specify both. Often, the manufacturer supplies a package redistributor–support plate combination.

2. Liquid flow must be prevented through the vapor passages. Risers are usually fitted with covers (Fig. 3.8a). Alternatively, a combination support plate–redistributor can sometimes prevent liquid downflow through vapor risers without using riser covers.

3. Guidelines 2, 5 to 8, and 10 in Sec. 4.10 are also applicable for redistributor riser covers. Exceptions to these are (1) flat hats are acceptable for redistributors, because a small amount of liquid dropping through the vapor risers can often be tolerated; and (2) when vapor distribution is not a major problem (e.g., column diameters smaller than 4 to 6 ft and a large number of vapor risers), uncovered vapor risers in a combination support plate–redistributor may be satisfactory. The top of such uncovered risers should not be more than 4 in below the bottom of the support plate above (305).

4. In very large columns (>20 to 30 ft), a redistributor may not be sufficient to mix the liquid adequately. In such cases, mixing can be improved by adding a liquid collector, from which the liquid is fed to the redistributor.

5. A combination redistributor–support plate which can also achieve good liquid mixing (212) is the support mixer redistributor (Fig.

3.8c). In this combination, the support plate also acts as a liquid collector. The collected liquid flows into a downcomer-like arrangement, which mixes the liquid and feeds it into the center of the redistributor plate below, behind a circular inlet weir. A potential problem with this redistributor is occlusion of perforations by migrating pieces of packing.

6. Rosette redistributors must be sealed to the tower wall in order to ensure adequate collection of wall liquid. A sealing strip is often included in the equipment provided by the manufacturer (305). If the column is constructed of flanged sections, a rosette redistributor can be mounted between flanges of adjoining sections.

7. A redistributor represents a flow restriction and may cause premature flooding, especially if packings or chips of packings find their way into the redistributor. Redistributors and supports should be carefully designed to prevent this.

   Special caution is required with wall wipers, since those are installed within the bed itself. A 1.5-in-wide wall wiper in a 2-ft ID column would reduce the column open area (and therefore its capacity) by as much as 25 percent. Nonstandard wall wipers should be avoided, and the rosette shape (Fig. 3.8b) should be preferred. The projections should be relatively narrow to avoid consuming excessive open area.

8. Distributors and redistributors are known to have been interchanged during installation (121), particularly when both are of similar design. Inspection prior to startup is necessary.

9. Separation of two immiscible liquid phases can be troublesome in some redistributors and promote irregularity in the makeup of the liquid supplied to the bed below. Special attention is required to redistributor design if liquid-phase separation is likely (386). Weir redistributors are least sensitive to this problem.

10. When a liquid product or pumparound stream is drawn from a redistributor, the area under the drawoff box must be properly irrigated. The author is familiar with one large-diameter (> 20-ft) column that had a redistributor containing a central 2-ft-wide liquid collecting draw box extending the full tower diameter with no irrigation underneath. The column performed extremely poorly, but mainly due to other more severe sources of maldistribution.

3.11   Flashing Feed Distributors

An intermediate liquid feed is almost always introduced into a redistributor. Introducing a vapor-containing feed via a distributor suitable for liquid feeds only can severely lower column efficiency, as demonstrated in one case history (76). When the feed contains vapor, a flashing feed distributor is used. Flashing feed distributors have two main functions:

1. Separate vapor from liquid, and either distribute the liquid evenly to the bed below or feed the collected liquid to a liquid distributor below.
2. Absorb and dissipate the forces exerted by the incoming feed.

The common types of flashing feed and vapor distributors are the baffle type (Fig. 3.9a), the vapor-liquid separator type (Fig. 3.9b), the gallery type (Fig. 3.9c), and the tangential entrance type (Fig. 2.2j). Some of the vapor distributors discussed in Sec. 3.12 are also sometimes used for flashing feeds, especially when liquid distribution to the section below is not critical (e.g., when it contains trays).

The baffle type (Fig. 3.9a) features an impingement baffle arrangement above an orifice distributor or a liquid collector. Feed is directed against the baffles through slotted piping, where velocities are lowered and vapor disengages from the liquid. The liquid can flow to an orifice-type distributor or may be collected and flow to a liquid distributor below (Fig. 8.1). This distributor is best suited for two-phase feeds that readily disengage and are nonfoaming (74), and for small towers (< 4 ft in diameter).

The vapor-liquid separator type (Fig. 3.9b) features an internal knockout drum directly above an orifice redistributor. It consumes more vertical space than the baffle type, but it has the potential of providing better liquid distribution to the bed below. This device is also most suitable for small columns (386).

The gallery type (Fig. 3.9c) is used for feeds in which liquid is the continuous phase. It features a gallery around the perimeter of the distributor above the distributor floor. The two-phase feed enters the gallery, where vapor separates from liquid. The liquid then underflows into a second distributor plate located below. A "tee" arrangement is often used at the feed inlet to force the feed to flow around the gallery. The gallery type is commonly used in foaming systems, such as cryogenic demethanizers and carbonate regenerators (74, 305), and in larger-diameter columns (386).

Careful design is required at the gallery region across from the inlet, as here two high-velocity streams meet. This meeting may be accompanied by excessive turbulence, splashing, and hydraulic pounding.

The feed entrance into the gallery also requires close attention. Severe hydraulic pounding by the entering vapor may cause mechanical

(a)

(b)

(c)

Figure 3.9 Flashing feed distributors. (*a*) Baffle-type distributor; (*b*) vapor-liquid separator-type distributor; (*c*) gallery-type distributor. (*Parts a, c Gilbert K. Chen, excerpted by special permission from* Chemical Engineering, *March 5, 1984, copyright © by McGraw-Hill, Inc., New York, NY 10020; part b reprinted courtesy of Norton Company.*)

damage in this area. There have been cases where substantial increases in the quantity of vapor in the feed during upset conditions flattened galleries. Such upset conditions should either be eliminated or taken into account in the mechanical design of gallery distributors. If mechanical damage cannot be prevented by strengthening the feed entrance area, it may pay to consider a tangential feed arrangement (below).

When a flashing feed distributor is installed between packed beds (rather than at the top of the columns), vapor risers should be equipped with hats to prevent liquid from the packed bed above from entering the risers. Guidelines 1, 3, and 7 in Sec. 3.10 are also applicable for flashing feed distributors.

Tangential feeds, often with helical baffles, are commonly used with high-velocity feeds in which the vapor phase is continuous and the liquid is present in the form of a spray (143). These arrangements are discussed in detail in Sec. 2.2 (arrangement *j*).

3.12 Vapor Distributors and Vapor-Distributing Supports

Vapor-distributing devices are typically located at or above a vapor feed, between a trayed and a packed section, or above the transition section where diameter changes. The following devices are used:

- A sparger pipe (Fig. 4.3*b*)
- A vapor distributor
- A vapor-distributing support (Fig. 3.10)

Sparger pipes are discussed in detail in Sec. 4.2. That discussion also applies to a sparger pipe introducing an intermediate feed.

Most vapor distributors are essentially chimney trays (Chap. 4) specifically designed to promote vapor distribution. A typical vapor distributor contains a large number of uniformly spaced vapor risers and

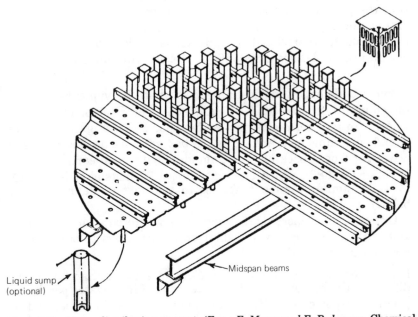

Liquid sump (optional)

Midspan beams

Figure 3.10 Vapor-distributing support. (*From F. Moore and F. Rukovena,* Chemical Plant and Processing, *Europe edition, August 1987. Reprinted courtesy of* Chemical Plant and Processing.)

has a relatively high pressure drop. A vapor distributor usually contains flow perforations for liquid downflow; alternatively, it may be equipped with downpipes. The diameter of each downpipe is restricted in order to avoid large areas with no risers. Total downpipe area must not be undersized (see Sec. 4.10, guidelines 14 and 15), and the downpipes must be adequately sealed to avoid vapor rising through them.

A vapor-distributing support (Fig. 3.10) is a flat perforated plate containing perforated vapor risers. Liquid descends through the floor perforations, while vapor rises through the riser perforations. The bottom portion of the vapor risers is unperforated, so that vapor is injected above the liquid pool on the plate. Sumps are optional and recommended (289) where liquid inventory is to be reduced. The vapor-distributing support combines two internals (a vapor distributor and a support plate) into one. Compared to a vapor distributor, this saves both vertical space and internals costs. On the other hand, obstruction of perforations by pieces of packing and possible liquid overflow into risers may make its vapor distribution quality somewhat lower.

Vapor-distributing devices should be installed whenever a high-velocity, unevenly distributed vapor flows toward a packed bed. In general, when the velocity head of this unevenly distributed vapor (e.g., vapor at the inlet nozzle) is of the same order as the pressure drop of the packed bed above, there is an incentive for installing a vapor-distributing device; when it is less than one-tenth of the bed pressure drop, a vapor-distributing device is unlikely to be beneficial. More detailed guidelines (289, 386) for distributing inlet vapor are

1. A vapor-distributing device is needed when the vapor F-factor at the column inlet exceeds $52.4\sqrt{\Delta P}$ (289). The F-factor is the inlet velocity in feet per second times the square root of gas density in pounds per cubic foot, and ΔP is the bed pressure drop in inches of water per foot of packing. The guideline updates an earlier version of this rule (386), which cited a value of 22 instead of $52.4\sqrt{\Delta P}$ (344). A similar comment applies to the next guideline (item 2 below).

   One case was reported (75) where vapor entered at an F-factor of 140 without a proper vapor distributor. Needless to say, performance was extremely poor.

2. If vapor enters the column at an F-factor smaller than $52.4\sqrt{\Delta P}$ (289) and the bed pressure drop exceeds 0.08 in of water per foot of packing, vapor distribution is unlikely to be troublesome, and no vapor-distributing device is needed. This applies to columns smaller than 20 ft in diameter (289).

3. If a high feed press: re drop is acceptable, the use of a sparger pipe (Fig. 4.3b) is recommended (289) when the vapor inlet F-factor ranges between 52.4 $\sqrt{\Delta P}$ and 81.2 $\sqrt{\Delta P}$. The use of a sparger pipe eliminates the need for a more sophisticated vapor-distributing device.

Kabakov and Rozen (183) measured severe gas maldistribution in 15-ft-diameter CO_2-amine absorbers containing 50-ft-tall packed beds. They did not report gas density, making it impossible to determine the F-factor at which the maldistribution was measured. The absorption efficiency in the columns experiencing maldistribution was roughly half the efficiency measured in another similar column that had a specially designed gas inlet sparger (183).

4. Pressure drop through the vapor distributor or a vapor-distributing support should be at least equal to the velocity head at the column inlet nozzle (386). Typically, a pressure drop of 1 to 8 in of water (289) is used for these devices. In one typical specific example (346), it was 4 ins.

A vapor distributor with insufficient pressure drop may be ineffective. A case has been reported (75) where poor column efficiency resulted from specifying a vapor distributor with lower pressure drop than recommended by the manufacturer. The vapor distributor (or vapor-distributing support) pressure drop can be minimized by entering the feed via a sparger pipe with the bottom quadrant removed (rather than perforated). Although such a sparger pipe will not be an effective vapor distributor, it will serve to break the incoming vapor jet and reduce its velocity head.

5. When a sparger pipe enters an intermediate feed, vapor jets must not impinge on redistributor liquid surface or other packed-column internals. Liquid surface agitation in redistributors, or mechanical damage to internals, can cause maldistribution, which can be detrimental to column efficiency.

6. Vapor-distributing supports should not be used with foaming systems (289).

7. A V-baffle (Fig. 2.2f, Sec. 2.2) is sometimes (albeit infrequently) installed as a primitive vapor or flashing feed distributor. If the feed contains liquid, the top and bottom plates of the baffle are often installed at a small downward slope. This arrangement is somewhat similar to the gallery distributor (Sec. 3.11), but with liquid falling off the edge rather than descending via perforations. If the liquid falls onto a distributor plate below, it can cause frothing, splashing, and waves, and is therefore not recommended. However, if the section below is insensitive to distribution (e.g., bottom sump or trayed section) this fall may be acceptable.

Experience with V-baffles is mixed. The author is familiar with columns in which the arrangement works well. On the other hand, an experience has been reported (299a) in which it did little to alleviate vapor maldistribution. With superheated feeds, it may lead to extensive coking. Coke growth on the back of such a baffle was reported (299a) in two entirely separate refinery FCC main fractionators (feed is highly superheated in this service). In one case, the coke grew right through the top of the grid bed above and needed to be dynamited out (299a).

8. Diffuser vanes are sometimes installed as a vapor distributor at the column inlet. The success of this arrangement depends on the vane design and the shape of the inlet vapor jet; the latter is a function of the column feed piping configuration. This arrangement can also be sensitive to plugging, coke buildup, and corrosion.

Chapter

4

Bottom Section and Column Outlets

The main considerations for the bottom section and column outlets are achieving the required phase separation, avoiding unfavorable interaction between different internals, and establishing adequate hydraulics. Failure to achieve these may cause premature flooding, outlet line choking, excessive entrainment, loss of efficiency, and mechanical damage. With liquid outlets, providing a satisfactory surge volume for downstream equipment is also an important consideration.

This chapter examines common practices of withdrawing streams from a column, focuses attention on internals used at the bottom of a column (an area which causes frequent headaches to troubleshooters), outlines the preferred practices, highlights the consequences of overlooking traps, and supplies guidelines for avoiding pitfalls and for troubleshooting column outlets and column bottom sections.

4.1 Bottom Feed and Reboiler Return Inlets

The space between the bottom tray or packing support plate and the liquid level is a prime potential troublespot in distillation and absorption columns. Numerous mishaps originating in this section have been reported (12, 49, 71, 107, 203, 231, 238, 255, 296, 375, 440). It is estimated (143, 207) that 50 percent of the problems in the lower part of the column are initiated in this space. Many of these problems are associated with the reboiler return (or bottom feed) inlet. Therefore, it is important to follow these guidelines.

1. Bottom feed and reboiler return inlets should never be submerged below the liquid level (see the Sec. 4.2 for the only acceptable ex-

ception to this rule). At times, it may appear attractive to sub-
merge this inlet in order to dampen a pulsating feed (12), or to
desuperheat the bottom feed, or to gain a theoretical plate. But this
is bad practice (Fig. 4.1a), mainly because the column of liquid
above the submerged inlet can vary in height, and under certain

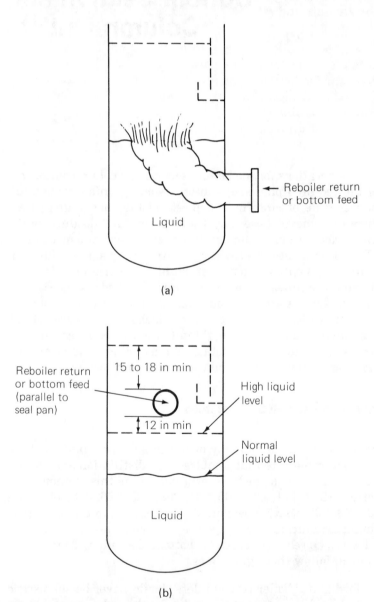

(a)

Reboiler return
or bottom feed
(parallel to
seal pan)

15 to 18 in min

High liquid
level

12 in min

Normal
liquid level

Liquid

(b)

Figure 4.1 Good and bad practices for reboiler return or bottom feed arrangements. (*a*) Bad practice, must be avoided; (*b*) good practice.

conditions, slugs of liquid and vapor can be blown up the col
and lift ("bump") trays off their supports or dislodge packing support plates. Inlets below the liquid level can also be responsible for
excessive entrainment and premature flooding. Experiences where
liquid levels above the reboiler return nozzle caused trays to lift off
their supports (49, 107, 231, 296) or induced premature flooding
and excessive entrainment (71, 208, 238, 375) have been reported
in the literature.

2. Bottom feed and reboiler return nozzles should not be too close to
   the maximum liquid level. The space between the bottom of the
   reboiler return (or bottom feed) nozzle and the maximum liquid
   level (Fig. 4.1b) should be at least 12 in (143, 192, 207, 208, 237,
   375). Failure to do this will promote turbulence on the liquid surface, erratic level control, and liquid entrainment into the rising
   vapor.

3. Bottom feed and reboiler return flows should not impinge on the
   bottom liquid surface. It is best to enter these flows in parallel to
   the liquid level. Impingement can occur when an internal pipe is
   "elled" down, or if the column inlet pipe slopes downward at the entrance. Arrangements to be avoided are shown in Fig. 4.2. Impinge-

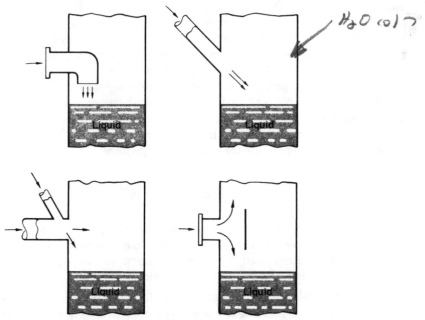

Figure 4.2 Practices to be avoided in bottom feed arrangements. (*From Henry Z. Kister, excerpted by special permission
from* Chemical Engineering, *May 19, 1980, Copyright © by
McGraw-Hill, Inc., New York, NY 10020.*)

ment on the liquid surface will produce consequences similar to those described in guideline 2 above. The author has experienced excessive entrainment problems following the addition of the sloped nozzle at the bottom feed of a column, as shown on the lower sketch on the left (Fig. 4.2).

4. Bottom feed and reboiler return flows should not impinge on the bottom seal pan, the seal pan overflow, or the bottom downcomer. The preferred arrangement is to introduce the feed parallel to the edge of the seal pan. If this cannot be done, the inlet pipe should be extended into the column and terminated with an inlet baffle, a horizontal tee, or another device that enters the vapor parallel to the edge of the seal pan (Fig. 4.1b). Vapor impingement on the seal pan overflow may result in liquid entrainment to the bottom tray. Impingement on the downcomer or seal pan can cause vaporization and lead to premature downcomer flooding, particularly if the feed is superheated. Bottom feed and reboiler return flows should also not impinge on level-sensing or other instrument connections. This is discussed in Sec. 5.3.

5. Reboiler return and bottom feed lines must be correctly sized. Undersized reboiler return lines can lead to liquid level backup to the reboiler outlet nozzle and may result in premature flooding. The author is familiar with an incident where a column flooded at 50 percent of its design rates because of an undersized reboiler line. Excessive inlet velocities may also induce turbulence and entrainment and have also been reported (98) to cause tray vibration, loosening of tray fasteners, and tray failure.

   A detailed sizing procedure for reboiler return inlets is discussed elsewhere (113). For bottom feeds, the criterion for sizing mixed and vapor feeds (Sec. 2.3, guideline 4) can be used (355). For reboiler return lines, it has been recommended (82) that the velocity in feet per second not exceed $\sqrt{4000/\rho_m}$, where

$$\rho_m = \frac{100}{\%vapor/\rho_v + \%\text{liquid}/\rho_L}$$

and ρ_v and ρ_L are vapor liquid densities in pounds per cubic foot. In some cases, an excessive inlet velocity can be accommodated by introducing the bottom feed or reboiler return via a vapor space sparger. This is discussed in the next section.

6. The top of the reboiler return or bottom feed nozzle should be located at least 15 to 18 in below the tray (or packing support plate) above (143, 207, 211, 237, 307) (Fig. 4.1b). Some designers (211, 237) recommend that the distance from the centerline of the nozzle

to the tray above be at least one tray spacing plus 12 in; others (143, 207) recommend that the distance from the top of the nozzle to the tray above be at least 12 in plus the nozzle diameter. One recommendation specific for packed towers is that the distance from the top of the nozzle to the packing support plate be at least 12 in plus half the nozzle diameter (289); another states that this distance should be made at least 1.5 times the nozzle diameter (75).

In large-diameter packed columns where I-beams support the packing support plate, a larger space between the reboiler return nozzle and the packing support plate may be required. The prime consideration here is allowing sufficient open area between the bottom of the beam and the high liquid level for adequate vapor distribution. Detailed discussion is in Sec. 8.2.

7. The section of column wall directly opposite a bottom feed (or reboiler return) nozzle is often prone to corrosion and erosion attacks. High inlet velocities, small column diameters, and corrosive chemicals are conducive to this problem. The author has experienced a case where metal was thinned and the column wall was eventually punctured opposite the bottom inlet of a 2.5-ft-diameter column. Installing an impingement plate on the wall will shield it from such corrosion and erosion attacks.

8. Tangential reboiler return or bottom feed nozzles, or geometries promoting tangential velocities, should be avoided (316). Tangential velocities will impart a swirl to the sump liquid, disturb the liquid surface, and promote vortexing. An exception to this rule is a well-designed and well-sized vapor horn nozzle (Fig. 2.2*j*), as described in Sec. 2.2.

4.2 Bottom Feed and Reboiler Return Spargers

Using a well-designed submerged sparger (Fig. 4.3*a*) is the only acceptable means of introducing bottom vapor feed below the liquid level (see guideline 1 in the previous section). The sparger is a perforated pipe which emits vapor into the liquid as bubbles so that slugging is avoided. It is important to monitor liquid level adequately and to shield the level-measuring device from the presence of bubbles. In one troublesome incident (440), loss of the liquid layer covering the sparger resulted in overheating of column internals; in that case, the system was designed to maintain a constant level, but the level was not monitored.

When the bottom feed or reboiler return enters the column with a high velocity, or when there is a concern about excessive turbulence at

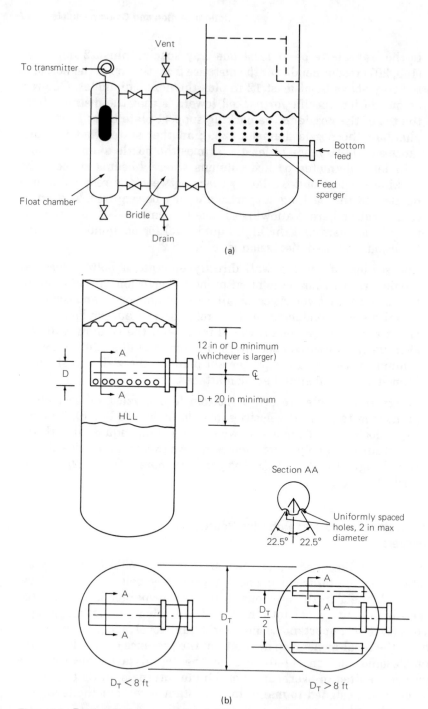

Figure 4.3 Bottom feed spargers. (*a*) Submerged bottom feed sparger; (*b*) vapor space (nonsubmerged) bottom feed sparger, with dimensions recommended by Moore and Rukovena (289).

the entry, it may be attractive to introduce it via a sparger located in the vapor space above the liquid level (Fig. 4.3b). The vapor-space sparger is primarily used with vapor-only bottom feeds. With mixed-phase bottom feeds or reboiler returns, the high-momentum liquid will tend to travel toward the closed end of the pipe, and preferentially issue there, while vapor will preferentially issue from perforations near the inlet to the sparger. While this is unlikely to be troublesome in most tray columns, some designers (289) recommend against using vapor space spargers for packed-column mixed-phase bottom feeds or reboiler returns.

In kettle and thermosiphon reboiler circuits, vapor space spargers are usually undesirable because of their high pressure drops. In one case (145), high pressure drop in a vapor-return sparger from a kettle reboiler caused liquid to back up above the reboiler return nozzle, which in turn led to premature flooding. The problem was solved by chopping off the sparger to reduce pressure drop and removing the bottom tray to permit vapor distribution.

Figure 4.3b shows some recommended dimensions (289) for a vapor space sparger. Drain holes are required if the bottom of the sparger is unperforated. Guidelines 2 and 4 in Sec. 4.1, and guidelines 1, 3 to 9, 11, and 12 in Sec. 2.4 also apply to vapor space spargers. In packed columns, it was recommended (289) to restrict perforation pressure drop to less than 1 psi.

Where pressure drop is critical, a sparger pipe with the bottom quadrant cut out (rather than perforated) is sometimes used, but at the penalty of inferior vapor distribution. Similarly, the sparger can be entirely eliminated and substituted by a "dog house" baffle parallel to the direction of fluid entry. This baffle is somewhat wider than the nozzle diameter and stretches from wall to wall parallel to the direction of the incoming fluid. The author is familiar with one experience where addition of a "dog house" baffle eliminated a packed-tower vapor maldistribution problem.

4.3   Liquid Outlets Pitfalls

The main consideration in the design of a column outlet is to achieve the required phase separation. For liquid outlets, providing the required surge capacity is important as well.

The presence of vapor in liquid outlet lines has been a major troublespot in distillation and absorption columns. Numerous case histories have reported premature flooding, column instability, and pump cavitation and erosion caused by presence of vapor in liquid outlet lines (12, 85, 208, 255); the author is familiar with a few additional cases. A simple rule of thumb states: "Either avoid the presence of vapor in a liquid outlet—or design for it."

Arrangements such as bottom sumps, chimney trays, side-draw drums, and surge drums are usually designed for avoiding the presence of vapor in the liquid outlet. On the other hand, downcomer trapouts are usually designed to allow for the presence of vapor in the outlet liquid.

The common mechanisms responsible for the presence of vapor in a column liquid outlet line are

• Insufficient residence time for vapor disengagement from the liquid. Liquid arriving at the pan or sump from which it is withdrawn almost always contains entrained vapor bubbles. Additional bubbles are generated by frothing (see below). These bubbles disengage from the liquid by a buoyancy mechanism. When sufficient residence time is not available for adequate bubble disengagement, vapor will be entrained in the liquid leaving the sump or pan. Residence time is discussed in the next section.

• Frothing (waterfall pool effect) is caused by the impact of falling liquid on the liquid surface in the draw pan or sump. The impact induces vapor entrainment into the liquid, in the same manner as a waterfall induces entrainment of air into the pool below it. Both the quantity of vapor entrained and the depth to which the vapor bubbles travel below the liquid surface are enhanced when large quantities of liquid fall into a small sump area, and when the vertical drop height is large. Frothing is therefore a far greater problem in tray column sumps (where liquid fall resembles a waterfall) than in packed-column sumps (where liquid fall is more like rain). Techniques to minimize the effects of frothing are discussed in the sections on bottom sumps and chimney trays.

• Flashing occurs when the liquid vapor pressure exceeds the static pressure in the outlet pipe. To avoid flashing, sufficient liquid head should be available at the sump, and the outlet pipes should be correctly sized. Sizing should take into account reasonable variations in outlet flow. Pump suction lines should be as short as possible and run without loops or pockets. One troublesome experience with flashing at the entrance to a side-draw line has been reported (85).

• Vortexing occurs because of intensification of swirling motion as liquid converges toward an outlet. This intensification results from conservation of angular momentum. Vortexing promotes the entrainment of vapor into the drawoff line. One troublesome experience of vortexing causing vapor entry into a liquid outlet line was reported (255).

To avoid vortexing, vortex breakers should be routinely installed. Vortex breakers introduce shear in the vicinity of the outlet, thus suppressing the swirl. While flat plates and simple crosses help in vortex prevention, radial vanes or floor-grating configurations are far more effective (316, 413a). The floor-grating type is most popular. For a high degree of effectiveness, the vortex breaker should be four times the outlet nozzle diameter, with a maximum size of one-third the vessel diameter (316, 413a). The height of the breaker above the outlet should be about half the nozzle diameter, with a minimum bottom clearance of several inches (316, 413a). Vortex breaking is discussed in detail elsewhere (316, 413a).

In vacuum towers, air may leak (e.g., through a flange) into a liquid outlet line. Minimizing the sources of leakage (e.g., flanges) and thorough leak testing prior to startup are important when the line is designed for avoiding the presence of vapor.

4.4 Residence Time

Providing adequate residence time is a primary consideration in arrangements that avoid vapor in liquid outlets. Such arrangements typically include bottom sumps, chimney trays, external side-drawoff drums, or surge drums. Sufficient residence time must be provided in the liquid-drawoff sump for one or more reasons:

1. To disentrain vapor contained in the sump liquid. Entrained vapor bubbles may lead to downstream pump cavitation and to "choking" in downstream pipelines.

2. To buffer downstream units from upstream and column upsets. This is most important when the sump is feeding sensitive units such as furnaces.

3. To buffer the column from downstream upsets. This is most important if the liquid is feeding a downstream unit by direct pressure, and the pressure difference between the two is small.

4. To give the operator sufficient time to take corrective action if common upsets occur (e.g., pump trips, level lost or gained too fast).

5. To provide sufficient settling time if two liquid phases such as hydrocarbons and water are also to be separated.

Residence time is calculated by dividing sump volume by the volumetric flow rate of liquid leaving the sump. In many instances, this definition is tricky to apply. The author recommends that the volume and liquid flow rates used in the above definition be those relevant to a specific purpose (Fig. 4.4), thus:

1. When the controlling consideration is satisfactory vapor disentrainment, the relevant volume is from the top of the liquid outlet nozzle to the normal liquid level. The relevant liquid flow rate is the total liquid flow rate leaving the sump. For example, if the bottom sump is not separated by a baffle from the reboiler compart-

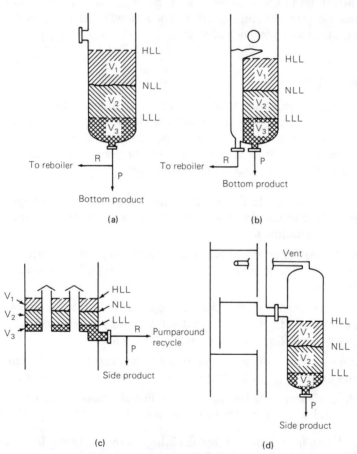

Figure 4.4 Residence time definitions. Residence time t_R given by:
For vapor disentrainment: $t_R = (V_2 + V_3)/(P+R)$ for a and c
$t_R = (V_2 + Y_3)/P$ for b and d
  For buffering upsets: $t_R = (V_1$ or $V_2)/P$ for all parts

ment and is used to supply liquid both to the reboiler and bottom product, the relevant liquid flow rate is the bottom product plus the liquid to the reboiler.

2. When the controlling consideration is surge volume to buffer the column or downstream equipment from upsets, the relevant liquid volume is from the low to the normal (or normal to high, depending on the nature of upsets expected) liquid level. The relevant liquid flow rate is only the liquid stream going to the next unit. In the above bottom sump example, the relevant liquid flow rate for buffering from upsets is the bottom product flow.

Guidelines for sump residence time are scarce in the published literature. Those given by Wheeler (420) for chimney trays are the only detailed list available (Table 4.1). Wheeler (420) did not state the residence time definition on which his guidelines are based. The author believes that their application with the above definition is reasonable for chimney trays, bottom sumps, and side-draw drums. Caution is needed when these guidelines are applied, particularly when the main consideration is to buffer upsets.

The author recommends that for adequate vapor disengageme it, liquid residence time should be at least 1 minute, based on the above definition. The author is familiar with experiences where sump residence times of the order of 20 to 40 s were insufficient to adequately disentrain vapor from liquid in sumps and draw pans of columns handling nonfoaming liquids. A greater residence time may be required

TABLE 4.1   Residence Time for Liquid in the Sump

Operating condition	Minimum residence time, min
Liquid is withdrawn by level control and feeds another column directly by pressure.	2
Liquid is withdrawn by level control and pumped away. Spare pump starts manually.	3
Liquid is withdrawn by level control and pumped away. Spare pump starts automatically.	1
Liquid is withdrawn by level control and feeds a unit that is some distance away or that has its instruments on a different control board.	5–7
Liquid is withdrawn by flow control.	3–5
Liquid flows through a thermosiphon reboiler without a level controller, to maintain a level in the sump.	1

SOURCE: From D. E. Wheeler, "Hydrocarbon Processing," July 1968, reprinted courtesy of Hydrocarbon Processing.

...e frothing is likely to be significant and where the system tends to foam.

A residence time smaller than 1 minute using the above definition may be satisfactory on chimney trays that are used for withdrawing pumparound liquid in nonfoaming services. Because of the large circulation rate typical in such applications, a 1-minute residence time may require an excessively tall chimney tray. Since pumps used in such applications typically develop low heads, they usually can tolerate some entrained vapor.

Under some conditions, it may be impractical or unattractive to provide sufficient residence time in the liquid-drawoff sump. Then, separate side-drawoff drums or surge drums that vent back to the column are added outside the column to provide the required residence time (Fig. 4.4d).

When vapor disentrainment is the controlling consideration, and it appears attractive to minimize the sump size, the liquid pipe downstream of the sump can be sized to provide a self-venting flow. In this case, the liquid residence time in the pipe contributes to the residence time requirement, and the sump itself can be designed for a smaller residence time. The high points on the self-venting lines should be vented back to the column.

A short-cut correlation (354) for sizing self-venting lines is shown in Fig. 4.5. The author has had a lot of favorable experience with this correlation. With foaming systems, pipe diameter should be increased in proportion to the foaming factor (355). Caution with applying Fig. 4.5 is required (355) when an extraordinary flashing flow situation is expected (e.g., when the liquid contains appreciable amount of a high-relative-volatility component).

4.5 Column Bottom Arrangements

Three different arrangements are commonly used for the column bottom:

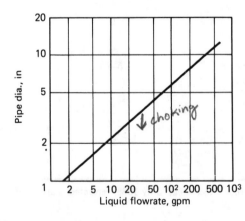

Figure 4.5 Correlation for self-venting flow. (*From A. Sewell, The Chemical Engineer, No. 300, 1975. Reprinted courtesy of the Institution of Chemical Engineers, UK.*)

1. *Unbaffled arrangement* (Fig. 4.4a): Here both bottom product and reboiler liquid are withdrawn from a common bottom sump.

2. *Baffled arrangement* (Fig. 4.4b): Here the space at the bottom of the column is divided into a bottom-drawoff sump and a reboiler feed sump by a preferential baffle. Typical arrangements are shown in Fig. 4.6.

3. *Once-through reboiler arrangement:* Here reboiler liquid is withdrawn from the bottom downcomer or from a chimney tray located above the bottom sump. Typical arrangements are shown in Fig. 4.7.

The unbaffled arrangement has the advantages of simplicity and low cost. It is preferred

- In small columns ($<$ 3 ft in diameter), where its simplicity is a major advantage and where baffles are difficult to inspect and maintain.

- With many packed columns. Here liquid raining from the bed above needs to be deflected into the reboiler side of a baffle, making a baffled arrangement more complicated, more expensive, and less economically attractive.

- With kettle reboilers, because bottom product is withdrawn from the kettle reboiler surge compartment and not from the column.

- With forced-circulation reboilers. Here the large reboiler circulation rate (compared to the bottom product rate) makes it difficult to achieve a steady liquid overflow across the baffle.

- With internal reboilers.

- In services where the feed consists mainly of lights and a small amount of residue, in order to avoid residue accumulation in the reboiler loop and thickening of material due to excessive residence times in the bottom sump.

Baffled or once-through arrangements are usually preferred with thermosiphon reboilers in large columns ($>$ 3 ft in diameter). Compared to the unbaffled arrangement, they offer the following advantages:

1. They provide an additional theoretical stage, or fraction of a theoretical stage.

2. They supply a constant liquid head to the reboiler. This is most important in vacuum systems, where a steady liquid driving head is essential for stable thermosiphon action (see Sec. 15.3).

3. They lower the boiling point of liquid supplied to the reboiler. This maximizes reboiler ΔT and its capacity and minimizes its fouling

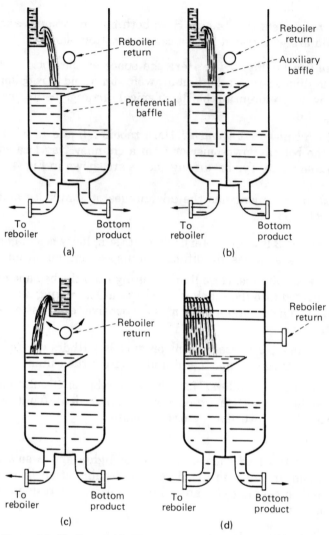

Figure 4.6 Preferential-baffle arrangements for bottom sumps. (*Henry Z. Kister. Excerpted by special permission from* Chemical Engineering, *July 28, 1980, copyright © by McGraw-Hill, Inc., New York, NY 10020.*)

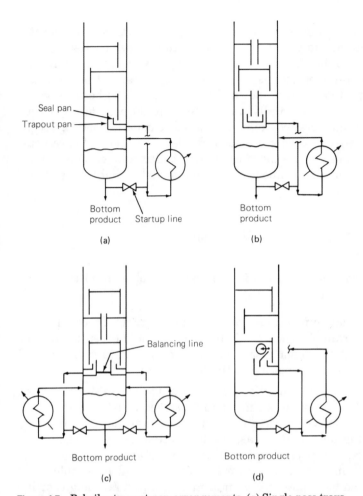

Figure 4.7   Reboiler trapout-pan arrangements. (*a*) Single-pass trays (once-through reboiler); (*b*) two-pass trays, single reboiler (once-through reboiler); (*c*) two-pass trays, two reboilers (once-through reboiler); (*d*) single-pass trays (recirculating reboiler).

tendency. This is a major consideration when the bottom liquid is a wide-boiling mixture.

4. They maximize bottom sump residence time when vapor disengagement is the main consideration.

The above benefits are most thoroughly realized with the baffle arrangement shown in Fig. 4.6*b* or with the once-through arrangements in Fig. 4.7*a* to *c*. With the baffle arrangements in Fig. 4.6*a*, *c*, and *d*, with packed columns, and with the trapout pan arrangement in Fig. 4.7*d*, benefits 1 and 3 above are only partially realized, because some liquid from the bottom tray (or from the packed section) can find its way to the bottom sump without passing through the reboiler.

Once-through reboiler arrangements can only be used when the quantity of liquid vaporized is less than 20 to 30 percent of the total amount of liquid descending down the column. This is because good thermosiphon action is usually limited to 20 to 30 percent vaporization. Once-through arrangements are therefore usually used for strippers or other low-boil-up applications. Baffled arrangements are usually used for most other thermosiphon reboiler applications. More detailed discussion of the above arrangements is available elsewhere (178).

4.6 Dos and Don'ts for Column Bottom Arrangements

Below are guidelines for avoiding operating problems with column bottom arrangements.

1. Column bottom sumps must provide sufficient residence time for vapor disentrainment (see guideline 4 for the only exception to this rule). Inadequate vapor disentrainment may lead to cavitation of the bottom pump or choking of the bottom line.

2. Frothing (waterfall pool effect) can entrain vapor into the bottom liquid, with consequences similar to those described in guideline 1 above. Conditions conducive to frothing are high liquid flow rates, liquid falling into small areas, shallow bottom sumps, large liquid fall heights, and poorly designed reboiler return (or bottom feed) spargers. In most other situations, frothing is seldom troublesome. Frothing can be suppressed by minimizing liquid fall height or increasing sump depth. An impingement baffle, or a layer of packing, close to the liquid level (and below the reboiler return nozzle), is sometimes used to absorb the impact of the falling liquid. A less common means of avoiding frothing is extending the bottom downcomer below the normal liquid level of the bottom sump (this, however, may cause premature flooding by downcomer backup; see guidelines 17 and 18 in Sec. 4.10).

3. Vapor disentrainment is not required in a sump or draw pan used exclusively to feed a thermosiphon reboiler, because the reboiler can handle some vapor bubbles in its liquid supply. It is important to size the reboiler sump for self-venting flow; otherwise vapor pockets may choke the reboiler sump. Figure 4.5 can be applied to determine the maximum allowable velocity through the reboiler sump. The pipe diameter on the y axis of the diagram is considered

as the equivalent hydraulic diameter (four times the flow area divided by the wetted perimeter) of the reboiler sump.

4. Completely different concepts are applied for bottom sumps containing thermally unstable materials. Here the prime consideration is to minimize the hot liquid residence time in order to avoid material degradation. The design in Fig. 4.8a is often recommended (68). With this arrangement, the liquid leaving the sump contains some vapor, and piping needs to be designed for self-venting flow. Commonly, the bottom liquid passes through a cooler which condenses the entrained vapor. If additional residence time is needed, or if the liquid is to be pumped, a surge drum is installed downstream of the cooler.

When the bottom sump does not supply liquid to a reboiler (e.g., when a reboiler trapout pan is used, or when the column has a bottom feed and no reboiler), the design in Fig. 4.8b can be used with thermally unstable materials. This arrangement eliminates the need for a surge drum and self-venting lines, and immediately quenches liquid reaching the column base.

5. Sometimes it may be desirable to have the outlet pipe elevated above the bottom of the sump in order to prevent solids or polymers from leaving in the column bottom (one example is shown later in Fig. 10.6a). This arrangement can be troublesome because it forms

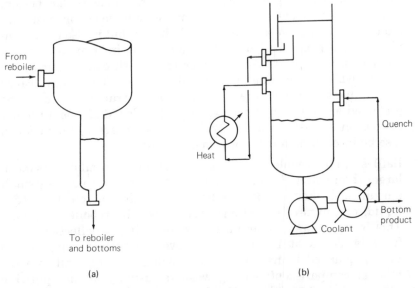

Figure 4.8 Common bottom sump arrangements for thermally unstable materials. (a) Sump liquid flows into reboiler; (b) sump liquid does not flow into reboiler.

a dead leg at the column bottom. It should be avoided altogether in services prone to water-induced pressure surges (Sec. 13.5), because the dead leg may trap water.

6. Flanges should be avoided inside the column skirt. These may leak and generate a flammable or toxic atmosphere inside the skirt. It is best to extend the nozzle to outside the skirt and install the flange there.

4.7 Dos and Don'ts for Preferential Baffles

When a baffled arrangement is used, the following additional guidelines apply:

1. Each compartment must be treated as a separate sump and designed accordingly.

2. Each sump must have its own drainage facilities. This is best achieved by installing an external valved dump line at a low point to interconnect the liquid outlet lines from each sump (see Fig. 15.4a). Alternatively, a number of holes can be drilled at the bottom of the baffle, but this gives slow drainage rates and susceptibility to plugging and is therefore not recommended. The author is familiar with cases when it took over a day to drain the reboiler sump via such holes.

3. Figure 4.6 shows preferential baffle arrangements for bottom sumps. Arrangement a has a simplicity advantage and is most common. Arrangement b performs slightly better, but is also more complex than arrangement a. Arrangement c is inferior to arrangement d because it forces the vapor to flow through a curtain of liquid while ascending to the first tray. This may cause entrainment or premature flooding (208). Arrangement d is the preferred arrangement for two-pass trays (Sec. 6.21 has a further discussion). A baffle design similar to that in arrangement b can also be incorporated in arrangements c and d.

4. Baffles can be troublesome when reboiler liquid circulation is much larger than bottom flow rate. In such cases, even a relatively small amount of leakage across the baffle, splashing of reboiler return liquid, or column fluctuations may starve the reboiler of liquid. This leakage can be minimized by installing a deflection baffle (usually sloped at 15° or more) above the bottom sump, which routes splashed liquid and liquid runoff down the wall into the reboiler compartment; adding washers (usually of the split-ring type) to the bolts of the baffle plate and the hatchway; gasketing; and seal-welding. In this type of service, it often pays to test the

reboiler compartment for leaks by filling it with water prior to startup.

5. When reboiler circulation is much larger than bottom flow rate, baffles can cause unstable bottom flow. Small variations in the reboiler heat input or column flow rates show up as large changes in reboiler sump overflow, and therefore in bottom sump level. Bottom level fluctuations cause bottom flow fluctuations, which may be troublesome if the bottom flow is the feed to another column or is used to preheat column feed. This problem is illustrated elsewhere (258) and is most severe when the bottom sump compartment is relatively small. Providing additional residence time in the bottom sump can minimize this problem.

4.8 Dos and Don'ts for Once-Through Reboiler Draw Pans

In tray columns, liquid to once-through reboilers (Fig. 4.7a to c), as well as liquid to many recirculating reboilers (Fig. 4.7d), is normally withdrawn from the bottom tray via a trapout pan. In packed columns, a chimney tray is usually preferred for this purpose. This trapout pan can often be troublesome. Successful operation requires close attention to the following guidelines:

1. Leakage at the trapout pan or bottom tray may starve the reboiler of liquid, increase the reboiler fractional vaporization, and reduce heat transfer. Leakage also bypasses liquid around the reboiler, reducing the mass transfer efficiency of the bottom stage. This problem is most severe at low liquid and at low vapor flow rates. Gasketing and/or seal welding, as well as making the bottom tray a leak-resistant valve tray or (if capacity considerations allow) a bubble-cap tray, may help minimize this leakage. Using a chimney tray instead of a trapout pan below the bottom tray can avoid weeping, but is also more expensive.

2. A dump line connecting the bottom of the column with the reboiler liquid inlet line should be provided (Fig. 4.7a to d). Without it, liquid will weep through the tray at startup (there is no vapor flow until the reboiler starts up), and no liquid will be available to commission the reboiler and initiate vapor generation.

3. An abrupt increase in heat to the reboiler may back up liquid to the first tray. The overflow arrangement in Figure 4.7a to d provides some protection against flooding in such circumstances.

4. The wall of the trapout pan should be 4 to 8 in higher than the wall of the seal pan (237). This prevents the liquid overflowing the seal

pan from bypassing the trapout pan. A case where failure to follow this rule caused a reboiler to be starved of liquid has been reported (237). Some other recommended dimensions are shown in Fig. 4.9.

5. Sufficient vertical height must be provided from the top of the trapout pan overflow weir to the bottom tray. The overflow pan liquid is usually less aerated and therefore denser than the downcomer fluid. This will cause greater downcomer backup.

6. If the column liquid flows to more than one reboiler using separate trapout pans (Fig. 4.7c), a liquid balance line is needed to ensure that liquid is equally distributed to each reboiler.

7. The trapout pan is most suitable for once-through reboilers (Fig. 4.7a to c). With a recirculating reboiler, the trapout pan can still be used, but its arrangement is more complex. Here the reboiler return nozzle needs to be located above the trapout pan, with liquid returning from the reboiler routed into the trapout pan and away from the seal pan. This is usually accomplished by a baffle (Fig. 4.7d) and/or by using a wide pan. Failure to properly route the reboiler return liquid into the trapout pan may starve the reboiler of liquid, and/or lower reboiler ΔT, and/or reduce the reboiler mass transfer efficiency.

For the above reasons, the sump baffle arrangements (Fig. 4.6) are usually preferred with recirculating reboilers.

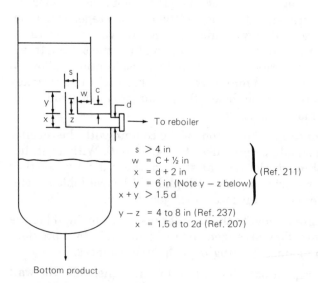

Figure 4.9 Dimensions recommended for arrangements in Fig. 4.7a to c.

4.9   Chimney Trays

Chimney trays (Fig. 4.10a, b) are used for withdrawing intermediate liquid streams from the column; in a packed tower, they are also used as liquid collectors or vapor distributors. Alternative devices used for liquid withdrawal are downcomer trapouts in tray columns, chevron collectors, and some redistributors in packed columns. Compared to these alternative devices, chimney trays have the following advantages:

1. They minimize liquid leakage to the section below. For this reason, they are preferred
   - When all liquid in a column section is withdrawn ("total drawoff").
   - When most (but not all) of the liquid is to be withdrawn, because excessive leakage will starve the draw off of liquid.
   - In low-liquid-load applications.

2. They provide greater residence times for vapor disentrainment, a greater surge volume, a better buffer against upsets, and smoother control. Therefore, they are frequently preferred
   - When the intermediate drawoff is pumped
   - When most of the liquid in a column section is withdrawn
   - When two-liquid phase separation is required.

3. In packed columns, chimney trays are the most effective liquid collection devices for high liquid flow rate services, for interreboilers, and for once-through reboilers. They are also sometimes used for collecting liquid from an upper bed for redistribution (when the redistributor is not self-collecting), and as vapor distributors.

4. In tray columns, chimney trays do not suffer from unsealing problems and are generally less troublesome than downcomer trapouts.

   The main drawback of chimney trays is that they consume more column height than alternative drawoff devices, resulting in a more expensive arrangement. Chimney trays are also relatively high-pressure-drop devices, which is a major disadvantage in packed columns operating in deep vacuum.

4.10   Dos and Don'ts for Chimney Trays

Detailed procedures for the design and operation of chimney trays are available in the published literature (138, 207, 420). The important factors and guidelines are summarized below:

1. The flow area of the risers is set by the allowable pressure drop. Too large an area wastes sump space; too small an area causes excessive pressure drop. A pressure drop of 5 in of water was recom-

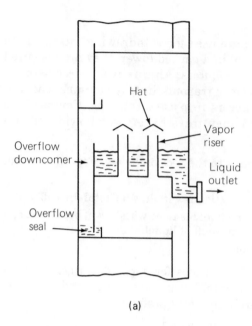

(a)

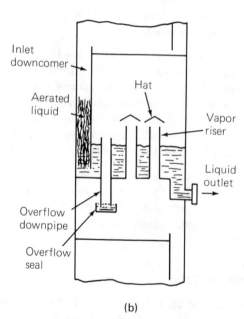

(b)

Figure 4.10 Chimney tray arrangements. (a) With liquid entering from a seal pan; (b) with a submerged inlet downcomer.

mended for chimney trays (420). A pressure drop of 8 in of water was recommended for packed-tower chimney trays in pressure services (>25 psig) in order to improve vapor distribution to the bed above (237). A method for estimating this pressure drop was detailed elsewhere (420). A lower pressure drop may be desired in packed columns operating under deep vacuum. In such cases, it has been recommended (237) that the chimney tray be designed to a pressure drop of 1 in of liquid. An alternative criterion (179, 307) recommends setting the risers' flow area at 15 percent of the tower area.

2. The number of risers must be large enough to ensure good vapor distribution. This is essential in packed towers, especially when the chimney tray delivers vapor to a short or a low-pressure-drop bed.

   An excessive number of risers should be avoided, because they will obstruct liquid flow and form a hydraulic gradient on the chimney tray. With short risers, this can provoke liquid overflow into risers. Caution is required to prevent any row of risers from restricting liquid flow. The hydraulic gradient can be estimated by techniques similar to those used for estimating hydraulic gradients on bubble-cap trays (48, 257, 319, 371).

   In larger towers, the hydraulic gradient can be minimized by using a split-flow arrangement, with two downcomers or draw sumps located at opposite ends of the chimney tray. In packed towers, this split-flow arrangement also assists in obtaining a more uniform vapor profile. The split-flow arrangement has been advocated (386) for large-diameter (>12-ft) packed columns unless the amount of liquid drawn is small.

   Risers may be round, square, or rectangular. Square or rectangular risers are usually less expensive to fabricate (144, 207). For good vapor distribution, rectangular risers are typically 4 to 6 in wide (237). In packed columns, risers should be uniformly spaced to maintain uniform vapor distribution (386).

3. The riser must be tall enough to ensure adequate residence time on the tray. A riser height of at least 12 to 18 in was recommended (74, 237). Further discussion is in Sec. 4.4. Where residence time is not a significant consideration (e.g., when the sole purpose of the tray is to serve as a vapor distributor), riser height as low as 6 in (237) can be used.

4. For total draw-off arrangements, it has been recommended to locate the top of the vapor risers 6 to 9 in above the maximum liquid level on the chimney tray (179, 307).

5. A cap or hat should be placed over each riser to deflect liquid raining from above, so that no liquid enters the risers and bypasses the chimney tray. The caps also improve vapor mixing by forcing it to move laterally.

All sides of each hat should either slope down, or be equipped with a drip lip (typically a 1-in-wide strip seal-welded to the edge of a flat hat and protruding and sloping downward) (143, 207, 421). The recommended slope is 15° (143, 207), although up to 45° slopes are sometimes used in leak-tight services (421). Alternatively, standard pipe caps may be used (420). The practice of using flat hats (without drip lips) is not recommended because they may not prevent liquid from running back underneath the hat and down into the chimney (207), and because they may lead to blowing the liquid running off the hat into the tray above.

Chimney trays equipped with rectangular risers that stretch from one end of the tray to another often use open-top gutters ("rainwater conduits") rather than hats (see Fig. 8.6b later). The gutters can be V-shaped or U-shaped, are mounted above the risers, and slope toward the short edges of the rectangles. Compared to hats, these reduce riser pressure drop and eliminate the downward velocity component of vapor leaving the chimneys, but at the expense of allowing some liquid to rain into the risers. This makes them less suitable for total drawoffs.

6. Flow area for the hat is set by the vapor velocity, which should be the same through the hat as through the riser. A method for determining this flow area is detailed elsewhere (420). An alternative criterion (207) is to make the peripheral area of vapor outlet at the hat 1.25 times the riser area. In services prone to pressure surges, the peripheral area should be further increased to at least twice the riser area (237), because surges often blow off the hats.

7. Hats should extend at least 1 in past the chimney (on all sides) to prevent liquid from entering the vapor risers (143, 207, 421).

8. The net flow area between each hat and the column shell must be sufficiently large to avoid excessive vapor velocities. Excessive vapor velocities may entrain liquid dropping from above, or fling this liquid on the shell, thus promoting corrosion.

9. In tray columns, vapor leaving the chimneys should not impinge on the downcomer feeding the tray. Failure to do this may cause vaporization in the downcomer.

10. Spacing between the top of the hat and the tray or packed bed above should be at least 12 in, but 18 in or more is preferred (74, 207). This is most important in packed towers, especially when

the chimney tray delivers vapor to a short or low-pressure-drop bed. This space is necessary to ensure adequate vapor distribution to the bed above. Additional spacing should be provided if support beams obstruct vapor mixing in packed towers. Section 8.2 has further discussion and a case history describing vapor maldistribution due to support beam obstruction.

11. It is desirable to place the draw-off nozzle in a portion of the tray floor that is lowered to form a sump. This lowers the height of liquid on the chimney tray by an amount equal to the nozzle diameter and reduces liquid leakage and the mass of liquid that the tray must support. The outlet nozzle should be flush with the sump floor to ensure adequate drainage. This practice is particularly fruitful in leak-tight applications, because it eliminates the need for weep holes.

12. Whenever possible, the drawoff sump should be positioned at the column shell to minimize internal piping. Internal flanges should be avoided in the draw lines, because these may leak and deprive the draw-off of liquid. One troublesome experience of such leakage has been described (237).

13. When the chimney tray is used for total drawoff, liquid leakage at its joints and support ring may be troublesome. The problem is amplified when liquid head on the chimney tray is high and when liquid flow rate is low. Seal welding of chimney tray sections has been recommended (237), and has been successfully implemented in several columns (231, 296, 421). Gasketing and sealants are sometimes used (145, 296, 386, 412), but they are often unsuccessful in harsh temperature or chemical environments (296, 412). A leakage test, similar to that conducted on bubble-cap trays (Sec. 7.12), is often specified for chimney trays. Techniques for troubleshooting leakage from drawoff pans are discussed elsewhere (231).

   Seal welding at the support ring is frequently disfavored because it hinders thermal expansion of the chimney tray. This leaves the support ring joints as a potential source of leakage. One technique for minimizing this leakage (231, 237) is welding a strip some distance from the column wall along the tray periphery (Fig. 4.11a) (alternatively, the chimney tray may be fabricated as a pan). A sloped wiper ring welded to the shell a short distance above the strip prevents liquid flowing down the wall from entering the support ring area. An even better technique (236) is covering the support ring with an angle iron, approximately 3 by 3 in (Fig. 4.11b). The angle iron is rolled to the tower diameter providing a snug fit at the column wall. One edge of the angle iron is seal-welded to the shell; the other is seal-welded to the tray, so

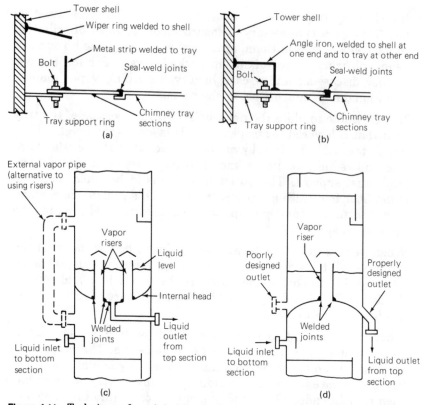

Figure 4.11 Techniques for minimizing leakage in intermediate draw-offs. (a) Strip and wiper ring; (b) angle iron; (c) internal head (downward bulging); (d) internal head (upward bulging). (*Part a, based on Process Design for Reliable Operation, Second Edition, by Norman P. Lieberman. Copyright © 1988 (May) by Gulf Publishing Company, Houston, TX. Used with permission. All rights reserved. Part b contributed courtesy of Norman P. Lieberman, private communication.*)

that the support ring joint sees no liquid. The angle iron is flexible enough to handle the thermal expansion. This technique has been successfully implemented in many installations (236).

A technique which can positively eliminate leakage is using a welded internal column head instead of a chimney tray for liquid collection (Fig. 4.11c, d). The vapor can travel from the column section below the head to the section above either through risers or via an external pipe. This technique is expensive, and its use is restricted to situations where no leakage below the collection point can be tolerated.

The internal head can be either downward-bulging (Fig. 4.11c) or upward-bulging (Figure 4.11d). An upward-bulging internal head may be difficult to drain. The liquid outlet shown in dashed lines on Fig. 4.11d provides improper drainage and should therefore be avoided. One incident was reported (4) where this outlet arrangement collected water, which later caused a pressure surge and tray damage in hot-oil service. The outlet arrangement shown in heavy lines should be used.

14. An overflow pipe (or downcomer), with an opening located below the top of the risers, is recommended for preventing liquid from overflowing the risers at high liquid levels (Fig. 4.10). The overflow pipe should be liquid-sealed at the bottom to avoid vapor rise through it. Two experiences have been described (57, 237) where failure to provide an overflow pipe caused liquid to overflow the risers and prematurely flood the column section above the chimney tray. In a third case (334), liquid overflowing the risers caused entrainment (Sec. 8.2).

In packed columns, liquid overflowing the chimneys may preferentially descend into some risers (often the peripheral), with vapor ascending through the others (often the central risers). This may cause severe maldistribution of vapor to the bed above.

15. Downcomers, downpipes (internal or external), and overflow pipes are often used to transport liquid from the chimney tray to the column section below. Undersizing these will cause liquid to overflow the chimneys, with the consequences described in guideline 14 above.

Liquid normally enters a chimney tray downpipe (or downcomer, or overflow pipe) close to the liquid level. The liquid is therefore usually aerated (unless sufficient residence time for vapor disengagement is available between the liquid level and the entrance to the downpipe), and the downpipes must be sized for self-venting flow (Fig. 4.5). When the residence time described above is very short (<10 s), and in foaming systems, it may be better to size the downpipes using downcomer-choking criteria (Sec. 6.16). Downpipes must be adequately sealed to prevent vapor from rising through them.

The author experienced one troublesome case, which was also reported by Lieberman (237), where liquid overflow through the chimneys caused a severe loss of efficiency in the packed section above. The chimney tray had undersized downpipes that were not liquid-sealed; either the undersizing or the lack of seal (or both) could have caused the overflow. Lieberman (237) suggests that the overflow led to entrainment and flooding, hence the loss in efficiency. However, subsequent pressure-drop measurements and other observations provided no supporting evidence for the existence of flooding, and the author believes that vapor maldistribution due to liquid overflow (guideline 14 above) caused the loss in efficiency.

16. When the chimney tray is used for liquid drawoff, liquid level on the tray should be controlled or maintained by a sufficiently tall overflow weir in order to avoid vapor in the outlet (unless downstream piping is designed to handle vapor).

17. In tray columns, the downcomer from the tray above the chimney tray can either end in a seal pan from which liquid spills onto the chimney tray (Fig. 4.10a), or can be extended below the liquid level of the chimney tray (Fig. 4.10b). Extending the downcomers below the liquid level minimizes frothing (waterfall pool effect) and splashing generated by the entering liquid, assists the action of the level controller, and eliminates problems with blockage of weep holes in the seal pan (Sec. 6.21). Minimizing the frothing effect is beneficial for chimney trays because the available residence time for vapor disentrainment is usually relatively small. Smooth level control is particularly important if two liquid phases are separated on the chimney tray by an interface level controller. The above reasons make extending the downcomers below the liquid surface attractive, but this technique may be troublesome (guideline 18 below).

18. If liquid on the chimney tray seals the downcomer from the tray above, particular care must be taken with the design of this downcomer (Fig. 4.10b). The liquid in the downcomer is aerated, while most of the liquid on the tray is degasified. The degasified liquid on the tray exerts a greater hydrostatic head than the column of aerated liquid in the downcomer. This produces additional liquid backup in the downcomer. If not allowed for, and if sufficient height is not provided, downcomer backup may exceed the spacing between the liquid level and the tray above, and lead to premature flooding. Troublesome experiences caused by this effect have been reported (179, 391); the author is familiar with one other. This problem is aggravated if two liquid phases are separated on the chimney tray.

19. In some services (e.g., refinery fractionators), vapor approaching the chimney tray is hotter than the chimney tray liquid. Heat will be transferred from the vapor to the liquid. If the vapor is condensable, some will condense on the bottom face of the chimney tray. The net result is analogous to leakage. The author is familiar with situations where refractory was installed on the bottom face of the chimney tray. In all these cases, steps were also taken to minimize leakage, making it difficult to independently assess the effectiveness of the refractory. For multicomponent, partially condensable vapor condensing on an uninsulated bottom face of a chimney tray (e.g., in a refinery fractionator), a typical heat transfer coefficient is 15 Btu/(h $\cdot$ ft^2 $\cdot$ °F) (237).

4.11   Downcomer Trapouts

Downcomer trapouts (Fig. 4.12) are mainly used for partial liquid drawoff from tray columns. They are preferred when the liquid withdrawn is only a small fraction of the column liquid, or when the drawoff feeds a once-through reboiler (Sec. 4.8). They are sometimes also used for drawing most or all the column liquid at an intermediate point, but are less suitable than chimney trays for this purpose.

Some of the main considerations in the application of downcomer trapouts are

- Downcomer trapouts seldom provide sufficient residence time for vapor disentrainment, and the venting process must be completed downstream of the column outlet. The vent-return nozzles must always be located above the tray's liquid level. Downstream piping must be designed for self-venting flow (Fig. 4.5).

- Sometimes an external drawoff drum (Fig. 4.4d) is installed at the liquid-drawoff elevation. The drum provides the required residence time for vapor disengagement and for smooth control and is vented back into the column. Such a drum is usually supported by the column.

- Liquid may weep through the tray openings and escape the trapping downcomer, especially when the vapor rate is low. This will starve the trapout from liquid and is particularly troublesome with total drawoffs or where most of the column liquid is withdrawn. For this reason, the author and others (237) strongly recommend against using downcomer trapouts for total drawoffs. A particularly troublesome case history has been reported (237) where several different trays were tried above a downcomer trapout, but none got close to achieving the desired drawoff rate. Total trapout was finally achieved by converting the downcomer trapout to a seal-welded chimney tray trapout.

- For similar reasons, in partial trapout services, it is desirable to use a leak-resistant type of tray above the downcomer trapout.

- Vapor may flow up the downcomer. A positive seal may be needed to prevent this, as discussed below.

4.12   Maintaining Downcomer Trapout Seal

The downcomer in Fig. 4.12a is likely to lose its seal whenever its liquid height drops below tray level. When the seal is lost, vapor from the tray ascends the downcomer, which may cause flooding, cycling, and/or poor separation. Downcomer unsealing by this mechanism is most likely to occur when the liquid drawn constitutes a large portion of the downcomer liquid flow, when the quantity drawn tends to fluctuate, and/or when excessive leakage takes place due to tray weeping or draw pan leaks.

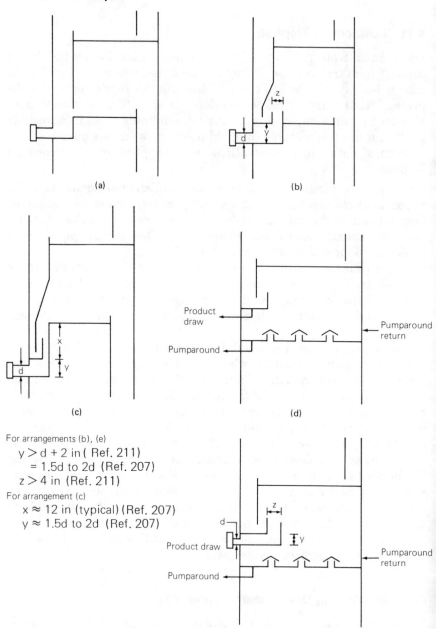

For arrangements (b), (e)
$$y > d + 2 \text{ in } (\text{Ref. 211})$$
$$= 1.5d \text{ to } 2d \ (\text{Ref. 207})$$
$$z > 4 \text{ in } (\text{Ref. 211})$$

For arrangement (c)
$$x \approx 12 \text{ in (typical)} (\text{Ref. 207})$$
$$y \approx 1.5d \text{ to } 2d \ (\text{Ref. 207})$$

Figure 4.12 Downcomer trapout pan arrangements. (*a*) No positive seal; (*b, c*) with positive seal; (*d*) with pumparound, no positive seal; (*e*) with pumparound and positive seal.

The arrangement in Fig. 4.12*b* does not suffer from unsealing, but it lowers the tray bubbling area. It may also induce excessive weeping through the inlet rows of tray perforations because of the downward momentum of liquid overflowing the inlet weir. Lowering the seal and drawoff pans below tray level (e.g., Fig. 4.12*c*) can overcome the weeping problem. The loss in bubbling area is usually minimized by using sloped downcomers (Fig. 4.12*b, c*). If the geometry and required liquid flow rates permit, the reduction of bubbling area can be entirely eliminated (Fig. 4.12*c*).

Arrangements 4.12*b* and *c* are more expensive than arrangement 4.12*a,* and are therefore only justified when unsealing is likely to be a problem. It is also important to ensure that weep holes are located at the bottom of each pan to facilitate proper drainage.

In large columns processing nonfoaming systems at low or moderate pressures, the draw pans shown in Figs. 4.12*b* and *c* are often enlarged to provide sufficient residence time for vapor disengagement, avoiding the need for downstream venting.

In refinery fractionators, product draws are often located above a chimney tray used for pumparound circulation (Fig. 4.12*d* and *e*). Arrangement *d* is analogous to arrangement *a* and can suffer from downcomer unsealing. An unsealing problem sometimes unveils when a refiner maximizes pumparound heat removal, because stepping up heat removal lowers the liquid overflow from the section above, thus increasing the fraction of downcomer liquid withdrawn as a side product. Unsealing can be avoided by using arrangement *e*, which is analogous to arrangements *b* and *c*.

4.13 Do's and Don'ts for Downcomer Trapouts

The following guidelines have been recommended for design and operation of downcomer trapouts:

1. Sump width should preferably not exceed the width of the downcomer in that location, in order to maintain interchangeability of tray parts (207). However, it should not be less than the nominal drawoff nozzle diameter plus 2 in (211).

2. Normal sump depth should be 1½ to 2 times the nominal nozzle diameter for partial-draw sumps (207, 211).

3. The distance from the bottom of the drawoff sump to the floor of the tray below should be at least 75 to 80 percent of the tray spacing (179, 211, 237, 307). Another designer (207) feels that this distance needs to be at least 60 percent of the tray spacing.

4. Downcomer area should not be decreased below 50 percent of the

top area if the downcomers are sloped and extend into the sump (207).

5. Whenever possible, downcomer trapouts and draw pans should be positioned at the column shell in order to minimize internal piping. Internal flanges should be avoided, as these may leak and deprive the drawoff of liquid. One troublesome experience with such leakage has been reported (237).

6. Downcomer trapouts may be troublesome with foaming systems. Chimney trays are often preferred in such services.

4.14   Chevron Collectors

Chevron collectors are used in packed columns as liquid collectors for partial drawoff or for feeding to a redistributor which is not self-collecting (e.g., a notched-trough redistributor). They are sometimes also used as total drawoffs, but are less suitable than chimney trays for this purpose. The chevron collector (Fig. 4.13) consists of evenly spaced chevron blades several inches high. Liquid collects at the bottom of the blades and runs into a draw pan. From there it can be taken out or fed to a redistributor.

Figure 4.13   Chevron collector. (*G. K. Chen, excerpted by special permission from* Chemical Engineering, *March 5, 1980, copyright © by McGraw-Hill, Inc., New York, NY 10020.*)

The chevron collector features a high open area for vapor flow, and is therefore a low-pressure-drop device. Pressure drop is usually less than 0.1 in of water (74). This type of collector is most suitable for vacuum applications. This collector also acts as a vapor distributor; accordingly, the distance required between the top of this collector and the support plate above is less than 12 in (74). The short vertical distance requirement of this device compared to the chimney tray often makes it less expensive than the chimney tray.

The disadvantages of chevron collectors are difficulty in handling large liquid loads and greater liquid leakage compared to chimney trays. They provide insufficient residence time for vapor disengagement, and the venting process must be completed downstream of the column outlet. Downstream piping must be sized for self-venting flow (Fig. 4.5).

4.15 Redistributor Drawoffs

In packed columns, liquid may sometimes be withdrawn from a redistributor. This drawoff method is suitable only when the liquid withdrawn is a small fraction of the column liquid. Compared to the chimney tray, this method consumes far less vertical height and is therefore less expensive.

Liquid is withdrawn from the redistributor either via an overflow pipe or using a submerged drawoff nozzle. When an overflow arrangement is used, the drawoff may frequently run dry, and flow to downstream units may be intermittent or unstable. When a submerged drawoff nozzle is used, excessive drawoff rates may lower the liquid level in the redistributor and cause poor distribution to the bed below.

Redistributors usually provide insufficient residence times for vapor disengagement (especially if an overflow pipe is used), and venting must be completed downstream of the column outlet. Downstream piping must be sized for self-venting flow (Fig. 4.5).

When this arrangement is used, caution is required to ensure that the drawoff does not generate large-scale maldistribution of liquid to the bed below (see Sec. 3.1) and that unirrigated areas do not form on top of the packing under the drawoff pipe or sump. This is most likely to be a problem when a submerged drawoff nozzle or sump is used.

4.16 Collector Box

When a small fraction (<6 percent) of a packed tower liquid is withdrawn, a special collector box can be installed within the packing

(386). Careful design is required to prevent blockage or interference of this box with liquid and vapor distribution. A channel or open-pipe box is sometimes used for this purpose.

4.17   Mechanical Dos and Don'ts for Liquid Outlets

Below are some mechanical considerations generally applying to liquid outlets:

1. Seal welding rather than gasketing should be applied for total drawoff pans. Gaskets may not provide a sufficiently effective seal under the operating conditions. For large-diameter towers and high-temperature services, expansion joints should be provided. If the service is also corrosive, a heavy-gage metal or a corrosion-resistant material should be specified for the entire trapout tray (or pan).

2. Areas in the column that are likely to be heavily loaded with liquid, such as drawoff sumps, require special attention. Thus, the design may call for trusses to be added to the structure.

3. Pipe supports should be located near the outlet nozzles so that the pipes are not supported by the nozzles. Pipe guides should be used to prevent pipes from swaying in the wind.

4. All liquid outlet areas must have low-point drains or weep holes to allow liquid drainage at shutdown. Recommended weep-hole practices are discussed in Sec. 7.11.

5. Any infrequently used drawoff lines should be isolated (and if possible, blinded) at the column to avoid a formation of a dead leg. Water or chemicals trapped in dead legs can freeze, plug, react or enter the column at an undesirable time. In one case (275), a maintenance worker was sprayed with liquid trapped in a plugged dead leg at shutdown.

6. When liquid reaching the sump or chimney tray contains small amounts of lighter, but insoluble liquid (or solids), or oil from leaking machinery, two liquid phases may separate in the sump. The lighter phase will have no way out unless skimming connections are provided. Typically, three 1-in valved connections are provided, one being at the same elevation as the normal liquid level, the others being 1 in above and 1 in below that level (421). When more thorough skimming is required, an external skimming surge drum is often used.

7. When liquid from a packed column is pumped, it is recommended to

install a screen upstream of the drawoff nozzle in order to protect pump impellers from migrating pieces of packings. Vortex breakers can often be designed to perform this function. This technique may also be beneficial for columns containing valve trays with a history of valves popping out of their seats.

4.18 Vapor Outlets

Vapor outlets are far less troublesome than liquid outlets. The prime consideration is avoiding the presence of entrained liquid droplets in the vapor streams leaving the columns. The degree of liquid removal is set by the downstream unit. If the vapor flows directly into a condenser, a relatively crude droplet separation can be tolerated, because at worst, this entrainment would slightly reduce column efficiency. However, if the vapor flows to a compressor or a turbine, liquid separation must be fine; otherwise the liquid droplets may damage the blades.

Design methods are discussed elsewhere (417) and are based on limiting the maximum diameter of droplets entrained in the vapor stream. Maximum droplet diameter can be reduced by supplying a knockout facility upstream of the compressor. This can be either an enlargement of the vapor space above the top tray, an in-line liquid separator, or a separate knockout drum. Mist eliminators are effective and can be installed above the top tray or in the drum. If the vapor temperature is higher than the worst ambient conditions, the lines from the knockout facility to the compressor should either be kept short and insulated or be provided with liquid-removal facilities.

If mist eliminators are used, they must be carefully designed, specified, and supported. Plugging of mist eliminator pads is not an uncommon problem. Occasionally, a dislodged part of a mist eliminator pad is sucked into the compressor rotor or becomes lodged in downstream piping (232, 239). These potential problems should be discussed with the manufacturer. The beneficial effects and the various types of mist eliminators available are extensively described elsewhere (101, 165, 270, 431–433).

Small quantities of liquid should always be expected in column vapor outlets. The origin of this liquid can either be entrainment from the column (fine droplets can pass even through mist eliminators and coalesce in the overhead line), or atmospheric condensation. Low points in vapor outlet lines should be avoided as these tend to trap and accumulate liquid. The accumulated liquid back-pressures the column, causing instability, erratic operation, and slug flow into the downstream unit; one case history where this occurred has been reported (203). Vapor outlet lines should be sloped (i.e., self-draining) either back into the column or into a downstream vessel. If sloping into a downstream vessel, the vessel should be able to handle at least a small quantity of liquid.

It has been recommended (354, 355) to size top vapor outlet lines for a maximum velocity of 60 ft/s for atmospheric and pressure columns, increasing to 200 ft/s as pressure is lowered below 10 psia, and to confirm that pressure losses between column and condenser are acceptable. In medium- and high-pressure columns (>100 psig), pressure drop considerations usually dominate and velocities well below 60 ft/s are normally used.

Chapter

5

Gravity Lines and Instrument and Access Connections

The prime consideration for gravity lines is achieving adequate hydraulics. Deficient hydraulics in gravity lines may cause column instability and mechanical damage. This chapter outlines considerations unique to gravity lines, highlights common problems, reviews the preferred practices, and supplies guidelines for avoiding and overcoming gravity line pitfalls.

The prime consideration for instrument connections is to avoid hydraulic interference in the column or impulse line, which would lead to erroneous measurements or instrument malfunction. False information supplied by instruments has been the cause of premature flooding, column damage, and poor separation in many columns. This chapter examines the preferred practices, reviews common pitfalls, and supplies guidelines for avoiding pitfalls with column instrument connections.

Provision of adequate access to column internals makes cleaning, inspection, maintenance, dismantling, and reassembly of the column internals easier and safer. The access requirements depend on the anticipated frequency of personnel entry into the column. This chapter reviews the preferred practices and supplies guidelines for avoiding unsatisfactory access to column internals.

5.1 Gravity Lines Pitfalls

Gravity flow is often used in bottom, reflux, side-draw, or feed lines. Unique considerations for gravity flow are described below, with specific reference to reflux lines. Considerations for gravity flow at other inlets and outlets are similar and can be directly inferred from those described. Overlooking these unique considerations is a frequent cause of column instability and poor performance.

In smaller columns, the overhead condenser is often mounted above the column, and reflux flows to the column by gravity. The reflux drum and/or reflux control valves are frequently omitted; sometimes, the bottom section of the condenser is used as a surge compartment.

The main considerations unique to gravity reflux arrangements are

1. Reverse flow of vapor from the column into the reflux line would interfere with liquid downflow, condenser action, and may cause hammering if the reflux is subcooled. A seal loop (Fig. 5.1a to e) is almost always used to avoid reverse flow. The low point in the seal loop needs a drain (normally closed).

2. It is best to locate the bottom of the seal loop at a lower elevation than the outlet rim of the reflux pipe (Fig. 5.1a, b, d). This prevents the seal loop liquid from siphoning back into the column. Arrangements c and e may suffer from siphoning unless an open high-point vent is installed to break that siphon (188). However, an open vent can be troublesome as described below (item 3).

3. A high point in the reflux line downstream of the seal loop should be avoided (if there is no seal loop, a high point should also be avoided). Arrangement 5.1a is best, with lines downstream of the seal loop sloping toward the seal loop to avoid a high point. If the reflux nozzle is at the column head, arrangement 5.1d is best, with the vertical inlet pipe sized for vapor upflow simultaneous with liquid downflow.

When an upward loop is not vented (Fig. 5.1b, c, e with a closed vent valve or no vent line), inerts may accumulate at the high point, causing intermittent siphon action and reflux flow oscillation. Cases were reported (68) where hot vapor was sucked back from the column into the high point and caused hammering upon contacting the subcooled reflux. This ruptured the reflux line and nozzle. The phenomenon is most troublesome in vacuum, where slight air leaks occur.

The high-point vent in arrangements b, c, e in Fig 5.1 vents inerts and serves as a siphon breaker. When the reflux piping downstream of the vent is large enough to permit vapor upflow simultaneous with liquid downflow, arrangements b and c become equivalent to arrangement d. A well-sized restriction orifice on the vent will minimize vapor upflow.

An undersized reflux line downstream of the vent, or an oversized vent line or restriction orifice can perform worse than no vent at all. At high reflux flow rates, vapor may be sucked from the vent line into the vertical leg downstream of the high point (158). At low

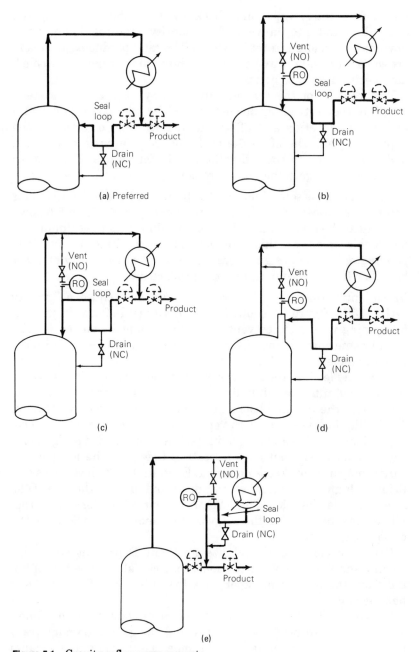

Figure 5.1   Gravity reflux arrangements.

reflux flow rates, vapor may backflow from the column up the reflux line and into the vent. The quantity and direction of vapor flow may therefore alternate with reflux rate, producing unsteady pressure drop and reflux flow. When the liquid is highly subcooled, hammering may also occur.

The location of the control valve (if any), the control system, length of the vertical leg downstream of the high point, and actual piping configuration also affect the quantity and direction of the vapor flow, and therefore the performance of arrangements *b, c,* and *e.* Attention to these details, and judicious use of nonreturn valves, can mitigate the above problems.

4. Arrangement *e* is sometimes used to liquid-seal the condenser. This arrangement does not suffer from the gravity cycle (item 5 below). Venting at the high point and vapor backflow from the column may be troublesome with this arrangement, as per item 3 above. Depending on piping design, there may also be a possibility of vapor backflow from the column into the product line.

5. Vapor entrainment into the reflux liquid may cause a "gravity cycle" (158). Figure 5.2 shows how a gravity cycle forms in the Fig. 5.1c piping with a closed vent. Although this phenomenon is most troublesome with arrangements *b* and c, it may also be encountered in arrangements a and d.

When the level in the condenser surge compartment is low, vapor is entrained into the reflux line (Fig. 5.2*a*). This increases flow resistance in the line and backs up liquid in the surge compartment. As the liquid level rises, less vapor is entrained (Fig. 5.2*b*). Some vapor, however, is still trapped in the line, especially if a high point exists. This vapor continues to restrict flow, and the level in the surge compartment keeps on rising. Eventually, the level rises sufficiently high to bar vapor entrainment and to clear the line of the remaining gas bubbles (Fig. 5.2*c*). A siphon is now formed. It rapidly drains the liquid in the surge compartment (Fig. 5.2*d*), and the cycle starts again.

Gravity cycles often destabilize a column. Depending on the geometry and violence of fluctuations, they can also cause loss of liquid from the seal loop, vapor backflow into the reflux line, and hammering.

Gravity cycles tend to be most severe in column bottom lines. These cycles can be quite violent, can dislodge and damage trays and packing support plates, and in some cases, even pull a vacuum deep enough to collapse a vessel (158). Gravity outlet lines should be surveyed with special care to ensure no gravity cycle can form.

To avoid a gravity cycle, it is best to ensure constant level in the

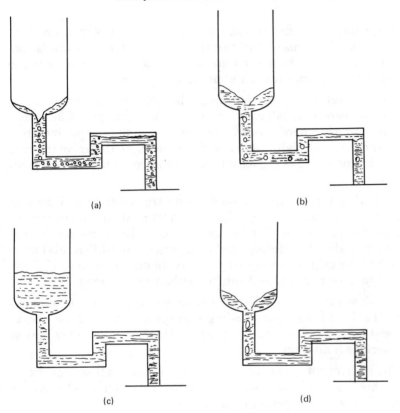

Figure 5.2   Gravity cycle formation.

surge compartment. This can be achieved either using arrangement *e* in Fig 5.1, or by using a level control loop. The vents in Fig. 5.1 can prevent the siphoning; the author eliminated one gravity cycle by adding this vent.

6. Some control systems may induce an instability known as "reflux cycle" (68) in gravity systems. This is discussed in Sec. 19.2.

5.2   Instrument Connections

Information supplied by instruments is vital for column control and operation. False information prompts wrong operator (or control loop) action, erratic column operation, and incorrect diagnosis of problems. Numerous experiences (7, 71, 97, 203, 210, 232, 237, 239, 255, 268) have been reported where faulty instrument readings caused, or directly contributed to, column problems.

General considerations important for instrument connections are:

1. The connection must be located where the desired measurement can be properly made. For instance, if liquid temperature is specifically desired, the measurement connection must be close to the tray or downcomer floor, and not in a vapor space.

2. The connection(s) must be compatible with the instrument and permit correct installation of impulse lines. Figure 5.3b and c shows a case where overlooking this guideline prompts incorrect readings (see guideline 3 in Sec. 5.3 for details). Defective impulse line piping is one of the most frequent sources of instrument problems.

3. A flowing fluid must not impinge on transmitters and connections. Impingement will produce incorrect readings and may damage the instrument. For instance, velocity head (converted into static head) will be interpreted by a pressure (or differential pressure) transmitter as a higher pressure. Impinged-upon level floats will move erratically and can be mechanically damaged.

4. The instrument must be accessible for operation and maintenance. If an instrument reading is suspect, a check of the instrument is required, often in a hurry, and inaccessibility may impose a serious limitation.

5. The instrument connection should not be too small so that it easily breaks or plugs; ½-in connections are often considered too small; 1 to 2 in is normally adequate.

6. Instrument lines should be kept as short and as free of elbows and bends as possible. Internal instrument lines should be avoided. This is most important when there is a tendency to plug. One experience has been reported (237) where overlooking this rule caused plugging and loss of level indication.

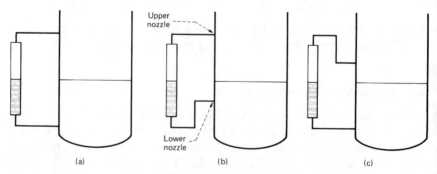

Figure 5.3   Piping arrangements for level glasses. (*a*) Correct location for glass; (*b*) glass below lower nozzle; (*c*) glass above upper nozzle. (*Henry Z. Kister, excerpted by special permission from* Chemical Engineering, *July 28, 1980, copyright © by McGraw-Hill, Inc., New York. NY 10020.*)

7. Having a single nozzle serve two (or more) instruments can lower costs. This, however, must not be achieved at the expense of reduced reliability. For instance, using the same nozzle to serve a pressure transmitter and a pressure alarm switch must be avoided, because both will become inoperative should the nozzle plug. An incident has been reported (239) where this led to serious overpressuring of a column. In another incident (237), a level indicator, a level gage, and a level alarm connected to the same reflux drum taps signaled erratically when a tap plugged.

8. Vibration-free mounting of transmitters is essential. Vibration can loosen the transmitter's bolts and/or lead to erratic signals.

9. Where ambient temperatures can fall below the freezing point of water (or of the process fluid), the possibility of liquid freezing in impulse lines must be averted. Extensive discussions are presented elsewhere (120, 266).

10. When duplicating instrumentation from an "identical" or "similar" unit, it is crucial to be alert to any differences, albeit small, as these may have a major impact on the location of instrument connections. In one peroxide-service incident (97), a small change in reboiler design, which was not matched by a change in instrument setting, resulted in an explosion.

11. When instrument connections are purged by an inert gas, the source of the purge gas and its pressure must be reliable. Backflow of column fluids when purge gas pressure suddenly dips can cause plugging, erroneous indication, and even a hazard. Further, the purge gas must be free of entrained solids; in one case (145), such solids blocked the purge gas restriction orifices, and the purge was lost.

12. When instrument connections are purged by an inert gas, adequate venting must be provided at the overhead condenser to prevent gas blanketing. In one incident (239), condensation was impaired after a nitrogen-purged instrument was added to the column. Gas-purged instrument connections should be avoided if the gas can be troublesome in the condenser or in downstream equipment.

5.3 Dos and Don'ts for Level Sensing

Proper design and installation of level sensors is crucial. The author's experience and others' (97, 107, 210, 234, 237, 375) has been that losing track of the true liquid level is a prime source of column upsets and damage (Secs. 13.1 and 13.2). Guidelines for level measurement connections in columns are described below. Further details of good design practices are described elsewhere (68, 234, 237).

1. A well-designed liquid level measurement assembly (234, 237) has two bridles (Fig. 5.4), one for the level transmitter, the other for level glasses. The latter bridle usually harbors the high and low level alarms. If the assembly consists of a lone bridle, all level measurements and alarms will give the same misleading reading when any line joining the column to the bridle plugs or freezes. The operator will have no means of knowing that the reading is incorrect until a major upset occurs.

   The use of stilling wells (Fig. 5.4) has been highly recommended (68, 237), particularly with float-type devices. The stilling well is an inexpensive device, often a 4-in piece of pipe. It dampens any oscillating signals that may originate from waves on the liquid surface; moderates the "dancing" of level floats, thus protecting them from mechanical damage; helps keep the level float free of dirt; and lessens the likelihood of nozzle plugging. A stilling well may be external (Fig. 5.4) or an internal baffle (68); the external arrangement is usually preferred.

2. With pressure and differential pressure transmitters, it is important to ensure that the measured head is the fluid head that the instrument is intended to sense. If a transmitter is mounted above a nozzle, the line from the nozzle to the instrument must be suffi-

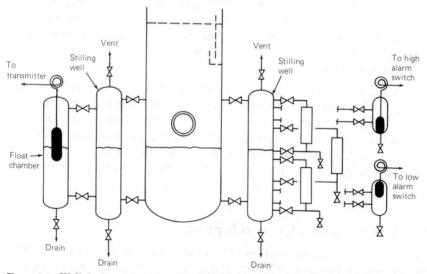

Figure 5.4   Well-designed liquid level measurement assembly. (*Based on Norman P. Lieberman, excerpted by special permission from* Chemical Engineering, *September 12, 1977, copyright © by McGraw-Hill, Inc., New York, NY 10020.*)

ciently large to permit proper venting and drainage. If a transmitter is mounted below a nozzle, a means for removing accumulated or condensed liquid must be devised (e.g., steam tracing). It is best to insulate lines from the nozzle to the transmitter in order to lessen the impact of ambient changes.

3. Level glasses should be mounted above the lower level-glass nozzle and below the upper level-glass nozzle (Fig. 5.3a). If part of the level glass is below the lower nozzle (Fig. 5.3b), the level reading will be incorrect when the liquid level in the column falls below the lower nozzle, because the glass will indicate the liquid level in its own loop. If the piping from the upper nozzle to the level glass has a high point above the upper nozzle (Fig. 5.3c), the level reading will be incorrect when the liquid level in the column exceeds the nozzle elevation, because the liquid will entrap the vapor at the high point. Hence, the arrangements of Fig. 5.3b, c must be avoided.

4. Level glasses should preferably be staggered (Fig. 5.4) with overlapping transparent sections (237). If the transparent sections do not overlap, the liquid level may "hide" between adjacent sections. Each level glass should have its own drain. Draining the glass and watching it refill is a valuable means of detecting plugging or freezing in the line joining the bridle to the glass. The drains also permit connection of nitrogen hoses for blowing back and clearing blockages.

   One designer (210) recommends against using level glasses in pressure columns that handle flammable or toxic liquids at temperatures higher than their normal boiling points. The same designer also recommends fitting level glasses with ball check valves which prevent a massive leak if the glass breaks (Fig. 5.5).

5. Caution is required when interpreting level measurements in reboilers, draw pans, downcomers, or other devices containing aerated liquid. Unless the aeration factor is known (which is rarely the case), these measurements can be misleading and are best avoided. In one case (56), troubleshooters were misled by a normal level indication in a downcomer trapout pan when the pan was flooded and backed up liquid to the trays above (Fig. 5.6). In that case, the level glass measured the clear liquid height in the section between its own tappings. In a peroxide-service accident (97), a level indicator in the reboiler liquid measured the aerated liquid height instead of the actual liquid height. A low liquid level was therefore unnoticed; this resulted in an explosion.

6. In aqueous systems, accumulation of a layer of oil or other low-specific-gravity organic liquid will cause a level instrument to indicate a level that is lower than actual. The oil may originate in

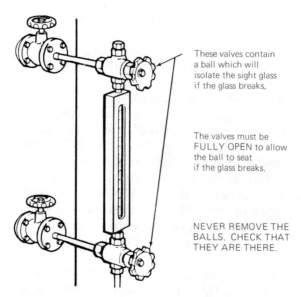

These valves contain
a ball which will
isolate the sight glass
if the glass breaks.

The valves must be
FULLY OPEN to allow
the ball to seat
if the glass breaks.

NEVER REMOVE THE
BALLS. CHECK THAT
THEY ARE THERE.

Figure 5.5   Recommended practice of installing ball check valves to isolate a level glass. (*What Went Wrong?* Second Edition, by Trevor A. Kletz. Copyright © 1988 by Gulf Publishing Company, Houston, TX. *Used with permission. All rights reserved.*)

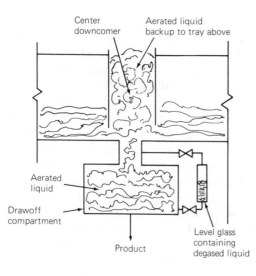

Center
downcomer

Aerated liquid
backup to tray above

Aerated
liquid

Drawoff
compartment

Product

Level glass
containing
degased liquid

Figure 5.6 Misleading liquid level measurement in a draw pan. (*Courtesy of Doug S. Bouck (Sohio), private communication.*)

leaking machinery or condensation of oily components in the tower. At times, a layer of oil may be seen in the level glass, but if the upper bridle or level glass tapping is above the liquid level, the oil will not reach the level glass. To restore proper level indication, the oil must be skimmed (Sec. 4.17).

A misleading level indication due to oil accumulation in the column base caused severe operating problems in many aqueous services. In several cases, liquid level in amine towers rose above the bottom feed inlet and caused column flooding, while the level instrument indicated normal level (238). In another case (238), a glycol regenerator performed poorly when the liquid level in a kettle draw compartment rose above the overflow weir, while the level instrument indicated normal level (238).

7. It has been recommended (67, 68, 255) to avoid orientation of level-measurement nozzles by angles greater than 90° from a vapor inlet or reboiler return nozzle, and to refrain from positioning these nozzles under the bottom downcomers. If the angle exceeds 90°, a shielding baffle should be provided in front of the measurement nozzle.

   Failure to follow this guideline may cause incorrect level reading because (1) the transmitter will interpret part of the vapor velocity head as higher static pressure at the top nozzle, and (2) an excessive quantity of liquid may enter the top level nozzle from the reboiler return or from the wall flow under the bottom downcomer. With level floats, failure to follow this guideline may also lead to dancing and mechanical damage of the float. In one troublesome experience (255), impingement of a vapor inlet on the level measurement connection caused erroneous level readings, and eventually breakage of a level float.

8. Connections for differential-pressure transmitters measuring level should not be placed across a restriction in the vapor space of the column (237). The transmitter will interpret the pressure drop across the restriction as a higher level.

9. Horizontal leg plugging may be a severe problem in fouling services. A number of techniques are often used to overcome this:

   - *Using nucleonic level detectors:* These usually employ gamma ray absorption or neutron backscatter techniques, but may also use gamma ray backscatter (Sec. 14.5). A case where a gamma ray absorption indicator solved a level control problem at the base of a column has been described (71).

   - *Purging back into the column* (Fig. 5.7a): This is effective in some services (237), but the purges are often unable to clear deposits from the bottom leg or liquid from the top leg. If the liquid is heat-sensitive, it may polymerize in the top leg. Purge rates must be regulated at a rate low enough to avoid affecting the measurement.

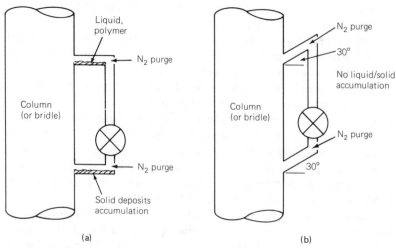

Figure 5.7 Transmitter arrangement for fouling services. (*a*) Purging back; (*b*) purging back with sloped nozzles.

- *Purging back into the column with angled nozzles* (Fig. 5.7*b*): The slope drains liquid from the top leg before it polymerizes, and resists accumulation of deposits in the bottom leg. This technique, however, may reduce the available level transmitter span.

Similar techniques may also be advantageous for minimizing leakage in toxic services.

10. In fouling services the bottom level-measurement nozzle should be located at least 6 in above the bottom of the sump, and the transmitter should not be mounted below the bottom nozzle.

11. Radioactive materials (e.g., from radioactive tracer tests) reaching the column base and/or concentrating there may interfere with the action of nucleonic level controllers. In one experience (210), this caused level buildup and subsequent column flooding. Radiography performed nearby may also affect nucleonic level indicators (210).

Additional guidelines for differential-pressure level transmitters on column-bottoms sumps have been presented by Buckley (67).

5.4 Dos and Don'ts for Pressure and Differential-Pressure Sensing

The following guidelines apply for pressure and differential-pressure measurement connections:

1. Column pressure is usually measured in vapor spaces at the reflux

accumulator, at the top of the column, and/or below the bottom tray (or packing). The reflux accumulator point is generally the most popular location for pressure control. The bottom point is the preferred pressure control location in many vacuum services where bottom pressure is most important. The bottom point, however, will be affected by the liquid head in case of flooding. The top of the column is a suitable location for pressure control in short columns, but it makes transmitter maintenance and testing difficult in tall columns and is therefore often avoided.

2. Pressure transmitters should be mounted above the measurement nozzle so that any condensed liquid can drain back. The line to the transmitter should be sufficiently large to permit proper drainage.

3. Guidelines 2, 7, and 9 for level connections also apply to pressure and differential-pressure connections.

4. Column differential pressure is usually measured by one of the following techniques:

   a. The most popular technique (Fig. 5.8*a*) mounts the transmitter at or above the top pressure connection. The leg from the transmitter to the bottom connection must be sufficiently large to permit proper drainage and venting, and well insulated to avoid excessive condensation that may in turn generate vacuum (and, therefore, an incorrect reading). The leg is often also steam or electrically traced to minimize condensation.

   The main drawback of this technique is that transmitter maintenance is performed at high elevation. If the top of the column is inaccessible, the technique should not be used.

   b. A transmitter mounted at the bottom nozzle (Fig. 5.8*b*) is another approach. Often, this is the most troublesome technique because of condensation and accumulation of liquid in the upper leg. Facilities must be provided to minimize condensation (i.e., good insulation) and to vaporize any condensed liquid (e.g., steam or electric tracing).

   To mitigate the condensation problem, the upper leg is often filled with a high-boiling-point liquid, with a zero suppression to allow for the weight of the liquid. The top of the upper leg is usually diaphragm-sealed to barricade the fill liquid from the column liquid. However, during service, the diaphragm seldom remains absolutely leakproof, especially under harsh chemical and temperature conditions. Leakage across the diaphragm will change the fill-liquid density or induce loss of the fill liquid when the differential pressure fluctuates. These effects are magnified when a diaphragm seal is absent. To minimize the effects of leakage, the fill liquid must be heavier then the column liquid and preferably insoluble in it. Trapping of air bub-

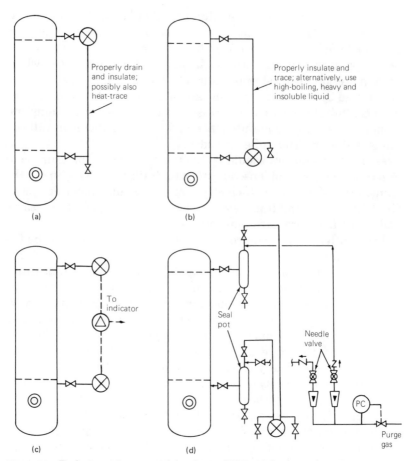

Figure 5.8 Techniques for measuring column differential pressure. (*a*) Transmitter located at top; (*b*) transmitter located at bottom; (*c*) two transmitters; (*d*) gas-purged system.

     bles while filling the upper liquid leg will also affect the fill-liquid density, and can be troublesome.

  c. A third method is two transmitters, one mounted at the bottom, the other at the top, with the difference between their readings giving the differential-pressure measurement (Fig. 5.8*c*). This is the least troublesome technique, but is inaccurate where the differential pressure is small compared to the absolute pressure of the system. The author would not recommend this technique when the pressure drop is less than 10 percent of the absolute pressure of the system.

  d. A gas- or liquid-purged system is a fourth approach. The transmitter can be mounted at the top or bottom pressure connection, even at a lower elevation. A well-designed gas-purge system is shown in Figure 5.8*d*. The purge is supplied via a pressure regulator. It then splits into two separate purge lines, one for each leg of the transmitter. Each line is equipped with its own rotameter, needle valve, nonreturn valve, and isolation valve.

Simplified versions of the Fig. 5.8*d* arrangement are common; some are described elsewhere (68). In some cases, the purge lines split downstream of a shared rotameter and needle valve. An even simpler system replaces the pressure regulator, rotameters, and needle valves by a restriction orifice in each line. Although the simplified systems save capital and can be made to work, the simplifications are achieved at the expense of more troublesome operation, and the author would not recommend them.

To avoid a measurement offset, the purge to each transmitter leg must be steady, low, and equal. This is readily achieved with the Fig. 5.8*d* system, but more difficult to accomplish with simpler arrangements. The pipe length downstream of the needle valve should be minimized to lessen pressure drop. In one simplified purge system in vacuum service (145), excessive pipe length downstream of the restriction orifice (which served for pressure letdown) and excessive purge flow led to erroneous measurements. The problem was rectified by resizing the restriction orifices for a smaller flow and relocating them close to the column taps.

Adequate pressure regulation is essential, particularly in vacuum services. If the pressure drop across the needle valves is too high, the valves will need to be manipulated at an almost-shut position, which will make flow regulation difficult and increase the potential for valve plugging. High-pressure sources of purge gas should be avoided whenever possible.

In vacuum systems, purge gas velocity increases as the column pressure is lowered from atmospheric. Continuous regulation is required during periods of column pressure changes (e.g., during startup) to prevent erroneous measurements and even overloading of vacuum jets (68). It has been recommended to avoid gas-purged systems when column pressure is less than 6 psia, unless the system is carefully designed and recalibrated after startup.

In high-pressure systems, the gas purge makes it difficult to correct pressure drop for static vapor head (see item 6 below), because of the changeable vapor composition in the transmitter vapor legs.

Seal pots are optional for any pressure-drop measurement

technique, but were specifically recommended for gas-purged systems (2). Seal pots allow drainage of liquid back to the column, trap column liquid that may surge into the transmitter lines, and dampen pressure fluctuations. It has been recommended (2) to make each seal pot an 8-in length of 2-in pipe.

Liquid condensation in the transmitter legs may be troublesome. Using high (but not excessive) purge rates, good insulation, and heat tracing can mitigate this problem. The purge gas used should be free of components that may condense out at the column temperature. If the purge gas contains solids, it is best avoided or at least properly filtered.

Gas purging should be avoided when inert blanketing in the condenser is a concern or when the gas can be troublesome in downstream equipment.

Liquid-purged systems suffer from most of the problems described in (b) above; in addition, they may require facilities for separating this liquid from the bottom product.

Because of the above difficulties, it is best to restrict the use of a purged system to services that can either readily tolerate the purge or to systems where transmitter lines are to be kept free from column vapors. This technique is most suitable for toxic or corrosive systems.

5. Valved connections should be available to enable measuring differential pressure across each section of column or each packed bed. Differential pressure is one of the most valuable measurements for identifying the nature and location of problems or bottlenecks in a column.

6. Differential-pressure measurements in pressure columns should be corrected by subtracting the static vapor head from the measurement. This can be achieved by taking a zero reading at the operating pressure or by calculation. This correction is especially important with packed columns, where the static vapor head can be much greater than the pressure drop at high pressures.

7. Local pressure gages located at the top of tall columns are not only a waste of money (they are rarely read) but may also be hazardous (a high potential for leakage) at a point which is rarely inspected. They should therefore be avoided.

5.5   Dos and Don'ts for Temperature Sensing

The following guidelines are recommended for column temperature measurement connections:

1. Temperature measurement connections should be strategically placed to cover zones of maximum temperature change (2).

2. When distilling heat-sensitive materials, a correct temperature indication near the bottom is critical. Under these conditions, thermowells are also most likely to foul, and consequently, read low. The low reading can mislead the operator or automatic controller into adding heat at a time when heat input needs cutting. In one peroxide service accident (97), this led to an explosion. In such services, additional temperature connections should be positioned a short distance above the bottom to permit cross-checks.

3. When the system distilled is fairly uncommon, temperature measurement connections should be placed in locations that can serve as alternative control points in case the design control tray is not the best control tray (98).

4. Temperature connections in regions where the temperature does not greatly vary (e.g., the pure product end of a column in a close separation) are a waste money and should be minimized.

5. It is best to measure liquid temperature; this is particularly important in services such as refinery crude distillation, where vapor temperature often differs from liquid temperature. Best measurement location is near the bottom of the downcomer (2) and close to its widest point.

    When the temperature measurement is used for control, speed of response considerations usually outweigh the accuracy consideration above. A control temperature located in the downcomer gives slow dynamic response because it relies on change in downcomer liquid composition, which lags behind changes in the vapor or in the active area (68, 370). It was therefore advocated to measure the control temperature in the vapor phase (370) or in the tray active area (68).

6. A liquid sensor below the bottom tray (or bottom supports) may read the vapor temperature when the level falls. This can mislead the operator or automatic controller into adding heat at a time when the heat needs cutting. In one incident (275), this induced overheating and a rapid exothermic reaction. Preventive measures include careful location of the temperature connection and/or the measures suggested in 2 above.

7. Thermowells are usually extended into the column. Care is required to ensure that they do not obstruct passageways or manways and become a hazard to personnel working inside the column.

8. Installing thermowells inside a bed of structured packing often requires drilling the packing. Once the packing is assembled, removal of drilling debris is difficult, and the debris remaining in the packing can promote channeling and maldistribution. The author is familiar with a case where a manufacturer strongly objected to drilling the packing in the field for this reason. It is, therefore, important to adequately assess the temperature-measurement needs and provide the manufacturer with this information well ahead of the assembly phase.

5.6   Connections for Sampling

With sampling connections, it is important to ensure that the fluid arrives at the analyzer inlet at the desired phase. This is most important for vapor samples. Vapor samples must contain no liquid droplets when leaving the column, because on vaporization they become a significant part of the sample volume and can prompt a misleading analysis. Another important consideration (particularly in packed columns) is to obtain a representative sample. An excellent extensive discussion on column-sampling techniques is available elsewhere (2). A valuable discussion on the best locations for sample connections is also available (268). An excellent technique for obtaining representative samples in packed columns was described by Silvey and Keller (366).

5.7   Viewing Ports

Viewing ports are expensive, increase the leakage potential, and may lead to a massive chemical release and hazard to personnel if the glass breaks. Viewing ports are therefore seldom justified, and should be avoided in hazardous, pressure, vacuum, and high- or low-temperature services.

In tray columns processing nonhazardous systems at near-ambient conditions, viewing ports may be desirable for troubleshooting, frequent testing, or operator training. For these purposes, they are very effective. One experience has been reported (55) where viewing ports led to identifying and solving a tray problem.

Viewing ports should be fabricated from armored glass of sufficient structural strength and high impact resistance. Attention should be paid to proper installation and leak prevention. An adequate light source for observing inside action must be provided. Often, two ports are installed at the same tray, one for the light source, the other for viewing. When column temperature is above ambient, mist may settle inside the port and obstruct viewing, and some type of demisting device may be required.

5.8 Manholes for Column Access

Entry into the shell of a distillation tower is made via manholes. These are usually fitted in the column so that each serves 10 to 20 trays (48, 177, 354). When the service is clean and noncorrosive, up to 30 trays or more may be served by one manhole. When frequent cleaning is anticipated, or if the trays are large and the process of removing them through the hole is slow, the smaller number above should be used. This enables multiple crews to work on removal or installation. If the column diameter is too small to admit personnel, cartridge trays (Sec. 7.13) should be used.

In packed columns, manholes should be positioned so that distributors, redistributors, and other internals can be accessed. Usually, manholes are located above each bed support plate, above the top distributor, and at the bottom of the column. Locating manholes above the support plates also permits removal of packing from the bed above.

In small-diameter packed columns (<3 feet), either handholes or flanged construction is normally used for access instead of manholes. Handholes usually have diameters of 8 to 14 in and are positioned at the same points as manholes in larger columns. The access offered by handholes is limited, and they are used only when little maintenance on the column is anticipated. In most cases, construction of small-diameter packed columns out of flanged sections is the preferred method of providing access to internals. Flanges are usually installed near the top distributor and just below each support plate. Although flanged construction is more expensive than handholes, it greatly lowers maintenance costs by improving access to internals and by making packing removal easier. To remove packing, a flanged section is simply tipped over after its top distributor and bed limiter are removed.

The inside diameter of the tower manhole affects the width of the tray manways and the number of pieces that are used to assemble each tray, distributor, or internal pan. Small manholes may not permit optimum design. In large towers, and/or when a manhole serves several trays, there is a strong economic incentive to minimize the number of pieces to be assembled or dismantled, and using large manholes becomes a necessity. Larger manholes are mandatory if protective clothing may need to be worn by personnel entering the column. Small manholes will impede entry of personnel wearing bulky protective clothing or breathing apparatus, and will also hinder emergency rescue efforts. Recommended manhole diameters vary from one designer to another, but are always in the range of 16 to 24 in. (48, 138, 177, 354).

Overly large manholes (> 24 in) should be avoided unless essential for admitting a person wearing bulky protective clothing (e.g., a breathing apparatus). Such manholes are expensive, hard to seal, difficult to open or close, and hard to carry by the davit and hinges. They enhance the injury risk to personnel opening or closing them, and locally increase the tray spacing requirement and the column height.

Often, tray spacings must be locally increased to be larger than the manhole diameter. For this reason, and whenever practicable, it is a good policy to install manholes in the space above the feed trays where the tray spacing is normally extended (for reasons given in Sec. 2.3). When this is done, care must be taken to ensure that feed nozzles and distributors do not impede entry by personnel into the column. One designer (48) recommends that tray spacing at the manhole be at least 36 in to provide adequate work space.

The method of installing the trays and normal tray-cleaning procedures should be considered when manholes are designed. If removal of trays for cleaning is unnecessary and trays are to be cleaned in place, the number and size of manholes can be reduced. Tray parts should be top-removable for columns larger than 3.5 ft in diameter (399), because bottom-removable trays can put workers in danger of having a poor foothold, incurring the risk of a long fall during installation. Similarly if packing removal is not anticipated to take place frequently, the number of manholes can be reduced.

Whenever possible, all manholes should be oriented in the same direction. It is also preferable that the manholes face the main accessway to the column. The aligned manholes will occupy a segment of the total tower circumference that should not be occupied by any pipe runs. This minimizes the difficulties in lowering tower internals to the ground (190).

To minimize possible damage to column internals, care should be taken not to locate the manhole in the downcomer seal area.

To enable access from the bottom sump (where the manhole is often located) to the bottom tray, rungs are often installed at the column shell. Internal ladders are also sometimes used for the above purpose, but these may corrode and become unsafe. Corrosion-resistant materials are recommended if ladders are to be used.

Finally, accessibility to the manhole should not be overlooked. Access platforms should be provided for all manholes 12 ft or higher above grade (190). The manhole door should open away from the ladder leading to the platform and should be capable of being opened fully. Although these recommendations may appear obvious, it is surprising how many times they are overlooked, and either tray removal is slowed because a manhole door cannot be fully opened or injuries are caused to personnel because a manhole door opens the wrong way.

6

Tray and Downcomer Layout

Designing trays and downcomers for distillation columns involves at least two stages: a primary (basic) design stage, and a secondary (detailed layout) stage. At the basic design stage, the following parameters are usually established:

- Vapor and liquid flow rates, operating conditions, and desired flow regime
- Tray diameter and area
- Type of tray
- Bubbling area and downcomer hole area
- A preliminary estimate of tray spacing and number of passes
- A preliminary tray and downcomer layout

Methods for determining these basic factors are discussed at length in most distillation texts (123, 193, 319, 371, 404, 409). Once these are determined, the basic design stage ends, and the secondary stage begins. In the secondary stage, tray and downcomer layout and other preliminary estimates are finalized. This secondary stage will be described in this chapter. It is possible to change the basic design parameters at the secondary stage, but such changes are usually small.

Unlike the internals discussed in previous chapters, poor tray layout of one- or two-pass trays rarely causes spectacular column failures such as flooding at 50 percent of design rates or extremely poor separation. Ill effects resulting from poor tray layout seldom extend beyond suboptimum design or performance; a moderate reduction in capacity, efficiency, or turndown; and some increases in capital or operating costs. Nevertheless, the troubleshooter must be familiar with tray layout pitfalls, not only for pursuing process improvements but also while attempting to identify the cause of a major failure. In the latter case, several theories are often voiced, and it is important to correctly rule out the less likely ones.

Tray layout discussions in this chapter emphasize sieve and valve trays, as these trays are most frequently encountered in industrial practice. Several of the considerations also apply to other tray types (e.g., bubble-cap trays). Considerations unique to bubble-cap trays were excluded from this chapter. The infrequent application of this type of tray in modern distillation practice argues against a detailed discussion here. A large amount of information on bubble-cap tray layout is available and is well documented in several texts (48, 257, 371, 409).

Unlike tray layout, poor downcomer layout is as likely to cause a spectacular column failure as several internals discussed in previous chapters. Adequate hydraulics is the primary consideration, and if not achieved, premature flooding may occur.

This chapter examines common practices for tray layout, including tray spacing, hole diameter, fractional hole area (and hole spacing), valve tray layout and valve selection, calming zones, outlet weirs, setting and changing the number of liquid passes, considerations unique to multipass trays, and preventing flow-induced vibrations. It also examines common practices for downcomer layout, including types of downcomers, downcomer width and area, downcomer sealing, clearances under the downcomer, inlet weirs, and seal pans. In addition, this chapter also considers unique baffles and other practices sometimes used on trays, including splash baffles, vapor hoods, reverse-flow baffles, stepped trays, and antijump baffles.

For each of these internals, the chapter outlines the preferred practices, highlights the consequences of poor practices, and supplies guidelines for troubleshooting and reviewing their designs.

6.1   General Considerations

The following general factors influence tray performance and, therefore, the desired tray layout. They need to receive attention at an early stage and guide the specification of tray and downcomer layout.

Flow regime.   The flow regime is the nature of the vapor-liquid dispersion on the tray. Most industrial trays operate either in the froth or the spray regime, depending on tray geometry and operating conditions. In the froth regime (Fig. 6.1a) liquid is the continuous phase and vapor is dispersed as bubbles in the liquid. In the spray regime (Fig. 6.1b), the phases are reversed; vapor is the continuous phase,

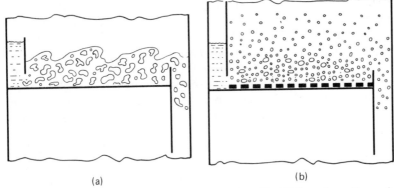

(a)   (b)

Figure 6.1 Flow patterns on trays. (a) Froth regime (liquid phase is continuous); (b) spray regime (gas phase is continuous). (*Henry Z. Kister, excerpted by special permission from* Chemical Engineering, *September 8, 1980; copyright ©, by McGraw-Hill, Inc., New York, NY 10020.*)

while liquid is dispersed as drops in the vapor. Some investigators (163) identify an additional liquid-continuous regime ("the emulsion regime"), in which the dispersion behaves as a uniform two-phase fluid. Since tray layout considerations for this emulsion regime generally resemble those for the froth regime, it is treated here as an extension of the froth regime.

The froth regime is the most common flow regime on industrial trays. This regime occurs at low and moderate vapor velocities and moderate and high liquid loads, conditions typical of pressure and most atmospheric distillations. The spray regime usually occurs at high vapor velocities and low liquid loads, conditions typical of vacuum distillation. Quantitative criteria for predicting the expected flow regime on the tray are described elsewhere (163, 193, 243, 244, 321).

Spray regime operation is desirable (321) for negative-surface-tension systems (i.e., where the mixture's surface tension decreases from the top tray toward the bottom tray). Froth regime operation is desirable for positive-surface-tension systems (surface tension increases from the top tray down), and when liquid entrainment needs to be minimized (321). In most commercial applications, vapor and liquid loading requirements override these desirability considerations and dictate the tray flow regime. For optimum tray performance, the tray layout must therefore accommodate the expected flow regime.

Capacity restriction mechanism(s). Column throughput is restricted by one of several different mechanisms. These include spray entrainment flooding, froth entrainment flooding, downcomer backup flooding, downcomer choke flooding, excessive entrainment and excessive pressure drop. Optimum tray and downcomer layouts vary with the mechanism that restricts column throughput. The nature of the various flooding mechanisms is discussed in Sec. 14.1.1. Most distillation texts and review articles (73, 123, 193, 243, 319, 371, 404, 409) contain quantitative criteria for predicting column throughput.

In the spray regime, column throughput is usually restricted either by entertainment flooding or by excessive entrainment. Excessive pressure drop may also restrict spray regime throughput in vacuum columns. Downcomer flooding and downcomer choke seldom restrict throughput in the spray regime.

In the froth regime, column throughput is usually restricted by flooding due to excessive entrainment, excessive downcomer backup, or downcomer choke. Excessive entrainment or excessive pressure drops may occasionally (but infrequently) restrict throughput in the froth regime.

Corrosion. The likelihood of corrosion and its potential effect on column internals must be reviewed. The consequences of operating with corroded internals, and the cost of their repair, including the cost of lost production, must be evaluated. The extent to which corrosion can be tolerated and/or inhibited must be defined. This definition affects several decisions to be made about tray and downcomer layout.

Fouling. The likelihood, type, degree, and consequences of fouling in the column must be evaluated, and the method of cleaning its internals should be established. Fouling may result either from extraneous material introduced in the feed, or from polymerization or decomposition inside the column (either under normal or abnormal operating conditions).

Fouling problems may be overcome by keeping the contaminants out, by using additives to inhibit fouling, by operating in a way that will minimize fouling, or by designing the column so that fouling does not significantly affect column performance. In any case, the effect of fouling of column performance, frequency of shutdowns, and method of cleaning have an important bearing on tray and downcomer layout.

Turndown. Turndown requirements should be critically examined. Generally, turndown ratios (ratio of normal operating throughput to minimum allowable throughput) of up to 3:1 incur a negligible cost penalty. In most cases, higher turndown ratios can be readily achieved at a relatively small cost penalty, as long as they do not exceed about 5:1. Beyond this, the cost penalty for enhancing turndown escalates. It is also important to appreciate that high turndown may turn out beneficial due to some unforeseen circumstance, and is often worth at least a minor cost penalty.

Simplicity. The simpler the tray and downcomer layouts, the less expensive they are, and the smaller the chance of fabrication, assembly, and installation errors. Tray design and fabrication are often highly computerized; any oddities [e.g., downcomers following the contours of the weir (128)] may require considerable hand work.

Tray and downcomer layout should be as uniform as possible and their components as interchangeable as possible throughout the column. Machinery employed by manufacturers is most efficient on long production runs (128), which minimizes costs. Further, a uniform layout can eliminate costly delays when replacing defective or worn-out parts, and it minimizes the chances of interchanging dissimilar parts during column assembly.

Layout changes within a column section (except at feed or draw trays) should be avoided. Layout changes from one section to another should only be performed when process conditions widely differ between the sections, and where there is a clear economic advantage for such changes. Even in these cases, layout of one section should be as similar as possible to that of the other section.

Others. Other considerations affecting tray and downcomer layout include efficiency, access for maintenance, and cost. These tend to be specific rather than general, and are therefore discussed in the relevant sections.

Tolerances. Tighter tolerances than necessary on column internals represent a significant waste of money and should be avoided. In particular, each specified tolerance should be considered in relation to its contribution to the functional or mechanical integrity of the tray. The magnitude of the various tolerances should be realistic and as large as feasible for the particular application. Often, it is desirable to leave tolerances to the discretion of the fabricator.

A tolerance can be expressed as a "plus-or-minus" value or a "maximum-to-minimum" value. For instance, if weir height is allowed to vary between 1 $15/16$ and 2 $1/16$ in, the tolerance can be expressed as a $\pm$ $1/16$ inch or as "maximum to minimum" of $1/8$ in. Note that for an equivalent variation, maximum-to-minimum tolerances are always double the "plus-or-minus" tolerances. When specifying tolerances, the relevant definition must always be stated, or the specified value may be misleading. Specifying maximum-to-minimum tolerances is often considered a less rigid practice; in the above weir height example it would permit weir height to be between 1 $7/8$ and 2 $1/8$ in, as long as the difference between the highest and lowest point does not exceed $1/8$ in.

For convenience and clarity, this book will use the plus-and-minus definition. The reader can convert this to the maximum-to-minimum tolerance simply by doubling the tolerance value. Tolerances for various tray items are discussed in the relevant sections. A summary of typical tolerances is shown later in Fig. 10.5.

6.2 Tray Spacing

Tray spacing in industrial columns ranges from 8 to 36 in. A prime factor in setting tray spacing is the economic trade-off between column height and column diameter. Enlarging tray spacing adds to the column height requirement, but it permits a smaller column diameter to be used. Usually, the costs of lengthening the column match the savings achieved from the corresponding reduction in column diameter. It follows that as long as the above trade-off is not constrained, tray spacing has a minor impact on column costs. It is then normally set to allow easy access for maintenance.

A tray spacing of 24 in is most common for columns 4 ft and larger in diameter (48, 211, 271, 272, 409). This spacing is sufficiently wide to allow a worker to freely crawl between trays. Some designers (272, 404, 409) recommend a tray spacing of 24 to 36 in. Tray spacing larger than 24 in make crawling between trays easier and are often desirable where frequent maintenance is expected (e.g., fouling and corrosive services) and in large-diameter (> 10 to 20 ft) columns. Deep tray support beams in large-diameter columns restrict the crawling space available and may also interfere with vapor movement across the tray, making larger tray spacing more desirable.

A tray spacing of 18 in is recommended for columns 2 ½ to 4 ft in diameter (48). Here crawling between trays is seldom required, because a worker can reach the column wall from the tray manway. For such columns, the recommended tray spacing is lowered to reduce the support problems associated with tall and thin columns (see item 8 below).

A tray spacing smaller than 18 in makes access for maintenance difficult. It is therefore generally not recommended, but is often used (257, 272, 409).

When the economic trade-off between column height and column diameter is constrained, optimum tray spacing is dictated by cost considerations, while access considerations assume a secondary role. Some situations where the trade-off is constrained are

1. *When column height is restricted:* Examples include a column that needs to be enclosed inside a building, a cold box, or another expensive shield (and perforating the roof is impractical or costly); a tall column that is to be erected in the vicinity of an airport (and must comply with a maximum height regulation); and a column 250 to

300 ft tall. Since it is almost always more economical to perform a service in a single column than in two, there is an incentive to select a tray spacing small enough to match the height restriction. Cryogenic air separation columns, which are enclosed in cold boxes, sometimes have a tray spacing as low as 4 to 8 in; C_2 and C_3 splitters, often containing 100 to 200 trays, are sometimes designed with tray spacing as low as 12 to 18 in (218).

2. *When column diameter is restricted:* Examples include small columns, where the diameter needs to be at least 2 ½ to 3 ft to permit personnel entry; an existing column used for a new service; or a hydraulically underloaded section of a constant-diameter column. When the diameter in such columns is hydraulically oversized, it becomes attractive to minimize tray spacing, thus utilizing excess capacity to reduce column height or increase the number of trays.

3. *When the column (or column section where maximum hydraulic loading occurs) contains very few (< 5 to 10) trays:* Here tray spacing has only a weak effect on column height, and savings due to diameter reductions outweigh the costs of lengthening the column. In such cases, it becomes attractive to enlarge the tray spacing (371).

4. *Columns operating in the spray regime:* These should have a tray spacing at least 18 in, and preferably 24 in or more (218, 321), to avoid excessive entrainment. Vacuum columns often operate in the spray regime and are correctly designed with a tray spacing of at least 24 in and often 30 in (257). The same applies to low-liquid-load services (< 2 gpm per inch of weir length).

5. *When froth regime operation is favored:* Lower tray spacing restricts the allowable vapor velocity, thereby promoting froth regime operation. Tray spacing of 18 in or less is sometimes advocated when froth regime operation is specifically desired (218).

6. *Systems with a high foaming tendency:* Here a tray spacing of at least 18 in, and preferably 24 in or more, is recommended (218, 354) to avoid premature flooding.

7. *Systems restricted by downcomer choke:* With such systems, increasing tray spacing does little to enhance capacity and there is an incentive to minimize tray spacing. Unfortunately, downcomer choke usually coincides with a high foaming tendency, and guideline 6 above overrules.

8. *Supporting tall and thin columns (high ratio of height to diameter) is difficult and expensive:* One designer (243) forecasts problems when this ratio exceeds 25. In such cases, savings from shortening

the column usually exceed the cost of enlarging its diameter, and there is an incentive to lower tray spacing (often to about 18 in).

6.3 Hole Diameters on Sieve Trays

Hole diameters on industrial trays range from 1/16 to 1 in. In selecting hole diameters, the following should be considered:

1. *The nature of the service:* Small holes are not suitable for fouling or corrosive services because they may block or partially block, leading to excessive pressure drops and premature flooding. Further, hole blockage often takes place in a nonuniformly distributed pattern, causing uneven vapor flows and lower tray efficiencies. Incidents where plugging of small holes resulted in premature flooding have been reported (126, 239). If the service is corrosive, large holes have two advantages: (1) the rate of change of hole area and tray pressure drop with time is much slower (192, 268); and (b) the allowable tray thickness is greater, and a greater degree of corrosion can be tolerated (317). In either corrosive or fouling services, 1/2-in or larger diameter holes are recommended (192).

2. *Hydraulics:* Small holes enhance tray capacity (218, 227, 243, 354), but this enhancement may be small. Small holes appreciably reduce entrainment in the spray regime and at low liquid loadings (21, 201, 204, 205, 227, 371), but this reduction is less pronounced at moderate and high liquid loadings (>3 gpm/in) in the froth regime (176, 201, 205). Small holes slightly lower the pressure drop (172, 218, 226, 227, 371). It is generally believed that small holes reduce weeping (175, 176, 218, 223, 226, 245, 281, 382, 437, 438), but this does not always apply (88, 218, 227, 245). Small holes promote froth regime operation (244, 321). Extremely small holes (< 3/32-in) may sometimes promote foaming (249).

3. *Mass transfer:* In froth regime operation, small holes provide better vapor-liquid contact, and hence higher efficiency (119, 172, 184, 247, 252, 273, 317, 326, 347, 438, 439). The effect is small, and often negligible (119, 184, 218, 273, 326, 409). In the spray regime there is some indication that large holes enhance efficiency (273, 347), although this may not always apply (326).

4. *Frothing:* In clean vacuum services operating in the froth regime, small holes (1/8 in diameter) have been specifically recommended for both entrainment and efficiency reasons (175, 257, 317).

5. *Turndown:* Small holes have a better turndown characteristic because they reduce tray weeping and increase tray capacity.

6. *Costs:* Trays punched with larger holes are cheaper because fewer

holes are required. Holes with diameters smaller than 3/16 in may require drilling, which is far more expensive than punching. As a rule of thumb, carbon-steel or copper alloy trays can be punched when the hole diameter is equal to, or greater than, the tray thickness; for stainless steels, the hole diameter must be 1.5 to 2 times the tray thickness (73, 172, 317, 371, 409).

7. *Errors in diameter:* In trays with small holes, a small error in hole diameter when drilling or punching has a greater effect on pressure drop, capacity, and turndown than the same error in a large hole.

8. *Installation:* Tray panels should be installed with the rough edge of the hole facing the vapor flow (Fig. 6.2a). This reduces the risk of injury to personnel working in the column or installing the trays (49, 123, 192). The practice has the disadvantage of slightly increasing the tray pressure drop and the advantage of slightly decreasing the weeping tendency. If pressure-drop considerations are of major importance, the rough edge of the tray should be ground smooth so as to be free of burrs (88, 317), as illustrated in Fig. 6.2b.

In general, large holes are recommended for fouling and corrosive services and for spray regime operation (provided tray spacing is sufficiently large in the latter case). In other applications, smaller punched

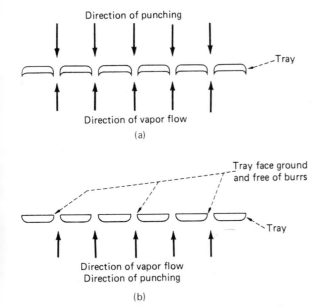

Figure 6.2 Recommended sieve tray hole-punching practices. (*a*) Normal; (*b*) pressure drop is important. (*Henry Z. Kister, excerpted by special permission from* Chemical Engineering, *September 8, 1980, copyright © 1980 by McGraw-Hill, Inc., New York, NY 10020.*)

holes are preferable, with $\frac{3}{16}$ inch diameter being a favorite general choice (73, 172, 192, 226, 243, 249, 257, 281, 317, 319).

6.4   Fractional Hole Area and Hole Spacing on Sieve Trays

Fractional hole area is the ratio of the total area of the tray holes to the tray bubbling area (bubbling area being column area less areas of unperforated regions such as downcomer, downcomer seal, and large calming zones). The number of holes is obtained by multiplying the bubbling area by the fractional hole area and dividing the product by the area of a single hole.

Hole pitch is the center-to-center hole spacing (Fig. 6.3). Holes

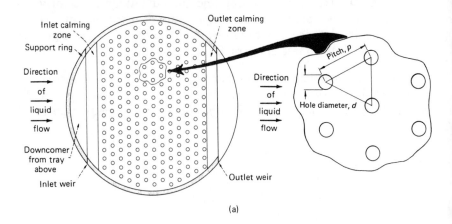

(a)

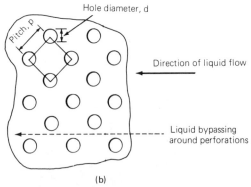

(b)

Figure 6.3 Typical layout of the components for sieve trays. (a) Recommended, with equilateral triangular hole pitch; (b) Square hole pitch (*Part a Henry Z. Kister, excerpted by special permission from* Chemical Engineering, *September 8, 1980, copyright © by McGraw-Hill, Inc., New York, NY 10020.*)

should be spaced on an equilateral triangular pitch (Fig. 6.3*a*), but sometimes a square pitch (Fig. 6.3*b*) is used. Liquid should always flow perpendicular to the rows. An equilateral triangular pitch minimizes liquid bypassing around perforations and affords a greater ratio of hole pitch to hole diameter for a given fractional hole area (Fig. 6.4). It is therefore preferred. The ratio of hole pitch to hole diameter is a geometric function of the fractional hole area and the hole pattern, given by

$$\text{Fractional hole area} = K(\text{hole diameter/hole pitch})^2 \qquad (6.1)$$

where $K = 0.905$ for an equilateral triangular pitch, and $K = 0.785$ for a square pitch. The relationship is shown graphically in Fig. 6.4.

Fractional hole areas on commercial sieve trays usually range from 0.05 to 0.15 (corresponding to pitch to diameter ratios of 2.5 to 4), with 0.08 to 0.12 normally considered optimum. Some designers (123, 382) suggest setting the fractional hole area relatively high for vacuum services. The optimization is usually between capacity and turndown. Both tray capacity and tray-weeping tendency increase with fractional hole area. When fractional hole area is increased beyond 0.08 to 0.10, capacity gains approach marginal returns, while turndown reductions continue to be substantial.

The following guidelines apply to fractional hole area and hole pitch selection:

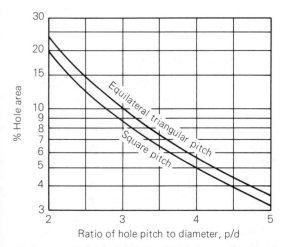

Figure 6.4 Graphical relationship between fractional hole area and hole pitch to diameter ratio (*From J. David Chase, excerpted by special permission from* Chemical Engineering, *July 31, 1967, copyright © by McGraw-Hill, Inc., New York, NY 10020.*)

1. When entrainment flooding limits column capacity and fractional hole area is between 0.05 and 0.08, an increase of fractional hole area of the order of 0.01 will enhance column capacity by about 5 percent (112, 218, 227, 319, 382). When fractional hole area exceeds 0.1, capacity gains from increasing fractional hole area are substantially lower (112, 218, 319, 382, 429), becoming negligible at high liquid loads (>6 gpm/in of outlet wein). Column capacity also increases with fractional hole area when entrainment, pressure drop, or downcomer backup limit column capacity. These increases are substantial when fractional hole area is less than 0.08 to 0.10, but are small at higher fractional hole areas. Column capacity is unaffected by fractional hole area when downcomer choke is the capacity limit (429).

2. In the spray regime, and at low liquid loads in the froth regime, increasing the fractional hole area reduces entertainment (201, 204, 205, 382). This reduction is particularly large when fractional hole area is less than 0.08, but it becomes small when fractional hole area exceeds 0.10. Entrainment is essentially unaffected by fractional hole area at moderate and high liquid loads (> 3 gpm/in) in the froth regime (176, 201, 205).

3. Increasing fractional hole area reduces pressure drop (176, 218, 226, 227, 371), and therefore also downcomer backup. These reductions are significant at low fractional hole areas, but become small as fractional hole area exceeds 0.10.

4. Increasing the fractional hole area enhances the weeping tendency (175, 176, 218, 226, 243, 245, 319, 371, 437, 438), and therefore lowers turndown.

5. High fractional hole area promotes froth regime operation (321).

6. There is some indication that lower fractional hole area enhances efficiency (218, 409, 429). The magnitude of this enhancement is uncertain; some reports suggest it is substantial (218, 429), while others imply a negligible or small effect (409). There is some evidence to suggest that substantial efficiency enhancement occurs mainly in the spray regime (273, 326).

7. When the column operates at turned-down conditions for lengthy periods, excess holes are often blocked off by blanking strips. This lowers the fractional hole area, thus decreasing weeping. Blanking strips should be installed from one wall to another, perpendicular to the liquid flow. The width of each strip should not exceed 4 in to avoid the formation of inactive regions (192). Alternatively, it has been recommended (73, 371) that the width of each strip not exceed 7 percent of the tray diameter in small-diameter (< 10-ft) columns

or 5 percent of column diameter in large-diameter (> 10-ft) columns. It is desirable to install at least four blanking strips per tray, and they should be scattered across the tray. The blanked area should not exceed a quarter of the fractional hole area (73, 371). Blanking strips should be fastened so that a tight seal is provided, removal is easy, and the perforated sheet is not distorted or buckled (371).

8. When setting tray pitch, it is preferable to select a standard punching pattern, used by tray fabricators, and to adjust the hole pitch accordingly. A nonstandard pattern increases the cost of trays. Following this practice rarely alters the desired fractional hole area substantially.

9. Tolerance on fractional hole area is typically ±3 to 5 percent (49, 371).

6.5 Valve Tray Layout and Valve Selection

A valve tray (Fig. 6.5) is a flat perforated plate, with each perforation equipped with a movable disk. The perforations and disks may be circular (Fig. 6.5a, b, d) or rectangular (Fig. 6.5c). At low vapor rates, the disk settles over the perforation and covers it to avoid liquid weeping. As vapor rate is increased, the disk rises vertically.

The vertical rise of the disk is restricted either by a cage (e.g., Figs. 6.5b, 6.6b) or by retaining legs attached to the bottom of the disk (e.g. Figs. 6.5a, 6.6a, d). The cage and retaining legs also prevent horizontal movement of the disk.

The following general considerations are important for valve trays:

1. Valve trays are proprietary, and it is always best to have the manufacturer specify and design the tray layout. The amount of information available on valve tray design is limited, although some is available in manufacturers' literature (138, 211, 307).

2. The manufacturer must be supplied with concise information on the corrosive, erosive, and fouling nature of the system, as well as the turndown requirements. Certain valve types and layouts are far better suited to cope with each of these problems than others. Lack of concise information can lead to a poor selection of valve type, and subsequent troublesome operation.

3. A mechanical problem often encountered in valve trays is wear or corrosion of either the retaining legs or the perforations that harbor the valve. The constant vertical movement of the valve, and spinning of circular valves in their perforations, impose fatigue

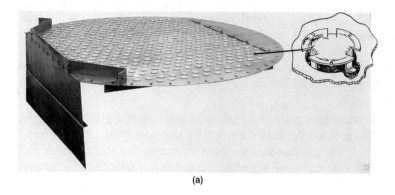

(a)

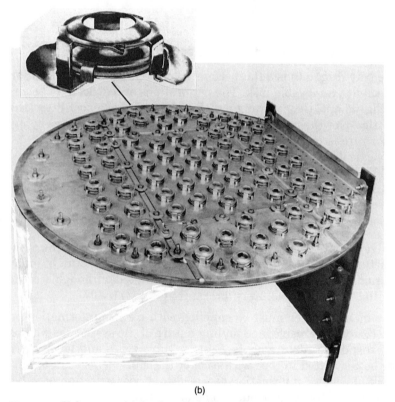

(b)

Figure 6.5   Valve trays. (a) A valve tray equipped with round uncaged valves; (b) a valve tray equipped with round caged valves; (c) a valve tray equipped with rectangular valves; (d) a valve tray containing both round uncaged valves and perforations.(*Parts a and d courtesy of Koch Engineering Company, Inc.; part b courtesy of of Glitsch Inc.; part c courtesy of Nutter Engineering.*)

Figure 6.5 *(Continued)*   (d)

stresses on its retaining legs and wear the perforations. It is not uncommon to have valves pop out of their seats.

In many instances, popping out of valves has no apparent effect on column performance, at least at high throughputs. The author is familiar with several situations in which operating personnel did not realize that valves popped out until the column was opened up for inspection or until popped-out valves damaged the column bottom pump. In all these instances, operation was at high throughputs; in some, the fraction of popped out valves was high.

Wear of the valve legs and orifices may also accelerate valve and orifice corrosion, especially in services that rely on

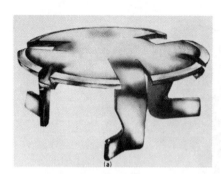

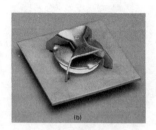

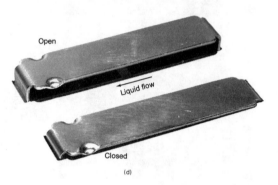

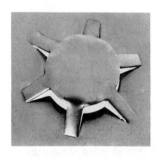

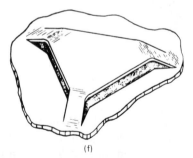

Figure 6.6   Types of valve units. (*a*) An uncaged round valve unit; (*b*) a caged round valve unit; (*c*) a caged round valve unit, featuring a contoured orifice; (*d*) a rectangular valve unit; (*e*) a fixed round valve unit; (*f*) a fixed triangular valve unit. (*Parts a and e courtesy of Glitsch Inc.; parts b, c, and f courtesy of Koch Engineering Company, Inc.; part d courtesy of Nutter Engineering.*)

passivating films to provide corrosion protection. The wear destroys the passivating film; one experience was reported in which valves were corroded to paper thickness and orifices were enlarged by this mechanism (38, 42).

The sensitivity of valve trays to wear can be minimized by adequate selection of materials and valve type. In the above case history, switching valve type from uncaged round valves to caged round valves eliminated the corrosion problem (42). The problem could have also been solved by changing tray and valve material (38). Spinning is often prevented by short horizontal tabs which protrude into the perforations which harbor uncaged valves; in caged designs, tabs are sometimes attached to the disk (Fig. 6.5b).

"Home-made" valves should be avoided because it is difficult to bend the retaining legs without excessively stressing the metal. The author is familiar with one experience where home-made valves lasted a very short time before popping out. Careful handling of valve trays prior to installation is important, because the retaining legs of valve units are easily damaged (Sec. 10.1).

4. Another mechanical problem often encountered in valve trays is "sticking" of the valve. Sticking occurs because of buildup of sludge and/or corrosion products while the valve is in contact with the tray floor. Sticking reduces the open area of the tray and can initiate premature flooding.

Sticking is inhibited by running the column at high throughput. High vapor velocities sweep foulants away from the valve base. The author is aware of cases where valve trays handled extremely sticky materials without sticking. In these cases, high-throughput operation was constantly sustained, and column run length was restricted by downcomer plugging. Successful application of valve trays in the highly fouling service of distilling alcohol from an aqueous sludge was reported (19). On the other hand, the author is familiar with cases of valve sticking while handling mildly sticky materials; in these cases, column throughput varied. Two experiences of valve tray fouling were reported by Harrison and France (150a); in one, a scaling problem was alleviated by replacing valve trays by sieve trays.

Valve sticking can be alleviated by proper selection of valve type. Several designs have the valves seated on turned-down nibs which hold the disk about 0.1 in (138) above the tray deck to prevent sticking (Figs. 6.5a, 6.6a, b, d). This clearance, however, may enhance the weeping tendency and somewhat reduce turndown. One experience has been reported (85) where weeping from this type of valve at low liquid rates did not permit adequate liquid circulation in a pumparound loop.

5. The most common types of valves (138, 211, 306) are shown in Fig.
6.6. The uncaged round valve (Fig. 6.6a) is perhaps the most popular type. Normally, the disk is flat (Fig. 6.6a) or flat-dome shaped (Fig. 6.5a); the legs are formed integrally with the disk, and the orifice on the deck is standard (sharp-edged). A more leak-resistant variation (138) uses welded legs instead of integrally formed legs; this eliminates the leak area at the legs. Another design variation, often called an *uncaged venturi valve* (Fig. 6.6c) replaces the sharp-edged orifice by a contoured orifice to reduce pressure drop and downcomer backup (138, 211).

The caged round valve (Figs. 6.5b, 6.6b) is more resistant to spinning, corrosion, wear, and popping out of its seat (37, 38, 42, 268). It was also reported to have a lower tendency to stick (37, 42, 211), a lower pressure drop (37, 42, 322), and a slightly lower efficiency (37, 42) than the uncaged round valve. The disk is usually flat-dome shaped (138, 211), and the orifice is sharp-edged, but a lower-pressure-drop (venturi valve) variation which uses a contoured orifice is also available (138, 211). Leak-resistant variations have disks that can completely close the orifice. In one leak-resistant variation, a small perforation is drilled in the disk to allow vapor passage at very low rates (39); in another variation, the orifice cover is a separate, lightweight plate located below the normal disk (138, 144). Cases were reported (237, 239) where the latter variation was successful in overcoming a turndown problem which occurred with standard valve units.

The uncaged, rectangular valve (Fig. 6.6d) eliminates valve spinning and is resistant to popping out. The pressure drop, capacity, and efficiency characteristics of this valve were reported (14) to be of the same order as those of the uncaged round valve. The long edge of the valve is parallel to the liquid flow (Figs. 6.5c, 6.6d). This minimizes the valve opening exposed to the approaching liquid flow and improves its leakage resistance (306). The legs are formed integrally with the rectangular plate, and the tabs which hold the valve above the tray deck to prevent sticking are located on the downstream end (Fig. 6.6d).

The fixed valve (Fig. 6.6e, f) is a low-cost stationary assembly which imitates the shape of a valve. The absence of the moving disk eliminates wear and sticking, but at the expense of lower turndown than other valve trays. The imitation disk can be flat-dome shaped (138, Fig. 6.6e), triangular (211, Fig. 6.6f), or rectangular (306).

A sieve-and-valve tray (Fig. 6.5d) contains a combination of venturi valves and sieve clusters in alternate rows. This tray has a low pressure drop while maintaining turndown equivalent to most valve trays (37, 38).

Further details on valve types, dimensions, and design variations are available in manufacturers' bulletins and in the open literature (37, 39, 42, 138, 139, 211, 306).

6. The valves should be heavy enough to prevent excessive opening at low vapor flow rates. If excessive valve opening occurs prematurely, weeping will result (46) and may drastically lower turndown. On the other hand, heavier valves incur greater pressure drops and are more expensive. As a compromise, manufacturers often employ two weights of valves in alternate rows on the same tray (38, 46, 84, 171, 243, 409). Typically, the light valves are fabricated from 16-gage metal, while the heavy valves are fabricated from 12- or 14-gage metal (144, 171).

A case history contributed by S. W. Golden (145) demonstrates a successful application of this technique. A trapout pan at the bottom of a coker debutanizer was feeding liquid to a once-through reboiler (an arrangement identical to Fig. 4.7a, but with an unpartitioned combined seal and trapout pan). At low rates, excessive leakage through the bottom valve tray starved the reboiler of liquid. The tray contained 16-gage uncaged circular valves with turned-down nibs. The problem was completely eliminated by replacing half of the bottom tray valves by uncaged 12-gage valves that seat flush with the tray floor. This modification stopped leakage to such an extent that the reboiler could be started up without operating the startup line (Fig. 4.7a), which was always needed in prior startups.

When the liquid flow path is relatively short (< 30 in), the row of valves closest to the tray inlet should contain light valves. If the inlet row contains the heavy valves, aeration at turned-down conditions may not begin until some distance into the tray. This shortens the effective flow path and may lower tray efficiency.

7. A tray containing too many valves is likely to perform as poorly as one in which valves which are too light (46) (6 above). In either case, the open area will be excessive at low vapor rates, weeping will occur, and poor turndown will result. In one case, the author physically observed heavy weeping from a valve tray which operated at 70 percent of its measured flood point. In another column containing valve trays (239a), efficiency severely dropped upon turndown to 50 percent of the normal production rate. In other cases (14a, 239, 296) improved efficiency and reduced tray leakage were reported after portions of the valves were removed and their openings blanked. An excessive number of valves is also likely to cause instability and channeling, as described in the next guideline.

8. Vapor channeling and instability at low vapor flow rates can be a severe problem in valve trays containing too many valves, particularly where valves are permitted to seat flush with the tray deck (138, 237, 294, 334, 444). Vapor channels through a few wide-open valves in a small aerated zone located in some intermediate position on the tray, with the remaining valves completely closed. On two-pass trays, one panel sometimes tends to be active and the other inactive, with activity switching back and forth. The nibs used to prevent sticking (item 4 above) also help to prevent this instability (138). Some modern valve designs do not experience this instability (138).

In one case, contributed by D. W. Reay (334), such channeling is believed to have induced severe corrosion of a two-pass valve tray located just above the feed zone in a 44-ft-diameter refinery vacuum tower. Two-phase feed entered via a feed distributor, and the flash vapor passed through a chimney tray before entering the valve tray. The valves were round, uncaged, and had turned-down nibs to prevent sticking. Five deck panels only, all in the same area, were very severely corroded to the extent that valve holes had enlarged and become knife-edged; some valves were severely thinned. All other deck panels suffered relatively minor damage although some 90 percent of all the valves had become displaced, mainly owing to leg erosion. If vapor had been uniformly distributed, velocity would have been too low to cause severe corrosion. Apparently, the vapor preferentially channeled through a relatively small number of valves, and the velocity at which corrosion is known to become severe was exceeded. Calculations showed that about 40 percent of the valves on the tray should have been blanked. Vapor maldistribution could have been initiated at the feed distributor or the valve tray, and its consequences were aggravated by valve channeling.

To minimize channeling, valve trays are designed to exceed a minimum unit reference (144). A unit reference is the ratio of the vapor rate to the vapor rate at which all valves are open. A minimum unit reference of 40, 60, and 80 percent is recommended for one-, two-, and four-pass trays, respectively (144). If the unit reference falls below the minimum, or if a higher unit reference is desired, selected valves can be blanked, valve density can be reduced, or the ratio of light to heavy units can be varied (138, 144).

9. Most valve trays are designed with up to 12 to 16 valves per square foot of bubbling area (bubbling area being column area less downcomer, downcomer seal, and any other large unvalved area). Typically, orifice and disk diameters are about 1½ and 2

in, respectively. The disk typically rises $3/16$ to $7/16$ in above the tray deck. Open area of fully open valves is typically about 10 to 15 percent.

10. Valve trays are prone to damage during commissioning, startup, and other abnormal operation. Seating of the valves restricts downward movement of vapor or liquid, and this restriction can be troublesome during abnormal operation. Adequate planning and operating procedures can circumvent these problems. This is discussed in detail in Chap. 11.

6.6   Calming Zones

It has been common practice to provide a blank area between the inlet downcomer or inlet weir and the hole field, and another blank area between the hole field and the outlet weir (Fig. 6.3). These blank areas are termed *calming zones.*

Inlet calming zones are used because the entering liquid possesses a vertical velocity component in a downward direction. This component causes excessive weeping and inhibits bubble formation at the first row of holes or valves. A calming zone 2 to 4 in wide is recommended to attenuate the effect of this downward component (73, 175, 192, 226, 281, 317, 409). With recessed seal pans, this problem does not occur and a calming zone 2 in wide is satisfactory (73, 172, 281). A method for estimating optimum calming zone width is available (73, 191). At least one designer (371) feels that an inlet calming zone may not be necessary.

Outlet calming zones allow vapor disengagement from the froth on the tray prior to liquid entering the downcomer. An outlet calming zone also increases liquid holdup on the tray, but at the expense of a greater pressure drop (83, 151). An outlet calming zone 3 to 4 in wide is useful for trays operating in the froth regime (172, 192, 226, 281). Excessive width for the outlet calming zone is to be avoided because it wastes space and promotes weeping and backmixing (90, 151). Some designers (73, 90, 371) feel that outlet calming zones are unnecessary.

When trays are operating in the spray regime, different considerations apply. The liquid does not enter the downcomer by flowing over the outlet weir. Instead, it enters when liquid droplets suspended in the vapor space descend into the downcomer (Fig. 6.1b). The closer the weir is to the holes, the easier it is for these suspended droplets to reach the downcomer. Further, the liquid pool on the tray floor near the outlet weir induces liquid backmixing onto the tray, which lowers tray efficiency (333). For these reasons, the outlet weir should be placed as close to the holes as possible for spray regime operation (90,

151, 192, 333). The minimum width of the outlet calming zone is often limited to about 2 in by the presence of a tray support beam (73).

6.7 Outlet weirs

Outlet weirs maintain a desired liquid level on the tray. In the froth regime (Fig. 6.1a), liquid enters the downcomer by flowing over the outlet weir, and the weir height directly sets the liquid level and holdup on the tray. This liquid level should be high enough to provide sufficient liquid-vapor contact time and good bubble formation. Tray efficiency increases with weir height in the froth regime (119, 175, 184, 218, 226, 243, 273, 326, 409), but for weirs 1½ to 3 in high this increase is often small (211, 218, 326, 409). On the other hand, the higher the liquid level, the higher the tray pressure drop (175, 226, 243, 371), downcomer backup (172, 226), entrainment rate (205), and weeping tendency (218, 226, 245, 371). Higher liquid levels also imply higher inventories, which is a distinct disadvantage if the liquid is hazardous.

For most froth regime applications, a liquid level ranging between 2 and 4 in provides the best value (88, 172, 192, 226). Weir height can be determined from the following equation (172):

$$(4 - h_{ow} - 0.5\Delta) \geq h_w \geq (2 - h_{ow} - 0.5\Delta) \tag{6.2}$$

where h_w = weir height, inches
h_{ow} = height of liquid crest over the weir, inches
Δ = hydraulic gradient, inches

Methods for estimating h_{ow} and Δ are described elsewhere (73, 172, 319, 371). Application of Eq. (6.2) leads to weirs 2 to 3 in tall for most froth regime services (123, 138, 243, 382, 413).

One exception to the above rule is where a long liquid residence time is necessary, e.g., when a chemical reaction takes place on a tray. A weir height of 3 to 4 in is common for absorbers and strippers (243), and up to 6 in can be used (138, 218). If the weir height exceeds 15 percent of the tray spacing, care should be taken to allow for the reduction in effective tray spacing when capacity limits are estimated. A second exception to the rule expressed by Eq. 6-2 is when tray spacing is very low (<12 in). Here, a weir height of 2 to 3 in significantly reduces the effective tray spacing, and therefore tray capacity. This imposes a more severe economic penalty than the efficiency loss effected by shortening the weir. In such cases, weir heights of ½ to 1 in are often set (218).

When trays operate in the spray regime (Fig. 6.1b), liquid enters the downcomer as a shower of liquid droplets, precipitating from the vapor space above the downcomer (90). Under these conditions, liquid holdup on the tray is independent of weir height (116); thus, a weir serves little purpose and can be eliminated entirely. However, even with columns designed to operate normally in the spray regime, it is good policy to provide outlet weirs because at low rates the column may be operating in the froth regime, and because spray regime entrainment increases as weir height decreases (21, 201, 202, 205).

Minimum recommended weir height is ½ in (73, 138), but ¾ to 1 in is preferred (123, 138, 211, 382). Low weirs (about 1 in) are frequently used in vacuum columns (123, 211, 226, 243, 382). Vacuum columns generally operate in the spray regime for most of their operating range (123, 441), and low weir heights are suitable. Weir heights of 1 to 2 in are generally used for other spray regime services.

Adjustable weirs (Fig. 6.7a) were common in early designs. Their intent was to provide flexibility for interchanging capacity and efficiency, but their effectiveness was generally limited. This design is no longer recommended because experience has shown that maladjustments of the weirs outweigh the potential benefits (123, 211, 399).

Swept-back weirs (Fig. 6.7b) are sometimes used at high liquid loads. They extend the weir length, which in turn lowers the effective liquid load (gallons per minute per inch of weir length), without changing tray or downcomer area. Swept-back weirs reduce tray pressure drop and downcomer backup, improve liquid distribution on the tray, and improve tray efficiency by inducing liquid flow into peripheral stagnant zones. However, the above improvements are usually small.

Perhaps the main application of swept-back weirs is for extending the length of side weirs of trays containing three or more liquid passes (Sec. 6.12). This equalizes the length of the side weir(s) to that of the center weir(s), which in turn equalizes liquid flow to the nonsymmetrical tray panels.

There are several variations of the swept-back weir design. The weir can be segmental (Fig. 6.7b) or semicircular. The downcomer wall can follow the contour of a swept-back weir (Fig. 6.7b) or it may be a straight vertical wall (Fig. 6.5d). The former type of wall is more expensive, but it provides more downcomer area and better utilization of tray space than the latter.

The tolerance on weir height is usually ± 1/16 or ± 1/8 inch (38, 48, 86, 177, 211, 257, 371, 399). Weir length should be designed so that at least a 1/4- to 1/2-in liquid crest is maintained above the weir (73, 88, 172, 382) to provide a good liquid distribution. The higher value should be used if the tolerance is ± 1/8 in. If the tolerance is too high relative to the liquid crest, the head of liquid above the weir will vary along the weir length, resulting in large flow variations

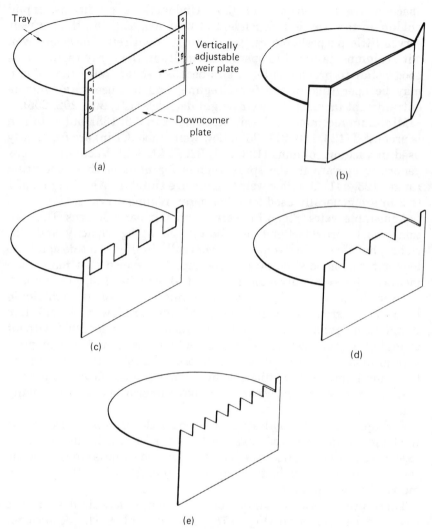

Figure 6.7   Unique outlet weir designs. (*a*) Adjustable; (*b*) swept back; (*c*) rectangular notch; (*d*) Intermittent triangular notch; (*e*) continuous triangular notch. (*Parts a, and c to e from Henry Z. Kister, excerpted by special permission from* Chemical Engineering, *September 8, 1980; copyright © by McGraw-Hill, Inc., New York, NY 10020.*)

from one end of the weir to the other, and the tray efficiency will be reduced.

At low liquid loads, a liquid crest of ¼ to ½ in may be difficult to achieve unless notched weirs (Fig. 6.7c to e) are used. Notches can be triangular or rectangular. The only notched weirs which are effective in the spray regime are those with deep (6 to 8 in or more) rectangular notches (Fig. 6.7c; often called "picket-fence weirs"). In this regime, most of the liquid is dispersed as drops suspended several inches above the tray floor, and these easily skip over low notches. Since at low-liquid loads the dispersion is usually in the spray regime, picket-fence weirs are the most popular type of notched weir. Picket-fence weirs are recommended whenever the liquid load is lower than 0.5 to 1 gpm per inch of weir (144, 211, 243). Others (123, 138) recommend their use when liquid load is lower than 0.25 gpm per inch of weir.

Weirs with triangular notches (Fig. 6.7d, e), or shallow (1 to 2 in) rectangular notches are only effective either when used together with a splash baffle or a vapor hood (discussed below), or when the dispersion is in the froth regime. Since at low liquid loads and high throughputs the dispersion is seldom in the froth regime, such weirs are mainly of value with splash baffles (144) or when notching is only required for turndown. Rectangular notches are usually preferred to triangular notches because they are less sensitive to out-of-levelness.

6.8 Splash Baffles and Vapor Hoods

A splash baffle (Fig. 6.8a) is a flat vertical plate parallel to the outlet weir and located a short distance in front of the weir. The bottom edge

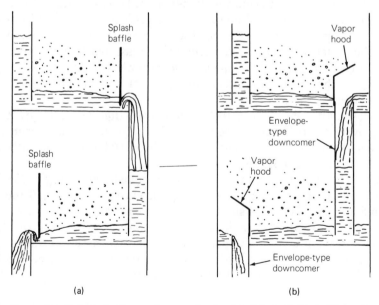

(a)   (b)

Figure 6.8 Splash baffles and vapor hoods. (a) Splash baffles; (b) vapor hoods.

of the baffle is a short distance above the tray floor, permitting liquid flow underneath.

Splash baffles are used in low-liquid-load services. The baffle backs liquid up onto the tray and increases its liquid holdup and froth height (31, 83, 374). The baffle also helps to prevent the tray from drying up and promotes froth regime operation at low liquid loads (374). In small-diameter columns (< 2 ft) it also prevents liquid drops formed at the tray inlet from being flung directly into the downcomer.

A splash baffle is recommended when liquid flow rate is less than 0.1 gpm per inch of outlet weir (144). One successful application of tailor-designed sieve trays with splash baffles at liquid rates ranging from 0.01 to 0.07 gpm per inch of weir has been described (374). Splash baffles should only be used at low liquid loads, because they restrict the downcomer inlet area and can lead to premature downcomer choke. A calming zone upstream of the baffle is sometimes used to minimize this problem (374). Splash baffles also increase tray pressure drop and froth regime entrainment (31).

Splash baffles should only be used with leakage-resistant trays. Leakage is a major problem at low liquid loads and is aggravated by the buildup of level induced by the splash baffle. When liquid flow rate is of the order of 0.1 gpm per inch of weir or less, even a small leak can be detrimental to column efficiency (374). Tray joints should be gasketed or seal-welded when using splash baffles. *Horizontal leakage* (i.e., leakage from the tray into the downcomer under or at the sides of the outlet weir plate) may render the splash baffle ineffective. Seal welding, or use of an envelope-type downcomer with its own weir, can overcome the problem. A successful application of the latter technique has been reported (374).

The splash baffle should be tall enough to avoid liquid droplets from skipping over its top. To minimize effects of out-of-levelness and to ensure adequate head over the outlet weir, the outlet weir is usually notched with shallow notches (Sec. 6.7).

The vapor hood is similar in function to the splash baffle (Fig. 6.8b). A vapor hood is used with an envelope-type downcomer (see Fig. 6.14c and d in Sec. 6.15). The hood shelters the downcomer from the shower of drops on the tray and forces liquid buildup on the tray until it overflows the weir. Application and performance characteristics of the vapor hood are similar to those of the splash baffle, and, likewise, it is only suitable for very low liquid-flow-rate services.

6.9 Internal Demisters

Demisters are sometimes installed in the intertray space to eliminate entrainment. Typical applications include some clean, low-liquid-load chemical services at close tray spacing (<20 in), and refinery vacuum towers where entrainment of metal-containing residue into the gas oil section is troublesome. Pilot scale tests (348) show that internal demisters substantially reduce entrainment and enhance efficiency. Two experiences were reported (55, 278) where internal demister installation resolved severe entrainment problems. Photographs illustrating the effectiveness of this technique were published (431).

Internal demisters are expensive, have a high plugging tendency, and their sections may disintegrate. The author is familiar with one incident where chunks of wire mesh from an internal demister tore loose and found their way to a draw pan above and to the bottom pump. For these reasons, they should only be installed in clean services experiencing entrainment problems that cannot be eliminated by alternate means. It is a good practice to install an internal demister only in the problem regions rather than above every tray in a column. In one case (278), it was found that demisters installed in the nonproblem regions did little to improve performance.

Typically, an internal demister in the intertray space is about 2 in thick and contains wire mesh. Attention must be paid to correctly specifying the mesh type, size, and materials. In a refinery vacuum column, it is also important to keep the mesh wet (eg, by spraying wash oil above the demister) in order to inhibit coking. When the service has a high plugging tendency, chevrons rather than wire-mesh may be the best demister choice (270).

6.10   Reverse-Flow Baffles

In a reverse-flow ("half-pass") tray (Fig. 6.9), liquid is forced to flow around a central baffle. Both the downcomer and the downcomer seal area are on the same side of the tray, and the liquid flow path is quite long. This tray is mainly suitable for low-liquid-flow-rate applications. Making the baffle at least twice the height of the highest calculated clear liquid height on the tray has been recommended (48) in order to avoid short-circuiting of liquid. For the same reason, leakage under the baffle should also be minimized.

6.11   Stepped Trays

Stepped (cascade) trays (Fig. 6.10) are sometimes used in large-diameter columns. They are used when there is a concern that an excessive hydraulic gradient will induce vapor to preferentially channel through the outlet portion of the tray. Since the hydraulic gradient is

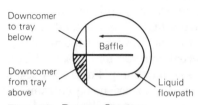

Figure 6.9   Reverse-flow tray.

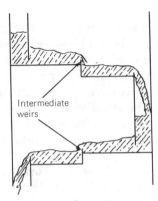

Figure 6.10   Stepped trays.

relatively steep on bubble-cap trays, and relatively flat on sieve and valve trays, stepped trays are mainly used with bubble-cap trays. At least one designer (409) recommends against stepping sieve or valve trays. The author is familiar with one troublesome experience with stepped sieve trays operating at relatively low liquid loads. Presumably, excessive weepage where liquid dropped over the intermediate weir deprived the downstream tray panel of liquid and caused the channeling effect the trays were trying to prevent. The net result was excessive entrainment and poor separation. There is little reported in the literature about the effectiveness of stepped trays.

Each step of the tray should be designed with the same hydraulic gradient, the same liquid head over the weir, and the same hole area (86, 88). Intermediate weirs should rise above the calculated liquid height immediately downstream (48, 86, 257, 371, 409); if submerged, they would not operate as true weirs.

6.12   Unique Considerations for Multipass Trays

Multipass trays lower the tray or downcomer liquid load by splitting the tray liquid into two or more paths. For instance, the use of two liquid passes instead of a single liquid pass will halve the liquid load. This enhances tray and downcomer capacity and lowers tray pressure drop, but at the expense of a shorter path length. Shorter path lengths reduce tray efficiency, and if extremely short, may be inadequate for accommodating tray manways. Trays containing more than two liquid passes may also be sensitive to liquid and vapor maldistribution because panels are nonsymmetrical. Once such maldistribution forms, it tends to persist throughout the trays below (88), causing a loss of efficiency and capacity (47).

Guidelines for setting the number of passes and minimizing operating problems with multipass trays are listed below:

1. There is an incentive to use as few passes as possible (88, 307). Increasing the number of passes decreases efficiency, increases the sensitivity to maldistribution, and increases costs.
2. Capacity gains from splitting liquid phases are substantial at high liquid loads, but negligible at low liquid loads. Most sizing practices (35, 38, 39, 123, 138, 211, 240, 246, 249, 307, 382, 404, 413) set the number of passes so that liquid loads do not exceed 7 to 13 gpm per inch of weir. The author feels that the upper end of this range (10 to 13 gpm) is optimum for most applications. This range is suitable when tray spacing exceeds 18 in; for smaller tray spacings, the recommended maximum liquid load is often lower (38, 211, 307). A detailed chart for setting the number of passes is available (38, 211). An earlier chart or set of guidelines (48, 172) was recommended by several designers in the past (48, 73, 172, 257, 319, 371, 409). These were based on bubble-cap trays and are not generally popular for setting the number of liquid passes on modern sieve and valve trays.
3. Short path lengths should be avoided. A path length smaller than 16 in is not feasible for internal manway installation (138); a path length 16 to 18 in is considered tight for this purpose. Path lengths smaller than 18 to 22 in may severely lower tray efficiency. One experience has been reported (237) where a refinery reboiled absorber was revamped by replacing single-pass trays by two-pass trays. Although the expected capacity gain was achieved, the reduction in flow-path length from 36 to 18 in caused a loss in efficiency large enough to justify reinstalling the original trays.

   To avoid excessively short liquid paths, several designers recommend that two-pass trays should only be installed when column diameter exceeds 4 to 6 ft (38, 48, 73, 138, 172, 211, 257, 319, 371, 409); three-pass trays should only be installed when column diameter exceeds 7 to 9 feet (38, 138, 211); and four-pass trays should only be installed when column diameter exceeds 10 to 12 ft (38, 138, 211).
4. Whenever possible, the number of passes should not exceed two. Trays containing a larger number of passes are prone to maldistribution among the passes. Neither the cause of this maldistribution nor its effect on efficiency is well understood, and prediction is difficult. If more than two liquid passes have to be used, it is best to follow Bolles' guidelines (47) for minimizing liquid maldistribution (guideline 7 below).
5. The author and others (238, 371) prefer avoiding an odd number of passes (e.g., three-pass trays) altogether. Their panels are far less

symmetrical than even-pass trays, which makes adequate liquid distribution particularly difficult to achieve. Nevertheless, it has been the author's and the industry's experience that when adequate liquid distribution is achieved, trays with an odd number of passes perform well.

6. When more than two liquid passes are used, it is essential to achieve adequate split of feed or reflux liquid among the passes. Similarly, tray irregularities that inadvertently interfere with liquid split to the passes (e.g., a nonsymmetrical variation in weir height due to construction error on even one tray) can initiate liquid maldistribution. Further discussion is in Sec. 2.4.

7. In a multipass tray, vapor distribution between the passes is largely determined by the hole area, while liquid distribution is largely a function of the outlet weir height and length. If the geometry of the passes is perfectly identical, the distribution of vapor and liquid is the same for each pass, and tray efficiency is uniform. This is readily achievable in two-pass trays, where each pass is identical to the other, but not when a larger number of passes is involved. For instance, in a four-pass tray, the weir length of the center passes differs from that of the side passes. Unless this effect is allowed for in the design, the L/V ratio will vary from pass to pass, with a resulting reduction in tray efficiency, as demonstrated by Bolles (47).

To minimize the effects of pass maldistribution on efficiency, the following guidelines (Fig. 6.11) were proposed (47):

a. *Providing equal vapor flow to each pass:* This is achieved by subdividing the column into equal-pass (bubbling) areas and providing equal hole area within each pass (alternatively, an equal number of valves or bubble caps). This practice is also recommended by others (179, 307).

b. *Providing equal liquid flow to each pass:* This can be achieved either by installing inlet weirs, so that even liquid distribution is established at the tray inlet, or by tuning the length of the outlet weirs. Swept-back weirs and picket-fence weirs may be ideal for the latter purpose. Other designers (179, 307) recommend tuning both the heights and lengths of outlet weirs to provide equal liquid flow in each pass. The author feels that judiciously varying the clearance under the downcomer from one pass to another may be another desirable alternative.

It is difficult to state whether the tray inlet or outlet geometry has a greater influence on liquid distribution between the passes. Some simulator tests (384) imply that at low liquid loads (1 to 2 gpm per inch of outlet weir), tray inlet geometry would be far more important. It is uncertain whether this con-

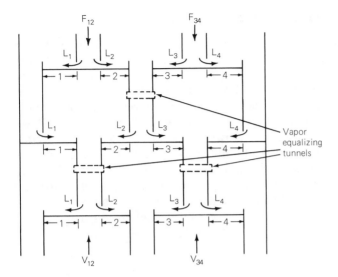

(A) Equal bubbling and hole area on each panel
$A_1 = A_2 = A_3 = A_4$
(B) Equal liquid flow to each pass
$L_1 = L_2 = L_3 = L_4$
(C) Vapor equalizing tunnels
(D) Evenly distributed feed
$F_{12} = F_{34}$; $V_{12} = V_{34}$

Figure 6.11 Prevention of maldistribution in multipass trays.

clusion can be extended to the higher liquid flow rates (above 6 gpm/in), and therefore, to the frothier dispersions normally encountered on multipass trays.

c. *Providing tunnels for vapor equalization between compartments:* The tunnels equalize vapor flow and also dampen liquid maldistribution (12). Other designers (12, 88, 179, 192, 307) also advocate the use of tunnels when the number of liquid passes exceeds two, and sometimes even with two-pass trays.

d. Guidelines 6 and 8 in Sec. 6.5, and Sec. 6.17 contain additional recommendations relevant to multipass valve trays.

8. When the number of passes is restricted, either by the short liquid path length (3 above) or by fear of maldistribution (4 and 5 above), liquid loads up to 20 gpm per inch of weir can and have been used (138). Careful sizing of trays and downcomers is required with such high liquid loads, since tray sizing correlations often require extrapolation beyond their recommended limits at such loads. Short tray spacings (< 18 to 24 in) should also be avoided in such applications.

6.13   Change in Number of Liquid Passes

Transition from one number of liquid passes to another is often required where a feed stream or a circulating reflux stream is introduced. Such transitions can be achieved by a number of methods, but care is required to ensure that the chosen method causes neither maldistribution nor flow restriction, nor interferes with downcomer sealing. One experience has been reported (12) in which a flow-restrictive design caused premature flooding. A design involving downcomer rotation by 90° and a notched-seal pan (Fig. 6.12) has the advantage of simplicity and has been recommended (143, 179, 192, 207, 307). Dimensions for this arrangement (143, 207) are described in Fig. 6.12. In addition, it has been recommended that the tray spacing at the transition tray should exceed the normal tray spacing by at least 1 ½ ft, and preferably 2 ft (138, 143, 179, 192, 208, 307).

6.14   Flow-Induced Tray Vibrations

Flow-induced tray vibration is a relatively infrequent occurrence that was observed in a number of large-diameter (5- to 25-ft) columns containing sieve and valve trays (62,108). The pressure of these columns ranged from deep vacuum to high pressure, although the majority operated under vacuum (62). The most severe vibrations were observed close to the weep point (62, 108). The vibrations caused fatigue cracking of trays, support beams, weirs, and tray-to-shell supports. Extensive and widespread fatigue cracking sometimes occurred within hours of operation at the damaging vapor rate. In one case, total internal collapse occurred; in another, shell cracking resulted (62). In individual cases, it is likely that secondary factors such as column acoustic or mechanical resonance increased the severity of the damage caused by flow-induced vibrations (62, 108, 329, 425).

Several mechanisms were postulated to explain the origin of flow-induced vibrations. Brierly et al. (62) proposed that vapor flow fluctuations occur when wet-tray pressure drop passes through a minimum as the hole velocity varies (Fig. 6.13). The existence of such a minimum under some conditions was experimentally confirmed by Wijn (425). Near the minimum, the vapor flow rate can fluctuate between two values with the same wet-tray pressure drop. Priestman and Brown (329–331) have proposed that fluctuations are initiated when vapor passes through the liquid as a pulsating jet. At low vapor rates, liquid reaches the rims of tray perforations and starts choking the

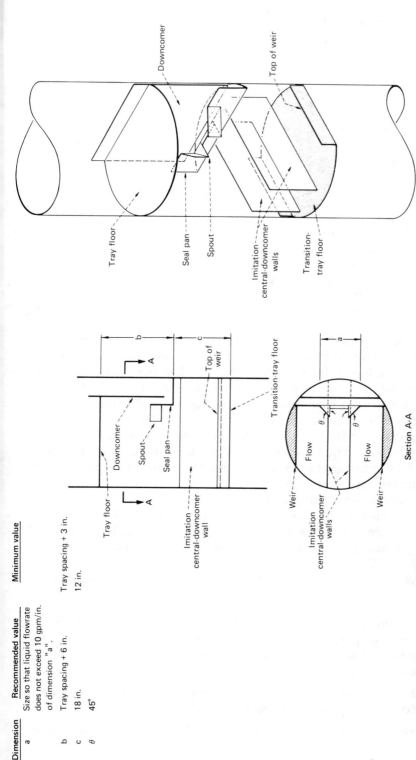

Dimension	Recommended value	Minimum value
a	Size so that liquid flowrate does not exceed 10 gpm/in. of dimension "a".	
b	Tray spacing + 6 in.	Tray spacing + 3 in.
c	18 in.	12 in.
θ	45°	

Section A-A

Figure 6.12 Arrangement for the transition tray. (*From Henry Z. Kister, excerpted by special permission from Chemical Engineering, December 29, 1980, copyright © by McGraw-Hill Inc., New York, NY 10020. Concept and dimensions based on L. Kitterman. Paper presented at Congresso Brasileiro de Petroquímica, Rio de Janeiro, November 8–12, 1976.*)

171

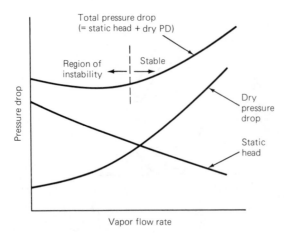

Figure 6.13 Region of instability where flow-induced vibrations may develop.

holes. This tapers the hole open area, and vapor hole velocity accelerates. Eventually, the vapor velocity reaches a critical value, approximately at a hole F-factor (hole velocity in feet per second times the square root of vapor density in pounds per cubic foot) of 16. At this value, a vapor jet forms and pressure drop rises, clearing the choke, and triggering off a pressure pulse. When a significant number of pulsating jets occur in phase, they may synchronize and cause pulsation across the tray. Wijn (425) proposed that the fluctuations result from a springlike behavior of the vapor volume below the bottom tray, dampened by tray pressure drop.

The mechanism of transforming the above fluctuations into vibrations capable of damaging trays is not clear. Amplification by acoustic or mechanical resonance is commonly postulated (62, 108, 329, 425). Some predictive models and correlations have been proposed (62, 108, 329, 330, 425).

Common techniques for avoiding flow-induced vibration damage are:

1. *Reducing the fractional hole area (62, 108, 329, 425):* This enhances the pressure drop and the dampening of oscillations. This technique has been powerful for overcoming flow-induced vibrations (62).

2. *Altering the stiffness of support beams:* This technique has been successful on a number of occasions (62).

3. *Keeping column vapor flows above those conducive to vibra-*

tions: This technique may limit column turndown; in some cases, high vapor flow rates are difficult to maintain continuously.

6.15 Downcomers: Function and Types

Passage of liquid from the top to the bottom of trayed towers occurs primarily via downcomers. Downcomers are conduits having circular, segmental, or rectangular cross sections that convey liquid from an upper tray to a lower tray in distillation columns. Different types of downcomers are shown in Fig. 6.14. The major differences are in the cross-sectional areas and in the slopes of the lengthwise extension.

The straight, segmental, vertical downcomer (Fig. 6.14*a*) is the type most commonly used in distillation columns. It represents good utilization of column area for downflow and has a cost and simplicity advantage over all others.

The circular downcomer, or "downpipe" (Fig. 6.14*b*), was widely used early in column history. It has fallen out of grace because it provides low downflow area and limited vapor disengagement space. Presently, downpipes are only used when liquid loads are extremely small and segmental downcomers are not suitable (e.g., in some alcohol rectification columns and glycol dehydrators, where liquid load is well below 1 gpm per inch of weir length). Two outlet weir arrangements are used with downpipes: a segmental weir (Fig. 6.14*b*) or a circular weir, the latter being simply a short, vertical extension of the downpipe above the tray floor. A circular weir is less expensive, but a segmental weir gives better liquid distribution on the tray and may be expected to enhance tray efficiency and lower pressure drop. For this reason, segmental weirs are usually preferred to circular weirs when column diameter exceeds 3 ft. Alternatively, two or more downpipes with circular weirs are used in larger columns to ensure adequate liquid distribution.

"Envelope" types of downcomers (Fig. 6.14*c* and *d*) are sometimes used in low-liquid-load applications. They are often used to satisfy the minimum downcomer width criterion or to minimize liquid leakage. In moderate- and high-liquid-load applications, minimum downcomer width and liquid leakage are rarely major factors, and this type of downcomer is seldom used.

The arrangements of Fig. 6.14*b* to *d* are generally not recommended for applications other than those where liquid loads are low.

Sloped downcomers (Fig. 6.14*e* and *f*) represent the best utilization of column area for downflow. They provide sufficient volume for vapor-liquid disengagement at the top of the downcomer without wasting the active area on the tray below. These downcomers are par-

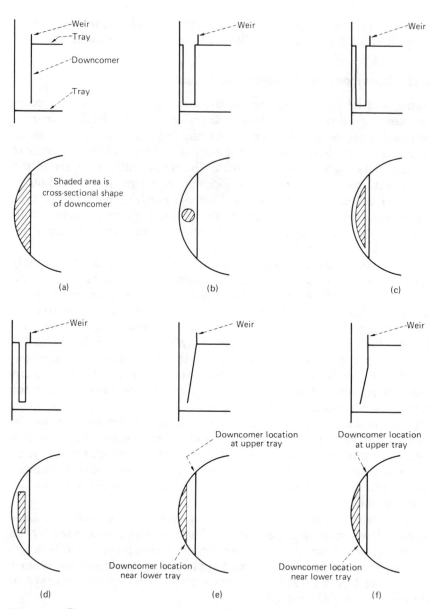

Figure 6.14 Downcomer types. (*a*) Segmental; (*b*) circular; (*c*) envelope (segmental); (*d*) envelope (rectangular); (*e*) sloped; (*f*) sloped. (*Henry Z. Kister, excerpted by special permission from* Chemical Engineering, *December 29, 1980, copyright © by McGraw-Hill, Inc., New York, NY 10020.*)

ticularly useful when vapor-liquid disengagement is difficult (e.g., foaming systems and high-pressure applications) and when the downcomers occupy a substantial portion of the tray area (e.g., high liquid loads). Although these downcomers are more expensive than straight downcomers, the active area savings usually more than offset the additional costs when downcomer area exceeds 20 to 30 percent of the tower area.

It is recommended that the ratio of the top area to the bottom area of sloped downcomers be between 1.5 and 2.0 (commonly, 1.7) (61, 123, 179, 192, 211, 226, 238, 307).

When the outlet weir is of the swept-back type (Sec. 6.7), a segmental vertical downcomer which follows the contour of the weir (Fig. 6.7b) is sometimes used.

6.16 Downcomer Width and Area

While insufficient tray area is one of the least common causes of poor column performance, there have been numerous instances of premature column flooding because of inadequate downcomer area. A downcomer must be sufficiently large to permit liquid downflow without choking. The fluid entering the downcomer is essentially the froth on the tray. If the friction losses in the downcomer and/or downcomer entrance are excessive, liquid will back up onto the tray and eventually flood the column. This is termed *downcomer choke*. A secondary function of the downcomer is to disengage vapor from the descending liquid. Vapor disengagement reduces the friction losses at the downcomer and minimizes recycling of vapor to the tray below.

The prime design parameter for avoiding downcomer choke is the downcomer top area. Downcomer width is a geometric function of downcomer area. With sloped downcomers, the downcomer bottom area is set according to the criterion in Sec. 6.15. This criterion permits the area near the bottom of the downcomer to be smaller than at the top, because near the bottom most of the vapor has disengaged and fluid velocity is lower.

Criteria for determining downcomer area are described below. The author recommends that the downcomer area be set large enough to satisfy all of these criteria except for liquid throw over the weir.

Sizing procedures. A detailed sizing procedure is normally used to determine downcomer area at the basic design stage. At the downcomer layout (secondary design) stage, this area is checked against other criteria (below). The detailed procedure may be based on one of these criteria. Detailed design procedures are described elsewhere (61, 138, 193, 211, 307).

Downcomer velocity. The maximum velocity of clear liquid in the downcomer needs to be low enough to prevent choking and to permit rise and satisfactory disengagement of vapor bubbles from the downcomer liquid. This is most restrictive in systems that have a high foaming tendency.

Values recommended for maximum downcomer velocity range from 0.1 to 0.7 ft/s. (35, 73, 175, 192, 226, 243, 249, 257, 272, 382, 409), depending on the foaming tendency of the system. Most recommended values are in the 0.3 to 0.5 ft/s range (35, 175, 226, 243, 249, 257, 382). These values were criticized for being conservative for nonfoaming low-pressure systems (73, 246, 396–398), and at least one source (138) claims that velocities up to 3 ft/s can sometimes be used in such applications.

Lockhart and Leggett (249) and Ludwig (257) presented similar sets of guidelines for maximum downcomer velocity as a function of system foaming tendency and tray spacing. Erbar and Maddox (272) presented an equation for maximum downcomer velocity as a function of physical properties. This equation presumably applies to light hydrocarbon systems similar to those encountered in gas processing plants. Lockett (243) derived plots comparing recommendations by a number of tray manufacturers for maximum downcomer velocities. The author incorporated the above with his experience into a single set of guidelines (Table 6.1). The author feels that the values shown in Table 6.1 are not conservative, and some may even be slightly optimistic. For a conservative design, a value from Table 6.1 can be multiplied by a safety factor of 0.75.

Residence time in downcomers. Sufficient residence time must be provided in the downcomer to allow adequate disengagement of vapor from the descending liquid, so that the liquid is relatively vapor free

TABLE 6.1   Maximum Downcomer Velocities

Foaming tendency	Example	Clear liquid velocity in downcomer, ft/s		
		18-in spacing	24-in spacing	30-in spacing
Low	Low-pressure (<100-psia) light hydrocarbons, stabilizers, air-water simulators	0.4–0.5	0.5–0.6	0.6–0.7
Medium	Oil systems, crude oil distillation, absorbers, midpressure (100–300 psia) hydrocarbons	0.3–0.4	0.4–0.5	0.5–0.6
High	Amine, glycerine, glycols, high-pressure (>300-psi) light hydrocarbons	0.2–0.25	0.2–0.25	0.2–0.3

by the time it enters the tray below. Inadequate removal of vapor from the liquid may choke the downcomer.

Two different definitions are used for downcomer residence time (371). The "apparent" residence time is the ratio of the downcomer volume to the clear liquid flow in the downcomer. The downcomer volume is based on the tray spacing times the average downcomer cross section. The true residence time is the ratio of froth volume in the downcomer to the frothy liquid flow in the downcomer. The true residence time can alternatively be expressed as the ratio of the clear liquid volume in the downcomer to the clear liquid flow. Different literature sources use different definitions; the definition adopted here is that of the apparent downcomer residence time. The author found this definition easier to apply, and to give a better correlation with the guidelines below [which were based on apparent residence times (49, 86)]. Further, the author found that applying the true residence time definition as outlined by some early sources (371) can lead to oversized downcomers.

Recommended values for downcomer residence time are based on Davies' survey (86) of flooding towers. The survey found that the apparent downcomer residence time of any of the towers that were flooding was 4 s or less. Based on this survey, Davies (86) and most subsequent designers (48, 88, 123, 172, 179, 192, 257, 371, 382, 409) recommend a minimum apparent residence time of about 5 s. This figure has been criticized for being conservative for low-pressure nonfoaming systems (73, 396–398). Some designers (49, 88, 123, 172, 179, 192, 271, 272, 409) recommend higher residence times for foaming systems. Bolles (49) and Erbar and Maddox (271, 272) presented similar guidelines for residence time as a function of the system's tendency to foam (Table 6.2). These guidelines are recommended (49, 192).

Liquid throw. Liquid throw (or jump) over the weir is the horizontal distance the liquid travels from the outlet weir before reaching the main body of liquid in the downcomer (Fig. 6.15).

TABLE 6.2   Recommended Minimum Residence Time in the Downcomer

Foaming tendency	Example	Residence Time, s
Low	Low-molecular-weight hydrocarbons,* alcohols	3
Medium	Medium-molecular-weight hydrocarbons	4
High	Mineral-oil absorbers	5
Very high	Amines and glycols	7

*The author believes that "low-molecular-weight hydrocarbons" refers to light hydrocarbons at atmospheric conditions or under vacuum. The foaming tendency of light hydrocarbon distillation at medium pressure (>100 psia) is medium; at high pressure (>300 psia) is high.
SOURCE: Bolles, W. L. (Monsanto Company), private communication, 1977.

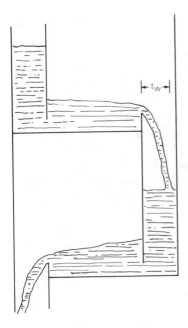

Figure 6.15 Liquid throw over
the weir.

In early designs, downcomer width was set so that the liquid throw
over the weir does not reach the column wall (48, 73, 88, 172, 257,
371). The purpose of this practice was to provide a pathway for vapor
disengaged in the downcomer to the tray above. Design criteria for adequate liquid throw are presented in the literature (48, 73, 88, 172,
257, 371). Since then, it has been demonstrated (396–398) that even at
very low liquid flow rates and wide downcomers, the liquid throw hits
the column wall, and that the above criteria were ineffective. These
criteria were based on clear liquid, while the downcomer liquid is aerated, and therefore travels longer horizontal distances.

Thomas et al. (396–398) observed that even when the liquid velocity
exceeded the design criterion, the downcomer inlet was not completely
closed. They pointed out that it may even be advantageous for the aerated liquid to hit the column wall, as it assists in breaking up the
froth, thus reducing the quantity of vapor entering the downcomer.
On this basis, liquid throw over the weir is not considered a limiting
factor (246, 371, 396–398), and is normally omitted from downcomer
sizing calculations.

Minimum downcomer width and minimum downcomer area. Reducing
downcomer area lowers column diameter or increases tray bubbling
area. This can lead to substantial cost savings when the downcomer is
large. As the downcomer becomes smaller, these savings diminish and
eventually approach marginal returns. When downcomer area is

small (less than 5 to 8 percent), there is little economic incentive to reduce it further. Further reduction of downcomer area is undesirable for one of the following reasons:

- As the downcomer becomes smaller, its width decreases faster than its length, thus converting the downcomer into a long and narrow slot. This geometry increases the friction resistance to liquid downflow and to upflow of disengaging vapor, but this increase is seldom taken into account by the previous sizing criteria. The previous criteria may therefore give optimistic area predictions for such situations.

- Small downcomers become extremely sensitive to foaming, fouling, construction tolerances, and introduction of debris.

- Smaller weirs associated with small downcomers distort the liquid flow pattern as it approaches the weir ("weir constriction effect"), which increases tray pressure drop (48, 257, 319, 371) and promotes the formation of stagnant regions near the tray periphery (243, 376). Such stagnant regions may be detrimental to column efficiency.

Downcomers smaller than 5 to 8 percent of the column cross-sectional area should therefore be avoided (73, 123, 144, 243, 246, 249, 409). Note that several of the cited designers expressed this rule as a minimum ratio of weir length or downcomer width to column diameter; these ratios can be geometrically converted into the stated minimum area ratio. An alternative recommendation advocated by some designers (61, 138) is to set the minimum downcomer area to either twice the area calculated using the normal design procedure or 10 percent of the column cross-sectional area (whichever is smaller).

If liquid flows are extremely small and the system is nonfoaming, circular or envelope downcomers (Fig. 6.14b to d) may be installed within the area subtended by a segmental weir (123). These should have twice the area calculated using the normal design procedure. Avoiding downcomer areas smaller than the minimum is most important in superatmospheric services and when there is a tendency to foam. The author has experienced several cases of premature downcomer choke flooding in superatmospheric columns whose downcomer areas were less than 5 percent of the tower area, even though downcomer sizing, downcomer residence time, and downcomer velocity appeared adequate. Such premature flooding occurred at rates as low as 40 to 50 percent of the design liquid loads.

Tolerance.   There has been at least one case (56, 57) where failure to meet the required downcomer tolerance contributed to premature column flooding. For large columns (> 10 ft in diameter) a downcomer width tolerance of ± ⁵⁄₃₂ to ± ¼ inch is recommended (38, 257). For smaller columns, a smaller tolerance is often justified, especially if the downcomer is small. One designer (38) recommends a tolerance of ± ⅛ in for columns 5 to 10 ft in diameter, and ± ³⁄₃₂ in for columns smaller than 5 ft in diameter. The author feels that since the critical variable is downcomer area rather than downcomer width, it is more appropriate to follow a practice similar to that for tray hole area (Sec. 6.4). Accordingly, downcomer area can be specified to a ±3 to 5 percent tolerance.

6.17 Antijump Baffles

Antijump baffles (Fig. 6.16) are often installed in the center and off-center downcomers of multipass trays in order to avoid a phenomenon similar to liquid throw (Sec. 6.16). Here the concern is that liquid jumping across the center downcomer, from one side to another, may cause excessive localized liquid buildup near the tray outlet, which may lead to premature flooding. Simulator tests (144) showed that antijump baffles can enhance tray capacity by as much as 25 percent in some instances.

In early designs, criteria for including antijump baffles were based on the liquid throw over the weir (48, 73, 88, 371). More recent publi-

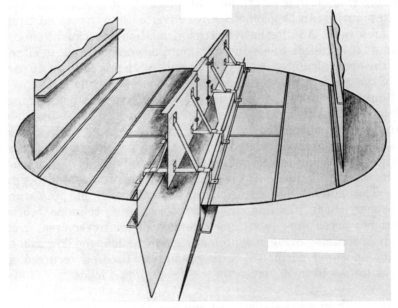

Figure 6.16 Antijump baffle. (*Reprinted courtesy of Glitsch Inc.*)

cations (138) attribute liquid jump to vapor expansion at the outlet weir that "pumps" (or gas-lifts) liquid over the weir. Such gas-lifting action is important in valve and bubble-cap trays, but is small in sieve trays. The gas-lifting mechanism therefore implies that antijump baffles are likely to be less effective with sieve trays.

Antijump baffles deflect the jumping liquid into the downcomer, as does the tower shell when the flow is toward the side downcomer. Antijump baffles also assist in breaking up the froth. The following guidelines have been recommended for antijump baffles:

1. The bottom edge of the baffle should be at the same elevation as the outlet weir (138, 409). Its top edge should be 11 to 20 in above the tray floor (138). Its length should be essentially the same as the weir length (138). Sealing at the joints or to the tower shell is not required. The baffle should be located at the center of the downcomer.

2. The center piece of the baffle should be easily removable and serve as a manway.

3. A criterion to determine the need for antijump baffles is presented elsewhere (138). Antijump baffles are not required when downcomer width exceeds 16 in (144).

6.18 Downcomer Sealing

One of the most common problems during startup of distillation columns, particularly those that operate at low liquid loads, is the difficulty of establishing a liquid seal in the column downcomers (49, 55, 192, 196, 203, 238). A case history of a column that could not be started up for a lengthy period because of this problem has been reported (196, 203). Another experience has been reported (203) where inadequate downcomer sealing was believed responsible for column instability. A downcomer sealing problem caused by incorrect installation of tray panels (145) is described in Sec. 10.9.5. Loss of downcomer seal has been reported to cause entrainment and separation problems in gas plant glycol dehydrators (238). A downcomer sealing problem can be easily eliminated at the downcomer layout stage by incorporating simple design features at minimal, or even no, additional cost.

During normal operation of distillation trays, the vapor flows through the tray perforations and liquid flows through the downcomers. The downcomer liquid seal prevents vapor from breaking into the downcomer, and the vapor hole velocity prevents the bulk of the liquid from weeping through the tray perforations.

At startup, this situation is reversed. Vapor tends to flow through both the downcomer and the tray perforations, while liquid tends to weep through the perforations rather than travel across the tray to the downcomer. To establish a satisfactory downcomer seal at startup, the liquid must:

1. Travel across the trays and outlet weirs and reach the downcomers (i.e., the vapor velocity through the trays must be high enough to prevent most of the liquid from weeping through the holes).
2. Travel into the downcomers (i.e., the initial vapor velocity through the downcomers must not be so high that liquid is prevented from descending through them).
3. Seal the downcomers (i.e., the downcomer backup must exceed the clearance under the downcomer).

A potential sealing problem can easily be detected by using a "startup stability diagram." The method of constructing such a diagram is discussed elsewhere (196). Some equations presented in the original publication (196) have been revised since then by the author. Revisions and updates to the equations for constructing such a diagram are described in Appendix B. A typical startup stability diagram is shown in Fig. 6.17. The lower curve illustrates the first criterion; the area below that boundary represents the range of conditions under

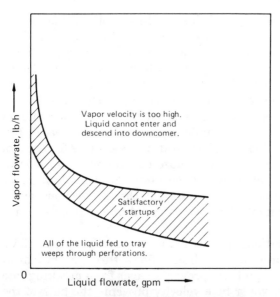

Figure 6.17 Typical startup stability diagram. (*Henry Z. Kister,* Hydrocarbon Processing, *February 1979. Reprinted courtesy of* Hydrocarbon Processing.)

which all liquid fed to the tray weeps through the perforations and none reaches the downcomer (the "dumping" region).

The upper curve illustrates the second criterion; the area above this boundary represents the range of conditions under which the vapor velocity in the downcomer is too high to allow liquid to enter and descend. The area between the upper and lower curves in Fig. 6.17 represents the range of conditions under which satisfactory startup can be achieved.

A potential sealing problem exists if (1) the vapor and liquid flow rates available for startup fall outside the area between the two curves, or (2) the area between the two curves is too narrow, and relatively small variations in flow rates during startup result in column operation outside the required sealing region.

Both curves in Fig. 6.17 (especially the upper curve) are very sensitive to some of the downcomer layout design parameters, particularly to the clearance under the downcomer, and the design of inlet weirs and seal pans. Figure 6.18 illustrates the effect of reducing the clearance under the downcomer, h_{cl}, from 1.5 to 1.0 in on the startup-stability diagram shown in Fig. 6.17.

Therefore, it is important at the downcomer layout design stage to determine whether a potential sealing problem exists. If it does, means of overcoming the problem should be sought. This is discussed in the following sections.

6.19 Clearance under the Downcomer

Three major factors govern the specification of clearance under the downcomer: downcomer pressure drop, the fouling and corrosive nature of the system, and downcomer sealing.

Downcomer pressure drop. If the clearance under the downcomer is too low, it may add substantially to the downcomer backup and consequently reduce downcomer capacity. Cases have been reported (61) where column capacity was increased by simply cutting 1 in off the bottom of the downcomer. Methods of estimating the backup caused by hydraulic losses through the opening under the downcomer are available in most distillation texts (48, 319, 371, 409).

To avoid excessive downcomer backup, the clearance under the downcomer is usually set so that clear liquid pressure drop at the downcomer outlet does not exceed 1 in of hot liquid (61, 172, 192, 249). Alternatively, some designers recommend outlet pressure drops not exceeding 1½ in of hot liquid (211), or liquid velocity at downcomer outlet not exceeding 1 to 1.5 ft/s (123,243), or area between the bottom of the downcomer and the tray floor one-third to one-half the area at

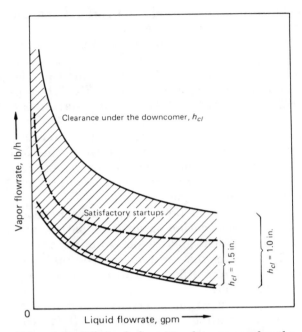

Figure 6.18 Effect of changing clearance under the downcomer on the startup stability diagram. (*Henry Z. Kister*, Hydrocarbon Processing, *February 1979*. Reprinted courtesy of Hydrocarbon Processing.)

the top of the downcomer (138). In most cases, these criteria give similar results.

When downcomer backup is critical, pressure drop of liquid leaving the downcomer is sometimes minimized by using a rounded downcomer outlet lip (Fig. 6.19). This is done infrequently, and usually only with highly foaming systems.

Fouling and corrosion.   If the service is a fouling one, dirt and polymer may accumulate under the downcomer and restrict the flow area. This may cause excessive backup, premature flooding, and maldistribution of liquid to the tray. Clearance under the downcomer should never be less than ½ in (38, 86, 123, 172, 192) in order to avoid blockage. It is best to avoid clearances smaller than 1 in, particularly if fouling may occur.

Downcomer sealing.   While pressure drop and fouling considerations set the minimum values for downcomer clearance, sealing considerations set the maximum value for downcomer clearance. If the service

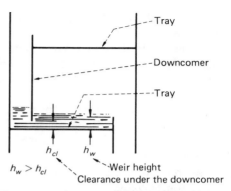

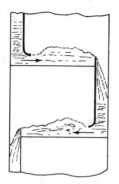

Figure 6.19 Downcomers with rounded outlet lips.

Figure 6.20 A common design practice of ensuring a positive downcomer seal. (*Henry Z. Kister, excerpted by special permission from* Chemical Engineering, *December 29, 1980, copyright © by McGraw-Hill, Inc., New York, NY 10020.*)

is corrosive, the area under the downcomer may become larger during service, and sealing it at startups may become more difficult.

Figure 6.18 shows the effect of the clearance under the downcomer on the startup-stability diagram. The smaller the clearance, the higher the upper curve in the diagram, and the easier the startup. Smaller clearances restrict vapor flow up the initially empty downcomer during startup. Consequently, the impedance to downflow is reduced.

The third criterion for satisfactory startup (Sec. 6.18) requires that downcomer backup exceed the clearance under the downcomer. A positive way of satisfying this criterion in froth regime operation is by setting the clearance under the downcomer h_{cl} lower than the weir height h_w (Fig. 6.20).

For froth regime operation, most designers (88, 123, 172, 211, 404) recommend a clearance about ½ in less than the weir height. Some designers (39, 123, 409) recommend making the clearance 0.5 to 0.7 times the weir height. The latter policy is often too restrictive, and the former is usually sufficient to ensure adequate sealing. In many applications, particularly at high liquid loads, even the former policy may propose clearances lower than the minimum required by the pressure drop criterion above. Increasing both downcomer clearance and outlet weir height can resolve this conflict, but is often undesirable and even self-defeating, because it increases tray pressure drop and downcomer backup (Sec. 6.7). It may be more satisfactory to set a downcomer clearance equal to or greater than the weir height. This practice relies on the tray pressure drop and tray liquid height to back up enough liquid in the downcomer to bridge the difference between the weir height and the clearance. The author has experienced several cases in which high-pressure and high-liquid load columns experienced no problems with downcomer clearances up to 1 in greater than the weir heights.

If a clearance greater than the weir height is desired, the author recommends that it be made as low as permitted by the downcomer pressure drop and fouling criteria; that the clearance not exceed weir height by more than 1 in; and that the clearance be made at least 2 in less than the clear liquid backup in the downcomer at minimum vapor and liquid loads.

In the spray regime (low liquid loads, low pressures; Fig. 6.1b), liquid holdup is essentially independent of weir height (204, 244). Therefore, the policy of setting a downcomer clearance lower than the weir height (Fig. 6.20) will do little to ensure satisfactory downcomer sealing. Trays operating in the spray regime rely on tray pressure drop and clear liquid height to back up enough liquid to seal their downcomer. This downcomer sealing consideration is surprisingly similar to that of columns where downcomer clearance exceeds the weir height, and the same guideline recommended above by the author applies.

Two additional considerations need to be taken into account in the spray regime. First, at reduced vapor rates, the flow regime may change to froth. Second, and most important, experience teaches that downcomer sealing is particularly difficult at low liquid loads (144, 179, 211), and there is a strong incentive to make downcomer clearance as low as possible.

Tolerances. The clearance under the downcomer should have an installed tolerance of $\pm$ ⅛ in (38, 179, 211, 257, 399).

6.20   Inlet Weirs and Recessed Seal Pans

Inlet weirs (Fig. 6.21a) and recessed seal pans (Fig. 6.21b) are primarily used for achieving a downcomer seal in cases where a potential sealing problem exists and clearance under the downcomer is limited by one of the design criteria previously cited (Sec. 6.19). These devices provide a positive seal on the tray under all conditions and ensure that the second and third sealing criteria (Sec. 6.18) are always satisfied. Sometimes it is argued that these devices improve liquid distribution to the tray, but this function is usually performed satisfactorily by the downcomer outlet (48, 172, 257, 404) and can rarely justify using either device. One exception is when the downcomer is circular

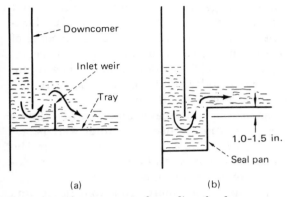

Figure 6.21 Arrangements for sealing the downcomer. (a) Inlet weir; (b) Recessed seal pan. (*Henry Z. Kister, excerpted by special permission from* Chemical Engineering, *December 29, 1980, copyright © by McGraw-Hill, Inc., New York, NY 10020.*)

(Fig. 6.14b); here an inlet weir is usually required for liquid distribution.

In high-liquid-load applications, an additional use of recessed seal pans is for increasing capacity of columns limited by downcomer backup, excessive pressure drop, or excessive froth height. The recessed seal pans permit using large clearances under the downcomer without resorting to taller outlet weirs, thus reducing downcomer backup. Often, the recessed seal pans permit outlet weirs to be lowered, which reduces pressure drop, froth height, and downcomer backup.

A recessed seal pan distributes liquid to the tray with an upward vertical motion rather than a horizontal motion containing a vertical downflow component. This results in better aeration at the inlet edge, increases both capacity and efficiency, and avoids precipitant weeping. Using a recessed seal pan can also be exploited to reduce the tray's inlet calming zone.

Inlet weirs and recessed seal pans have the disadvantage of creating areas of stagnant liquid, where sediment, dirt, and polymer can build up. Such buildup may restrict the downcomer outlet flow area and cause premature flooding. For this reason, these devices should be avoided in fouling services (88, 144, 257, 268, 404, 409). Inlet weirs and recessed seal pans are also more expensive than flat seal areas, but often by a small amount (61). The cost difference between inlet weirs and recessed seal pans is usually small (73, 317).

The inlet weir has several disadvantages which are not shared by the recessed seal pan. Liquid overflowing the weir has a strong downward component. This requires a larger calming zone at the tray inlet in order to eliminate precipitant weeping at the inlet row of perforations, and therefore wastes tray space. An inlet weir uses up some of the downcomer height and often increases downcomer backup (in comparison, the recessed seal pan "extends" the downcomer). The liquid seal provided by the inlet weir is usually shallower and therefore less reliable than that provided by a recessed seal pan.

Because of the above considerations, a recessed seal pan is almost always preferred to an inlet weir, and the use of an inlet weir is seldom recommended (48, 73, 123, 138, 192, 371, 404, 409). Perhaps the only exception occurs at low liquid loads when liquid leakage is a major consideration and/or when circular downcomers are used. Recessed seal pans support greater liquid heads, are more prone to leakage, and are more difficult to make leak-tight than inlet weirs, while downcomer backup is seldom a limiting factor under these conditions. At least one designer (144) prefers inlet weirs to recessed seal pans for such services. Should an inlet weir be considered necessary, the following guidelines have been recommended (86, 88).

1. Inlet weir height should equal the clearance under the downcomer but be less than that of the overflow weir. Excessive inlet weir height will lead to excessive downcomer backup and excessive weep through the inlet row of perforations and should therefore be avoided. If a positive downcomer seal is required, the inlet weir needs to be higher than the clearance under the downcomer, but this may cause a reduction in downcomer capacity.

2. Inlet weirs should not be notched.

3. The horizontal distance between the downcomer and inlet weir should not be less than the clearance under the downcomer.

4. Drain holes should be drilled through the bottom of the inlet weir to enable liquid drainage at shutdowns.

5. Waste of horizontal space can be minimized by sloping downcomers into the inlet weir area (Fig. 6.22). The area at the bottom of the slope should not be less than one-half to two-thirds the area at the top of the downcomer. In the inlet weir area, it is best to equalize upflow and downflow areas when using sloped downcomers (73).

6. Inlet weirs should be provided with at least two slots ¾ by 1 in. This is recommended to aid in flushing out trapped sediment (257).

When recessed seal pans are included on the trays, the following guidelines are recommended:

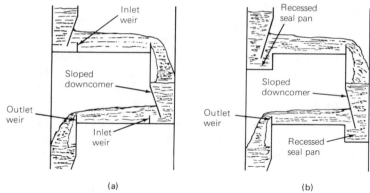

Figure 6.22 Inlet weir and recessed seal pans with sloped downcomers.

1. The bottom of the downcomer should extend about 1 to 1½ in below the tray level to provide a good liquid seal (73, 179, 317, 409).

2. The cross-sectional area for liquid flow at the bottom of the downcomer (between the downcomer and the bottom of the seal pan) and the upflow area of the seal pan should be as close to the same as possible to avoid restricting liquid flow. It has been recommended (73) that a seal pan depth of 4 in be normally used, but this should be increased to 6 in for high-liquid-load applications. A deeper seal pan and a greater clearance between the bottom edge of the downcomer and the pan floor may also be justified when some sedimentation is expected.

3. Recessed seal pans have a high tendency to leak. Some attention is required to minimize this leakage. A weld-in pan is often used where leakage is critical (179).

4. Guidelines 4 and 5 for inlet weirs also apply to recessed seal pans.

6.21 Seal Pan Below the Bottom Tray

The downcomer descending from the bottom tray of a tower (or from the tray above a chimney tray) must be sealed to prevent vapor upflow through it. A case where improper sealing of chimney tray downpipes could have been responsible for poor column performance is described in Sec. 4.10, guideline 15. The guidelines for recessed seal pans also apply to seal pans below the bottom tray (or above a chimney tray). Additional considerations are

1. The clearance between the bottom downcomer and the seal-pan floor should minimize pressure drop and restriction to fluid flow.

This clearance should be greater than the normal clearance under the tray downcomers, and should be at least 2 in, because the bottom seal pan is one of the most sensitive areas for solids accumulation.

2. Submergence of the downcomer within the bottom seal pan should be approximately the same as the clearance between the bottom of the downcomer and the seal-pan floor. One designer (307) recommends that the submergence be made at least 2 in.

3. The horizontal distance between the bottom-seal-pan overflow weir and the downcomer bar should be somewhat larger than the clearance under the downcomer in order to minimize pressure drop.

4. Adequate spacing must be provided between the bottom tray and the seal-pan overflow weir. This distance should equal at least one tray spacing. The distance between the bottom tray floor and the seal-pan floor should be 6 in larger than the normal tray spacing (143, 179, 192, 207, 208, 211, 237, 307). The additional distance helps to prevent flooding at the bottom tray. An incident where premature flooding was attributed to insufficient distance between the bottom seal pan and the bottom tray has been reported (12).

5. Bottom seal pans should be oriented so that the reboiler return nozzle does not impinge on the liquid emerging and falling from the seal pan. This is discussed in detail in Sec. 4.1.

6. With multipass trays, it is best to design the bottom tray with side outlet downcomers (307). This simplifies the introduction of vapor to the bottom of the column, giving a cheaper and less troublesome arrangement.

   Using side downcomers below the bottom tray is especially recommended when a preferential baffle (Sec. 4.5) divides the bottom sump into a reboiler compartment and a bottom compartment. Here, an overflow weir arrangement (Fig. 4.6d) is commonly used to direct the liquid from the bottom downcomer(s) into the reboiler compartment. This arrangement works far better with side downcomers than with center downcomers. The author is familiar with an experience where a column with a center downcomer below the bottom tray failed to achieve its separation efficiency because liquid from the bottom tray overflowed the higher weirs (Fig. 4.6d) into the bottom compartment.

7. Special attention must be given to ensure proper drainage of the bottom seal pan in fouling services. One experience has been reported (296) where plugged weep holes trapped water, which later caused a pressure surge and tray damage. Recurrence was successfully eliminated by installing a drain pipe from the seal pan to below the bottom sump liquid level. Such a drain pipe must be sized so that it neither plugs nor backs up liquid to the first tray. Guidelines 15 and 18 in Sec. 4.10 are particularly important.

Mechanical Requirements for Trays

The mechanical requirements for trays affect both their mechanical integrity and their performance. Inadequate catering for these requirements may lead to accelerated corrosion, damaged internals, excessive leakage, hazards to personnel working inside the column, difficulty in cleaning and maintaining trays and internals, and excessively prolonged startups and shutdowns. Some of these factors may in turn cause poor efficiency, premature flooding, or rapid deterioration of performance in service.

This chapter reviews common practices concerning materials of construction, thicknesses, supports, fastening, levelness, thermal expansion, manways, leakage, drainage, and other mechanical aspects of tray design. It outlines the preferred practices, highlights the consequences of overlooking traps, and supplies guidelines for avoiding pitfalls and for troubleshooting the mechanical details of trays.

For reasons described in the previous chapter, the discussions emphasize sieve and valve trays and exclude considerations unique to bubble-cap trays. Those unique considerations are detailed elsewhere (48, 86, 257). Nevertheless, several of the considerations discussed here also apply to several other tray types, including bubble-cap trays.

7.1 Materials of Construction

The main factors affecting the choice of construction materials of tray parts are:

- Compatibility with the chemicals processed (e.g., toxic materials must be avoided in columns processing food or beverages).
- Compatibility with the column materials of construction.

- Anticipated rate of corrosion.
- Degree of confidence in the predicted rate of corrosion.
- Procedure and expected frequency of cleaning, particularly if cleaning is to be done by high-pressure water jets or abrasive blasting.
- Initial cost of the materials.
- Effect of decline in the tray performance on plant operation.
- Cost of lost production due to a possible need to retray.
- Maintenance cost.

Bolles (48, 371) and Molyneux (288) provided rough guides for tray costs as a function of their construction material. Although these guides were prepared over 30 years ago and are based on bubble-cap trays, they roughly match the author's experience with recent sieve tray costs. These guides (48, 371) are listed in Table 7.1

In general, the effects of corrosion are least severe in large-hole sieve trays. In valve trays, the caged type is less sensitive to corrosion than the "uncaged" type. These factors are discussed in detail in Chap. 6. The effect of corrosion is generally more severe on bubble-cap trays than on sieve or valve trays (123).

7.2 Thickness of Tray Parts

Thickness of tray parts is usually determined by the corrosion- and erosion-resistance requirements and material used. Thicknesses are usually specified in sheet-metal gages.

Minimum deck thickness is usually 14 gage (0.0747 in) for corrosion-resistant alloys and 10 gage (0.1345 in) for carbon steel (88, 137, 211, 399). The latter typically provides a corrosion allowance of $5/64$ in for the top side or bottom side (137). A corrosion allowance of $1/16$ in is recommended for major support beams constructed from carbon steel; none is usually required for alloy construction (399).

The greater the tray thickness, the lower the dry-tray pressure drop in sieve and valve trays (88, 226, 257, 409). It is commonly believed

TABLE 7.1   Relative Tray Costs as a Function of Construction Material

Material	Relative tray cost
Carbon steel	1
Type 410 (11–13% chrome)	2
Type 304 stainless (18-8)	2.5
Type 316 stainless (18-8 Moly)	3.5
Monel	4

TABLE 7.2   Minimum Recommended Sheet Metal Thicknesses for Tray Parts

	Gage	Inch
Major support beams	7	0.1793
Minor support beams		
Nonferrous a.1d alloy	12	0.1046
Carbon steel	10	0.1345
General construction		
Nonferrous and alloy	14	0.0747
Carbon steel	10	0.1345
Downcomer bars		
Alloy	—	$3/16$–$1/4$
Carbon steel (noncorrosive service)	—	$1/4$
Carbon steel (general service)	—	$1/4$*–$3/8$

*Plus corrosion allowance.

that thicker sieve trays have a higher weeping tendency (226), but some contradicting experiences (438) were also reported.

When tray thickness is determined for sieve trays, hole diameter should be kept in mind. A thick plate may require larger holes because of the restrictions on hole punching (Sec. 6.3).

Minimum thicknesses recommended for other tray parts (211, 399) are given in Table 7.2.

In services where leakage is to be minimized and pressure surges tend to occur, it is often recommended to specify heavier gage for chimney trays and trapout pans (232).

7.3   Support Rings

A ring that is welded circumferentially around the shell usually is used to support the tray and, frequently, the tray-support beams (Fig. 7.1).

A support ring that forms a closed ring will help stiffen the column to withstand a high external pressure. However, this construction reduces the effective downcomer area at the top and can cause premature flooding if not allowed for in downcomer design. Unless the downcomer has excess area, it has been recommended (48, 86) that the support ring should not extend into the downcomer area.

Whenever possible, tray holes should be avoided in that part of the tray located over the support rings. Having tray holes over the rings reduces the effectiveness of the support, particularly in corrosive services. A support ring should be installed level. Adequate tray levelness cannot be achieved with tilted support rings.

Support rings should be installed by the tower fabricators, who are

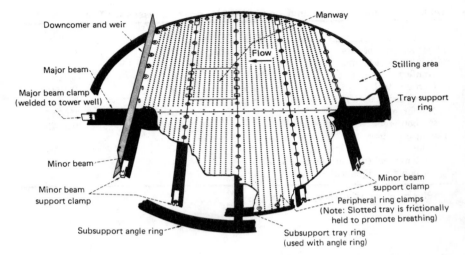

Figure 7.1 Mechanical supports for and assembly of sieve tray. (*F.A. Zenz, in P.A. Schweitzer* (ed.) Handbook of Separation Techniques for Chemical Engineers. *Copyright © 1979, by McGraw-Hill, Inc. Reprinted by permission.*)

able to accurately roll their rings to the "as built" internal diameter of the towers. Others, including tray fabricators, must roll a larger ring and then tailor it to the tower (399), an unnecessarily expensive procedure.

Support-ring designs, thicknesses, and widths vary from one fabricator to another. Support ring width increases with column diameter, and is usually between 1½ and 3½ in. Detailed guidelines for support ring widths and thicknesses are presented elsewhere (38, 211, 399). The thickness of the support ring must include a corrosion allowance for one side only.

7.4 Support Beams

The support beams prevent tray deflection under load from exceeding the specified requirements, and the beams support service personnel without having the tray acquire a permanent deformation.

In large-diameter (> 10 to 12 ft) columns, one or more major support beams (Fig. 7.1) are required, and these also support the minor support beams. In smaller-diameter (< 10 to 12 ft) columns, major support beams are often omitted, and trays are entirely supported by the support rings and minor support beams. In some cases, tie-rods are used for minimizing tray deflection.

Support beams should be located parallel to the liquid flow on the tray. This is recommended in order to minimize vapor crossflow (86, 88, 138). In large columns (> 10 to 12 ft), the major support beams should be installed parallel to the liquid flow, and the minor beams normal (i.e., at right angles) to the liquid flow (138).

Structural members should not extend deep enough to affect tray action. In order to minimize vapor cross flow, it is a good practice to limit the maximum vertical height of beams normal to liquid flow to about one-fourth of the tray spacing. In one case history contributed by L. Kitterman (206a), violating this practice led to severe froth oscillations on 10-ft-diameter trays of a nitric acid absorber. The trays were at 12-in spacing. Each tray was supported by two beams, 6 ins deep each, oriented parallel to the liquid flow. The average froth height reached the bottom of the beams. Observations through viewing ports indicated that beams divided the tray into three cells, with most of the bubbling taking place in the cell with the lowest liquid level. The liquid would then violently move from the high-liquid-level cell to the low-liquid-level cell, reversing the cell actions. This generated violent back-and-forth liquid oscillations perpendicular to the beams.

The support beams should carry the highest likely distributed load on the trays without exceeding the maximum allowable deflection. Maximum deflection is usually $\frac{1}{8}$ in (48, 137, 138, 211, 257, 288, 371, 399) for small-diameter trays (up to 8 to 10 feet). For larger trays, the maximum allowable deflection is greater, and a value of $\frac{1}{1000}$ to $\frac{1}{720}$ of the tray diameter has been recommended (48, 211, 288, 371, 399). Beam deflections must be determined in both the uncorroded and corroded states.

The highest distributed load likely is usually taken as the tray weight plus the weight of 2.5 to 5 in of water distributed across the tray (86, 138, 211, 399), or plus 1 ft of water for the seal-pan area (211, 399). A higher value should be used if the column is to be washed with a denser solution during preparation. Typical values of beam depths for different tray diameters were presented by Davies (86). Cambered beams can be used to reduce the beam size, particularly in large-diameter columns (86, 138).

Beams must also be designed to support work crews standing on the tray, or other equipment and tray sections that may be stacked on the tray. Usually, trays are designed to support a concentrated load of 200 to 300 lb/ft^2 at any point and at ambient temperature without overstressing (137, 138, 211, 399).

Tray supports must prevent trays from being lifted by upflowing vapor. Trays are usually designed to withstand an uplift pressure of about three times the tray pressure drop at maximum vapor and liquid loads (49) or 0.25 psi (107) (normally, whichever is larger).

Frequent pressure surges, such as those occurring when slugs of water are introduced into a hot column or into circulating reflux streams, may cause tray damage unless "explosion-proof" trays are used. Such trays are designed to withstand a load of 600 lb/ft^2, or more, from either top or bottom without yielding (138). Note that this provides a much greater uplift resistance than normal trays. Lieberman (237) cites one technique for enhancing uplift resistance by welding a bracket to the tower wall under the tray support ring. A horizontal tierod and shear-clip arrangement is used to nonrigidly attach tray decks to this bracket.

Support beams must be installed level, or tray levelness may be difficult to achieve, especially in large columns. Support beams are often equipped with levelling screws (288) to permit on-site levelling. A less satisfactory alternative is using slotted holes on the joint of a major support beam to its shell support brackets. Slotted holes, however, are often used on the joints of minor beams to the major support beam to allow levelling.

Downcomers are sometimes used to mechanically support a portion of the tray. If used for this purpose, they should be designed with sufficient mechanical strength to avoid buckling.

If severe tray vibration is expected, beams should be designed for the fundamental resonant frequency of the tray panel. Trays may be tied together by rod arrangements to raise their resonant frequencies.

7.5 Fastening of Trays, Downcomers, and Support Beams

Trays may be clamped (Fig. 7.2a) or bolted (Fig. 7.2b) to their supports. The overlap of the tray plate on the support is usually ¾ to 1 in (86). Similarly, downcomer panels are usually clamped (Fig. 7.2c) or bolted (Fig. 7.2d) to the vertical downcomer support bars. The downcomer support bars are welded to the shell. The fastening of downcomer panels to their supports should permit some on-site adjustment for ensuring correct setting of clearances under the downcomer.

Downcomer support bars should be compatible with commercial downcomer adapters. Downcomer adapters allow downcomer enlargement in a future revamp without the need to weld new downcomer support bars to the shell. These adapters are clamped to the existing downcomer support bars and to the new downcomer panels (274), and their use can dramatically lower the cost of a future revamp. Downcomer enlargement is one of the most effective and most popular revamp techniques in foaming, high-liquid-load, and high-pressure services.

Major support beams are usually bolted to support brackets (Figure 7.3). The support brackets for major support beams are welded to the

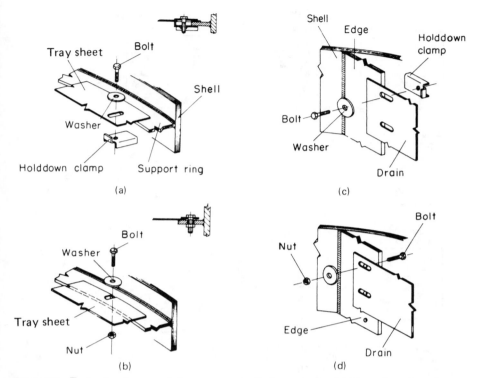

Figure 7.2 Fastening trays and downcomers to their supports. (a) Tray clamping; (b) tray bolting; (c) downcomer clamping; (d) downcomer bolting. (R. *Billet* Distillation Engineering, *Chemical Publishing Company, New York, 1979. Reprinted courtesy of Chemical Publishing Company.*)

column shell. Typical clamping arrangements for minor support beams are shown in Fig. 7.3.

The cost of tray assembly and installation rises with the number of tray parts. Tray assembly often costs as much as the trays themselves (128). Increasing bolt or clamp spacing reduces assembly costs, but also enhances the leakage tendency and renders the trays less secure. It is best to have the tray manufacturer specify fastener type and fastener spacing and to avoid imposing restrictive "maximum bolt or clamp spacing" specifications. Any specific concerns about tray leakage, or any unique considerations pertaining to the mechanical strength of the tray assembly, should be communicated to the manufacturer. In such cases, a maximum spacing specification (commonly about 6 in) may be justified.

It is a common (and desirable) fabrication practice to supply loose nuts, bolts, and clamps and to have these fastened inside the column.

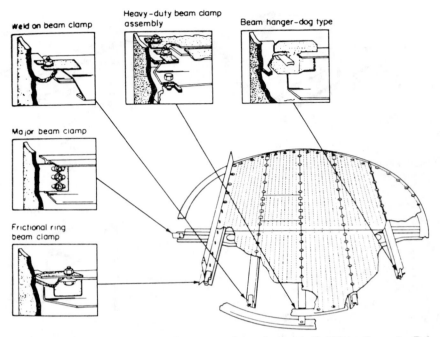

Figure 7.3 Fastening support beams to the column shell. (*F.A. Zenz, in P.A. Schweitzer* (ed.) Handbook of Separation Techniques for Chemical Engineers. *Copyright © 1979, by McGraw-Hill, Inc. Reprinted by permission.*)

An alternative is to have the tray fabricated with all nuts welded to the tray pieces. This alternative saves both assembly costs and installation time, and the savings easily pay for the added fabrication costs (128, 274). It is estimated (274) that a loose-bolted tray takes 15 percent more time to install than one which has no loose bolting. Often, a loose-bolted tray assembly requires two assemblers (one to screw the bolt and the other to hold the nut), while only one assembler is required when the nuts are welded. Despite the cost and time disadvantages, loose-bolting is almost always preferred to welded nuts, because it enables easy removal of corroded or worn parts in the future. Welded nuts are only justified in clean, noncorrosive services, and when installation time is critical. Sometimes, loose bolting is specified between the tray floor pieces, and welded nuts are specified for the downcomer and tray inlet areas (274).

Bolts, nuts, and clamps should be made from materials having good corrosion resistance so that they can be easily unfastened after being in service. This is particularly important for bolting at tray manways. Bolts, nuts, and clamps should be to the manufacturer's standards if acceptable (137). Bolts or clamps should be accessible from the top side of the tray. This will make their removal easier.

Special care should be exercised in systems where previous experience has indicated tray vibration problems. When tray vibration is expected, bolted-through tray connections should be used, along with lock washers or self-locking nuts, rather than frictional fastening.

Bolts ⅜ in in diameter were recommended and generally used on trays (211, 274). Larger bolts are recommended for major support beams (211).

7.6 Tray Manways

Tray manways (Fig. 7.1) allow maintenance workers and inspectors to travel from one tray to another. Manways should be sufficiently large to allow workers to rapidly climb out through them (in case of emergency) without cutting themselves. However, manways must not be too large because the larger they are, the heavier they become and the greater the risk of injury to the workers removing them. Further, the manways should be small enough to be removable through the column manholes. For convenience, manways are usually rectangular. Some designers recommend a minimum manway size of 12 by 16 in (211, 399). The author feels that manways should be at least 16 by 20 in, and preferably larger. The author is familiar with one incident in which a worker cut himself while quickly climbing through 16 by 20 in manways during a false emergency. At least one designer (73) recommends a minimum manway size of 18 by 24 inches. The manway panel weight should not exceed 65 lb (211, 399). Manways should be top- and bottom-removable.

Partial vertical alignment of manways is recommended (211, 399) for easy passage of lights, hoses, and communications and to allow personnel to stand erect. Full alignment of manways is not recommended because it increases the distance that a worker or tools can fall freely down the column.

For multipass trays, one tray manway is required for each tray pass because the central downcomers restrict access from one side to the other.

Some manufacturers (140, 213) market quick-opening types of manways, in which a latch replaces bolting as a means of keeping the manway in place. These offer advantages of reducing shutdown time and labor (140), but little has been published about their field performance.

7.7   Thermal Expansion

Tray temperature often varies from one set of operating conditions to another (e.g., it is higher when a distillation column is pressured than when it is depressured). Often, the tray temperature is not the same as the shell temperature. Tray design should allow for thermal expansion of tray sections. Failure to do so may result in tray buckling or beam warping.

To avoid thermal-expansion problems, one must (1) provide adequate fastener spacing or slotted holes; (2) establish satisfactory clearances between the trays and the shell, [½ in per 10 ft has been recommended (86)]; and (3) refrain from welding chordal members such as downcomer panels, beams, or tray sections at both ends to the column shell.

7.8   Tower Roundness

Neither the tray nor the tower is a perfect circle. Usually, both tower and tray are round to within ± 0.5 to 1 percent of their respective nominal diameters (38, 86, 192). The normal practice is to allow a clearance of about 1 in between the tray and the shell (86, 138). More detailed guidelines are presented elsewhere (138).

7.9   Tray Levelness

In tray design calculations, we assume that the tray is perfectly level. In practice, this is impossible. Deviations from the horizontal (unlevelness) can affect performance in one or more ways:

1. Sieve tray weeping may be expected to increase with tilting, but tests have shown that this effect is only important when the column operates at liquid flow rates [<5 gpm per inch of weir (250)].

2. Out-of-levelness can promote vortexing in bubble-cap trays. This effect is particularly evident in bubble caps that operate with a low slot seal, such as atmospheric and vacuum columns (55, 257).

3. Unlevelness may lead to variation of vapor load across the tray, and a consequent loss of efficiency. However, data surveyed by Lockwood and Glausser (250) showed that this effect is of little significance.

4. Unlevelness may lead to maldistribution of liquid flowing across the outlet weir because of variations in the liquid crest along the weir length. This occurs mainly when the liquid crest is low in froth regime operation.

5. Unlevelness may induce liquid maldistribution, and thereby lower efficiency in multipass trays (177). Specifically, it has been stated (211, 399) that a series of multipass trays tilted in the same direction will suffer a severe efficiency loss.

6. Unlevelness due to column sway (e.g., when located on an offshore platform) can induce wave motion on trays (323), variation in tray liquid depth, and loss of downcomer liquid seal (154). It has been stated that tray columns are unsuitable for offshore plants (154, 323), but little published test data are available to substantiate this. The author experienced one onshore tower whose top section swayed considerably in strong winds with no noticeable deterioration in performance, and is familiar with other similar experiences. However, this sway is far less severe than what can be anticipated offshore.

In summary, unlevelness may be detrimental to the performance of trays containing more than two liquid passes, and it should be minimized in such services. Tray unlevelness due to column sway on offshore plants may also be detrimental to column performance, and trays are best avoided in such services until more experience with swaying trays becomes available.

Unlevelness may significantly affect the performance of trays operating at low liquid flow rates, and trays that have a low liquid crest and/or a low slot seal (bubble caps). Such features are common in vacuum columns. In most other cases, some unlevelness is not detrimental to tray performance, and a tilt even two or three times the recommended tolerance can often be tolerated.

Factors involved in a tray being unlevel are:

1. *Tray-level tolerance:* It is desirable to specify the highest tolerance that will not affect tray performance. Column costs increase significantly as this tolerance diminishes. In one particular case, reducing the levelness tolerance by $\frac{1}{32}$ and $\frac{1}{16}$ in (from a $\pm\frac{1}{8}$-in tolerance) increased the column cost by 2 and 4 percent, respectively (250).

   Many designers generally recommend a tolerance of $\pm\frac{1}{8}$ in for most services (38, 48, 86, 88, 211, 250, 317, 371, 399, 404, 409). This is probably unnecessarily fine in view of the previous discussion and of other factors that can affect tray levelness.

   The effects of unlevelness appear to depend more on the slope of the tray than on its vertical deviation. Therefore, it is logical to specify a diameter-dependent tolerance (i.e., maximum tolerable slope) rather than a "flatness" tolerance. For this reason, Ludwig's guidelines (257) are recommended: $\pm \frac{1}{8}$ in for columns smaller than 3 ft in diameter, $\pm \frac{3}{16}$ in for columns between 3 and 5 ft in diameter, and $\pm \frac{1}{4}$ in for columns larger than 5 ft in diameter.

Care should be taken with columns whose performance is likely to be affected by unlevel trays. For such cases, smaller tolerances should be specified. A tolerance of $\pm \frac{1}{16}$ in has been recommended for vacuum services (317).

2. *Vertical alignment:* The tower initially is set in its foundation so that its top may be within 1 or 2 in from the vertical. Closer alignment is of little value because the tolerance on the straightness of the column-shell axis is often greater then the vertical alignment. In a column 120 ft tall and 12 ft in diameter, this causes the top trays to be up to $\frac{1}{8}$ in out of level (86).

3. *Foundation settlement:* Column foundations often settle somewhat unevenly on their substrate (i.e., a minor "Tower-of-Pisa" effect occurs).

4. *Skirt alignment:* The skirt base support ring is not perfectly even. This unevenness affects the vertical alignment of the column, and therefore also the tray tilt.

5. *Column bowing:* Joining all the sections from which the column is fabricated is not perfectly even. In addition, some bowing of the column occurs prior to installation, while the column is horizontal. Either factor may affect the straightness of the column-shell axis, and therefore also the tray tilt.

6. *Unlevel supports:* Unlevel support rings or support beams will cause trays to tilt. Correct installation can minimize this effect. Techniques for permitting on-site levelling are discussed in Sec. 7.4.

7. *Tray deflection:* Uneven tray deflection under load and uneven corrosion of the tray deck can induce unlevelness.

8. *Wind loading:* Deflection under a high wind load may arise. However, the effect of wind load on tray levelness is usually small (250) and does not persist.

9. *Thermal expansion:* If the column is uninsulated and trays are installed in the morning on a sunny day, one side of the column may become hotter than the other. In a 100-ft column, this can add up to $\frac{1}{4}$ in of tilt to the top tray (250). During normal operation, thermal stresses vary with the time of day and the directions of sunshine and wind; this affects tray tilt.

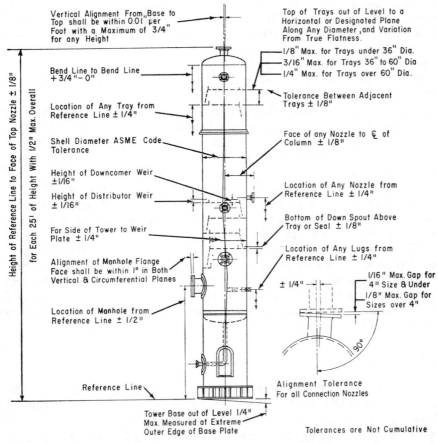

Figure 7.4 Typical column fabrication tolerances. (*From Applied Process Design for Chemical and Petrochemical Plants*, Volume 2, Second Edition, by Ernest E. Ludwig. Copyright © 1979 by Gulf Publishing Company, Houston, Texas. Used with permission. All rights reserved.)

7.10 Fabrication Tolerances

Figure 7.4 illustrates typical tolerances used for column fabrication (257). Excluded are most tray and downcomer tolerances; these are depicted later in Fig. 10.5.

7.11 Tray Drainage

When the column is shut down, some liquid is trapped on bubble-cap trays and certain valve trays, and at low points such as seal pans and inlet-weir areas. Weep holes allow this liquid to drain.

In services processing water-insoluble materials, weep holes also prevent the accumulation of small quantities of free water at low points such as seal pans. The free water may otherwise dissolve acidic impurities and cause severe corrosion at the low points. In one instance (296), water trapping following weep-hole blockage resulted in a pressure surge and tray damage.

Weep holes usually range in size from ¼ to ¾ in (48, 73, 86, 237, 248, 257, 371, 409). Small holes are preferred in clean, noncorrosive services but should be avoided in fouling or corrosive applications. The recommended hole area is 4 in^2 per 100 ft^2 of tray area (48, 257, 371, 409). This hole area will drain a column containing 50 bubble-cap (or other leak-tight) trays with 4-in weirs in about 8 hours (48, 371). Alternatively, the required weep-hole area can be derived from the required drainage time (usually a few hours) by using Bolles' (48, 371) or Lockhart's (248) procedure. Excessive hole area will cause leakage, and should be avoided, especially in draw pans; one case in which this was troublesome was reported (231).

When weep holes are provided in bubble-cap or valve trays, it is good practice to locate them as close to the outlet weir as possible. Hence, if liquid weeps during normal operation, it descends as close as possible to the inlet of the tray below. However, at least some holes should be located at potential low spots on the tray to permit complete draining of the column.

7.12 Leakage

The main current use of bubble-cap trays is for minimizing tray leakage or achieving high turndown. In either case, tray joints must be leak-tight if the tray is to achieve its objectives. An experience where joint leakage caused excessive entrainment, tray dryout, and poor efficiency in bubble-cap trays has been reported (55). Minimizing leakage is also important on liquid outlet pans and chimney trays, especially when used for total drawoff (see Sec. 4.10).

To prevent or minimize bubble-cap tray leakage, it is essential to pack its joints with gasket material. Gasketing substantially increases tray installation and maintenance costs, and may be troublesome. A joint whose gasket deteriorated during service may leak a lot more than a nongasketed joint. Gasket deterioration may be caused by mechanical wear (e.g., due to thermal expansion) or by chemical attack. Gasket bits or installation debris from packing of joints often come loose and may plug column internals. Adequate selection of gasket materials and correct installation are imperative. It is best to avoid gasketing whenever possible, but in leak-tight applications gaskets are often difficult to avoid (unless seal welding is used; Sec. 4.10). When neither gasketing nor seal welding is used, adequate metal thickness, finishing, and bolting should be provided to minimize joint leakage.

To ensure practically no leakage, a leakage test is commonly conducted upon installation. In this test, the weep holes are plugged and the trays are filled with water to the top of the weir. The rate of fall of water level is then measured. It has been recommended that this rate should not exceed 1 in per 20 minutes (48, 86, 257, 371, 409). If the leakage is not critical, a more liberal tolerance can be adopted (48). This test is not always conclusive, because a tray that leaks badly under test conditions may perform satisfactorily at the design temperature and vice versa.

In valve and sieve trays, no positive liquid head is maintained on the tray, and liquid leakage is seldom a problem. For this reason, no gaskets are generally required and leakage tests are not carried out. Occasionally, however, a leakage problem may occur in low-liquid-rate applications, and gasketing may be required (237).

7.13   Cartridge Trays

When column diameters are smaller than about 2½ to 3 feet, a person cannot enter the column to install, inspect, and maintain the trays. Because of this limitation, a 2½- to 3-ft trayed column is often installed, even if a smaller diameter is hydraulically sufficient. Alternatively, either packings or cartridge trays can be used, and column diameter reduced below 2½ ft.

Cartridge trays (Figure 7.5) are a bundle of trays and downcomers held together on a rod assembly. Each bundle is called a cartridge and is typically 10 to 12 ft long for convenient shipping and installation (306, 308). The first cartridge is slipped into the column until it reaches its supports. Each successive cartridge is stacked on top of the previous cartridge. Over 70 trays can be installed in this manner (306, 308). The top head of the column needs to be flanged to facilitate installing and removing these trays.

The two main disadvantages of cartridge trays are leakage around the circumference and their relatively high cost. A vertical metal seal ring is often installed around each tray to minimize leakage. Successful operation at liquid rates lower than 1 gpm has been claimed with a sophisticated variation of this sealing technique (308).

Since cartridge trays are not fixed to support rings (like normal trays) they can be easily uplifted. Kitterman (206b) shows that the weight of the trays does little to resist uplift. The pressure drop sufficient to uplift a bundle is set by the shell friction, and this can be small. Uplifting has been troublesome on many occasions. In one

Figure 7.5 Cartridge fixed-valve (V-Grid) trays, 23-in ID assembly. (*Reprinted courtesy of Nutter Engineering.*)

tower (206*b*), one bundle of trays separated from another, leaving an unsealed downcomer between the two; this caused premature flooding. In another tower that had gasketing where two bundles met, separation of bundles resulted in several gasket pieces in the bottom of the tower (206*a*).

In order to prevent uplift, adequate holddown of cartridge trays is essential. Holddown can be achieved by either fixing the top tray at a flange or by using rods fixed to the top head of the tower.

Cartridge sieve, valve, bubble-cap, and other types of trays are commercially available. Further information can be found in the manufacturer's literature (e.g., 308).

7.14 Tower and Tray Specification

Specification sheets are vehicles for transferring and documenting information pertaining to column design. These sheets are used for transmitting process and design requirements to column designers, tray designers, or manufacturers. They are also used by column designers, tray designers, or manufacturers for submitting design details back for review.

Specification sheets vary from one company to another. In most cases, these sheets attempt to maximize the information transferred, present this information as coherently as possible, and minimize paperwork. In most cases, these sheets are used together with a separate reference mechanical specification.

Typical column and tray specification sheets are shown in Tables 7.3 to 7.5, respectively. The specification sheet in Table 7.4 (429b) was developed by FRI to cater to the needs of a large number of companies. When tray design is to be performed by others (e.g., the manufacturer), this is the best sheet to use. The spec sheet in Table 7.5 may be more appropriate when the user wishes to set several of the tray layout dimensions rather than let others set these. When supplying information to designers or manufacturers with the Table 7.5 spec sheet, only items dealing with process data and design requirements need to be filled. Other items will be completed by the designer or manufacturer. Filling in other details is usually interpreted by the designer as a specification or preference. For instance, if the user enters "2 in" under "weir height," the designer or manufacturer will attempt to incorporate a 2-in weir in the tray layout.

TABLE 7.3. Typical Column Specification Sheet

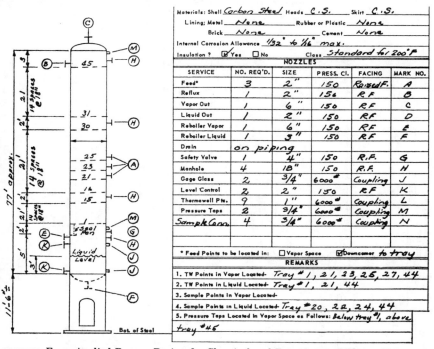

Client _____
Job No. _____
Item No. _____

Plant Location _____
Inquiry No. _____
Service _____

Engineer _____
Date _____

Tray No. 1= Top/Btm

Sectfon (Name/Description)

Tray Numbers Included

Loading at Actual Tray No.

Number of Trays Required

NORMAL VAPOR TO:

Rate, lbs/h

Density, lbs/ft3

ACFS

Mole Weight

Viscosity, cP

Pressure, psia

Temperature, $^\circ$F

Design Range, % of Normal

NORMAL LIQUID FROM:

Rate, lbs/h

Density, lbs/ft3

GPM

Mole Weight

Surface Tension, dynes/cm

Viscosity, cP

Temperature, $^\circ$F

Design Range, % of Normal

Item No. _____

Section (Name/Description)

Tray Numbers Included

Service _____

PERFORMANCE REQUIREMENTS:

Max. δP per Tray, mmHg

Max. % Jet Flood

Max. % DC Liq. Velocity

Max. DC Backup Clear Liq, in.

Derating Factor

Purpose for Derating
(Foaming, System, Safety)

MECHANICAL REQUIREMENTS:

Tower Diameter, inches

Number of Passes

Tray Spacing, inches

Type of Tray

Hole/B Cap Diameter, inches

Deck Material/Thickness

Valve/B Cap Material

Hardware Material

Support Material/Thickness

Total Corrosion Allowance

Vessel Manhole I.D., inches

MISCELLANEOUS:

Solids Present: Yes / No   Flashing Feed: Yes / No

Anti-Jump Baffles: Yes / No / Vendor Preference

Recessed Seal Pans: Yes / No / Vendor Preference

Specify Equal   Bubbling Areas / Flow Path Lengths   per pass

Design Load: _____ PSF with _____ inch deflection at _____ F.
or   Standard: 30 PSF with 1/8" at 300 F.

TABLE 7.5   Typical Tray Specification Sheet

Job No. _____ Ref. Drwg: _____ Ref Spec: _____	Tray Data Sheet Sheet 2 of 2		Service: _____ Column No: _____ Plant: _____		
Tray No.					
Design Case					Note No.
Tray Details					
Weir Height, S/C Inch					
Length of Each Weir, S/C Inch					
DC Width S/C Inch					
DC Clearance S/C Inch					
Special Features					
(Yes/No/Optional)					
Seal Pan					
Inlet Weir					
Antijump Baffles					
Splash Baffle					
Mechanical Details					
Tray Gauge					
Material Deck/Valves					
Corrosion Allowance, Inch					
Support Materials					
Bolting Materials					
Pressure Surge Tendency					
Uplift, ΔP psi					
Manhole ID Inch					
Manway Size Inch X Inch					

NOTES:

No.	Date	By	Approved	Revisions		
					Data Sheet No.	Rev

TABLE 7.5   Typical Tray Specification Sheet (*Continued*)

Job no. _____ Ref drwg _____ Ref Spec _____	Tray Data Sheet Sheet 1 of 2		Service: _____ Column no: _____ Plant: _____			
Tray no.						
Design Case						Note no.
Vapor to Tray						
Temp °F						
Pressure psia						
Z						
MW						
Density lb/ft^3						
Viscosity cP						
Flow rate ACFS						
Flow rate lb/h						
Liquid from Tray						
Temp °F						
MW						
Density lb/ft^3						
Viscosity cP						
Surf. Tension dyn/cm						
Flow Rate gpm						
Flow Rate lb/h						
Design Basis						
Column ID ft-in						
Tray Type						
% Flood, Allow/Calc						
% DC Froth, Allow/Calc						
ΔP per tray, Allow/Calc Inch Liquid						
Min Oper Rate, % Des at Des L/V						
DC Velocity, Allow/Calc ft/s						
Weir loading, Allow/Calc gpm/in						
Fouling/Sticking Tendency						
Corrosion Tendency						
Foaming Tendency						
System Factor						
Tray Layout						
Tray Spacing inches						
Liquid Passes per tray						
Bubbling Area ft^2						
DC Top Area ft^2						
DC Bottom Area ft^2						
No of holes/Valves						
Hole diam, Inch/Valve type						
Fractional Hole Area						

No.	Date	By	Approved	Revisions	Data Sheet No.	Rev

Internals Unique to Packed Towers

A "cutaway" of a packed column, highlighting arrangement of various internals, is shown in Fig. 8.1. Internals used for distribution, redistribution, and introducing feeds or reflux into packed columns were described in Chap. 3. Chapters 4 and 5 examined internals and practices which are generally common to tray and packed columns, including column bottom sections, column outlets, liquid collection devices, gravity lines, and connections for access and instruments. This chapter completes the coverage of packed-column internals and practices by reviewing those that have not been previously described, including packing supports, support structures, hold-down plates, bed limiters, and column verticality.

The primary function of packing supports is to adequately support the packings. The primary function of bed limiters and hold-down plates is to prevent packing fluidization. These functions must be achieved without adversely affecting liquid distribution or column hydraulics. This chapter outlines the preferred practices regarding these internals, highlights consequences of poor practices, and supplies guidelines for troubleshooting and reviewing designs of packing supports, hold-down plates, bed limiters, and column slant.

8.1 Packing Supports

Packing supports must perform the following functions:

1. Physically support the packed bed

2. Incorporate sufficient open area in order to permit unrestricted flow of liquid and vapor

3. Avoid downward migration of pieces of packing

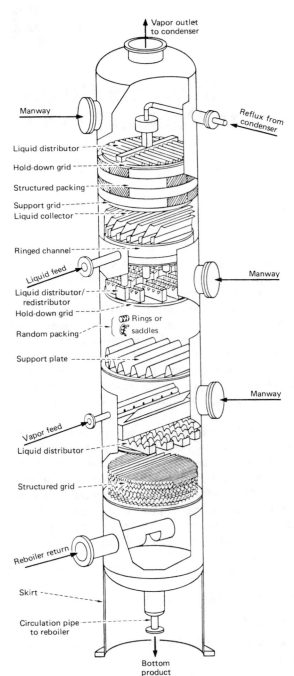

Figure 8.1 A "cutaway" of a packed column, highlighting arrangement of various internals (*Gilbert K. Chen, excerpted by special permission from* Chemical Engineering, *March 5, 1984; copyright © by McGraw-Hill, Inc., New York, NY 10020.*)

Packing supports therefore must provide both a high fraction of open area and the required mechanical strength.

The weight supported by the packing support is the weight of the packed section above it under flooded conditions. Allowance must be made for absorbing the shock of pressure surges, deterioration due to corrosion, and any additional weight not separately supported (such as a hold-down plate or redistributor resting on the packings).

Early designs were flat plates with 15 to 25 percent open area. Vapor and liquid passed countercurrently through the same openings, and a substantial hydrostatic head developed on the plate (Fig. 8.2a). This caused premature flooding of the column. The problem was aggravated by pieces of packing occluding some of the openings.

Open areas provided by modern packing supports are usually of the order of 70 to 100 percent of the column cross-sectional area (74, 257, 305). The open area depends on the materials of construction and the column diameter; some common designs in ceramic, carbon, and plastic have open areas smaller than 65 percent (214, 257, 305).

When the open area of the packing support is relatively small, it may bottleneck column capacity. A useful clue can be obtained by comparing the fractional open area at the support to the fractional open area of the packing (237, 268, 319). If the former is significantly lower, premature flooding may initiate at the support and propagate through the packing. Situations where this has occurred have been described (237, 268, 386). One designer (111) recommends an open area of at least 80 percent and preferably 100 percent of the column cross-sectional area for metallic applications.

In order to avoid downward migration of pieces of packing, the support openings must be smaller than the packing size. For circular openings, $\frac{1}{2}$-in holes have been recommended (111). The openings should be evenly distributed. A wire-mesh screen can be installed above the support, but this practice is not recommended because it reduces the open area of the support. In one reported case (237), adding such a screen resulted in premature flooding. A better alternative is to stack about a foot of larger packings right above the support (237). These larger packings should be stacked and not dumped, or they may easily mix with the smaller column packings.

In mildly corrosive applications it often pays to fabricate the packing support from stainless steel, even when carbon steel is satisfactory for the packing. This prevents localized corrosion that can puncture holes in the support, resulting in downward migration of packings. One service where this technique is commonly applied is in natural gas plant glycol-regeneration stills (84).

In services prone to pressure surges (some examples of such services are listed in Sec. 8.2), the resistance of a packing support to mechan-

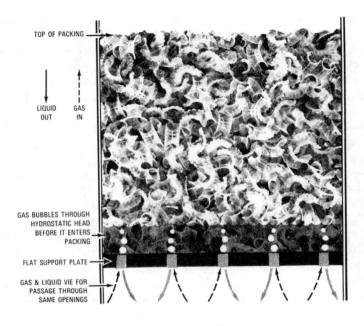

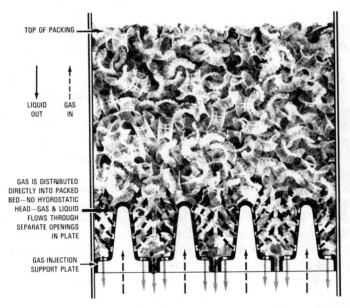

Figure 8.2 Action of flat and gas-injection support plates. (*a*) Flat support plate; (*b*) gas-injection support plate. (*Reprinted courtesy of Norton Company.*)

ical damage is a prime consideration. A damaged support is likely to lead to downward migration of packing pieces.

In services where coking may occur, the dripability at the bottom of the supports is an important consideration (299). The supports should be designed so that liquid drips off; liquid that hangs there tends to form stalactites.

The common types of packing supports are:

Gas-injection support plates (Fig. 8.3). These have separate openings for vapor and liquid. Vapor issues from the side openings, and liquid flows through the bottom openings, thus avoiding buildup of hydrostatic head (Fig. 8.2b).

Standard metallic and plastic gas-injection support plates provide open areas approximately 100 percent or more of the column cross-sectional area, are capable of handling liquid loads to 50 to 100 gpm/ft^2, and are available for column diameters of 1 ft or larger (142, 212, 305). Most standard units can handle packings 1 in or larger (142, 305), but smaller packings can often be accommodated with special design (305). At normal operating rates, pressure drop is usually less than 0.25 in of water (142, 212, 305, 386) and seldom exceeds 0.75 in of water (212). Ceramic units have a lower open area, usually about 50 percent or slightly more of the column cross-sectional area (305).

The large open area and the separation of vapor from liquid passages are major advantages which make gas-injection support plates most popular. In metallic and plastic applications, this support plate is the least likely to become a capacity bottleneck (142, 212).

Figure 8.3 Gas-injection support plate. (*Reprinted courtesy of Norton Company.*)

In large-diameter columns, gas-injection support plates have the additional advantage of minimizing the number of midspan beams required for supporting the support plate or grid. In metallic application, alternative support types require two to three times as many support beams as the gas-injection plate (74). The author and others (74, 111, 166) recommend specifying gas-injection support plates whenever possible in metallic and plastic random packing applications.

Grid supports (Fig. 8.4).   Grid supports consist of vertical bars arranged on one or two different planes (Fig. 8.4a and d, respectively). Grid supports can often provide as good a performance as (in ceramic applications, even better than) gas-injection supports.

Grid supports are generally less expensive than gas-injection supports and can provide open areas as high as 70 percent of column cross-sectional area in ceramic applications (305) and 95 to 97 percent in metallic applications (74). Grid supports are commonly used for structured packing, where gas-injection supports are usually unsuitable, and where most of the disadvantages listed below do not apply.

One disadvantage of grid supports is that pieces of packing may occlude some of their open area. This may build a hydrostatic head in the packing and bottleneck column capacity (similar to the phenomenon depicted in Fig. 8.2a). In ceramic services, a common practice for avoiding this bottleneck is to stack one or two layers of large (4 to 6 in) cross-partition rings or grid blocks on the grid bars (Fig. 8.4b, c). These layers form a high-capacity region at the support grid, which can accommodate liquid buildup without initiating column flooding. These layers also prevent smaller pieces of packing from reaching and occluding the grid openings. The stacked rings or grid blocks should be set tight and wedged in place with shards so that they do not shift around (302). Compared to cross-partition rings, grid blocks afford more open area at the support and easier installation (303), and are therefore often preferred.

Stacking a few layers of packing or grid blocks on the support can alleviate plugging when fouling is confined to or concentrated at the foot of the bed. The stacked layers eliminate pockets where deposits can accumulate and permit washing of scale by the downflowing liquid. In one case (170), stacking rings in the bottom foot of an acetylene plant compressor aftercooler wash column more than quadrupled column run length.

Grid supports often provide insufficient open area with small packing sizes. The space between adjacent bars, and therefore the grid

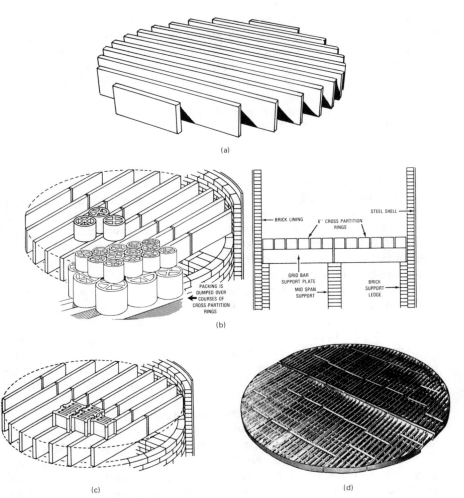

Figure 8.4 Grid supports. (*a*) Ceramic grid bar supports; (*b*) stacking large cross-partition rings on ceramic grid bar supports; (*c*) stacking grid blocks on ceramic grid bar supports; (*d*) metallic grid supports (*Parts a, b, c reprinted courtesy of Norton Company. Part d Gilbert K. Chen, excerpted by special permission from* Chemical Engineering, *March 5, 1984; copyright © by McGraw-Hill, Inc., New York, NY 10020.*)

open area, diminishes rapidly as packing size decreases, while the open area occlusion problem (above) escalates. Most standard designs of grid supports suit 1½-in or larger packing (142, 305).

Grid supports (particularly those whose bars are mounted in a single plane) are more likely to experience packing migration downward than gas-injection supports. Stacking larger packing directly above the support plate is often practiced to alleviate this problem. Migration is most pronounced with ring packings whose edges are not fused together, because the edges can easily penetrate the grid openings.

Grid supports with sharp edges should be avoided with plastic packing, because pieces of packing tend to extrude through the openings and be cut by the sharp edges, thus blocking the support (74, 144). This problem is most severe in high-temperature applications; in one reported case (26), a 20-ft bed lost 19 ft of its plastic packing through the support grid.

Structured packing. A layer of structured packing can sometimes be used to support a bed of random packings. This technique can add a fraction or even a full theoretical separation stage to the bed. The structured packing may also improve vapor distribution.

The structured packing selected for this purpose must have a large open area, a capacity at least as high as the random packings, and, when applicable, a resistance to fouling and corrosion at least equivalent to the random packing. The structured packing must also have sufficient mechanical strength to support the bed, and small enough openings to prevent packing migration. Grid and corrugated types of structured packings with large open areas are commonly used as random packing supports. This practice is most popular in short beds (e.g., refinery vacuum towers), where a relatively small layer of structured packing is sufficient to provide the required mechanical strength.

The use of the grid-type of structured packing to support random packing is also popular in situations where fouling tends to occur at the foot of the bed. Grid packings have a high fouling resistance and serve to eliminate regions which would otherwise be prone to plugging. On the other hand, structured packings (including grids) are difficult to clean and may require frequent replacement when used for the above purpose.

Corrugated supports (Fig. 8.5). Corrugated supports are used for light-duty support (typically up to 400 to 450 lb/ft^2), in small-diameter ($<$ 2 to 3 ft) columns (74, 212, 305). In very small columns ($<$ 2 ft in diameter) they are capable of supporting heavier loads (212). Corrugated supports are available in metal or plastic (305) and are the least expensive supports (74). Open area is relatively low (74), typically about 80 percent (305). The tendency of packing to nest between corrugated bars is high, which may further limit the open area. Because of the low open area and the common liquid and vapor passages through the bars, corrugated supports have a greater tendency to build up a hy-

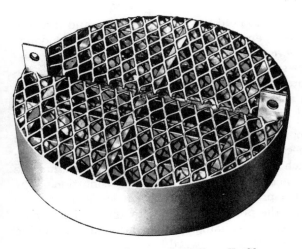

Figure 8.5 Corrugated support. (*Gilbert K. Chen, excerpted by special permission from* Chemical Engineering, *March 5, 1984; copyright © by McGraw-Hill, Inc., New York, NY 10020.*)

drostatic head (as depicted in Fig. 8.2a) than grid supports. Because of the above problems, use of corrugated supports should be restricted to light beds operating at low liquid and vapor loads.

Support-mixer redistributors.   These are discussed in Sect. 3.10.

Vapor-distributing supports.   These are discussed in Sec. 3.12.

8.2   Support Structure

In small columns, packing supports often rest on a tray support ring. The following considerations are important:

- Packing supports should be securely fastened to the support ring. One designer (386) recommends clamping support plates to the support ring at both ends for columns greater than 6 ft in diameter. Packing supports tend to be dislodged during pressure surges and upsets (e.g., those caused by slugs of water entering a hot hydrocarbon column, relief valve lifting, liquid level rising above the reboiler return nozzle, compressor surges, or instability of liquid seal loops). A dislodged packing support may either compress and damage the packing above or spill packing downward, or both. The problem is most severe with plastic packing. These supports are usually fabricated from nonmetallic materials which are difficult to fasten. The plastic packing is too light to hold the supports down, and the supports often have a small open area, and, therefore, a high tendency to lift. The author has experienced a number of cases where gas-injection and grid-type support plates in plastic packing service dislodged and disintegrated each time the column bottom seal loop experienced significant instability. The packing supports rested on, but were not fastened to, a number of support lugs.

- In corrosive environments, sections of the packing support are often glued to each other because of the difficulty of fastening corrosion-resistant resins. In such instances, it is important to ensure that the adhesive holds under actual and upset operating conditions. The chemicals processed, hot process temperatures, overheating during startup and upsets, and the presence of solvents and oxidizers may adversely affect the adhesives, causing disintegration of the packing support and spillage of packing downward.

- Resting the packing supports on lugs rather than a bona fide support ring is sometimes practiced as a cost-cutting measure. This arrangement generally provides weaker seating for the supports than a support ring, and may become troublesome if one or more of the lugs break or corrode.

- In extremely corrosive services (such as sulfuric acid towers), where large-diameter towers are used with ceramic supports, brick arches or piers are often used as the support structure (386).

- The support ring width should not be excessive, or it may restrict the column capacity. Strigle (386) gives the maximum recommended support ring width as a function of column diameter for columns between 4 and 12 ft in diameter. This width, W, in inches, can be calculated from $W = D/4 + 0.5$, where D is the tower diameter in feet. For small columns (< 4 ft in diameter), support rings sometimes restrict column capacity (386).

The maximum load that can be supported by a packing support plate or grid resting on a tray support ring decreases as column diameter increases. For instance, a typical standard stainless steel gas-injection support plate can support up to 1000 lb/ft^2 in a 4-ft column, but only half that load in a 9-ft column (212, 305). Greater loads can be handled either by fabricating the packing support out of thicker metal (often also at the expense of reducing open area) or by using one or more midspan beams.

Midspan beams are normally I-beams installed perpendicular to the support plate beams or grid bars (Fig. 8.6). With gas-injection support plates, midspan supports are usually required in columns larger than

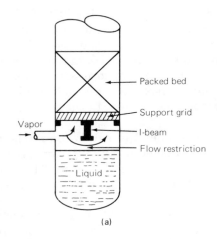

(a)

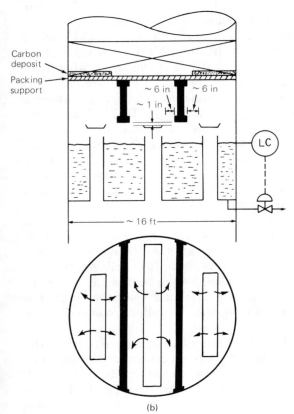

(b)

Figure 8.6 Maldistribution caused by I-beam support. (a) At column base; (b) at a chimney tray. (*Part a based on* Process Design for Reliable Operation, *second edition, by Norman P. Lieberman. Copyright © 1988 (May) by Gulf Publishing Company, Houston, Texas. Used with permission. All rights reserved. Part b courtesy of D. W. Reay (BP Oil), private communication.*)

9 to 12 ft in diameter in metallic applications (74, 166, 212, 386), and columns larger than about 4 ft in diameter in plastic and ceramic applications (305, 386). The sidewall of the column must be sufficiently rigid to support the point loading imposed at the beam seats (386). In very large columns, (> 20 ft in diameter) support trusses (e.g., as shown later in Fig. 8.8b) are sometimes used instead of I-beams. Trusses are lighter and enable a free flow of vapor in the section under the packing supports.

Midspan I-beams can be deep, sometimes 1 to 2 ft. When the bottom of the I-beam is too close to the liquid level, it may interfere with the distribution of vapor entering the packed bed (Fig. 8.6). This vapor maldistribution may lower the efficiency of the column and cause premature flooding.

I-beam interference at the column base (Fig. 8.6a) can be prevented either by installing two feed inlets, one on either side of the I-beam, or by providing sufficient area under the I-beam for vapor cross flow. For a single feed inlet, it has been recommended (237) that the vertical distance between the bottom of the I-beam and the high liquid level be set such that the calculated pressure drop for half the feed vapor flowing under the I-beam is less than 10 percent of the anticipated pressure drop in the packed bed above.

I-beam interference can be just as troublesome in the space above a chimney tray. In one case history contributed by D. W. Reay (334), this interference is believed to have led to severe vapor maldistribution in a refinery vacuum tower (Fig. 8.6b). The maldistributed vapor profile was displayed as a carbon deposit on the surface of the bottom packing. The deposit formed an annular ring about 5 ft wide that extended about 1 in into the bed. In that case, liquid was known to overflow the chimneys for several months because of an incorrect location of level tappings. This overflow caused liquid entrainment. Some entrained droplets ultimately carbonized on the base of the bed. Had the vapor profile been uniform, entrainment (and therefore deposit laydown) would have been more uniform. It is believed that vapor from the side chimneys was blocked by the beams and preferentially ascended around the periphery. If liquid overflow (down the risers) had been uneven, the maldistribution could have been further aggravated.

Techniques recommended for preventing I-beam interference at the column base also extend to chimney trays. It is best to provide sufficient open height for vapor cross flow between the top of the chimneys and the bottom of the I-beam (Sec. 4.10). If positioning the I-beams close to the chimneys cannot be avoided (e.g., in a revamp, where vertical space is limited), the chimneys should be designed to provide a uniform vapor profile to the bed without cross flow under the beam. This may be difficult when more than one I-beam is present.

8.3 Hold-Down Plates and Bed Limiters (Retaining Devices)

Hold-down plates (Fig. 8.7a,b) are used with ceramic or carbon random packing to prevent fluidization and restrict packing movement, which may chip or break packing particles. A hold-down plate rests directly on the packing. It is usually designed to support the top 5 ft of packing (257) and weighs roughly 20 to 30 lb/ft^2 (74,386).

Hold-down plates are not used with metal and plastic packings to avoid crushing the metal or compressing the plastic. To prevent fluidization with these random packings, bed limiters (Fig. 8.7c,d) are used. Bed limiters do not rest on the packing; instead, they are fastened to the column wall by means of a support ring or bolting clips. Alternatively, they can be suspended on tie-rods from the liquid distributor.

Random packing fluidization during even relatively minor upsets (such as column flooding) is common when a retaining device is absent. In large-diameter columns, packings tend to fluidize at random spots rather than uniformly over the entire top surface. Fluidization

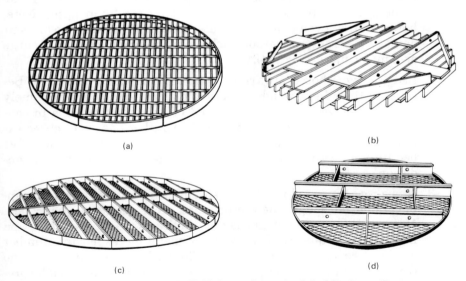

(a)

(b)

(c)

(d)

Figure 8.7 Retaining devices. (a, b) Hold-down plates; (c, d) bed limiters. (*Parts a to c reprinted courtesy of Koch Engineering Company, Inc.; part d reprinted courtesy of Norton Company.*)

may occur with metal, ceramic, and carbon packings and is most troublesome with plastic packings. Following the upset, packing particles settle unevenly, promoting maldistribution, which in turn reduces column efficiency and/or capacity. Fluidized packing particles frequently end up in distributor pans or troughs, causing further restriction and maldistribution. Fluidized packing particles sometimes reach the overhead system, and from there flow into the downstream plant. In the case of ceramic and carbon packings, fluidization is likely to chip and break packing particles.

With structured packings, fluidization is unlikely, but in large-diameter columns sections of the packing may be dislodged during upsets. To prevent this, hold-down bars are often used. These are bars fixed at the column shell and are perpendicular to the packing layers. In large-diameter towers, it may be desirable to use bona fide bed limiters rather than hold-down bars.

Guidelines for the design and operation of hold-down plates and bed limiters are listed below:

1. Hold-down plates should always be specified for ceramic and carbon services. The author has experienced extensive breakage in a bed of ceramic packing that did not have a hold-down plate. Others (74, 144, 302, 320) also share the above recommendation, stating that a hold-down plate is extremely economical considering the damage that may occur without it during a column upset. Note, however, that even hold-down plates may be insufficient to prevent damage to ceramic or carbon packing in the event of pressure surge or a major upset (74, 144).

2. Bed limiters should always be specified for plastic packings. Because of their light weight, pieces of plastic packing are extremely easy to fluidize and carry over, and often cause problems in condensers or downstream equipment. The author has experienced extensive carryover of plastic packing from towers that did not have bed limiters. Others (74, 144, 320) also share the above recommendation.

3. It is a recommended practice to specify bed limiters in metallic applications, although their economics are sometimes questioned. Metal packings are not as easy to fluidize as plastic packings, and neither the frequency of carryover of metal pieces nor the number of pieces carried over is as high. The author has experienced no carryover with metal packings in the same column in which he previously experienced extensive breakage with ceramic packing (in either case, no retaining device was used, while upsets occurred from time to time). The author, however, is familiar with experi-

ences in which metal packings settled unevenly, with some ending up in distributor troughs, following an upset. Another source (237) described cases of massive carryover of metal packings in a refinery fractionator that experienced pressure surges.

Bed limiters should be specified in metallic applications whenever carryover of packings may damage or seriously interfere with downstream equipment (e.g., rotating machinery), whenever surges that may induce massive carryover of packing occur, and whenever upsets are frequent. In other metallic services, the consequences and likelihood of packing carryover should be weighed against the cost of the bed limiter plus the cost of the additional column height required to accommodate the bed limiter. The author and some designers (74, 144, 299) feel that bed limiters should be routinely specified, while others (104) recommend their use when the packing pressure drop exceeds 0.5 to 0.75 in of water per foot of packing.

Note, however, that a bed limiter may provide insufficient protection against massive carryover of packing in the event of a pressure surge. A case has been reported (237) where a bed limiter disintegrated during a pressure surge. Techniques for preventing disintegration are discussed in guideline 8 below.

4. In metallic applications, a mesh fixed to the bottom of the distributor is sometimes used instead of a bed limiter. The mesh openings are made as large as possible, but sufficiently small to prevent passage of packing particles. Compared to a bed limiter, the mesh is less expensive and reduces column height requirements. It is effective for preventing packing carryover, but it is ineffective for preventing uneven settling of packing particles following an upset.

5. The retaining device must have at least the same capacity as the packed bed. It has been recommended (111, 305) that retaining devices provide an open area of at least 70 percent of the column cross-sectional area.

6. The openings of a retaining device should be small enough to avoid passage of pieces of packing. If this requirement cannot be met without an appreciable reduction in open area, and the service is nonfouling, a lightweight mesh should be installed under the plate. The author recommends routinely specifying this mesh backup to retaining devices in all nonfouling, noncorrosive services, because some pieces of packing often penetrate even through small openings. The mesh openings should be large enough to resist plugging, avoid reducing the open area, and avoid interference with liquid distribution. The mesh should be avoided in plugging services and fabricated from the appropriate materials in corrosive services.

7. The retaining device should not interfere with liquid distribution or be a good distributor. Wide beams or bars are likely to form unirrigated areas underneath, and retaining plates containing these should be avoided. Bed limiter support rings can form unirrigated areas near the column wall, and should either be avoided altogether (386) or should be of minimum width. Bed limiters are best either clipped to the tower wall or fixed to the supports of the distributor (or redistributor) above (346, 386). Combination distributor and bed limiter devices are commercially available (e.g., 212) with gravity distributors. These devices usually minimize interference of the limiter with liquid distribution and use a shared support.

8. Pressure surges may exert forces far greater than standard bed limiters are designed to tolerate, causing structural damage to the limiter and the packing (237) (systems typically prone to pressure surges are described in Sec. 8.2). It is important to ensure that the column designer is aware of the surging potential of the service, and to examine the operating procedure in order to avoid, or at least minimize, the frequency and magnitude of pressure surges. In large-diameter columns, some sections of a bed limiter may be easier to dislodge than others. Spot-welding sections of the bed limiter in place can help keeping it together (237). Another technique is to use vertical tie-rods to attach the bed limiter to the packing support grid (237; Fig. 8.8a). In one refinery vacuum tower, a common truss structure was used to support both the support plate and the bed limiter (147; Fig. 8.8b). The latter technique, however, is likely to be expensive and may only be justifiable for large-diameter columns which contain shallow packed beds.

9. With structured packings, hold-down bars are best oriented at right angles to the bricks. Bars should be arranged so that no brick is permitted to move upward.

8.4 Packed-Column Verticality

No column is perfectly vertical. Several reasons for column slant were described in Sec. 7.9. In this section, the effect of slant on packed towers is reviewed.

Since liquid flows vertically down, it preferentially flows toward one wall if the column is slanted. Some poor wetting on the upper side, as well as channeling, may also develop. All contribute to efficiency loss.

A number of pilot-scale investigations have been reported on the effect of a fixed slant on random and structured packing efficiency (162, 283, 323, 390, 428, 434). In all cases, slanting lowered packing effi-

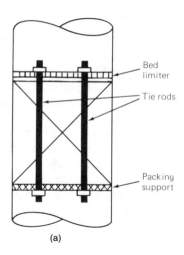

(a)

(b)

Figure 8.8   Techniques for improving bed limiter resistance to pressure surges. (a) Tie-rods for attaching the bed limiter to the packing support; (b) midspan truss supports both the bed limiter and the packing support plate. (*Part a based on* Process Design for Reliable Operation, *second edition, by Norman P. Lieberman. Copyright © 1988 (May) by Gulf Publishing Company, Houston, Texas. Used with permission. All rights reserved. Part b from Kenneth Graf,* Oil and Gas Journal, May 20, 1985. *Reprinted courtesy of* Oil and Gas Journal, May 20, 1986.)

228 Distillation Operation

ciency. Efficiency reductions in most tests were of the order of 5 to 10 percent per degree of inclination (162, 283, 323, 390, 434), but reductions as high as 30 to 50 percent per degree of inclination were also noted in some cases (283, 323, 428). One test showed a somewhat smaller efficiency reduction with structured packings compared to random packings at the same inclination (162). The tolerance recommended on slant in a column containing structured packings is about 0.2 to 0.5° (168, 428).

The effect of a variable slant, such as that experienced under barge motion conditions, was also studied in pilot-size columns (162, 323, 390, 434). It was shown that static slant can be regarded as the lower limit of column efficiency, and the motion diminishes the loss in column efficiency compared to static slant. In most cases of random motion, it was shown that the efficiency loss, compared to the vertically positioned column, is less than 10 percent (162, 390, 434).

Poor distributor performance due to motion may escalate the efficiency loss. Pilot tests (323) showed a smaller efficiency loss with pipe distributors than with orifice-trough distributors. Inclination, liquid sloshing in the distributor, and liquid overflowing the trough may all contribute to poor performance of some distributors under motion conditions.

8.5 Packed-Column Diameter Tolerances

With random packing, the tolerance on column diameter is similar to that for tray columns (refer back to Fig. 7.4). With structured packing, tighter tolerances are often specified to ease installation. Typical diameter tolerances for columns containing structured packings (168) are shown in Table 8.1.

TABLE 8.1 Typical Column Tolerances for Structured Packings

Column diameter	Diameter tolerance		Circumference tolerance	
	Wire mesh, in	Corrugated sheet, in	Wire mesh, in	Corrugated sheet, in
8 in to 2 ft, 6 in	5/64	1/8	5/64	
2 ft, 6 in to 4 ft	5/32	1/4	1/8	5/32
4 ft to 6 ft, 6 in	1/4	+ 3/8 − 1/4	5/32	3/16
6 ft, 6 in to 13 ft	5/16	+ 1/2 − 1/4	1/4	
13 ft to 20 ft	+ 3/8 − 5/16	+ 3/4 − 5/16	3/8	
> 20 ft	+ 5/8 − 5/16	+ 1 3/16 − 3/8	5/8	

NOTES: These are plus and minus tolerances; Some dimensions are rounded off upon conversion from the metric system.

SOURCE: *G.V. Horner, The Chemical Engineer Supplement, September 1987, reprinted courtesy of the Institution of Chemical Engineers, UK.*

Distillation
Overpressure Relief

A column and its auxiliary system undergo several changes through-out their lives—revamps, different feeds, changing reboilers, new product specs, not to mention changes to control system, control valve internals, control valve limiters, restriction orifices, and valves. Each one of these changes, even the minor ones, may have a large effect on column relief requirements. Failure to recognize this effect is likely to expose the column to the danger of overpressure and the possibility of a catastrophic failure. It is therefore essential for any designer, oper-ator, or engineer implementing column modifications to be familiar with the considerations involved in sizing distillation overpressure re-lief. The impact of any modification on the overpressure relief require-ments must always be carefully analyzed prior to implementing the modification.

This chapter reviews the common practices of relieving the column to avoid overpressure, outlines several considerations and factors that affect overpressure, highlights the pitfalls of undersizing and over-sizing relief devices, and provides guidelines for avoiding pitfalls.

9.1 Causes of Distillation Overpressure

Common possible causes of distillation column overpressure are listed below and illustrated in Fig. 9.1. The list is not comprehensive. Sources and possible causes of overpressure vary from column to col-umn, and each column must be thoroughly studied to determine its own overpressure sources. The list below can best serve as an initial checklist or a starting point. Overpressure in distillation columns may result from any one, or a combination of, the following conditions (9, 10, 60, 293, 351, 414).

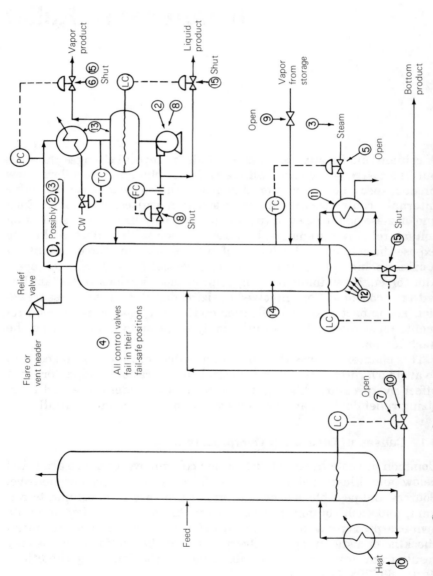

Figure 9.1 Common sources of column overpressure (numbers correspond to list in text). Note: open/shut signifies an undesirable valve condition which may induce overpressure.

A. *Utility failure*
   1. Loss of coolant (cooling   water, refrigerant, air fan, air louvers, pumparound coolers, feed cooler)
   2. Loss of electric power
   3. Loss of steam
   4. Loss of instrument air

B. *Controller failure (alternatively, human error which opens controller fully the wrong way):*
   5. Failure of steam controller
   6. Failure of pressure controller
   7. Failure of feed controller (each feed should be considered, including stripping steam)
   8. Failure of reflux (or pumparound) controller (or pump)

C. *Extraneous sources*
   9. Valve opening to an external pressure source
   10. Loss of heating in an upstream column (upstream column will dump, which will simultaneously increase feed rate and fraction of lights entering the column)
   11. Failure of exchanger (e.g., reboiler) tube
   12. Exterior fire

D. *Internal sources*
   13. Accumulation of noncondensables
   14. Chemical reaction
   15. Closed column outlets

E. *Transient sources (Including pockets of water in a hot tower, steam hammer, and internal explosions)*

9.2 Strategies for Determining Distillation Relief Requirements

Distillation relief devices are usually designed to protect the column and auxiliaries from overpressure due to any single cause (10, 414). The probability of two *unrelated* failures occurring simultaneously is considered remote and is not normally designed for (10, 414). Codes, regulations, or company policy may dictate a more conservative practice.

The word "unrelated" above is key. API's RP521 (10) defines causes to be unrelated if "no process, mechanical or electrical linkages exist among them, or if the length of time that elapses between possible successive occurrences of these causes is sufficient to make their classification unrelated." When one failure causes another, the events are related. For instance, if reboiler heat is controlled by a differential pressure controller, a cooling water failure will cause the steam controller to open fully. In this case, cooling water failure and failure of the steam controller are related, and the relief device must protect the column against overpressure when the two occur simultaneously. Two catastrophic overpressure accidents were reported (414) to result from failure to recognize the interrelation of the events in the above example (414). Another example of related events is an exterior fire and failure of a closely located air condenser. The air may be heated by the flames or the condenser may trip on automatic thermal cutoff. In either case, the fire will induce a complete cooling failure.

Distinguishing related from unrelated events requires a close examination of the events that will occur in case of each failure. The analysis yields a set of premises which defines these events. These premises must be based on correct assumptions regarding the behavior of plant and instrumentation. Credit for favorable events must only be applied for events that can be relied on to occur. This is discussed in detail in Sec. 9.4.

Once a set of premises is available for each failure, the list of possible overpressure causes can be narrowed down. A possible cause can be eliminated from the list if it is certain that its relief requirement is lower than or identical to the relief requirement of another source. For instance, when column pressure is controlled by manipulating cooling water to the condenser, failure of the pressure controller may have identical consequences to coolant failure. In this case, failure of the pressure controller can be eliminated from the list. Another example is a column whose heat-input control valve and all feed control valves fail shut, while cooling is likely to continue normally during an instrument air failure; in this case, the relief load is likely to be small (if any) upon instrument air failure, and this cause can be eliminated from the list.

Once the list is narrowed down, the relief capacity for each of the remaining failures must be determined separately. At this stage, the premises must be reexamined, and any credits must be carefully reviewed. The desired location of the relief device must be determined, since this will have some effect on its capacity. Common practices for setting relief device capacity, credit pitfalls, and preferred location are discussed in subsequent sections. Examples of the premises and calculations are presented elsewhere (293, 351). Each failure will lead to a different relief requirement. The largest requirement sets the size of the relief device.

The size of the relief device must be carefully examined to ensure that it is not excessive. A grossly oversized relief device is not only expensive, but it may cause damage to column internals, excessive discharges, and may even lead to lower discharge rates than a properly sized device. Guidelines for examining relief device size and techniques for preventing oversizing are discussed in Sec. 9.7

An alternative shortcut strategy sometimes used is to size the relief device for the gross overhead vapor from the tower (60, 351). Often, the gross overhead vapor rate at 90 to 95 percent of flood is used (60). Adjustments are made for unfavorable factors that may occur at the time of failure, while credits are taken for factors that can be relied on to reduce discharge rate. The strategy is discussed in detail elsewhere (60).

The "gross overhead vapor" strategy has been stated to be less accurate than the detailed calculation strategy recommended above (351). Perhaps an even greater disadvantage of the gross overhead vapor strategy is its inability to predict a situation where column relief requirement exceeds the column capacity. A corrective action, which may be required to reduce the relief requirement, may thus be overlooked. The gross overhead vapor strategy, however, is much quicker than the detailed calculation strategy. In most cases, the detailed calculation strategy is advocated (9, 10, 293, 351, 414).

In summary, the recommended strategy for determining column relief requirements is

1. Identify overpressure sources

2. Distinguish related and unrelated sources

3. Size the device

4. Examine the device for oversizing

9.3 Common Practices for Determining Relief Rates

Common practices for determining relief rates are described below. Good engineering judgment must be applied when considering them. Blind adherence to these (or other) practices must be avoided, as circumstances vary from column to column. As in previous sections of this chapter, these practices are best used as an initial checklist or a starting point. The relevant codes, regulations, standards, and company policy guidelines must be followed. Detailed equations (9, 293) and worked examples (293, 351) are available elsewhere.

Pressure, temperature, and composition

The relieving pressure upstream of the relief device is the set pressure plus allowable overpressure (9, 293). The allowable overpressure is set by the applicable code (e.g., 11), and depends on whether the set pressure is equal to or is lower than the allowable working pressure of the column (9). Often, the relieving pressure is 10 to 20 percent higher than the set pressure; a more detailed set of guidelines was presented by Mukerji (293).

The relieving temperature upstream of the relief device is usually the equilibrium temperature of the relief mixture at the relieving pressure (293, 351). The composition used is that of the relieving mixture. It is important to adequately account for gas compressibility and deviations from ideality. The calculations are discussed in detail by Mukerji (293) and in API's RP520 (9).

Relief vapor rate

The relief vapor rate depends on the cause of overpressure. Common practices for determining this rate for various failures are outlined below. Credits and some debits taken in the calculations are excluded from the guidelines below. These are examined separately in Sec. 9.4.

Loss of coolant.   This failure frequently sets the column relief requirements. A common practice (9) is to set the required relief capacity equal to the total incoming steam and vapor, plus that generated therein under normal operation.

Loss of electric power.   This is another failure that frequently sets the column relief requirements. A common practice (9) is to study the installation to determine the effect of power failures and to set the required relief capacity for the worst condition that can occur. All electrically driven equipment, such as pumps, compressors, and fans (including those in the site cooling water system or steam supply system), may fail, and so will electronic controllers and computer control equipment.

Loss of steam.   As for electric power (9).

Loss of instrument air.   In a well-designed system, all valves usually fail shut except for coolant and reflux valves that fail open. Unless the column significantly deviates from this practice, the relief requirements are likely to be small, if any.

Failure of steam controller.   A common practice with steam reboilers (293) is to assume the steam valve is wide open, the steam pressure in the reboiler is the same as in the steam supply line, and that reboiler area remains constant (condensate is removed as soon as it is formed). The process side is usually assumed to have the same temperature rise (outlet minus inlet) as in usual operation.

Failure of pressure controller.   When the tower pressure controller manipulates the rate of cooling, a pressure controller failure is usually equivalent to coolant failure. When the controller manipulates the vapor product rate, a common practice is (10) to set the relief requirement to the flow rate of uncondensed vapor.

Failure of feed controller.   A common practice (10) is to set the relief requirement equal to the difference between the maximum feed flow and the normal column outlet flow at relieving conditions, assuming that other valves in the system stay at their normal operating positions.

When the column is fed from a higher-pressure process unit (e.g., another column), the possibility of feed dryout (i.e., loss of liquid level in the upstream unit so that column is fed with vapor instead of liquid) also needs to be considered. The column relief requirement is commonly set equal to the peak vapor flow through the feed line, assuming the feed valve is wide open (10).

Failure of Reflux Controller.   A common practice is to set the relief requirement equal to the column internal vapor rate to the top tray. In case of a side reflux or pumparound, the relief requirement is commonly set equal to the difference between vapor entering and leaving the section (9).

Valve opening to an external pressure source.   Usually similar to feed controller failure.

Loss of heating in an upstream column.   Usually similar to feed controller failure, but with a lighter feed composition.

Failure of exchanger tube.   A common practice (9) is to set the relief requirement to allow for steam or heating fluid entering from twice the cross sectional area of one tube.

Exterior fire.   A common practice is to follow API's RP520 standard (9) and to set the relief requirement according to the amount of heat absorbed during the fire. The surface area wetted by the liquid is considered effective in generating vapor when it is exposed to fire. It is reasonable to assume (9) that the wetted surface is based on the total liquid in the bottom and in the trays, included within a height of 25 ft above grade (or above any level at which a sizable fire can be sustained). Equations for calculating the absorbed heat and further discussion are available elsewhere (9, 10).

Accumulation of noncondensables.   Usually similar to coolant failure.

Chemical reaction.   A common practice (9) is to set the relief require-ment equal to the estimated vapor generation from both normal and uncontrolled reaction conditions. Note that normal over-pressure relief will be too slow to respond and will provide insuffi-cient protection against a rapidly accelerating reaction (e.g., a "runaway"). Rapid relief before temperature and pressure rise to exponentially accelerating levels is a requirement for coping with runaway reactions (10). This relief is usually achieved by quench-ing the reaction zone by injecting cool liquid or gases; some experi-ences where quenching was used for arresting runaway reactions have been described (131, 275).

Closed outlets.   A common practice (9) is to set the relief requirement equal to the total incoming steam and vapor plus that generated therein under normal operation.

Transient sources.   The response of normal overpressure relief devices is usually too slow to provide effective protection against these sources. In some cases (e.g., a pocket of cold water entering a hot tower), the relief requirement depends on the pocket size and is there-fore extremely difficult to estimate (10). Normal overpressure relief is usually not designed to deal with these sources (10, 45).

9.4   Relief Capacity Pitfalls: Credits and Debits

Credit can be taken for a circumstance that is certain to occur during a relief situation and that will act to lower the column relief load. Credits are applied for reducing the calculated relief vapor rate, and thus the size of the relief device. Extreme caution is required in decid-ing whether credit should be taken for a given circumstance. If a credit is taken for a circumstance which cannot be relied on during a relief situation, the column may be overpressured.

The API RP521 standard (10) and Bradford and Durrett (60) present an excellent evaluation of various factors often credited, and also present pitfalls of these practices. Other guidelines and recommenda-tions have also been presented by several other authors (9, 10, 293, 351, 414). Many of these considerations, supplemented by this au-thor's experience, are summarized below. As in previous sections of this chapter, it is emphasized that these considerations must not be blindly adhered to; they best serve as an initial checklist or starting point. Codes, regulations, and company policy guidelines must be fol-lowed.

The following credits are commonly considered when sizing distillation relief capacity. Some of the items discussed are illustrated in Fig. 9.2

Instrumentation.   Standard API RP521 (10) recommends "that any automatic control valves, which are not under consideration as causing a relieving requirement and which would tend to relieve the system, (are assumed to) remain in the position required for the normal processing flow." Although this policy may appear conservative (it usually is), it is justified. Controllers are often operated on "manual"; at other times, they are tuned extremely slowly. Taking credit for controller action may lead to overpressure if the controller does not respond fast enough.

Similarly, credit is seldom taken for cutoff switches or trips that may correct a relieving situation (60, 414), as these may fail to operate. One possible exception is when a well-designed system comprising a family of (redundant) cutoff switches is installed (60, 414). This is discussed in Sec. 9.7.

Manual bypasses.   A manual bypass around a control valve may affect the column relief load. For instance, an open bypass around the reboiler control valve can increase the heat input into the column during reboiler controller failure (Fig. 9.2a).

Since the bypass can rarely be relied on to be in the favorable position at the time of failure, a conservative practice is normally followed. The bypass is usually assumed to be fully (or at least partially) open if this will increase the relief requirements; the bypass is usually assumed to be shut if opening it will decrease the relief requirements.

Upstream and downstream units.   With automatic control valves remaining in their position (above), flow through them will depend on the pressure difference between the column and the upstream or downstream unit. In several cases, a single failure may affect both the column and the connecting unit. For instance, a site cooling water failure may raise the pressure in an upstream unit (e.g., another column), which will force a greater flow from the upstream unit into the column. Similarly, an increase in pressure in a downstream unit (e.g., another column) may hinder product flow out of the column to that unit. A conservative practice is usually followed (Fig. 9.2b). The simultaneous failure of the column and the connecting unit due to a single cause is accounted for only if it is likely to increase the column relief requirement.

Operator response.   A commonly accepted time range for operator response is between 10 and 30 minutes, depending on the complexity of

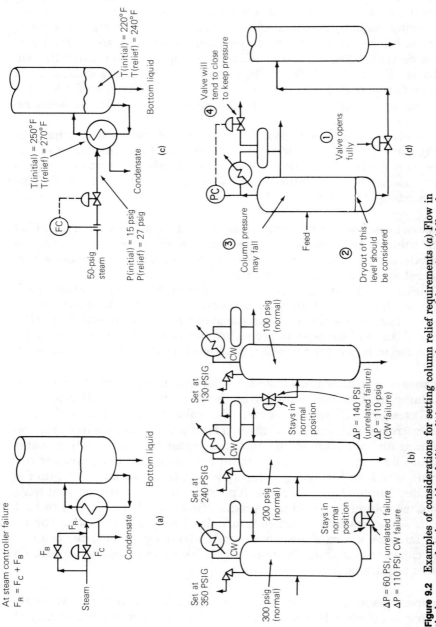

Figure 9.2 Examples of considerations for setting column relief requirements (*a*) Flow in the bypass needs to be considered; (*b*) conditions commonly assumed for setting middle column relief requirements; (*c*) No pinch will occur in this example; (*d*) considerations for feed dryout during feed controller failure (for setting relief requirement for column on the right).

the plant (10). Operator response can therefore rarely be relied on if overpressure can occur within a shorter time lag. Even when sufficient time is available, it is questionable whether the operator response will be correct and/or effective. This will depend on the operator's experience, training, the availability of detailed instructions, and human factors. In most cases, no credit is taken for operator action.

Cooling. Some cooling may continue at the time of a coolant or similar failure. The cooling that can be relied on to occur can be accounted for as credit toward the relief requirement. Each situation must be analyzed separately. Some common situations are discussed below:

1. No credit is normally taken for the residual coolant in the exchanger following a coolant failure, because this effect is time-limited (10).

2. If the process piping is unusually large and bare, some credit may be taken for the effect of heat loss to the surroundings (10). A hot, windless summer day is usually assumed in calculating this credit.

3. If the column is cooled by two or more unrelated coolants, failure of each coolant can often be considered independently. For instance, if air-cooled and water-cooled condensers share the column-cooling duty, credit for the air cooling can often be taken when cooling water fails. The above credit, however, may not apply when site power or steam failure is considered; here both the cooling water pump and the air cooler fan may simultaneously fail due to loss of power or steam. Failure of the reflux pump during a power failure may also back liquid into the condenser and stop condensation.

4. In many cases, feed subcooling provides a substantial portion of the cooling to the column. This subcooling credit is usually not taken into account because of a possible operator action. When cooling is lost, the operator may shut off the feed until cooling is reestablished. Loss of cooling and loss of feed, therefore, become related events (225).

5. With air coolers, louver closure is considered a total failure (10). Upon fan failure, or a fan drive (e.g., power or steam) failure, a credit is often taken for natural convection effects. This credit is usually 20 to 30 percent of the normal duty of induced-draft condensers. Forced-draft condensers have a considerably weaker chimney effect, and the credit taken is usually 10 to 15 percent of their normal duty. The above natural draft credit may not apply if a fire occurs near the cooler.

6. Vapor blanketing should be assumed to occur in a partial condenser if the vapor product route fails (60). For instance, if the re-

lief valve in Fig. 9.3 is located in position A or B, and the product compressor fails for any reason, inerts will accumulate in the overhead system, will rapidly blanket the condenser, and will interrupt condensation. Cooling credit, however, may apply if the relief valve is located in position C, since this would prevent vapor blanketing.

Reboiler temperature pinches.   As column pressure increases from normal operating pressure to the relieving pressure, column temperature rises. This may reduce the temperature difference (or create a temperature "pinch") in the reboiler. The pinching that can be relied on to occur can be credited toward the relief requirement. Each situation must be analyzed separately. Some common situations are:

1. The temperature of the heating medium often rises with column temperature. For instance, when steam to the reboiler is flow-controlled, steam pressure (and condensate temperature) will rise until it reaches the supply-line pressure (Fig. 9.2c). Credit can only be taken either when the steam pressure reaches the supply-line pressure, or when control valve capacity is reached.

2. When the column is reboiled by a fired heaters, temperature pinches rarely (if at all) occur (60).

3. As the column pressures up, boiling may temporarily cease or be reduced, and light components may dump into the column bottom. The bottom level will rise, and the level controller will act to open and dispose the heavy material, leaving the light material as reboiler feedstock. The expected sequence of events should be examined; if lights may migrate to the bottom of the column, less credit or no credit can be taken for pinching (60). This consideration is most significant with once-through and baffled arrangements (Sec. 4.5), because liquid from the column to the reboiler bypasses the bottom sump, and when reboiler surging (Sec. 15.4) is likely (60). If the bottom sump is unbaffled (Sec. 4.5), and its holdup lasts about 10 to 15 minutes, the effect of lights is likely to be small, and credits for pinching are often justified (60).

Heat input controller failure.   Some credits (and debits) often need to be considered when analyzing this (and sometimes other) failures. These include:

1. Allowance for a clean reboiler. The reboiler duty may be two to three times higher with a clean reboiler because of a higher heat transfer coefficient. If not accounted for, the column may be unprotected against overpressure when the heat input controller fails and the reboiler is clean.

2. A control valve limiter, a short length of narrow-bore pipe, or a restriction orifice in the steam line to the reboiler can be used for reducing the relief requirement for heat input controller failure.
However, these devices can only be relied on if it can be assured
that they stay in place and remain unchanged. A short length of
narrow-bore pipe is better than a restriction orifice (210) because it
is more difficult to remove. Alternatively, the restriction orifice can
be tack-welded to the flange of a short spool piece so that the pipe
cannot be reassembled without the spool piece (110). The instrument and safety valve registers should be marked in the appropriate locations to warn against changing these devices (210).

Decomposition.   As column pressure rises, so does column temperature. When the materials distilled are heat-sensitive, decomposition
or thermal cracking of the liquid may set in. Decomposition is likely to
yield gaseous products which behave as noncondensables. Decomposition will therefore increase the relief requirement and may reduce the
credit that can be allowed for cooling. The rate of decomposition may
be particularly high upon a heat input controller failure. This problem
is most severe when the reboiler is a fired heater or one that has a
high temperature difference.

Subcooling.   A subcooled reflux condenses some of the vapor rising up
the column. This condensation will not persist during coolant failure
because the reflux drum will either pump dry quickly, or reflux will be
reduced by the drum level controller. Credit for subcooling is not
taken toward the relief capacity when coolant or reflux fail (60).

Feed controller failure (Fig. 9.2d).   When considering feed dryout (i.e.,
loss of liquid level in the upstream unit so that column is fed with vapor instead of liquid), and if column volume is considerably larger
than the upstream unit, credit may be taken for the fall in pressure in
the upstream unit (10). Caution is required to allow for the action of
the pressure controller and the feed to the upstream unit, as both will
tend to counteract the fall in pressure.

Exterior fire.   API's RP520 standard (9) allows credit for adequate fire
insulation in its equation for heat absorbed during a fire. Credit toward the relief requirement is usually not taken for other fire-
protective equipment such as water sprays. Sprays are essential for
keeping the vessel cool, and they prevent premature bursting due to
overheating, but they are not considered reliable enough to warrant a
credit toward the relief requirements (9).

Consideration must be given to the possibility of column controls and auxiliary equipment failing during a fire. This may induce a coolant, power, or controller failure related to the fire.

Special consideration is required for exterior columns and auxiliaries that may contain unstable compounds (e.g., peroxides, nitro compounds, hydrocarbon oxides, acetylenic compounds, etc.). Here an external fire may cause overheating and polymerization, which in turn can lead to a runaway reaction and a decomposition explosion. These reactions will be related to the fire. Five major ethylene oxide column explosions caused by this sequence of events are cited in Ref. 209a. At least one involved a fatality, and in several the column was destroyed with column fragments travelling a long distance.

9.5 Locating Column Relief Devices

Relief valves, bursting disks, and major vents are best located at the top of superatmospheric columns (9, 45) (or in their overhead system), upstream of the condenser (location A, Fig. 9.3). The converse applies to vacuum services, where the vacuum-breaking device should normally be installed at the bottom of the column (192, 207).

The above strategy is usually preferred because the trays and the liquid on them may severely restrict vapor downflow toward a low-placed relief device. This consideration is most important for columns containing valve trays. The vapor downflow can generate excessive downward pressure differentials across the trays (see Sec. 11.1, guideline 2, and Sec. 11.2, guideline 2), and these can cause severe tray damage. Harrison and France (150 a) show a photograph of tray damage due to a sudden loss of vacuum from the top of a tower.

In addition, should the column become flooded, a low-placed relief valve may end up discharging large quantities of liquid. Liquid discharge may cause problems in downstream relief headers and may impede the ability of the relief valve to reduce column pressure.

When the column relief device discharges to atmosphere, a liquid discharge may be even more hazardous. In some cases, flammable liquid discharged from column relief valves caused fires (45). The flame produced when a liquid discharge fires is far longer and more dangerous than that of a gas discharge. If the liquid flashes, an explosive vapor cloud may form and detonate. For columns relieving to the atmosphere, location A is the safest and most desirable environmentally.

An alternative location, which is less commonly advocated, is at the vapor space in the bottom of the column, just below the bottom tray or packing supports (location B, Fig. 9.3). It has been argued (414) that the bottom location prevents trays from being uplifted when discharge rates are excessive, and it avoids the possibility of overpressuring the column base if the trays are heavily plugged (414). The author agrees that the bottom location has considerable merit where heavy plugging is expected. In one troublesome experience (131), pressure drop increased suddenly due to plugging, and a low-positioned relief valve blew repeatedly, thus effectively preventing overpressure. A relief device at the top to the column would have been ineffective in that case.

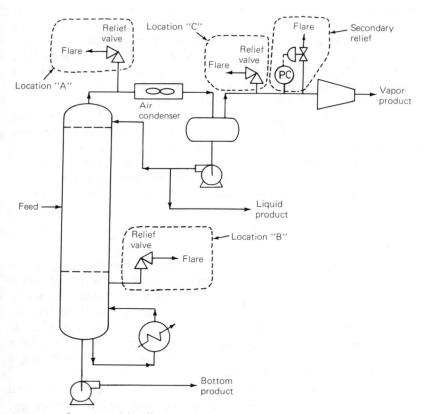

Figure 9.3 Location of distillation relief valves.

On the other hand, tray uplift prevention can (and should) be effectively done by alternative means (see Sec. 9.7), and should not be achieved at the expense of the disadvantages of a low placing of the relief device.

When a partial condenser is used, there may be an incentive for locating the relief device on the reflux drum (Location C, Fig. 9.3). This location positively assures continuous sweeping of noncondensables out of the condenser, thus maximizing condensation during the relief period. This in turn minimizes the rate requirement, temperature, and duration of the discharge. This location is best suited with air-cooled condensers, where natural draft cooling continues even during fan failure. This location must be avoided when the discharging vapor is hot enough to boil the coolant (e.g., cooling water) during a cooling water failure. The reflux drum must be large enough to avoid liquid discharge. Even then, the possibility of liquid discharge upon reflux drum overfilling (Sec. 13.3, also Sec. 13.2) remains.

The vapor product system in Fig. 9.3 is one example of a system that favors location C (60, 369). If the relief valve is upstream of the condenser (location A), the air condenser will be inert-blanketed and become completely ineffective during a power failure. A relief valve on the reflux drum (location C) will maximize condensation. Location C also offers easier access to maintenance and a shorter blowdown line.

9.6   Protection against Relief Valve Base Plugging

In some fouling services, plugging at the relief valve base may barricade the valve from the column, thereby making the valve inoperative. In one case (239), a plugged column relief valve did not lift while column pressure rose 150 psi above the relief setting; fortunately, the column did not rupture. Whenever the relief valve base may plug, it must be protected. Plugging protection is usually accomplished by a steam or an inert gas purge at the relief valve base, or by fitting a bursting disk upstream of the relief valve.

Steam or inert gas purging at the valve base may require large and costly purge rates, or may contaminate the top product. Inert gas purging may also gas blanket the condenser. A steam-purged relief valve in position A (Fig. 9.3) is best located in the overhead line rather than at the top of the column so that any condensate drains into the reflux drum. In one case (239), condensate draining back into the column during short outages induced pressure surges upon column restart. In services prone to water-induced pressure surges, using nitrogen rather than steam can also circumvent this problem.

Fitting a bursting disk upstream of the relief valve is often advocated (239), but it can be troublesome if the disk leaks. The space between a bursting disk and the relief valve may then pressurize, and the disk would not "see" a high column pressure and would not burst. To overcome this, the space between the disk and the relief valve must be continuously adequately vented to a gas disposal system (e.g., flare). Atmospheric venting of this space is often troublesome, because it may lead to a discharge of hazardous chemicals or pollutants. Valving such an atmospheric vent to prevent regular discharges is unreliable even when using a pressure gage, a pressure alarm, or an excess flow valve (210).

Other techniques sometimes used to prevent plugging at the relief valve base include (98a) using a liquid seal, steam tracing, and heating (from outside). The last two are particularly useful if the plugging can be caused by freezing. The reliability and effectiveness of any specific technique for a specific service must be carefully evaluated.

9.7   Reducing Distillation Relief Discharges

Oversizing column relief devices is commonly practiced, usually as a means of allowing for errors in the sizing estimates. Gross oversizing of these devices must be avoided as it may cause excessive depressuring rates.

The consequences of excessive depressuring rates are described in detail in Sec. 11.2. These include uplifting trays, packings, and retaining devices off their supports; flooding the column; and gas lifting of liquid ("champagne bottle" effect). Flooding and gas lifting of liquid often results in a discharge of a vapor-liquid mixture instead of vapor. Ironically, these consequences may restrict the discharge rate that oversizing is attempting to enhance. The author is familiar with the following incidents:

1. Excessive flow through a bursting disk in the overhead line blew off the top tray.

2. Excessive flow through a relief valve caused major liquid carryover, either due to flooding or due to gas lifting. The liquid caused a mess in the vent header.

3. Excessive flow through a relief valve lifted an entire bed of ceramic packings, which later came crushing down. Very few pieces of packing survived the crash.

Avoiding oversized relief devices also minimizes the environmental nuisance of a discharge, safety hazards of atmospheric discharges, and lowers the cost of the relief device and the vent system. These considerations are most important if the column contains toxic, flammable, or noxious chemicals.

The following are recommended for determining the incentive for reducing the discharge rate.

1. A flooding calculation should be performed assuming that vapor flow through the column equals the relief device discharge rate, relief pressure, and the normal liquid rate (or a higher liquid rate that may be experienced during an upset that can cause a discharge. Note that the relief valve discharge rate will be the same regardless of the cause of overpressure). If the calculation indicates that flooding is likely to occur, there is an incentive to reduce discharge rates.

2. A check should be carried out to determine whether pressure drop under discharge conditions (see 1 above) is high enough to uplift trays, packings, or packing retaining devices. If it is, there is an incentive for reducing discharge rates or increasing the uplift resistance of the trays.

3. The impact of reducing the discharge on the environment, safety, and vent header costs should be considered. If significant, there is an incentive to reduce discharge rates.

4. The list of techniques for reducing discharge rates (below) should be reviewed. Often, a simple and low-cost technique can drastically reduce the discharge rate requirement.

5. The cost incentive for reducing discharge rates is greater when the relief devices are large.

A suitable technique for reducing column relief requirements must only be implemented if column safety is not compromised. Several common techniques are listed below; in many applications, some of these may actually enhance column safety. The suitability of any of the techniques to a specific service must be examined critically and thoroughly. In examining this suitability, statutory and regulatory codes as well as company practices must be followed. The following techniques are often applied for reducing column relief requirements (Fig. 9.4).

1. *Increasing column mechanical design pressure:* This is one of the most effective techniques in cases where credit can be taken for a temperature pinch (see Sec. 9.4). When a close-boiling mixture is distilled, and the reboiler operates at a relatively small ΔT, a modest increase in column pressure can drive the reboiler ΔT to zero

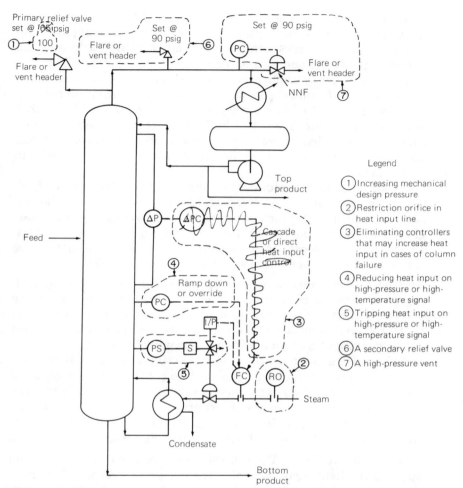

Figure 9.4 Techniques for reducing distillation relief discharges. See text for explanation.

(i.e., to a pinch). This can effect a severalfold reduction in the column relief requirement.

This technique not only lowers the discharge rate but also the probability, frequency, and duration of discharge, because the difference between the relief pressure and the normal operating pressure is enhanced. In atmospheric and low superatmospheric pressure services, a higher pressure rating of the column also permits higher throughput (80, 195 197–199) and flexibility (93) during normal operation.

The main disadvantage of this technique is that in many cases (particularly with wide-boiling mixtures), its effectiveness in reducing relief discharge rate may be limited. This technique can also be expensive when column pressure exceeds 100 to 150 psig.

However, in many applications, the reduction in vent header and protective relief equipment, together with the lower probability and frequency of discharge, can justify this technique. The author has experienced one case with a high-pressure, close-boiling system where the reduction in vent header and protective equipment cost was by itself sufficient to pay for the required increase in column design pressure.

In many instances, column shell thickness is independent of column pressure below 100 to 150 psig. In this pressure range, the cost penalty for increasing the mechanical design pressure is often marginal and can be easily justified. It is then a good practice to increase the mechanical design pressure to the maximum that can be accommodated without increasing shell thickness (93).

2. *Restricting heat input to the reboiler:* This can be achieved by adding a restriction orifice or a short length of narrow-bore pipe in the heat input line to the reboiler, or a valve limiter to the heat input control valve. Alternatively, a pressure controller can be installed in the heat input line (in addition to the normal heat input controller). Restricting heat input is most effective when heat input failure sets column relief requirement. Two cases have been reported (60, 414) where adding such a restriction orifice drastically reduced column relief requirements. Precautions are required to assure that the restriction orifice is not removed or resized in the future without resizing the relief device. Further discussion is in Sec. 9.4.

3. *Control modifications:* The control system can be modified to avoid opening the heat input control valve in case of a failure. Two incidents have been reported (414) in which the column automatic control system opened the heat input control valve during a coolant failure. In each case, the relief requirement was substantially greater than it would have been had the heat input controller stayed at its initial opening. In both cases, heat input was controlled by column ΔP (Fig. 9.4). Problems may also occur when heat input is controlled by bottom or reflux drum level.

Figure 9.5a shows another control system that can increase heat input to the reboiler if the bottom pump fails. Pump failure will interrupt column vapor flow, the column will dump, and the temperature controller will increase the furnace fuel. Unless a reliable trip system (discussed below) is installed, the furnace will overheat. In one incident (239), resumption of circulation caused rapid vaporization, which resulted in a pressure surge that dislodged

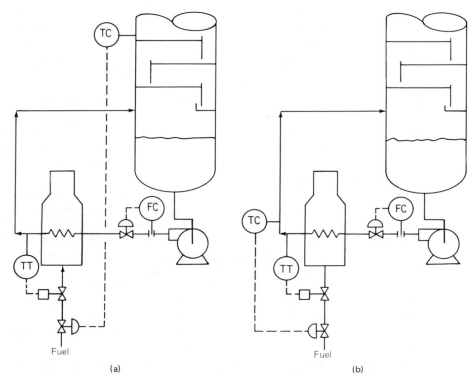

Figure 9.5 Fired-heater temperature controls. (a) Temperature controller will increase heat input upon pump failure. (b) Temperature controller will lower heat input upon pump failure.

trays. Note that this type of problem, while most likely to occur with a fired heater, can also occur with other forced-circulation reboilers. Figure 9.5b shows an alternative control system that tends to reduce the heat input upon bottom pump failure. Unfortunately, locating the temperature controller as shown in Fig. 9.5b is not always desirable for control reasons (Sec. 18.2, 18.3). In both schemes, a high-reliability trip system (on high temperature or low flow) is required to positively prevent furnace overheating, but the system in Fig. 9.5b will provide better protection should the trip fail.

Modifications of the control system to reduce heat input upon failure can help reduce the probability, frequency, and duration of discharges, but cannot be relied on for reducing the discharge rate requirements (60). In one case, such a controller was tuned too slowly to be effective (60); in other cases, it may be operated on manual control.

4. *Tripping heat input to the reboiler:* The use of a single temperature or pressure cutoff cannot be relied on for tripping heat input (60); it may not function when the failure occurs. Two incidents have been reported (239), one where a pressure trip failed to operate and avoid overpressure, the other where a temperature trip behaved similarly. The use of two temperature or pressure cutoff switches, with one of them sufficient to trip the heat input, is more reliable, but is generally not recommended because it is likely to induce spurious trips. One satisfactory system that can often be relied on for reducing the relief requirements (60, 414) includes three or more (redundant) switches, with a "voting" system. This may be expensive, but the expense can often be justified.

If a number of cutoff switches are used, they must be installed on separate nozzles and be completely independent. Otherwise, if the nozzle is plugged or the nozzle block valve inadvertently shut, all switches will simultaneously become inoperative. In such cases, they cannot be relied on to lower column relief requirements. The switches must also be regularly tested to be reliable.

The main disadvantage of even the best of the techniques of tripping heat input to reboilers is that they are generally somewhat less reliable than relief valves. They may also incur spurious trips, and at times cause unnecessarily large upsets. There are also situations for which their effectiveness may be limited.

It has been recommended (225) that if a trip system is used instead of a relief valve, it should be designed for a reliability 10 times that of the latter. Guidelines for evaluating this reliability are described elsewhere (225).

5. *Providing secondary relief at a lower pressure:* A relief device is usually sized to accommodate the failure that is expected to generate the largest relief discharge rate. This failure may rarely occur, and even when it does, quick operator intervention may reduce the discharge rate. However, once the relief device is sized for this failure, it will discharge this flow rate each time it opens.

Providing a secondary relief device, at a pressure of about 10 psi less than the main relief device, can reduce the probability, frequency, and duration of discharge through the main relief device to such an extent that it would practically never open.

The secondary relief device is usually sized to a much lower discharge rate than the primary. The secondary relief device can be a safety valve, a bursting disk, an automatic pressure-controlled vent valve, or a high-pressure switch that opens an automatic vent. If the secondary device is a relief valve or a bursting disk, credit for its rate of discharge can be taken when sizing the primary relief device.

The practice of using multiple relief devices at staggered settings is one of the most desirable applications of this technique (10). This practice reduces relief valve size, thus reducing leakage, valve chatter, and seat damage. It also lowers the reactive thrust at relief. Further discussion is available elsewhere (10).

A secondary relief device on the reflux drum may be particularly useful for sweeping noncondensables out of a partial condenser during a failure (e.g. Fig. 9.3) if the main relief device needs to be located upstream of the condenser (Fig. 9.3, location A). The principles are discussed in more detail in Sec. 9.5

If the column has a total condenser operating close to its maximum capacity, a secondary relief device, preferably an automatic high-pressure vent, is most beneficial. In one column that did not contain such a device, the relief valve lifted each time a significant amount of lights entered or accumulated in the column (239).

In some cases, a secondary relief device can discharge into a low-pressure system (e.g., fuel gas). This minimizes product loss. An automatic pressure-controlled vent is most suitable for this purpose. This technique can only be applied when the secondary discharge rate is small enough not to upset the low-pressure system. The secondary relief device must be carefully sized accordingly.

The main limitation of the secondary relief techniques is that some probability of experiencing the full relief load still remains.

6. *Miscellaneous techniques are effective for reducing the probability, frequency, and duration of column discharges:* These include improved operator instructions and operator training; smooth operation; regular testing of switches, alarms, trips, and controller action; a strategy of keeping column pressure at the minimum possible (either by judicious trimming or by automatic or computer control); provision of additional alarms, instrumentation, or computer controls. None of these techniques is considered sufficiently reliable to reduce the relief device discharge requirements. Other techniques may be effective for a specific type of failure, e.g., improved fireproofing of the column base, or installing a spare turbine-driven reflux or cooling water pump which automatically cuts in case of electric failure.

9.8   Emergency Depressuring Vents

If column metal overheats beyond its safe working temperature, the column may burst at the overheated area at pressures lower than the relieving pressure. Overheating may occur when the column wall is exposed to an external fire. The internal liquid will cool the metal and provide some protection against overheating, but it will do so only where the wall is wetted (e.g., the bottom sump or immediately above the tray). Metal in the column vapor spaces is not protected by internal liquid cooling and may overheat and rupture.

Water spray and deluge systems, as well as fire insulation, are common techniques for overheating protection. Various techniques are also used for preventing the initiation and spread of fire. These constitute a portion of the plant fire prevention system, and are outside the scope of this book. Some discussion is available in API's standards (9, 10).

One technique for overheating protection which is often incorporated in column design is the provision of a locally and/or remotely operated emergency depressuring vents. An overpressure relief valve will not depressure the column; it would merely ensure that the column relief pressure is not exceeded. If there is a danger of rupture (or a massive leak) at a pressure lower than the set relief valve pressure, the emergency depressuring vent is operated. The vent must be protected against fire, discharge to a safe place, and be operable for the duration of the emergency.

Emergency depressuring vents are commonly sized to reduce column pressure to 50 percent of the column design gage pressure within approximately 15 minutes (10, 45). This criterion is based on the resistance of column wall to rupture when its thickness exceeds or equals 1 in. The required depressuring rate depends on the wall thickness, metallurgy, and rate of heat input from the fire. Several alternative criteria for setting the depressuring rate are discussed by API's RP521 standard (10). One popular alternative is sizing the vent to reduce column pressure to 50 percent of the column gage pressure or 100 psig, whichever is lower, within 15 minutes (10, 79).

In order to achieve the required depressuring rate, the emergency depressuring vent must remove any vapor generated during the emergency. This includes liquid boiled by the fire and liquid flash due to pressure reduction. The calculation often assumes that other vapor-generation sources (e.g., feed, reboil) are ceased during the fire (10). Allowance must be made for the change in vapor density upon depressuring. The sizing procedure is detailed elsewhere (10, 79).

Column Assembly
and Preparation
for Commissioning

The column assembly and precommissioning period is most critical for assuring trouble-free operation. This is the last chance to detect any design or fabrication errors prior to startup. Flaws remaining hidden are likely to bring about poor performance and even mechanical damage. The prestartup cost and effort of rectifying flaws is often negligible, becoming enormous following the startup. In addition, new faults incurred during this period are extremely difficult to identify. Unlike the fabrication and design phases, which are usually well-documented by drawings, specifications, and correspondence, few (if any) records are kept of the assembly and preparation phase, or of differences between the "as-built" column and its drawings. If the column performs poorly, there is often a scanty basis for suspecting an assembly error. Nevertheless, a decision to shut a column down and reinspect it often hinges on this basis. A premature shutdown is extremely costly, and may turn out most embarrassing if it fails to cure the problem.

This chapter reviews the common assembly, installation, and inspection practices; outlines the preferred practices; and highlights the consequences of poor practices. It also supplies guidelines for avoiding pitfalls during assembly and preparation for commissioning.

10.1. Preassembly Dos and Don'ts for Tray Columns

Inadequate preparation of trays prior to installation may prolong the installation period and may adversely affect column performance. The guidelines below (192, 268, 274) can help avoid these problems.

1. It is important to ensure that adequately detailed installation drawings are available prior to assembly.

2. The tray manufacturer should be required to identify all parts clearly and to pack them separately for shipment.

3. Tray parts should never be removed from the crates prior to installation. Premature removal can lead to rusting, dusting, or loss of tray components. The crates should be stored in a dry, covered area.

4. Valve tray panels should never be shipped or placed "legs to legs" or "cap to cap" in order to prevent interlocking of valve units. Panels with interlocking valve units are extremely difficult to separate without damaging the valves (274).

5. Use of masking tape as flange covers should be banned. In one incident (203), erratic reboiler action resulted from a piece of masking tape left in a reboiler flange. Plastic flange covers are better, because these must be removed before bolting.

6. It is a good practice to order about 10 percent spares on nuts, bolts, or clamps in case some become damaged or lost. In some fouling or corrosive services, a larger percentage of spares is often stocked. In such services, spare trays are sometimes justified in order to minimize downtime (268).

7. Construction supervisors should be made aware of the functions of column internals and of any unique requirements of the service. For instance, if leakage is to be minimized, the construction crew should be made well aware of this need. They should also be alerted to the common installation traps that deserve specific attention (Sec. 10.9).

8. A mock-up tray installation outside the column prior to assembly is a valuable training tool for familiarizing the installation crew with tray parts and the installation procedure (268).

9. Before any work inside the column commences, it is essential to take steps for preventing small parts such as nuts and bolts from finding their way into downstream equipment, such as pumps, heat exchangers, and control valves. In one case (145), a column bottom pump frequently lost suction because a leftover piece of rope ladder reached its inlet and lodged there. Temporary plugs in the column base and some drawoffs can effectively prevent such incidents. It is important to positively ensure that these plugs are removed prior to startup.

   Alternatively (or in addition), temporary strainers can be installed in outlet lines, especially those feeding pumps. Strainers alone are less effective than plugs, because some debris can pass through strainers or damage strainer elements by impact and then pass through them. In one case (364) strainers broke due to blockage by debris, and pieces of strainer casings damaged the pumps. Strainer casings should be of adequate mechanical strength to withstand pump suction when fully blocked.

10.2 Preassembly Dos and Don'ts for Packed Towers

Inadequate preparation of packing prior to assembly may prolong the installation period and adversely affect column performance. The guidelines below can help avoid these problems. Guidelines 1, 3, and 4 primarily apply to random packings; guidelines 2 and 5 to 7 apply both to random and structured packings.

1. New packings often have a thin oil coating. The oil may be lubricants used in the packing press or an oil film used to inhibit packing corrosion during shipping or storage. When carbon steel packings are transported by sea, an oil coating is often essential (e.g., 148). This oil coating may inhibit the formation of liquid film on the packing surface, particularly in aqueous systems. Some lubricants may also cause foaming in high pH aqueous systems. The oil is also a fire hazard during hot work or hot commissioning/startup operations.

   It is important to be familiar with the nature of the oil and to adequately plan for removing it. It is best to seek the manufacturer's advice on the preferred removal procedure. Alternatively, the manufacturer can be requested to use a water-soluble lubricant in the press, which can be washed prior to startup, or to degrease the packing with solvent after pressing. Premature removal of the oil may cause corrosion, and should be avoided.

2. Packings should be stored in a dry, covered area. Packings may corrode or oxidize rapidly if left standing in the rain, or may collect dust if oil-coated. Plastic packings may also be attacked by ultraviolet trays, and should be protected from sunlight.

   Drums used for packing storage should be cleaned free of foreign material that may chemically attack packing or stick to packing surfaces and later inhibit liquid film formation, or cause undesirable effects (e.g., foaming). Oversized containers are hazardous to workers lifting them, and should be avoided.

3. Both new or reused ceramic packing should be screened to remove broken pieces. There were cases (219, 220) where up to 40 percent of the ceramic packings were damaged during transportation. Experiences of damage to ceramic packings during service have also been reported (34, 74, 145, 203, 257, 349). Figure 10.1 shows a few samples of ceramic saddles fresh from shipment. The breakage (Fig. 10.1a) consists mainly of chipping at the corners

(219). The large pieces can still be used, but the chips are likely to lower column capacity and increase its pressure drop. Screening must be performed and watched carefully, otherwise it may cause more particles to break than it removes. In one case (219), it was found necessary to pick chips out by hand. Figure 10.1*b* shows the size nonuniformity of saddles of a single nominal size from a fresh shipment.

4. A spare packing volume of about 10 percent (in case of ceramic packings, about 20 percent) should be ordered. The packing volume supplied is normally based on the volume of the supply containers. The packings are usually supplied in 1- or 2-ft^3 boxes or 25-ft^3 shipping containers; when these are emptied into the column, the total packed height may fall short of the specified height (219). The difference may be due to underfilling of boxes, unfilled space near the walls of the box, interlocking of packing particles, compression of packing when the column is filled, and dry packing techniques. The author is familiar with cases where this caused significant delays to column startup.

Instead of ordering spare packings, the user can specify that the manufacturer supply enough packings to fill the bed to the required height. The manufacturer will then allow for the spare volume in the price. If this procedure is adopted with ceramic packings, it is most

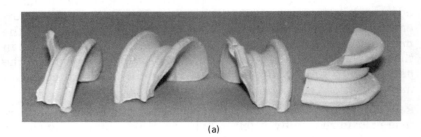

(a)

(b)

Figure 10.1 Samples of ceramic saddles fresh from shipment. (*a*) Chipped saddles; (*b*) nonuniform saddles. (*John G. Kunesh, excerpted by special permission from* Chemical Engineering, *December 7, 1987; copyright © by McGraw-Hill, Inc., New York, NY 10020.*)

important to allow adequately for any breakage occurring past the point where the manufacturer's responsibility ends.

5. When the service contains reactive chemicals, the resistance of packing samples to a chemical attack should be lab-tested under simulated process conditions. Specifically, the ability of plastic and ceramic packing to weather a chemical attack depends on their composition and texture, and these vary from manufacturer to manufacturer. In one case (349), lab tests were performed over several days on samples of ceramic packings from different suppliers under simulated hot potassium carbonate (hot pot) absorber-regenerator conditions. The tests revealed wide variations in the rate of loss of silica from the packings. With one sample, silica loss was so fast that the packing was branded unsuitable for the service. Packing deterioration in service can adversely affect packing performance or foul and plug the column and heat transfer equipment.

It is best to request manufacturers to submit samples and to have these tested before selecting a supplier. Any poor test results should be discussed with the manufacturer. A desirable (or acceptable) performance specification should then be included in the purchase order. Samples of the final shipment should be retested to ensure compliance with the specification.

6. If packing is to be installed in a column that previously contained trays, it is best to remove all internal support rings to within 3/8 in of the column shell (74, 299a). Horizontal support rings left in the column would interfere with liquid distribution and reduce the available open area, thus reducing column capacity and possibly its efficiency.

Strigle (386) states that when repacking with random packings, tray support rings need only be removed when they occupy more than 10 percent of the column cross-sectional area. Accordingly, the maximum acceptable support ring width depends on column diameter as follows (386).

Column diameter, ft	Acceptable ring width, in	Excessive ring width, in
4	1¼	1⅝
6	1⅞	2⅜
8	2½	3¼
10	3⅛	4
12	3¾	4¾

One case has been reported (346) where random packings performed well in a 13.5-ft-diameter column even though the tray support rings were not removed.

With structured packing, it has been recommended that the support rings always be removed prior to packing installation (168). Tests in a 3-ft-diameter column (285) containing 14 ft of structured packings showed a 10 percent efficiency loss and a 40 percent rise in pressure drop when 2-in support rings were left in. In larger columns, the support rings are likely to occupy a smaller fraction of the column area, and smaller pressure drop rise can be expected. On the other hand, the efficiency loss may escalate in taller beds due to distorted distribution profiles.

Vertical bars (e.g., downcomer supports) are generally far less detrimental to distribution and open area but would hinder structured packing installation. Usually, these are only removed if they are likely to interfere adversely with the new internals or with liquid flow through them (386). Some designers (299a) prefer to always have these removed.

In order to minimize dead liquid pockets and irregularities at the wall, it has been recommended to internally blind unused nozzles (299a) in the packed zone. Item 6 in Sec. 10.4 describes techniques for removing tray rings.

7. Guidelines 1 to 3 and 5 to 9 in Sec. 10.1 also apply to packed columns and their internals.

10.3. Safety Precautions for Work Inside the Column

Working in the confines of the column is hazardous. Appropriate safety precautions are required to minimize the hazards. Several of these precautions are required by law (405); OSHA standards are expected to be out soon. Some general precautions for installation, removal, inspection, and cleaning work performed inside the column are listed below. These are based on references 5, 7, 8, 131, 153, 185, 192, 210, 268, 274, 405, and 416, supplemented by the author's experience.

1. The officer responsible for safety must set the safety requirements. Before permitting personnel entry the officer must

   - Approve the work procedures and emergency plans of action
   - Check the safety equipment
   - Ensure workers are acquainted with all potential hazards and are trained in averting them
   - Ensure that work will be performed under qualified supervision
   - Be satisfied that the column is safe to enter

   The officer must be versed in the relevant statutory codes, the company safety regulations, and with the hazard and safety precautions unique to the service. The officer must not approve personnel entry unless satisfied that all hazards have been considered, all the statutory and company regulations are complied with, and all the required precautions are taken.

When satisfied that all is safe, the officer can issue a written permit for entry into the column. The permit should identify the job site, indicate date and duration, and specify all safety requirements. A qualified supervisor should accept responsibility for policing these requirements. A copy of the permit should be available at the job site and another in the control room. The permit should be valid for no longer than one shift, or at the most, one day. Extensive guidelines for the format of such permits are described elsewhere (416).

2. The column must be positively isolated from any chemicals or service lines. This includes water, nitrogen, and compressed air lines. The column must also be positively isolated from heating and cooling sources (e.g. reboiler, condenser). Isolation by valves is not satisfactory, as these may leak or be inadvertently opened. Likewise, heat exchanger tubes may leak and cannot be relied on to provide positive isolation. Any lines connecting to the column must be blinded, slip-plated, or fully disconnected prior to personnel entry. This includes any lines containing heat transfer media to and from the column heat exchangers.

   Any valves isolating the column, including those that need to be closed while blinds or slip plates are being fitted, must be locked shut with a padlock and chain or a similar device (210). It is a good practice to post a "Danger: workers in column" notice at the isolation. Blinds and slip plates should be installed as close to the column as possible, and on the column side of any isolation valves, in order to avoid trapping liquid between the blind and the column (210, 275). Liquid trapping in the bonnet of an isolation valve, which was blinded on its other side, has been reported to have caused at least one fatal explosion (275).

   Vent and blowdown lines are best fully disconnected rather than slip-plated or blinded (210), because slip plates and blinds in these lines are easy to overlook when returning the column to service.

   Good blinding and unblinding practices must be followed. These are detailed in Sec. 11.4 and in Ref. 210.

   Any electrical equipment connected to the column should be locked out and any radiation sources should be removed or shielded. The mere removal of fuses is unsatisfactory, as fuses can easily be (inadvertently) replaced. Ignition sources should be removed from the vicinity of the column. Any exceptions (e.g., electric lighting equipment) should be carefully examined and clearly stated on the entry permit.

3. Emergency procedures for evacuating people from inside the column and for rescuing any person injured inside must be prepared and approved by the officer responsible for safety. The work crew should be well-trained in implementing this procedure. First aid and standby medical aid should be available.

4. The atmosphere inside the column must be analyzed to contain 19.5 to 23.5 percent oxygen, less than 10 percent of the lower explosive limit (LEL) combustibles, and less than the threshold limit value (TLV) of each toxic chemical (185, 405) prior to column entry. These oxygen and combustibles limits are expected to tighten in the future (185).

   Care should be taken to ensure the analysis is correct. The sample should be taken from inside the column (e.g., using a sampling tube) and not from near the manhole. The testing should be repeated at regular intervals to ensure that a safe atmosphere is being maintained inside the column. The analytical instruments must be well maintained and frequently tested.

   Adequate ventilation and proper and reliable air supply to all work spaces must be assured whenever people are inside the column. Any unreachable pockets should be assumed to contain the normal column chemicals until inspected and/or tested (210). Any deposits should be carefully examined, as they may release adsorbed gas. If such deposits are found, they should be kept wetted, inerted, or waterlogged until cleaned.

   Ventilation is usually supplied by a blower or fan, forcing air from the bottom up. Exhaust (induced-draft) fans generate the risk of a combustible mixture passing through the impellers, and should therefore be avoided. Air intake must be from clear atmosphere. Air supply must be continuously maintained while people are inside the column.

5. "Sticking a head in" can be as hazardous as column entry (210), and must only be permitted when entry is permitted.

6. Any materials (e.g., small quantity of solvent for weld-testing) deliberately introduced into the open column, as well as their containers, should be critically examined. Their evaporation may generate a hazardous atmosphere inside. The possibility of container leakage or breakage should be considered and adequately allowed for.

7. The appropriate protective clothing must be worn inside the column at all times to avoid injury. Setting protective clothing requirements should consider hazards of toxic, corrosive, and flammable substances, as well as the presence of dust and other irritants.

8. Adequate and reliable lighting must be provided inside the column. Alternatively, workers should be supplied with easy-to-carry and powerful flashlights. Where relevant, the lighting equipment should be suitable for use in combustible atmospheres.

9. A proper communication system must be devised. A three-way system with a connection to the ground, tower manhole, and work area inside the tower is considered most efficient (274). A person must never be inside the column unless another person stands by at the manhole. The standby person must always be aware of the exact location of the person inside, and be trained in emergency procedures.

10. A means must be provided for ascending or descending from the manholes to any level at which internals are installed.

11. The routine of internals installation or removal must be planned so that the risk of injury to workers ascending or descending through the shell or manways is minimized.

12. The risk of accident due to dropping nuts, bolts, or tools must be minimized. This is most important when two or more crews are working in the column at the same time at different levels. Loose articles in the vicinity of the upper manholes must be eliminated. Ropes and cables going into the tower must be secured in place. Guideline 10 in Sec. 10.5 elaborates on the preferred practices when more than one crew works inside the column.

13. Work crews must be warned to support themselves on the support beams rather than on tray sections if the column has been in operation in corrosive service. Some of the corroded tray sections may have not retained sufficient mechanical strength to support the crews.

14. Installation of bottom-removable trays can put workers in danger of poor foothold and long fall. The use of movable platforms to minimize this risk has been recommended (268).

15. If welding is to take place inside the column, the officer responsible for safety must ensure absence of any flammable materials, and compliance with any additional requirements (e.g., presence of fire extinguishers). He should also ensure that personnel entry to areas affected by sparks is properly barred. In addition, he should spot-check the welding to ensure that smoke and fumes do not become too thick to be hazardous. When satisfied that all is safe, the officer can issue a written permit for hot work inside the column. The permit should be valid for no longer than one shift, or at most, one day.
Tests to ensure that the atmosphere near the welding area re-

mains nonflammable and nontoxic should be carried out before and while welding is in progress.

Any deposits must be thoroughly cleaned from the welding area, because they can decompose and release toxic or flammable gases upon heating. Columns containing heavy oils or tars can rarely be fully cleaned and welding inside or outside them is hazardous unless special techniques (210) are used.

16. Any heavy lifts should be completely avoided, or at least minimized. Heavy lifts can be exceptionally hazardous in the narrow work area inside the column.

17. A small movable bulkhead is required inside the column (274). It is set on a support ring above the heads of the workers. The workers should shelter under the bulkhead while tray pieces are lifted or lowered, in case the line fails.

18. A safety line must be in place at all times (274).

19. Whenever entry is not authorized (e.g., before column atmosphere is tested), the manholes should be blocked with barriers to ward off unauthorized entry (210).

20. The work crew must be strongly warned against joking about bolting the column up or about an emergency when someone is inside. The author is familiar with one case when such a joke caused injury to a worker as he was rushing out.

21. Consideration must be given for the provision of adequate and accessible sanitary facilities. This is most important for work on tall columns, and where the work is performed by multiple work crews. The author knows of one case where lack of such facilities was troublesome.

10.4 Removal of Existing Trays and Packings

Inadequate removal of existing trays or packings can prolong the shutdown, mechanically damage equipment, and endanger workers. The guidelines below can help minimize the above problems:

1. Data should be collected on the likely conditions of the internals, such as degree of fouling, corrosion, toxicity, and potential troublespots. This information can be gathered from previous inspection reports or by questioning personnel operating similar columns. This information is essential for preparing the proper tools, procedures, work and personnel schedules, and safety equipment.

2. Dimensions of tray panels and other internals should be carefully reviewed to determine whether they can be removed through

the manhole. If not, the old trays may need to be burned out. In one case (336), the need to oxy-cut tray parts extended the time required from tray removal in two large towers from 5 to 9 days.

If the old trays are to be burned out, some special safety equipment (e.g. good forced-air circulation for removing fumes, fire extinguishers inside the column, explosivity monitoring, water flooding, tarps) may be required. Measures may also be required to prevent insulation damage due to overheating. If the tower wall is fitted with a liner for corrosion prevention, some flammable material may be trapped behind the liner, and burning trays out is best avoided.

3. If the deposits inside the column are likely to be pyrophoric, all trays should be covered with water before dismantling. The water level should be progressively lowered to just below the tray that is being dismantled. A newly exposed tray and column wall section should be immediately cleaned.

4. Special safety precautions are required when removing damaged or bumped trays. Usually, pieces are left hanging and even those in place would not support a worker (274).

5. In corrosive services, it is a good idea to have various sizes of sockets on hand (274), as several bolt heads will be corroded.

6. If the old tray support rings and downcomer support bars are to be removed, it is best to cut them approximately ⅜ to ½ in from the shell (74, 168, 274, 299a, 336, 386). Complete removal (i.e. grinding flush) is time consuming, expensive, and rarely justified.

Support rings are usually hot-cut (see guideline 2 above for precautions), but cold-cutting techniques are also used. In one case (334), cutting tray rings down to 9/64 in using plasma-arc machinery consumed roughly 1.5 labor-hours per foot of ring or bolting bar removed. This figure includes tray deck removal and is typical for refinery towers about 20 to 30 ft in diameter. The labor-hour consumption is likely to escalate if hot-cutting is performed in the crammed conditions of a much smaller column. In another case (342), cold-cutting with a high-pressure abrasive water jet gave superior cutting quality and speed compared to hot-cutting. With water-jet cutting, column out-of-roundness may lead to an uneven cutting distance between the jet nozzle and the column wall (342).

If new support rings are to be installed, they can be installed about 1 in above the old rings; new downcomer support bars can be notched around the old downcomer support bars (274).

7. Some experiences have been reported (274) on blowing trays out by explosives when trays are highly plugged or coked. Little information has been published about the effectiveness of this technique. It is infrequently used and not always successfully (274).

8. High-pressure water jetting and hand-cleaning are the common methods of cleaning trays. Water jetting is reported to work well with sieve and valve trays, but not with bubble-cap trays (274). It is also reported to work well for cleaning the shell and support rings when the trays are removed (274).

9. Random packings are usually removed from the column by opening a manhole or handhole located right above the packing supports and letting the packing pieces roll into a collection drum. This operation must be performed carefully with ceramic or carbon packings to avoid breakage.

10. To speed up removal of random packing, suction equipment is often connected to the manhole (or handhole) via a wide flexible tube. It is important to ensure that this operation does not damage packing particles and that the impact of packing particles does not damage the tube. This operation is most commonly practices with plastic packings and should be avoided with ceramic or carbon packings where it may cause breakage.

11. When randomly packed columns are constructed of flanged sections, packing removal is usually performed by dismantling the top distributor (or redistributor) and retaining device, then lifting the packed section (including the support plate) and tipping it over. Extreme caution is required when the column contains ceramic or carbon packings, as packing pieces may break if tipped from heights exceeding about 2 ft.

10.5. Tray Installation

Correct tray installation aims at minimizing installation errors and installation time while following safe installation practices. Installation safety was discussed earlier (Sec. 10.3). Practices prior to installation which help achieve the above installation objectives are in Sec. 10.1. Practices which help achieve the above objectives during the installation period are described below.

1. The design or commissioning engineer should maintain close contact with the work crews. Often an installation problem is solved in a manner that can adversely affect column performance due to lack of proper communication.

2. The design or commissioning engineer should carry out periodic spot checks to ensure that trays and components are correctly installed, and inform the construction supervisors of any features that can adversely affect column performance.

3. Rough handling of valve tray sections must be avoided, as it can damage valve legs. This can cause valves to stick shut, stick open, or fall out when the column is placed in service.

4. The length of downcomer panels should be carefully adjusted to set the required clearance under the downcomer. One useful technique for achieving this (274) is having wooden blocks cut to the required dimension to act as spacers.

5. Downcomer panels should be installed on the wall side of the downcomer support bars, so that the weight of the liquid tightens the joints, and so that bowing out of the downcomer panel is minimized.

6. Any work that can be carried out on the ground is easier and safer than inside the column. It is therefore best to preassemble each piece on the ground as it comes out of the crate. Clamps, seal plates, washers, nuts, and bolts should be thus preassembled to each part (274). Care should be taken to ensure correct preassembly.

7. Keeping the preassembly line ahead of the installation can minimize the time spent by the installation crew inside the column and can speed up the installation (274). As soon as one tray is in the tower or at the manhole, the next tray should be hoisted to the platform and another tray should be preassembled on the ground, ready to hoist. Parts of only one tray should be passed on at a time, so that when the tray is installed, no pieces are left.

8. Major support beams should be installed first. The top of the beam should be flush with the tray support rings. Shims are often added to accomplish levelness.

   Generally, tray installation is started at the bottom and proceeds upward. When passing tray pieces to the work area, they should be sent in the order they go in. The preferred order (274) is downcomers first, underdowncomers next, followed by side-tray floors, center-tray floors, and finally the manways.

9. Bolts and clamps that fix tray panels to support beams should be fastened before those that fix the panels to the support ring. This minimizes the impact of support ring out-of-levelness on tray levelness. The manway should be temporarily set in place before tray panels are fastened in order to ensure it adequately fits in.

10. In large-diameter columns, when shutdown time is critical, and if manhole location permits, two tray installation crews are often used simultaneously inside the column. This practice suffers from the hazards of falling objects and/or poor ventilation and should be avoided whenever possible. If it has to be used, precautions are required for minimizing the risks.

Usually, a wooden bulkhead is installed just below the middle manhole and securely wedged in position. The first tray above the manhole is then installed. If no manhole is available between the middle and bottom manholes, half the bulkhead is then removed to enable ventilation and passage of tray pieces to the section below. Trays are then installed from the bottom up by one crew and from the middle up by another crew in the usual manner. If an intermediate manhole is available between the middle and bottom manholes, the entire bulkhead is best left in place, and tray pieces passed to the section below through the intermediate manhole. Having the entire bulkhead in place improves personnel protection from falling objects, but requires separate ventilation for each column half.

11. If seal welding of a tray or part of a tray is required, it is best to weld after tray installation is complete (274). However, this requires sufficient tray spacing (at least 24 in) and good ventilation to disperse fumes.

10.6. Dry versus Wet Random Packing Installation

A column can either be wet-packed (Fig. 10.2a) or dry-packed (Fig. 10.2b,c). When the column is wet-packed, it is filled with water following the installation of the bottom support plate and the bottom stacked rings (if any). Packing is gently poured from just above the water level and is "floated" to the bottom. The water cushions the fall and promotes randomness of settling. When the column is dry-packed, it is filled by dumping the packing from a certain height above the top layer of packing.

Wet packing minimizes breakage, compression, and mechanical damage to packings and supports. Wet packing also maximizes randomness and column capacity and minimizes the required number of packing pieces and the column pressure drop. In Billet's experiments (38) with 1½ -in pall rings in a 20-in-diameter column, changing the packing technique from dry to wet increased column capacity by about 5 percent, reduced the number of packing particles by about 5 percent, lowered column pressure drop by up to 10 percent, and had negligible effect on efficiency. The author also observed a 5 percent capacity improvement when the packing technique of a 3-ft column was changed

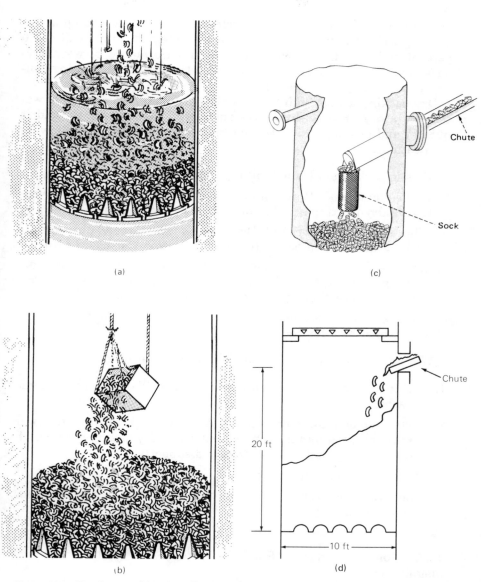

Figure 10.2 Random packing installation techniques. (*a*) Recommended wet-packing technique; (*b*) recommended dry-packing technique; (*c*) the chute-and-sock method of column packing; (*d*) packing technique promoting hill formation. (*Parts a and b reprinted courtesy of Norton Company. Part c Gilbert K. Chen, excerpted by special permission from* Chemical Engineering, *March 5, 1984; copyright © by McGraw-Hill, Inc., New York, NY 10020.*)

from dry to wet. Others (257) reported cases in which changing the packing technique from dry to wet lowered pressure drop by 50 to 60 percent and even more. The difference in performance between dry-packed and wet-packed beds is attributed to the greater packing density generated by dry packing.

Dry packing avoids high hydrostatic heads and introduction of water into a dry process, minimizes rusting of metal packings, and is quicker and less expensive. Dry packing is preferred in the following applications:

1. Plastic packings. Plastic typically floats on water.
2. Large-diameter (>10-ft) columns. Here the cost and speed advantage of dry packing are the controlling considerations.
3. Where downtime is critical. Dry packing is faster, and also permits packing installation before the column is ready for leak testing.
4. When only the upper sections of a column contain random packings. The expense of filling the entire column with water in order to wet-pack the upper sections is rarely justified.
5. When either the presence of water, or some corrosion to the packing, cannot be tolerated.

Wet packing is preferred in the following applications:

1. Packings constructed of breakage-prone materials, such as ceramic or carbon.
2. Small-diameter columns (<2 to 3 ft), where it is difficult to lower boxes of packing into the column and where dry dumping from high elevations may damage packings and supports.
3. Where column capacity is critical.
4. Where there is no clear advantage to dry packing. Wet-packed beds perform better.

10.7. Dos and Don'ts for Random Packing Installation

Poor packing installation may cause maldistribution, excessive pressure drop, and loss of efficiency, capacity, or both (34, 257, 303, 439). Poor packing installation may also damage packings and supports. It is therefore imperative that packing be properly installed.

Good column performance can be achieved with either the wet-packing or the dry-packing technique. Both techniques have pitfalls. Failure to avoid these traps, rather than a poor choice between the wet- and dry-packing method, is usually responsible for column mal-

functions. Some guidelines for avoiding these pitfalls are described below. Guidelines 1 to 3, 7, 8, 10, 13, and 14 apply to both dry and wet
packing; guidelines 4 to 6 and 9 apply only to dry packing; while
guidelines 11 and 12 apply only to wet packing.

1. The manufacturer's advice on packing procedure should be
   sought, reviewed, and followed. Any deviations should be discussed and agreed on with the manufacturer.

2. Formation of "hills" of packing (i.e., buildup of layers sloping
   away from the dumping region) must be avoided. Packings should
   be spread as evenly and as randomly as possible. One experience
   has been reported (34) where forming a "hill" caused severe
   maldistribution and lowered efficiency and capacity. Figure 10.2d
   illustrates the technique used in that case. Pilot-scale experiments by Billet (38) also indicate that hill formation during packing installation leads to a column efficiency lower than when the
   packing is uniformly spread.

3. When building each horizontal layer of packing, it is best to start
   from the edges and work toward the center (257). This is believed
   to reduce the tendency of liquid to accumulate at the tower wall
   (257).

4. Pieces of packing should not be pressed into place. Pressing the
   packings may increase bed density and may cause maldistribution
   and high pressure drop.

5. In large-diameter (>4-ft), dry-packed towers, it is often necessary
   for workers to stand on top of the bed to spread out and level the
   packing pieces. The workers should not stand directly on the packing. Plywood, or other rigid material, with an area of 4 ft^2 or
   greater should be used to stand on (74). With plastic packing, it is
   best to avoid this practice altogether.

6. Care should be taken to positively ensure that plywood and other
   foreign objects are removed and not buried in the bed. These have
   caused trouble in the past (74).

7. One frequently recommended technique is to dump the packings
   from buckets lowered into the column (Fig. 10.2b). These buckets
   should be emptied at several points, starting from the edges. In
   some instances, a worker is lowered into the column with the
   bucket and pours the packings out at a height of 6 to 10 ft above
   the top of the bed. Another variation of this technique (302, 320) is
   to construct a simple frame that will hold a shipping carton of the
   packing, with a trip cord attached to the frame. The carton is lowered into the column, and the packings dumped onto the bed by
   tripping the cord.

8. An alternative technique often practiced (74, 257) is the chute-and-sock method (Figure 10.2c). A sheet metal cylinder or cone (the "sock") is used to dump the packing at several points, starting at the edges. This method evenly spreads the packing, and breaks the packing fall height. Although this technique may be somewhat inferior to that in Fig. 10.2b, it is faster. This technique is often preferred in large columns (74).

9. For dry-packed towers, pieces of packing should not be allowed to fall a great distance onto the bed surface. With ceramic or carbon packings, the fall height should not exceed 2 ft (74, 257, 302, 303, 320). With metal and plastic packing, greater fall heights can be tolerated. Fall heights not exceeding 10 ft have been recommended for metal and plastic packings (303, 320); the author also prefers this practice. One source (74), however, feels that with metal packings, a fall height of up to 20 ft may be permissible.

Special care is required when introducing the first 2 to 3 ft of packing into the bed. Excessive fall height can damage the support plate. Special care is also required when plastic packing is loaded in near-freezing conditions. The impact resistance of plastic often drops at low temperature, and it is prone to breakage; the fall height should not exceed 4 ft under these conditions. In one case (206a), about 20 percent of the polypropylene rings were damaged when dropped from a height of 23 ft onto a steel support at about −15°F.

10. Dropping packings directly from a column manhole should be avoided, as it is likely to promote hill formation and excessive fall heights. Instead, the techniques suggested in guideline 7 or 8 above should be used. One source (74), however, feels that dropping packings from the manhole may be permissible in small-diameter (< 6-ft) columns as long as packing spread is visually inspected to avoid unevenness and as long as excessive packing fall heights are avoided.

11. Wet packing of towers requires that column shell and supports be designed to withstand the full hydrostatic head.

12. For wet-packed towers, at least 4 ft of water should be kept above the surface of the packing at all times (302, 303). Preferably, the water level should be up to the loading manhole. The former technique should be applied in combination with guideline 7 above.

13. When different packing sizes or types are used in different column sections, precautions must be taken to avoid interchanging.

14. When loading metal or ceramic packing into a tower, the ground area surrounding the tower should be out of bounds for personnel

entry and should be wired off. Falling packing particles are hazardous; the author is familiar with one accident where a falling
piece of sharp-edged metal packing caused a serious injury to a
passer-by.

10.8 Some Considerations for Structured Packing Installation

Relatively little has been published in the literature about preferred
installation practices for structured packings. The author has heard
several horror stories about poor performance due to heavy-footed stepping on structured packings, or failure to achieve snug fit of packings to
the tower wall, or excessive compression of some wire-mesh packings in
order to achieve a snug fit. The author is also familiar with one case of
poor column efficiency resulting from a wide annular open gap left between a structured packing and the column wall.

The following procedure has been recommended (168) for structured
packing installation.

1. Any welding of internals such as packing supports, redistributors,
   etc., is best completed before packing is installed (168). This eliminates possible packing damage or fouling by welding slags. If this
   is impractical, the packing and internals beneath the welding region should be shielded using tarpaulins and the like.

2. Sections of the packing support grid are best fastened only loosely
   until all parts have been assembled. This permits adjustment of the
   overlap of the rim of the support grid and the support ring prior to
   final clamping and tightening (168).

3. The packing is usually supplied in sections, or "bricks." Typical
   brick height is 8 to 12 in. A brick can be as long as the wall-to-wall
   distance (389) in smaller columns, but to minimize the possibility
   of damage it is better to have a number of bricks make up this distance in the central portion of the column (Fig. 10.3). The bricks
   should be lowered into the column using ropes. Metal hawsers and
   chains can damage the packings (168).

4. Any wall wipers should be bent outward prior to bricks placement
   (168).

5. A typical procedure for installation of the packing bricks (168, 342)
   is depicted in Fig. 10.3. The bricks are fitted in starting from the
   sides and using a slide plate (or "shoe horn") to fit center pieces. In
   the Fig. 10.3a example, brick 1 on the left is first inserted. Bricks 2
   and 3 on the left are then inserted and pressed against brick 1.

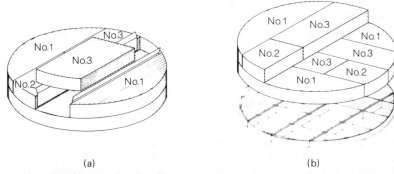

(a)   (b)

Figure 10.3 Structured packing installation. (*a*) Installing "bricks" within a layer; (*b*) layer orientation in the column (*G. V. Horner*, The Chemical Engineer Supplement, *September, 1987, reprinted courtesy of the Institution of Chemical Engineers, UK.*)

Brick 1 on the right is then installed, and finally bricks 2 and 3 on the right are inserted.

Column out-of-roundness or other irregularities may make it necessary to reduce or increase the width of some center blocks. This is accomplished by removing or adding a few strips of the packing material (342). The author is familiar with cases where strip removal was needed due to a tight fit.

6. The bricks of the next layer of packings have flow channels running at 90° to those in the layer beneath (Fig. 10.3*b*). Otherwise, the installation procedure is identical.

7. Tight fit of structured packings may result in a "growth" of the overall packed height; in one case (342), a 3-inch growth occurred. The additional height is often sawed off. The manufacturer's advice should always be sought, as sawing may damage the packing and cause premature flooding.

Because of the limited amount of knowledge publicly available, it is best to have the manufacturer supply an installation team. If impractical, the manufacturer should be requested to provide detailed installation instructions, which must be strictly adhered to. The manufacturer should also be requested to supervise the installation, or at least provide thorough hands-on training to the installation supervisors. Traps mentioned above should be considered, and, where relevant, steps should be taken to avoid them.

10.9. Prestartup Inspection

In general, any detail in the column that differs from the design and fabrication drawings may develop into a potential source of trouble.

The considerations covered in the previous chapters can provide a basis for evaluating the effects of departures from the original drawings on column performance. The discussion below emphasizes some common traps that deserve particular attention, either because of the severe consequences of overlooking them, or because of their high regularity of occurrence in practice.

The discussion below excludes column inspection for identifying corrosion, fatigue, and conditions that lead to deterioration of equipment mechanical integrity. These inspections should be performed separately by personnel specifically trained in these disciplines. The inspection emphasized here should be carried by the operations or process personnel.

10.9.1 Strategy and timing

The earlier a fault is discovered, the less expensive and less time consuming it is to rectify it. It is therefore recommended that the inspection be carried out simultaneously with the assembly of new internals. Field experiences which stress the importance of this strategy have been reported (295). Similarly, an existing tower should be inspected as early as possible after an entry permit is issued, and when feasible, before cleaning and disassembly (286). This has the added advantage of supplying data on fouling, corrosion, and internal damage. A second, briefer inspection may be required after cleaning and maintenance.

It has been recommended (286) that tech service personnel are most suitable for performing inspections. The author shares this recommendation. Besides helping to balance the turnaround workload (operations personnel being usually the busiest during these periods), tech service engineers generally have a good feel for equipment fluid flow and mass transfer considerations and for consequences of deviations from design dimensions. Perhaps even more important, column inspection is a most invaluable training exercise, and provides tech service engineers with experience essential for improving their design skills.

One of the best tools for column inspection is a checklist. Table 10.1 illustrates the type of checklist preferred by the author for new internals. For each satisfactory item, a tick is entered, for each unsatisfactory item, a number is entered and a comment is spelled out on a separate sheet. Items not applicable are left blank. Design values and relevant tolerances are entered in the right-hand columns prior to inspection. Any alternative checklist can be satisfactory as long as it is sufficiently detailed and the user is comfortable with it.

In a revamped or modified column, it is important to critically examine the modification areas and to identify the liquid and gas pas-

TABLE 10.1   Column Inspection Checklist

Column	Inspector									Date	
	Top								Bottom		Design
Tray no.	1	2	3	4	5	6	7	8	9	10	Value ± Tol
Orientation of internals Passage obstructions Possible impingement Instrument location											
Downcomer clearance Weir height Downcomer width Hole diameter No. of holes/valves Valves secure											
Tray levelness Tray spacing Tray/DC materials Fastening materials Trays/DCs secure Bolting firm Absence of debris Internal damage											

Inlet weir/seal pan
  width
Inlet weir/seal pan
  depth
Weep hole area
Weep hole diameter
Riser-hat clearance
Riser diameter
Riser number
Gaskets: tightness
          cutting											

Notch width Notch depth Screen size Others											

sageways. At least two cases have been reported (12, 57) where following a modification, there was no route open to the liquid to reach the section below.

10.9.2   An inspector's "survival kit"

Miller (286) has proposed an "inspector survival kit." The author wishes he saw this prior to some of his column inspections, and added one last item to the list. The list includes:

- A high-intensity flashlight.
- A bound pocket notebook and two pencils and pens.
- A short (6-in) steel ruler and a 6- to 10-ft tape measure. The ruler can also be used as a deposit scraper.
- A crayon marker. Yellow shows up best. The marks should be waterproof.
- Sample envelopes for sampling deposits.
- An unbreakable pocket mirror for looking around corners.
- An autofocus pocket camera. Photographing flaws, critical equipment items, and corroded and plugged regions provides management with factual information essential for immediate decision making. It also supplies a handy record for future troubleshooting.
- Templates for measuring repetitious internals.

10.9.3 Materials of construction

The importance of specifying the correct materials of construction was discussed in Chap. 7. It is equally important to make sure that each internal is installed according to these specifications. If lower-grade materials are arbitrarily substituted (even in only one section of the column), severe corrosion problems may occur. It is not uncommon to find that all internals except a very few are fabricated from the correct materials, but these few are usually sufficient to cause problems.

Particular attention must be given to nuts and bolts. Often, those used are of an inferior grade. These corrode in service and cannot be undone when required. One experience illustrating the above comments has been reported (295).

If different materials are used in different sections of the column, special care must be taken to ensure that no interchanging of materials occurs between sections. If the column has previously been in service, its internals should be inspected for corrosion and operation damage. The materials of any parts that were changed should also be checked.

10.9.4 Measurement of internals

All internal dimensions must be checked, but special emphasis must be given to ensure that:

1. Dimensions of small magnitude, such as clearances under downcomers, weir heights, notch dimensions, seal pan widths, distances between downcomers and inlet weirs, downcomer widths (for narrow downcomers), clearances between chimney tray or redis-

tributor risers and their hats, have been correctly set. In each of these examples, small errors can cause large deviations in flow area, which may severely restrict or channel the fluid flow. Clearances under downcomers, in particular, must be closely checked, since incorrect settings of these distances are among the most common assembly errors (85). In one incident (150a), clearances that were specified as 1 in were installed as ⅝ to ¾ in. The narrow clearances trapped scale and debris; this in turn led to premature flooding.

2. Narrow openings, such as clearances under the downcomer, seal pan widths, and distances between the downcomer and inlet weir, have not been blocked on any tray. Often, a fabrication error will cause this. In one case (206b), an inlet weir intended for the top tray was located five trays below, effectively closing off the downcomer.

3. Tray hole area is consistent with the design value. Care should also be taken to ensure that the holes are of consistent diameter, and that blanking strips (Sec. 6.4) are correctly positioned. The author is familiar with a case where one tray with a grossly diminished hole area was enough to prematurely flood an entire column. Hole, riser, and/or weep hole areas should be checked in a similar manner in distributors, redistributors, parting boxes, and chimney trays.

4. Dimensions around the bottom seal pan area are correct. Checks in this area are often difficult to implement because access to the bot-

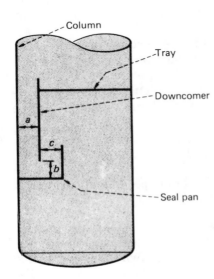

Figure 10.4 Clearances to be checked for a typical bottom seal-pan arrangement (*Henry Z. Kister excerpted by special permission from* Chemical Engineering, *February 9, 1981; copyright © by McGraw-Hill, Inc., New York, NY 10020.*)

tom seal pan is usually awkward. Figure 10.4 illustrates the clearances that must be checked (*a, b,* and c). It is also important to check that the downcomer plate going into the bottom seal pan is solidly supported. If this plate bends under liquid force, liquid flow will be seriously impeded and flooding of the bottom tray may result.

5. Screens, mist eliminators, packing support and hold-down plates, bed limiters, and vortex breakers have the correct opening size. The author is familiar with one experience in which a mist eliminator with excessively large openings was installed and was ineffective for checking entrainment.

6. Items where levelness is critical are indeed level. As a general rule of thumb, all items that contain notches and all gravity distributors and parting boxes must be thoroughly checked. It is a good practice to also check levelness in pressure distributors. Since tray levelness is normally less critical on one- and two-pass trays (Sec. 7.9), and tray-by-tray levelness checks are tedious, it is often satisfactory only to spot-check their levelness.

Figure 10.5 depicts the tolerances recommended in earlier chapters for various tray and column parts. All these should be carefully checked. The inspector should be aware of the possible adverse effects of departures from the recommended values; those were described in sections discussing the relevant parts.

Detailed measurement of internals can be a long, tedious process in tall trayed columns. One recommended shortcut is to use templates for measuring repetitious internals (e.g., clearances under downcomers). Another shortcut which is often satisfactory is to carry out detailed measurements on the first 10 to 20 trays, and then carefully eyeball the others with periodic spot checks. However, it is important to ensure that the dimensions of all internals on all trays are at least carefully looked at. There have been cases (e.g., 102) where irregularities on a single tray were sufficient to bottleneck an entire column.

Measurement of internals is particularly important for those dimensions that change from one section of the column to another. Assembly errors can often result in internals being interchanged. If these are not detected, severe hydraulic problems can arise. One experience has been described (295) in which the rectifying trays were interchanged with the stripping trays.

If the column was previously in service, and no changes were made, internals measurements need only be spot-checked to ensure absence of corrosion effects.

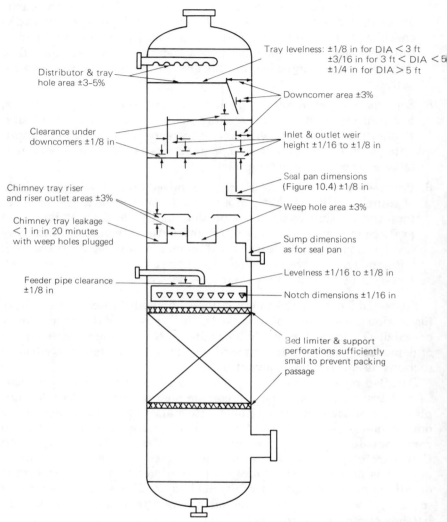

Tray levelness: ±1/8 in for DIA < 3 ft
±3/16 in for 3 ft < DIA < 5
±1/4 in for DIA > 5 ft

Distributor & tray hole area ±3-5%

Downcomer area ±3%

Clearance under downcomers ±1/8 in

Inlet & outlet weir height ±1/16 to ±1/8 in

Seal pan dimensions (Figure 10.4) ±1/8 in

Chimney tray riser and riser outlet areas ±3%

Weep hole area ±3%

Chimney tray leakage < 1 in in 20 minutes with weep holes plugged

Sump dimensions as for seal pan

Feeder pipe clearance ±1/8 in

Levelness ±1/16 to ±1/8 in

Notch dimensions ±1/16 in

Bed limiter & support perforations sufficiently small to prevent packing passage

Figure 10.5 Some dimensions that should be inspected and their typical installation tolerances.

10.9.5. Location and orientation of internals

One of the most common installation errors is incorrect orientation of internal pipes, baffles, and other removables. Often, these are installed in situ by people who have little understanding of column operation.

Figure 10.6 is an example of the correct and incorrect ways to assemble the inlet bend of a liquid drawoff line. The incorrect arrangement (Fig. 10.6*b*) caused vapor rather than liquid to escape through the bottom pipe (192). Hence, it is important to ensure that such pipes and distributors have not been installed upside down or sideways in the assembled column.

Figure 10.7 is another example of a correct and incorrect installa-

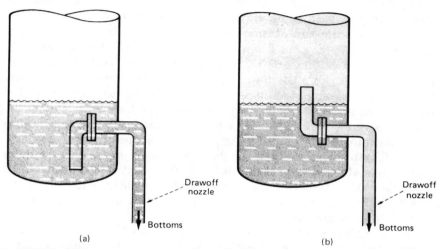

Figure 10.6   Orientation of a bottoms drawoff. (*a*) Correctly installed (liquid in drawoff), (*b*) incorrectly installed (vapor in drawoff). (*Henry Z. Kister excerpted by special permission from* Chemical Engineering, *February 9, 1981; copyright © by McGraw-Hill, Inc., New York, NY 10020.*)

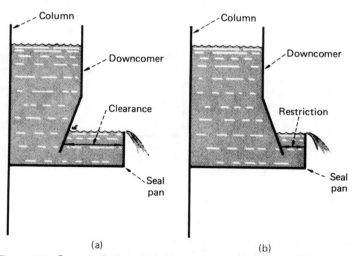

Figure 10.7   Incorrectly installed downcomer causes premature flooding. (*a*) Correct; (*b*) incorrect (*Henry Z. Kister excerpted by special permission from* Chemical Engineering, *February 9, 1981; copyright © by McGraw-Hill, Inc., New York, NY 10020.*)

tion. Here, a sloped downcomer was installed back to front (203). The incorrect arrangement (Fig. 10.7b) resulted in a restriction at the bottom seal pan, which produced cyclic flooding in service. Errors of a similar nature can easily occur in installing other baffles and internals.

Figure 10.8 is a third example of correct and incorrect installation. Here, tray panels in a low-liquid-rate amine contactor were rotated 180° to the desired orientation (145). The inlet weirs were therefore ineffective for sealing the downcomer (See Sec. 6.20). Vapor flowed up the downcomer and interfered with liquid descent, causing excessive liquid carryover from the tower. To minimize the carryover problem, liquid rate was reduced to 20 percent of the design rate. This provided poor tray washing; the valve trays plugged, and frequent cleaning became necessary. After the fault was finally corrected and normal liquid rates reinstated, the plugging problem disappeared.

In a fourth example (206b), bubble caps were installed under the tray panels; this column flooded at 30 to 40 percent of design. In a fifth example (206b), the downcomer clearance was about 7 to 8 in at the feed tray due to miscommunication, and premature flooding (due to lack of downcomer seal) resulted.

Liquid maldistribution in a packed tower can be initiated if a bed limiter, hold-down plate, or hold-down bar is poorly oriented relative to the distributor or redistributor above. With structured packings, these internals need to also be properly oriented relative to the top layer of packings.

A very common fault is the incorrect location or orientation of in-

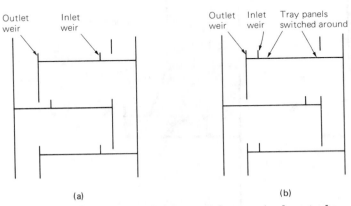

Figure 10.8 Incorrectly installed tray panels cause inadequate downcomer sealing (a) Correct; (b) incorrect. (*Courtesy of Scott W. Golden, Glitsch Inc., private communication.*)

struments. These may be installed at the wrong nozzles or in a stagnant region, and consequently give incorrect indications.

When the location and orientation of internals are being checked during inspection, note should be taken of the possible interference of internals with instruments or other parts. Often, different internals are shown on different drawings, and interferences between these remain unappreciated until the parts are assembled.

10.9.6. Fastening of internals

Trays and downcomers should be firmly bolted to their supports. Failure to do so may result in excessive deflection, flow restriction, leakage, and deficient mechanical integrity.

Firm fastening of downcomers and panels under downcomers is most important. If an inlet weir or recessed seal pan is present and the downcomer is not tightened properly, the downcomer may bow toward the weir under pressure from the downcomer liquid. This may cause a flow restriction at the downcomer outlet. Improper fastening of downcomer panels, or areas beneath downcomers, or installation of downcomer panels on the tray side (rather than the wall side) of the downcomer supports are likely to cause excessive leakage. Leaks in the area beneath the downcomer are most detrimental to efficiency since the leaking liquid completely bypasses two trays. This area also has the highest leak potential because of the high liquid head. The above could not be emphasized more for leaktight services.

Cracks and crevices should not take up more than 2 percent of the tray area (49). If they do, excessive weeping and/or channeling may result.

Trays and all internal parts should be properly secured and tightened. Otherwise, they may loosen during operation or lift during pressure surges. In one case (206b), poor tightening caused an entire grid bed to fall to the bottom of the column; in another (206b), poor tightening resulted in loud banging noise from an operating column. With valve trays, it is important also to ensure that valves do not easily come out of their orifices. If they do during inspection, they may do the same in service.

Particular care should be taken to see that the nuts and bolts at the drawoff trays (particularly for total drawoffs) are properly tightened. At least one experience (85) has been reported where poor column performance resulted from failure to tighten the drawoff box nuts and bolts.

In leaktight services, it is important to confirm that washers and gasketing were installed as specified. In one case (145), leaving out gaskets on a total-drawoff chimney tray during a revamp rendered the column inoperable. Gaskets must be carefully checked to ensure they were properly cut and are securely held by the joints.

10.9.7. Cleanliness of internals

The column should be inspected for proper cleanliness and absence of debris. While cleaning, guidelines 1, 3, 4, 8 in Sec. 10.4 and guideline 9 in Sec. 10.1 are important. Chemical cleaning is sometimes practiced; this is discussed in Sec. 11.6.

During the inspection, attention should be given to nozzles, downcomers, seal pans, and distributors. Debris commonly left in columns can consist of shipping covers, working tools, nuts, bolts, gaskets, rags, paper, masking tape, and beverage containers. Each can cause severe operating problems if not removed. Some troublesome experiences have been reported (98, 203, 268).

Temporary plugs are often used during construction and cleaning at column outlets (to prevent small parts such as nuts and bolts from entering outlet lines) and in weep holes (for leak testing). These plugs must be removed prior to startup. If the column has previously been in service, weep holes must be cleared of possible plugging; in one case (296), weep-hole plugging resulted in tray damage.

When the column was previously in service, inspection by backlighting may reveal internal blockages in mist eliminators and similar internals, even when the device appears visually clean (335).

Plugging and erosion in spray distributor nozzles that have been in service often occur at the nozzle spirals (109) and may be invisible from outside. It is best to water-test these distributors during the shutdown; such a test will also reveal internal leakage. However, this is not always practical. Alternatively, a sample of nozzles (about one in five) should be dismantled and inspected. Even better, these nozzles can be rigged and water-tested outside the column.

10.9.8. Final inspection

Safety considerations permitting, manhole doors should be shut immediately upon completion of column inspection and should only be kept open from then on while work (e.g., tray reassembly) is being performed inside. This is important in order to keep rain, sand, dust, and animals out of the column. There have been instances (e.g., 238) where animal carcasses lodged in delicate internals and caused premature flooding.

It is important to maintain a continuous watch of activities around the column during the period preceding the final bolting up of the column. This should be coupled with spot checks taken at random. At this time, the manways are reassembled and the column is prepared for commissioning. Common errors during this period are (1) manways left loosely placed and unbolted to the trays, (2) debris reintroduced, and (3) bottom baffle hatchways put into place but left unbolted. In one experience (255), poor column performance resulted from a failure to detect (during inspection) that a bottoms hatchway had not been bolted. The author has experienced a case where a few manways were left sitting on nearby tray decks, leaving a large gap in the tray floor; surprisingly, the column still functioned. Others (145) were less fortunate, and experienced poor separation or excessive entrainment due to an identical problem.

Prior to startup, it is also important to check for proper functioning of column instrumentation. Several useful guidelines have been described elsewhere (131, 150a, 357).

11

Column Commissioning

Once the column is bolted up (either after initial construction or after a shutdown), a number of operations are performed to prepare it for startup. This phase is often referred to as *column commissioning*. The main objectives of column commissioning are to clear the system of undesirable materials, to test the column and rectify potential problems, and to take preventive measures against performance deterioration (e.g. due to fouling).

Most commissioning operations are performed using readily available liquids and gases, such as air, nitrogen, steam, water, or oil. Using such fluids may appear straightforward, but this appearance is deceptive. Column hardware is seldom designed to cater specifically to commissioning operations. The commissioning operations must therefore be tailored to suit the limitations of the available hardware. Poor commissioning practice is a common cause of accidents, column and tray damage, and premature bottlenecks.

Columns containing valve trays are particularly vulnerable to poor practices during commissioning, startups, and other abnormal operations. The reasons are elucidated in this and the next chapter. The author knows of one plant that banned valve trays altogether, merely because of repeated tray damage during commissioning and startup. Such action is unfortunate, because it forfeits many important economic benefits that these trays can offer. In the vast majority of cases, awareness of the potential problems and avoiding the common commissioning and startup traps can entirely eliminate incidents of valve tray damage.

This chapter reviews the common practices of column commissioning, outlines the preferred practices, highlights several of the pitfalls, and supplies guidelines for avoiding these pitfalls. Several of the considerations extend beyond column commissioning and apply to other abnormal operations (e.g., relief valve lifting, shutdowns, and startups).

11.1   Line Blowing

A common prestartup practice is to pressurize the column with air or nitrogen. The column then serves as a "vapor reservoir" for blowing lines connected to the column, to remove construction debris. The following guidelines are recommended (Fig. 11.1):

1. Line blowing can cause vapor flow rates that exceed the design flow rates through the column. The resulting pressure differentials across the trays or packing may be far greater than the design values. In turn, the forces exerted on trays and supports by the high pressure drops can exceed the allowable stresses. These forces may permanently deform trays and/or supports or fluidize and carry over packings.

   This problem is best avoided by roughly calculating the expected flow in each line before blowing it. If likely to be excessive, either the flow rate (or rate of column depressuring) should be monitored and properly constrained, or blowing the line from the column should be avoided.

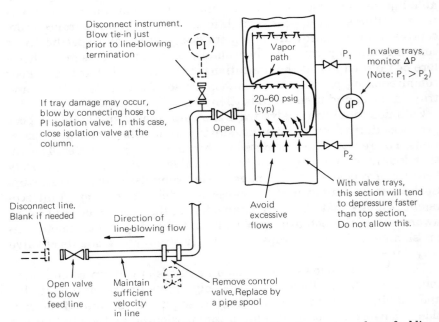

Figure 11.1   Good line-blowing practices, illustrated with reference to a column feed line.

2. Special care must be exercised with columns containing valve trays. In other types of columns, vapor can freely flow upward or downward through fixed apertures as well as through the downcomers (Fig. 11.2a). In valve trays, the valves seriously restrict downflow (i.e., they act as nonreturn valves that close for vapor downflow, Fig. 11.2b). Excessive pressure drops, which may cause serious tray damage, can therefore occur at relatively low downward flow rates in valve trays. This problem is most severe if the downcomers and/or clearances under the downcomers are small.

Blowing bottoms lines and reboiler lines from a valve tray column can be troublesome and is best avoided, because this operation relies entirely on downflow through the trays. Blowing intermediate lines can also be troublesome, especially if the downflow restriction of the valves causes the column section below the line to depressure faster than the section above the line (Fig. 11.1). Careful planning, adequate monitoring, and proper flow rate restriction are essential.

3. Sometimes it may appear attractive to blow lines into the column, and later clean the column. This practice is generally not recommended, and must be avoided where a feed distributor, a seal pan drawoff, or a vortex breaker is present. In each of these cases, debris can be blown into restrictive areas, and large pieces (particu-

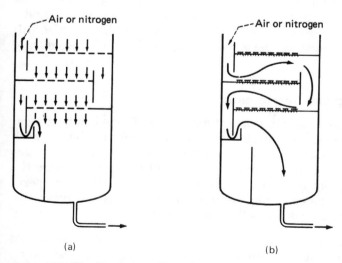

(a)   (b)

Figure 11.2 Blowing bottom lines from the column. (a) Sieve trays; (b) valve trays (*Henry Z. Kister, excerpted by special permission from* Chemical Engineering, *April 6, 1981. copyright © by McGraw-Hill, Inc., New York, NY 10020.*)

larly rags, plastic sheets, and the like) may be deposited in the column internals. Such debris can later restrict the flow passages when the column is in service.

4. All instruments, except for those that are directly required for monitoring the line-blowing operation, should be disconnected or at least blocked off during line blowing to avoid debris entering the impulse lines. With the impulse-lines disconnected, the tie-ins of the instruments to the lines should be blown free of debris just before blowing a line is terminated.

5. Any control or trip valves should be removed and replaced by pipe spools prior to line blowing.

6. While excessive velocities through the column must be avoided (1 above), sufficiently high velocities through the lines are essential to effectively remove debris. Blowing at velocities of 200 ft/s or higher has been recommended (295).

   Alternatively, it has been recommended (182a) that the F-factor ($u_s\rho^{0.5}$ where u_s = superficial velocity and ρ = fluid density) of the gas or liquid used for line blowing is at least as great as, and preferably 25 percent greater than, the maximum F-factor anticipated in the line (under normal or upset conditions, whichever is greater). The basis of the latter rule and a sample calculation are detailed elsewhere (182a).

7. The valve used for throttling during line blowing with air or gas should be as close to the column as possible. Since gas density decreases and its velocity increases upon throttling, this practice will maximize the length of line subject to the higher velocities during line blowing.

11.2 Pressuring and Depressuring

Pressuring and depressuring operations are performed during commissioning, startups, and shutdowns. They are used in order to check the column for pressure retention and leaks, remove air or inert gas prior to startup, free the column of gas at shutdowns, or prepare the column for entry by personnel. Pressuring and depressuring operations are also performed during venting and when safety valves open. The guidelines below address all these situations.

1. Pressuring and depressuring should be carried out at a controlled rate that is sufficiently slow to avoid tray damage. The reasoning for this guideline is identical to that of guideline 1 in Sec. 11.1. Some additional considerations unique to preventing excessive depressuring rates upon outage of vacuum generation equipment are discussed in Sec. 12.4.

2. Pressuring should be started at the base of the column and progress in an upward direction for atmospheric or superatmospheric columns containing valve trays. Progressing in the downward direction will severely restrict the allowable pressuring rate. Similarly, in vacuum columns, the valve that is used to draw vacuum should be at the top of the column. This guideline is most important when the downcomers and/or their clearances are small or when the column contains liquid. The reasoning for this guideline is identical to that of guideline 2 in Sec. 11.1.

One incident has been reported (364) where an 8-ft column equipped with valve trays was pressured from the top while it was circulating liquid. This caused the top 12 trays to be deflected 6 in downward in the middle and loosened manway covers. The loose covers were later displaced by upflowing vapor when normal operation commenced; this in turn caused premature flooding. The corrective action was to pressure the column from the bottom.

3. Location of relief valves, bursting disks, and major vents should be carefully reviewed. Normally, these should be at the top of superatmospheric columns (or in their overhead systems). Conversely, the vacuum-breaking device should normally be at the bottom of the column. This is important for avoiding damage to trays and for achieving the required relief rates. Detailed discussion is in Sec. 9.5.

4. Avoiding excessive vapor flow rates during depressuring is especially important when the column contains liquids. Excessive vapor flows may cause flooding and gas lifting of the liquid (a "champagne bottle" effect), resulting in a liquid discharge into the relief header and possible damage to column internals. The likelihood and consequences of excessive vapor flows should be investigated while sizing vents and relief devices (Sec. 9.7).

5. When a column is located between intermediate stages of a compression train, the effect of compressor surging on column internals should be considered. There were several incidents where demisters (206b) or trays were dislodged or damaged by back pressure during a compressor surge. The back-pressure problem is particularly severe with valve trays. Bypassing the column during compressor startup, using sieve trays or structured packings, or providing antisurge controls often overcomes this problem.

11.3 Purging

When the column separates combustible or hazardous materials, it is usually purged with an inert gas prior to startup to remove air. It may then be purged with the process gas to remove the inert gas. The reverse steps are performed at shutdown.

Nitrogen is the most common inert gas used for purging. Steam, carbon dioxide, and other inert gases are also used for purging. It is best to follow column steaming at startup by a nitrogen purge, because atmospheric condensation of the residual steam can pull a vacuum which may suck in air. The converse applies at shutdown. It is best to purge the column with nitrogen prior to steaming to eliminate combustibles. Further discussion is in Sec. 11.8.

Purging often requires installation of special purge lines or purge connections. It is a good policy to have such lines disconnected (or at least blinded) during normal operation. One rule of thumb (237) recommends sizing purge lines to deliver a purge gas volume four times the volume of equipment to be purged over a 10-hour period.

The purity of the inert purge gas must be tested before use and should also be spot-checked during use. An inert gas containing significant oxygen or combustibles may be hazardous and/or ineffective. Negligible concentrations of oxygen or combustibles will speed purging and reduce inert gas consumption. In one extreme case (7), columns were purged by 93 percent oxygen due to an error by the nitrogen supplier. Had the purge gas been tested, the several explosions and internal fires that followed could have been prevented.

Two techniques are commonly used to purge a column. The "sweeping" technique admits the purge gas near the inlet to the unit. From there, it flows from one vessel to another. The "pressuring and depressuring" technique isolates the column, then repeatedly pressures and depressures it a number of times. Either technique is adequate if correctly applied. The pressuring and depressuring technique is more effective for the column because it reaches pockets that will be bypassed by a sweeping gas. Sweeping is more effective for pipe runs. For maximum effectiveness, the pressuring and depressuring technique can be supplemented by sweeping column piping and dead-end connections.

The key consideration is to properly and systematically purge each portion of the system. Isolation valves in the column piping should be opened and remain open during the purge period to avoid trapping of liquid or undesirable components in their bonnets. If the column is to be shut down, checks should be made for plugging and obstruction. Several accidents occurred because of liquid or undesirable material being trapped by pipe plugging; Refs. 7 and 275 have some typical examples.

Drain and vent valves must be opened intermittently so that no dead pockets are left unpurged. When purging nears completion, analyses are performed at each vent or drain. Purging continues until the concentration of air, combustibles, or toxic gas does not exceed a desired value at any vent or drain. Some common test devices are described in Amoco's booklet (5).

The purging steps should be logged, including times at which each vent and drain was opened, and any analytical results (153). Coloring a column process and instrumentation diagram (P&ID) at the completion of each purging step is also helpful.

11.4 Blinding and Unblinding

Shutdown blinds and/or slip plates ("spades") are usually installed in all lines which leave or enter the column in order to positively eliminate leakage of material into the column when air is introduced. During commissioning and startup, these blinds and/or slip plates are removed. Other blinds ("running blinds") are often installed at the same time to prevent leakage of undesirable materials into or out of the column during normal running.

Blinding and unblinding is tedious, and errors can easily occur. Improper blinding or unblinding caused several major accidents involving loss of life, explosions, fires, and toxic releases. Less dramatic consequences of improper blinding include creation of hazardous conditions, line blockages, or serious startup delays. The author has experienced several cases in which improper blinding was troublesome. Considering the above, it is imperative to correctly blind and unblind the column. The following guidelines apply:

1. All statutory and/or company regulations or codes relevant to blinding or unblinding must be strictly adhered to.

2. The sequence of blinding and unblinding must be carefully planned prior to startup or shutdown. The point in the startup-shutdown program at which each blind is to be inserted or removed should be clearly defined in the operating procedure (153).

3. The possibility of reverse flow into the column must be considered when preparing the blinding schedule. Reverse flow into columns at shutdown has caused numerous accidents, some of which were fatal. In one accident (210), product from storage flowed into a reflux drum and then into the column through a leaking valve (Fig. 11.3a). In another case (210), toxic gas leaked from the blowdown header back into a column through a leaking vent valve, killing an operator who was draining the tower (Fig. 11.3b). In a third case (169), an explosion occurred in a shut down column containing flammable gas when an inert gas purge rather than blinding was relied on to keep air out. Each of these accidents could have been prevented had the valves been blinded.

Likewise, the possibility of reverse flow from the column into

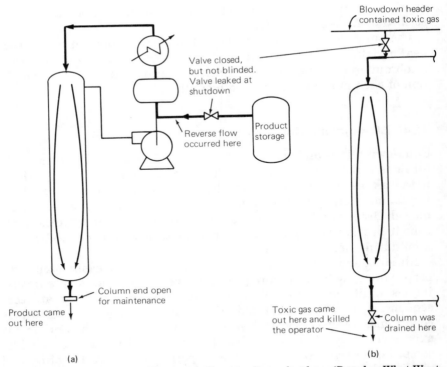

Figure 11.3   Accidents caused by reverse flow at column shutdown (*Based on* What Went Wrong? *second edition, by Trevor A. Kletz. Copyright © 1988 by Gulf Publishing Company, Houston, Texas. Used with permission. All rights reserved.*)

utility lines upon startup must be considered. In one case (237), propane leaked into the steam system through a steam-purge connection and was released from a fire-suppressant steam-purge nozzle. Similar nondistillation accidents have also been described (210). Such accidents can be avoided by disconnecting or blinding these lines.

4. A checklist of all the blinds to be installed or removed should be made. Whenever a blind is installed or removed, the operator witnessing the operation or removal should mark the date and time and initial the checklist (153).

5. The pressure and temperature ratings as well as the material specifications of each blind should be checked to ensure compatibility with the column or line. There were cases of slip plates that corroded right through (210; Fig. 11.4a). In other cases, slip plates bowed due to excessive pressure (210; Fig. 11.4b).

6. Blinds should be properly tagged, and the tag identification entered on the checklist (item 4 above). Tags should be at least 5 in

(a)

Figure 11.4 Importance of specifying correct pressure, temperature, and materials for blinds. (*a*) A slip plate left in position for many months and corroded right through. (*From* What Went Wrong? *second edition, by Trevor A. Kletz. Copyright © 1988 by Gulf Publishing Company, Houston, Texas. Used with permission. All rights reserved.*)

long when line diameter is 6 in or less, and at least 6-in long when line diameter exceeds 6 in (210). Insulation should be inspected to ensure that tags have not been lagged. A sound and unambiguous tagging practice should be followed; further discussion is elsewhere (210).

7. Extreme caution must be exercised when cracking flanges for blinding and unblinding (153, 210). The specific precautions must follow the safe handling procedures of the chemicals involved, and any statutory or company requirements.

   Cracking flanges can be particularly hazardous when the upstream isolation valve is plugged. Deposits trapped in the valve seat may inhibit valve closure. The deposits may loosen and break free once the flange is cracked. In one accident (7), this sequence of events was followed by a release of high-pressure liquid butane, after bolts in the housing of the bottom pump suction screen were loosened (without blinding). The release fired and exploded.

Figure 11.4 *(Continued)* Importance of specifying correct pressure, temperature, and materials for blinds. *(b)* A slip plate bowed by 470 psig pressure *(From* What Went Wrong? *second edition, by Trevor A. Kletz. Copyright © 1988 by Gulf Publishing Company, Houston, Texas. Used with permission. All rights reserved.)*

8. The line should be thoroughly surveyed prior to blinding or unblinding to ensure absence of hazardous conditions prior to cracking the flange.

9. Blinds whose removal or installation is potentially more difficult or hazardous should be clearly identified. Their removal or installation should be witnessed by a supervisor.

10. Cracking flanges must not be authorized without a written permit issued by a supervisor responsible for safety. The permit must list any specific precautions unambiguously. Breakdown of communication has caused several accidents during blinding and unblinding. In some cases (7), explosions resulted.

11.5 Leak Testing

After purging, and before introducing the process gas, the column is leak-tested. The most common leak-testing technique is pressuring the column up with inert gas, with all vents and drains closed, and monitoring the rate of pressure loss. At the start of the test, the pressure is usually about 50 to 60 psig, or about 10 psi below the safety valve setting, whichever is lower (5, 153). All joints, flanges, drains, and vents are inspected using soap solution, a sound sniffer, or an ultrasonic leak detector. When leaks are detected, bolts are tightened. If the leak persists, it may be necessary to depressure the system and correct the fault (e.g., replace a gasket). If air can leak into the column during the repairs, the column must be repurged. The test continues until no further leaks are detected and the column satisfactorily holds pressure.

Workers must be aware that nitrogen and other inert gases are hazardous and can quickly and easily cause asphyxiation. People working near leaking flanges have been affected (210); a burst of nitrogen can cause a fatality. Detailed discussion is presented elsewhere (210).

Alternative techniques involve hydrostatic testing of the column or pressuring it up with steam. Steam or water make leak detection easier, but these fluids are more troublesome to use than nitrogen. Hydrostatic testing requires that the column and its support structure can handle the hydrostatic loads; the disadvantages of steaming are described in Sec. 11.8. Testing with these fluids is conducted prior to purging and may require dehydration following the test.

In vacuum columns, it is best to first perform the leak test described above at just below the maximum permissible pressure (23). Leaks are generally easier to detect and repair when the column is under pressure. Next, the normal operating vacuum is pulled on the column. The vacuum equipment is then turned off and isolated from the column, and the rate of pressure rise is monitored. The air leakage rate can be inferred from the rate of pressure rise. Purging often needs to be repeated following a vacuum leak test in order to remove air that leaked in during the test.

The leak detection methods described above for pressure leak testing also apply for vacuum leak testing. More sensitive methods use an easily detectable tracer gas. Excellent detailed reviews of leak testing and leak detection techniques are available elsewhere (315, 345).

Water or other liquids should be adequately removed from the column prior to a vacuum leak test. Evaporation of liquids may cause a rise in pressure during the test.

In a complex column (e.g., a refinery vacuum tower), it may be beneficial to conduct the test by sections (5, 345). For instance, the tower is tested first, then the connecting lines are tested one at a time by opening block valves.

Once inert gas purging and leak testing are completed (in either a pressure or a vacuum tower), the column is usually depressured to a small positive pressure (5 to 10 psig) until the plant is ready to proceed with the startup.

11.6 Washing

Columns are often washed during a startup or a shutdown for one or more of the following reasons:

- To remove scale, mud, solids, and/or corrosion products
- To uncover leaks and check the operation of pumping systems prior to startup
- To cool the tower
- To dissolve undesirable material such as acid, caustic, polymer, oil, or sticky deposits at shutdown
- To coat column internals with a protective film that will inhibit corrosion, polymerization, or undesirable reactions during operation
- To remove water or other undesirable materials prior to startup

A water wash is effective when washing is performed for the first three reasons, and sometimes the fourth reason, above. A chemical wash is used when the reason for washing is one of the last two, and sometimes also the fourth above. When none of the above reasons is relevant, it is best to refrain from washing the column.

Some guidelines for washing columns are

1. Introduction of scale and dirt must be avoided. There have been cases (295) where the wash water contained a significant amount of solids, such as sand particles. These particles may deposit inside the column.

2. Washing may pick up debris resting in the column piping and deposit them inside the column or distributors. In one case (203), a welding rag lodged in a reflux distributor led to excessive liquid carryover from a column. The rag was transported by water from the column piping into the reflux distributor, where it remained stuck.

   It is best to water-flush (or gas-blow; Sec. 11.1) the water circulation lines before a prestartup wash. During flushing, lines should be broken at the column inlet to prevent debris from entering the column. For an effective flush, water velocities of 12 ft/s have been recommended (295). During the wash, strainers should be used in the water circuit, especially upstream of distributors or other delicate internals.

3. Washing may shake loose rust and scale adhering to the column walls or internals. In one case (148), this caused persistent poststartup plugging of filters and pump strainers in an absorber-regenerator system. In other cases, the particles may deposit in and later plug delicate internals. Strainers should be used during

the wash as per guideline 2 above. Prior to completion, the wash solution should be tested for particulates. Rinsing should be repeated until the particulates disappear.

4. Corrosive water must be avoided. If the column is made of stainless steel, the water should be low in chlorides (131). The author is familiar with one plant where steam-water operations were discontinued in stainless steel columns after it was found that chlorides in the water caused metal deterioration. Ideally, the water should contain a corrosion inhibitor. The corrosion inhibitor should be carefully selected, because several commercial inhibitors promote foaming in subsequent operation.

   Water should be drained from the column as soon as washing is complete. Drying by hot gas (Sec. 11.9) is often recommended (131). These measures are important for preventing corrosion, and in cold climates, also freezing.

5. When the column is heavily plugged with muddy or sandy material (e.g., natural gas glycol dehydrators which may plug due to sand or drilling mud), water is usually first run from the vapor outlet line down until it comes out of the bottom clear. Water is then run from the bottom up, so that it overflows the column, until the overflow is clear (238, 239). Washing from the bottom up should not precede washing from the top down, because it may lift trays off their supports. Excessive water flow, pressure circulation, or excessive pressure must be avoided when plugging is heavy, as these may damage trays.

6. Water washing is sometimes used (7, 23) as the last step of cooling a column, often from temperatures of 200°F or lower down to ambient. The wash water (or a cool wash liquid) should not be admitted unless the column is sufficiently cool and "heat pockets" are absent. In one case (23), premature introduction of water caused tray damage. Guidelines 1 to 4 in Sec. 11.7 also apply to washing hot columns.

7. Washing has been specifically recommended for newly packed columns (105) as a means of clearing construction debris, cleaning the dust off the packings, and removing small quantities of construction grease or oil, which may otherwise promote foaming. Some grades of polypropylene also tend to promote foaming in high-pH applications; it has been recommended (167) to wash these with caustic prior to startup.

8. When washing is used for preparing the column for personnel entry, it is important to avoid trapping undesirable materials in valve bonnets, piping, and dead pockets. All valves should be opened, all dead pockets and valve bonnets should be flushed, and

all drains operated. In some cases (5, 153), and only when column foundation and structural strength permit, the column is entirely flooded with water to remove undesirable materials.

9. Caustic or acid must be removed as thoroughly as possible from the system prior to the commencement of a water wash; otherwise, a violent reaction may occur and damage column internals.

10. Water used for process washing or flushing can carry air into the system. This must be avoided where explosive mixtures can form. There have been cases (4) of detonations in drums because of this.

11. In some cases, particularly with a regenerative absorption system, the washing solution is circulated through several columns (or other units) simultaneously. When this is practiced, it is imperative to appreciate that the wash solution will transport materials from one part of the system, and may accumulate them in another. Several incidents have been described (276) where circulating wash water absorbed small quantities of sparingly soluble gases (natural gas, hydrogen, nitrogen) in a pressurized absorber, and later desorbed them in a low-pressure regenerator and its connecting piping (Fig. 11.5). After some time, these gases concentrated in the regenerator. With natural gas and hydrogen, this caused several explosions (276); in the case of nitrogen, this caused a suffocation hazard in an open column (276). The system should be carefully examined to ensure that transportation of material by the washing solution does not generate such undesirable effects.

It is often desirable to gas-purge the column and its piping following a water wash. This will clear pockets of undesirable gases desorbed from the water. In one case (276), such a pocket caused an explosion a day after the wash was completed.

12. The possibility of reaction of wash chemicals with deposits in the column should be considered. The route of venting or disposing of any hazardous reaction products must be adequately planned. In one incident (7), hydrogen sulfide was liberated to the atmosphere following reaction of an acid used for washing with deposits in a caustic scrubbing system.

13. Chemical wash agents should be thoroughly removed from the column and its piping after the wash, particularly if the washing agent has the potential of reacting, interacting, or causing foaming during normal column operation. The possibility of wash agents remaining in dead pockets and later entering the column during operation should be considered.

If a hazardous or an explosive interaction may result, the use of

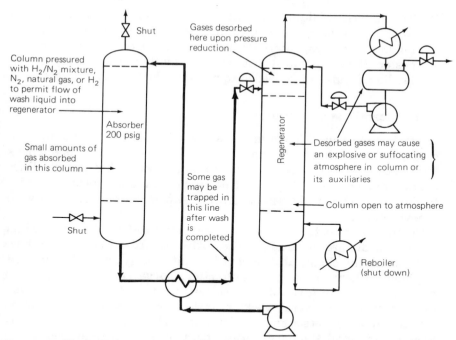

Figure 11.5 Wash of absorber-regenerator system where accidents can occur (heavy lines denote path of circulating wash liquid).

the wash agent should be avoided altogether. In one case (275), a small amount of caustic left over from cleaning in a dead pocket entered a vacuum still that was processing an acetylenic alcohol and caused a fatal explosion.

14. A chemical wash introduces the potential hazard of chemical leakage into utility systems and/or into lines connected to the column. Special attention must be given to adequate blinding. In one instance (7), caustic used for washing backflowed into a steam line while steam was being vented to atmosphere, causing a wide area to be sprayed with caustic.

15. If the wash solution is heavier than water, a check must be made to ensure that the tray or internal supports can handle it.

16. If the chemical wash is performed by an outside contractor, the operation must be carefully supervised by plant personnel. The contractor's equipment must be examined, and the contractor must be made aware of any safety requirements.

17. Excessively fast drainage of the wash liquid may induce a vapor gap problem. This is described in detail in Sec. 12.9.

18. The quantity of liquid charged into the system prior to the start of washing needs to be sufficiently large to accommodate liquid holdup in sumps, internals, and piping. In one amine absorber system (206*b*), pump suction was lost by lack of inventory. Regular liquid makeup is needed to accommodate solution losses.

11.7 Steam-Water Operations

A practice often used for column testing is to operate the column on a steam-water mixture. The main purpose of this operation is to check heat exchanger and column performance, observe that gravity lines operate properly, and check for leaks and malfunctions prior to the introduction of chemicals. This operation also acquaints operators with a new column.

Steam-water operation is often recommended (98), especially when the chemicals are hazardous or corrosive, in order to eliminate hazards before the chemicals are admitted into the system. The long list of potential heat exchanger malfunctions surveyed by Gilmour (134) argues strongly in favor of this recommendation. Further, if the test can be performed several weeks prior to startup, costly delays can be averted by early detection of faulty items. However, this operation is troublesome, and should generally be avoided in nonhazardous services. If used, caution is required with the operating procedure in order to avoid structural damage to the column. Such damage may arise from

1. *Creation of a vacuum:* When the test operation is completed or interrupted, steam and water are present in the column. The steam will condense because of atmospheric cooling, and a vacuum may be created. If the column is not structurally designed to withstand vacuum, and is not adequately protected against vacuum, it may implode. In one reported incident (275), a still imploded when cold water was added to cool it after steaming.

   To avoid column implosion, it is important to ensure that when the operation is completed or interrupted, a vent of adequate capacity is left open. The vent capacity should allow for the possibility of a sudden thunderstorm. Introduction of cold water into a hot still should be either avoided, or constrained and allowed for when sizing the vent.

2. *Imposition of excessive differential pressure:* When steam comes into contact with a large quantity of cold water, it will rapidly condense. The pressure at the contact zone will plunge and induce a rush of steam from above or below into the contact zone. This rush will impose a large downward-acting differential pressure across

the trays above, and a large upward-acting differential pressure across the trays below. If any of these differential pressures exceeds the allowable stress of the trays, tray damage may result. Trays above the contact zone will be deformed downward; those below will be deformed upward or lifted off their supports. A typical symptom of this problem is extensive tray damage near the contact zone, diminishing further up or further down the column. The author is familiar with incidents where this occurred at the feed point (Fig. 11.6). The downward-deformation problem is most severe in valve trays, particularly when the downcomers and/or downcomer clearances are small, for the reason explained in Sec. 11.1

Excessive differential pressures usually occur either when water is introduced before steam, or when the water flow rate is abruptly stepped up. It is therefore essential to introduce steam first, and then add the water slowly, always ensuring that not all the steam is being condensed (61, 207). Preheating the water can help and often eliminate this problem.

3. *Water hammer:* The differential pressures in guideline 2 above may also generate very high local steam velocities. The fast-moving steam will pick up slugs of water and fling them against anything that interferes in its path. The resulting water hammer can damage column internals and pipe fittings. Preventive measures are as per guideline 2 above.

4. *Thermal Stresses:* Rapid chilling of hot column shell or internals by cold water may generate excessive thermal stresses. These

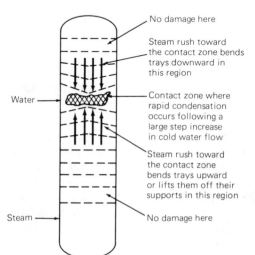

No damage here

Steam rush toward the contact zone bends trays downward in this region

Water

Contact zone where rapid condensation occurs following a large step increase in cold water flow

Steam rush toward the contact zone bends trays upward or lifts them off their supports in this region

Steam

No damage here

Figure 11.6 Tray damage caused by a large step increase in cold water rate during a steam-water operation.

stresses cause fatigue of the column or tray metal. In packed columns, this chilling effect can partially destroy ceramic packing (105). Steam–cold water operation is best avoided with ceramic packings. Other preventive measures are as per guideline 2 above.

5. *Overheating:* The ability of the column, its insulation, its internals, and its auxiliaries to withstand the expected temperatures should be carefully reviewed, especially if the steam is superheated. The column metal safe working temperature must not be exceeded.

   Steam-water operation can induce excessive thermal expansion which the column or its piping may be unable to accommodate. In one instance (210), a large distillation column was made in two halves, linked by a large vapor line containing a bellows. Steaming the line elevated one end of the bellows excessively above the other; the line was not designed for steaming. Steam-water operation is best avoided with plastic packings, because excessive temperatures can melt the plastic.

6. *Inadequate capacity:* Pretest checks must be carried out to ensure that the column's operating capacity and relief capacity are adequate for the expected steam and water flow-rates. Excessive flows must be avoided (see Sec. 11.1).

7. To prevent reboiler fouling, and accumulation of solids in the column during steam-water operations, an adequate water blowdown must be purged out of the column base.

8. The column and its piping should be steamed to warm it up prior to the commencement of a steam-water operation (see Sec. 11.8).

9. Guidelines 1 to 4 in Sec. 11.6 also apply to steam-water operation.

11.8 Steaming

A column may be steamed during commissioning to drive the air out, heat the column up, clear blockages, or leak-test the column. At shutdown, a column is often steamed out to strip out and drive out residual chemicals and to prepare the column for personnel entry. Steaming may also be practiced for cooling very hot (> 400°F) columns (7, 23,). Steaming the column can be troublesome and should be replaced by nitrogen purging whenever possible. However, nitrogen is usually ineffective for heating up the column, vaporizing heavy liquids, or clearing blockages, and steaming is sometimes necessary. Often, steam also has a cost advantage and enables a quicker purge than nitrogen (237). The following guidelines apply for steaming a column:

1. Precautions recommended for steaming the column are identical to guidelines 1 to 6 in Sec. 11.7. These problems are far less troublesome with steaming than with steam-water operations, because the only sources of cold in the column are atmospheric cooling and the initial temperature of the internals.

   The most critical periods for steaming the column are when steam is initially introduced into a cold column and when the steaming is terminated. In either case, rapid condensation may occur. Steam must therefore be initially introduced extremely slowly, until the column warms up.

2. Once steaming is completed, the column should be opened to let air or inert gas in on shutdown, or placed under gas pressure (inert gas such as nitrogen or fuel gas, or process gas) at startup. The alternative of leaving small steam purges to keep the column pressure up at the conclusion of steaming is often unsatisfactory and resulted in creation of vacuum in some instances (3). Unless the column is fully open to the atmosphere, so that no vacuum can be created, a pressure of at least 5 to 10 psig should be kept in it (3).

3. Following the conclusion of column steaming, all low points in the system must be drained. Water pockets may become troublesome during column operation, or may freeze, damage, or corrode drains and pipes.

4. Before steaming is completed, the effectiveness of the operation should be checked. Usually, indication is by temperature measurement or by carefully "feeling" lines. Any lines or sections which are considerably cooler than their neighbors imply presence of other material (e.g., inerts), and require venting or purging before steaming is terminated. When removing air (or other gases) is critical, a laboratory analysis rather than temperature measurement should be performed (see Sec. 11.3). It has been recommended (5) to continue steaming at least until there is a free flow of dry steam from all vents for 20 to 30 minutes.

5. Unless hazardous chemicals are present, key drains and vents should be opened before the commencement of steaming. Drains should be kept open until only dry steam is vented. They should then be opened intermittently to remove additional condensate.

   When hazardous chemicals are present, the column is often initially steamed to the disposal header (e.g., flare header). If steaming is sufficient to clear these chemicals, the vents and drains to the disposal header are then closed, and the atmospheric vents and drains are opened as steaming progresses.

6. Before steaming is commenced, the water side of condensers and coolers should be drained and vented. Instruments that may be damaged by steam need to be isolated. Pumps whose seals may be sensitive to heat need to be bypassed. Relief valves that may discharge air during steaming should be diverted away from the flare system.

7. Steaming should be avoided when the column may contain an explosive mixture, because steam may generate static electricity (5). If flammable materials are in the column, it is best to purge them with an inert gas before steaming the column.

8. Steaming should be avoided when an acid solution may form in the column upon condensation.

9. Steaming may accelerate stress corrosion cracking in non-stress-relieved carbon steel columns containing residual alkaline solution. Experience gathered in a large number of amine absorbers (337) links a high incidence of stress corrosion cracks to this shutdown procedure. Both the hot temperature during steamout and the concentration of residual alkaline pockets by evaporation contribute to this problem. In such services, it is best to water-wash the column to remove residual alkaline prior to steamout (337).

10. If the column may contain combustible or pyrophoric deposits or liquids, it should not be open to the atmosphere at the conclusion of steaming at shutdown. Instead, the column should first be cooled (e.g., by an inert gas), and only then opened. While this procedure is a good practice in all columns, it is imperative when the column contains internals with large surface area per unit volume [e.g., structured packing (389), mist eliminators; to a much lesser extent also random packing]. A considerable amount of deposits can adhere to the extended surface; these internals trap a substantial amount of heat, and their pressure drop tends to be low. Air ingress into such a hot system will generate a large "chimney effect," and at the hot temperature, the deposits may ignite. The author is familiar with one incident in which a bed of carbon steel structured packings burnt after air was allowed into a column shortly after steaming was completed. The author also heard of mist eliminator fires due to a similar procedure.

A more detailed discussion of several of the above guidelines, as well as some additional recommendations, is available in Amoco's booklets (3, 5).

11.9   Dehydration (Dryout)

Application of a special column dehydration (dryout) technique is usually required when the column needs to be bone dried. The common techniques used are:

1. *Dehydration by liquid circulation:* This technique is almost always used in hot services (> 200 to 250°F). It has the added advantage of heating up the column at the same time. The circulation needs to be performed at a positive pressure so that water can be drained from low points.

   This technique may be unable to bone-dry the column at temperatures lower than about 150 to 200°F, because water may dissolve or emulsify in the liquid and the temperature may be too low to strip water out. In one case, a temperature of at least 275°F was required for adequate drying (23). Details of this technique are discussed in Sec. 11.10

2. *Dehydration by total reflux distillation:* This technique is suitable for warm services (100 to 200°F) and is similar to liquid circulation; but at the same time, water is stripped from the liquid. Similar to dehydration by liquid circulation, this operation needs to be performed at a positive pressure. This dehydration procedure can be slow and tedious. Further details are discussed in Sec. 12.8

3. *Dehydration by solvent washing (e.g., glycol, acetone, methanol, propanol):* This technique is effective when the solvent is hygroscopic. Compared to other techniques, solvent washing is often expensive. It is commonly used in chilled and cryogenic services as well as in some vacuum services, where other techniques may not be suitable. In other cases, it is generally economical either for small columns, or when the solvent can readily be regenerated. Guidelines 1 to 4 and 9 to 18 in Sec. 11.6 extend to this technique. Some additional discussion is also available in Sec. 13.10

4. *Dehydration by purging with hot gas (air, inert gas, or superheated steam):* This technique evaporates water from the column. Water pockets in connecting pipes can be blown out using the purge gas (see Sec. 11.1). Dehydration can be performed at slight positive pressure or under vacuum (if practicable). A positive pressure is required for blowing water out of connecting lines, while vacuum promotes evaporation of water in the column. The operation can be initiated under pressure, and water pockets blown out, before switching to vacuum to enhance evaporation. The gas temperature

should be as high as possible without exceeding any equipment limitations or risking fire. The author knows of one bed of oil-coated steel random packing that caught fire upon hot air dehydration. Flame air heaters should be avoided, as flame impingement can damage column internals.

Dehydration by purging with hot gas is fast, but seldom achieves complete water removal. In one case (23), numerous water-induced pressure surges occurred at startup due to incomplete dryout using this technique. When bone-drying is required, this technique is best used for initially removing the bulk of the water. The rest is removed by one of the above techniques, such as liquid circulation (23, 54).

5. *Dehydration by purging with dried (or desiccated) gas (e.g., air, nitrogen, or natural gas):* The gas absorbs and evaporates water from the column. This technique is similar to the hot gas purging technique, but the gas is dried before entering the column. It is commonly used for dehydrating cryogenic columns, and can only be applied when a drier is available for startup. Figure 11.7 shows a typical block diagram.

To maximize water absorption, the gas temperature and the flow at the inlet to the column (or plant) should be as high as possible without exceeding any equipment limitations. The dehydration can

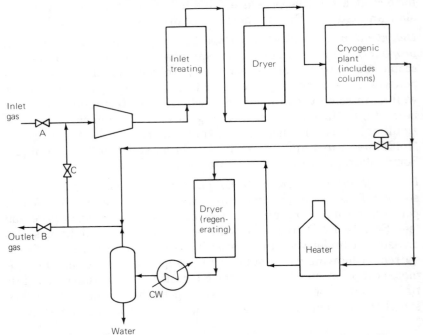

Figure 11.7 Dehydrating by purging with dried gas. Open circuit: valves A, B open; C shut. Closed circuit: valves A, B shut; C open.

be carried out using an open or closed circuit (Fig. 11.7). A closed circuit must be used if excessive water in the outlet gas is intolerable. A closed circuit is also preferred when the inlet gas has a high moisture content or is of limited availability. When a closed circuit is used, it pays to maximize pressure and minimize temperature at the knockout drum, since this is the only water outlet in the system. A more detailed discussion, specifically related to cryogenic gas plants, is available elsewhere (16).

Prior to dehydration, low points, vessels, heat exchangers, and pipes should be properly drained, as any water left in the system would prolong the dehydration process and reduce its effectiveness. Periodic purges and checks should also be conducted at these points during dehydration. All bypasses and vent lines should be adequately purged. When properly performed, dehydration by purging with dried gas can effectively eliminate water from columns (16).

11.10 Liquid Circulation

Liquid is circulated through the column in order to heat up the column or to flush pockets of undesirable materials (e.g., water from a hydrocarbon column; chemicals from a previous campaign) prior to feed introduction. The circulation liquid must be compatible with the chemicals which are normally in the column. Some key considerations are described below and illustrated in Fig. 11.8. Depending on the nature of the circulation, several of the guidelines in Sec. 11.6 also apply to liquid circulation.

When used for heating, circulating liquid is heated (e.g., in a startup heat exchanger) before entering the column. The liquid warms up column internals and also undergoes some evaporation, which partially replaces the inert gas in the column by a condensable vapor. Both actions hinder cooling of the feed when it later enters the column. The heating temperature must be kept low enough to avoid boiling, degradation, or decomposition of the circulating liquid, and high enough to prevent freezing or excessive viscosity.

The circulating liquid must remain fluid without solidifying or becoming excessively viscous even at the lowest temperature expected at the column internals before liquid circulation is commenced. This is especially important in winter startups in cold climates. It may pay to warm up the column internals and piping by passing hot air or steam through the column just before liquid circulation is started.

Alternatively (or additionally), a less viscous circulating liquid can be used. It is often possible to use a low-viscosity, freeze-resistant liquid in the initial circulation steps, and to replace it by a compatible high-boiling, decomposition-resistant liquid in the hotter steps.

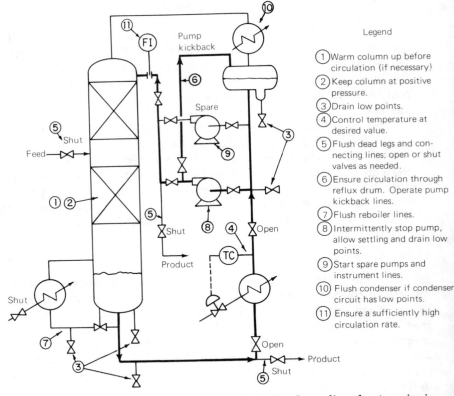

Figure 11.8 Some key considerations for liquid circulation (heavy lines denote main circulation path).

When liquid circulation is used for column dehydration (in addition to heating), it is common to perform the heating and dryout in steps of cold, warm, and hot circulation. The cold circulation ($\cong 100°F$) sweeps water to low points, from where it must be drained. All the drains must be operated until no more water comes out; any plugged drains must be unblocked. During the warm circulation, the temperature is gradually raised to about 180°F. This step melts any residual ice and vaporizes remaining pockets of water.

Pumps (both regular and spare) must be started up to flush out suction and discharge lines. Heat exchangers, side product lines, blowdown and emergency lines, and any connections used intermittently should be flushed and drained during this step. Circulation should be established and maintained in pumparound lines (296). Instruments should be flushed and be fully operational before this step is complete. If the column operates at temperatures lower than about 200°F, the warm circulation completes the heating and dryout.

When the column operates at temperatures greater than about 200 to 250°F, and when positive removal of water is necessary (e.g., services prone to pressure surges), a final hot circulation step is required. This step vaporizes and removes the last traces of water from the system. Vaporized water will collect in the cooler parts of the system (e.g., reflux drum boot) and must be drained from there. During hot circulation it is imperative to prevent refluxing of water back to the system. It is equally imperative to prevent hot circulating liquid from entering the cooler parts of the system where water condenses. The author is familiar with a case where overlooking the latter precaution caused a pressure surge of the same nature as the hot circulation is attempting to prevent. Any jumpover lines carrying circulating liquid to the reflux system (such as that shown in Fig. 11.8) should be disconnected, and the circulating liquid should enter the column at a lower point and be pumped by a separate (not the reflux) pump.

A common troublespot during hot (sometimes even during warm) circulation is loss of suction on the circulating pump (5). This results from vaporization of water at the pump. Repeated stops and starts of the pump and appropriate manipulation of pump discharge and vent valves are important.

If the circulating liquid is likely to form an emulsion with water (e.g., certain oils, especially during cold circulation), it may pay to periodically interrupt circulation and let water settle out of the emulsion. Circulation can then be resumed.

Dryout by circulation should be performed under positive pressure (5, 296); otherwise, water from drains will be sucked back into the column when a drain is opened. Circulation for dryout must be carried out at a rate high enough to ensure adequate sweeping of water to the low points. Dryout is considered complete when column temperature exceeds 250 to 350°F (or the maximum desirable temperature), no more water comes out of drains (especially the drain on the reflux drum boot), and when pump suction is steady.

Amoco's booklet (5) provides an excellent detailed discussion on dryout of refinery columns (and other refinery equipment); several valuable ideas from this discussion have been included above. API's booklets (7) contain some very useful guidelines for refinery vacuum towers dryout.

11.11 Solvent Testing

A "solvent" is a relatively safe fluid whose properties are close to those of the process fluid. Solvent testing is sometimes used as an alternative to steam-water testing (Sec. 11.7). The purpose of solvent testing is similar to steam-water testing: to check heat exchangers and column performance, observe that gravity lines operate properly, check for leaks and malfunctions prior to the introduction of chemicals, and familiarize the operating team with the column. Compared to steam-water testing, solvent testing usually affords a far closer simulation of actual process conditions, a wider and more meaningful range of tests, and is less troublesome. However, solvent testing may be expensive, and is generally avoided unless the process fluids are hazardous or difficult to handle.

An excellent discussion on solvent testing is available elsewhere (131). Some of the important considerations are summarized below:

1. Caution is required when selecting the solvent. The closer it is to the process fluid, the more meaningful and less troublesome the test is. The solvent must also be nonreactive and far less hazardous than the process fluid. The author has experienced cases where columns handling reactive unsaturated hydrocarbons were tested and started up by using analogous straight-chain hydrocarbons. The testing proved very fruitful.

2. The quantity of solvent required must be carefully calculated, making allowance to losses. Purchasing excessive quantities of solvent may be expensive.

3. Pretest checks must be carried out to ensure that the column operating capacity, relief capacity, safe working temperature, and internals supports are adequate for the planned test conditions.

4. The expected test conditions should be marked on a copy of the column flowsheet. Any deviations from the expected conditions should be thoroughly investigated.

5. The column should be commissioned prior to the introduction of solvent in a similar manner to its commissioning prior to introducing the process fluid.

6. Following the test, the solvent must be thoroughly removed. If the solvent material is acceptable in the product, without violating quality specifications, and is unlikely to interfere with column operation, removal may not be necessary.

7. A major advantage of solvent testing is that it permits testing the reliability of instrumentation, alarms, trips, and emergency systems. To achieve meaningful results these systems should be fully operational during the test. The installation of any additional instrumentation (e.g., temporary differential pressure recorders for flood-point determination) should be considered for obtaining better results. The tests should be carefully planned and their results

logged and documented. The column and its auxiliaries should be deliberately operated near their limits at the tests. It is worth keeping in mind that data collected in the solvent test may become extremely valuable for resolving future operating problems.

8. Test planning should recognize the value of familiarizing the operating team with column operation under conditions where mistakes are neither too costly nor too hazardous. It is often beneficial to schedule additional tests just for the sake of operator training.

11.12 Miscellaneous Considerations

Several important considerations for column preparation and testing do not adequately fit under any of the previous headings. Some are outlined below:

1. All the relevant statutory and regulatory requirements and company practices for handling the chemicals must be strictly adhered to during commissioning.

2. Freezing of water or residual chemicals prior to startup or following a shutdown can damage equipment and introduce hazards because of plugging. Water should be drained from all low point drains at a shutdown. All drains should be checked for plugging.

3. If pyrophoric deposits are anticipated, the areas where these deposits are likely to accumulate should be kept wet or oiled upon shutdown.

12

Column Startup
and Shutdown

The column's ability to weather foreseen and unforeseen difficulties is severely tested during startup and shutdown. Failure to recognize and eliminate pitfalls or hazards associated with these operations has caused catastrophic accidents, sometimes with injuries or loss of life; damage to columns and their internals; and/or prolonged periods of downtime. It is therefore imperative to recognize and eliminate (or at least minimize) the pitfalls associated with column startups and shutdowns.

This chapter discusses several general pitfalls inherent in distillation startup and shutdown operations. Hazards and pitfalls specifically related to the nature of the chemicals processed (e.g., toxicity, flammability), the simultaneous startup/shutdown of the column with other units (e.g., reactors, furnaces), or to the overall startup/shutdown policy (e.g., operator training procedures) are outside the scope of this distillation text. The safe handling procedures of the relevant chemicals, as well as the overall startup/shutdown policy, must be followed to cater for the latter hazards and pitfalls.

This chapter reviews common column startup and shutdown practices, outlines those which are generally preferred, highlights the consequences of poor practices, and supplies general guidelines for avoiding the hazards and pitfalls generally inherent in column startup and shutdown.

12.1 Strategy of Distillation Startups and Shutdowns

Column startup usually consists of the following steps:

1. Commissioning. These operations are discussed in detail in Chap. 11. Commissioning operations clear the system of undesirable ma-

terials and test it in preparation for introducing the chemicals. The demarcation between "commissioning" and "actual startup" is normally arbitrary, depending on the system and on subjective judgment.

2. Final elimination of undesirable materials.
3. Bringing the column to its normal operating pressure.
4. Column heating and cooling.
5. Introducing feed.
6. Starting up heating and cooling sources.
7. Bringing the column to the desired operating rates.

The sequence of steps performed at startup typically follows the above order, with some deviations. Such deviations are often desirable (even necessary) to suit the system and available hardware. The best sequence of events must be judiciously determined considering the chemicals, the system, and the practices generally preferred for distillation startup. The latter practices are discussed in the following sections.

Similarly, column shutdown usually consists of the following steps:

1. Reducing column rates
2. Shutting down heating and cooling sources
3. Stopping feed
4. Draining liquids
5. Cooling or heating the column
6. Bringing the column to atmospheric pressure
7. Eliminating undesirable materials
8. Preparing the column for opening to atmosphere

The above sequence of steps is typical of shutdown operations. Other comments made above regarding the sequence of startup steps extend also to the sequence of shutdown steps.

Distillation startup/shutdown must incorporate several measures which are common to startup/shutdown of most plant units. These measures are generally considered a portion of the overall plant startup/shutdown policy, and are discussed in detail elsewhere (e.g., 127, 131). These measures include

1. Preparing adequate operating, startup, shutdown and maintenance procedures.

2. Ensuring the startup/shutdown team consists of personnel who possess all the skills likely to be required.

3. Adequately training the startup team, supervisors, and operators.

4. Proper startup/shutdown planning.

5. Securing any raw materials, equipment, and spare parts required.

6. Developing adequate procedures for last-minute modifications, safety checks and audits, prestartup inspections and startup/shutdown recordkeeping.

7. Developing individual tasks and objectives and ensuring these are well understood by members of the startup team.

8. Preparing checklists for each phase of the startup/shutdown.

12.2 Useful Startup/Shutdown Lines

Column piping often contains lines whose primary purpose is to minimize material loss or speed up startups and shutdowns. While such lines may be extremely useful on such occasions, leakage across their valves during normal operation may cause severe operating problems. Further, during normal operation these lines become dead legs in which freezing, plugging, or corrosion may occur. It is therefore a good practice to ensure that such lines are isolated, blinded, and left free of any hazardous or undesirable materials during normal operation.

Startup/shutdown lines increase the complexity of the system and open up new routes for chemicals to travel back and forth through the plant and to reach locations where they are hazardous or undesirable and from where they would otherwise be barred. It is therefore imperative to engineer those lines carefully, and to examine thoroughly the possibilities and consequences of materials traveling through them into other units, both under normal and abnormal operation. The effects of material travel on the relief capacity of the relevant units should also be considered. Some examples of the potential consequences of undesirable material travel are in Sec. 12.11. It is a good policy to restrict installation of startup/shutdown lines to those absolutely necessary.

Some of the more common lines useful at startups and shutdowns are (Fig. 12.1):

1. *Pressuring up lines:* These enable bringing the column to pressure with desirable materials prior to starting up. These are discussed in detail in Sec. 12.3

2. *Column bypass lines:* These are essential if the column is in a com-

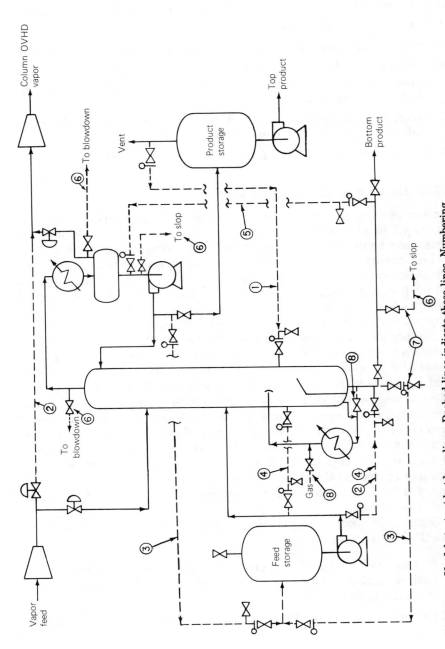

Figure 12.1 Useful startup/shutdown lines. Dashed lines indicate these lines. Numbering matches numbering in the text.

pressor train and may be sensitive to surging. Column bypasses can also be useful for overcoming sealing problems (Sec. 12.10), for tackling some column emergencies, and for starting up a column train.

3. *Lines for recycling top and/or bottom product to the feed tank:* These are useful for minimizing the amount of off-spec material produced. They are often essential when starting up a column train which is started one column at a time.

4. *Lines for introducing the feed at the reboiler inlet sump:* These lines are useful for avoiding excessive temperatures at the column bottom. When column heating is performed solely by reboil action, they can also speed up column heating (i.e., they eliminate the condensing action of a cold feed in the bottom section).

5. *Liquid circulation lines:* These are required when liquid circulation is performed at startup (Sec. 11.10). Often, a jumpover from the column bottom to the reflux line is needed, but other lines may also be required. In a refinery crude fractionator, it has been recommended (237) to install the jumpover line to the return line of the uppermost pumparound and to size it for 20 percent of the net distillate product rate.

6. *Process vents, high point vents, and low point drains:*

7. *Liquid dump lines:* These can help speed up column draining.

8. *Reboiler startup lines:* These are discussed in Sec. 15.4.

12.3 Startup/Shutdown Considerations: Pressure Distillation

In pressure distillation, it is generally a good policy to bring the column close to its normal operating pressure prior to liquid feed introduction. Pressuring up is best performed using one of the condensable feed components. At shutdown, all liquid should be drained out of the column before depressuring. The discussion below expands on these rules.

A pressurized liquid, whose temperature exceeds its atmospheric boiling point, will chill upon flashing to atmospheric pressure (Fig. 12.2b). If the flash temperature is lower than the minimum safe working temperature of the column metal, a brittle failure of the column metal may occur. Pressuring the column up before liquid introduction raises the flash temperature when feed is introduced.

Even if the safe working temperature is suitable to handle the flashing liquid, pressuring prior to liquid introduction is still the recommended startup technique. This procedure eliminates the large

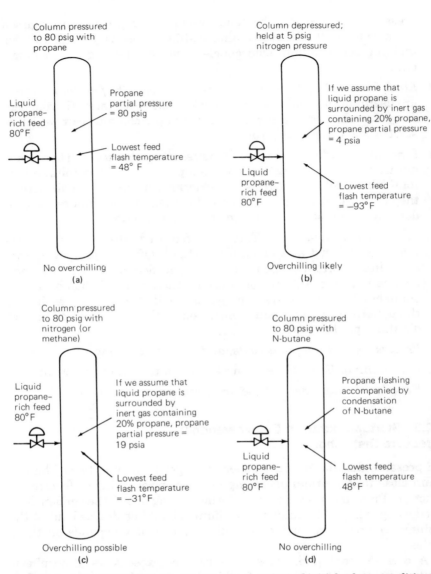

Figure 12.2 Lowest feed flash temperature for a variety of initial column conditions. Propane-rich liquid feed, metal minimum safe working temperature = − 20°F, assumed in all cases. Column pressured with (a) 80-psig propane, (b) 5-psig nitrogen, (c) 80-psig nitrogen, (d) 80-psig butane.

thermal stresses that can be caused by liquid flashing to excessively low pressure. Such thermal stresses can cause unnecessary joint leaks, metal fatigue, and shorten the column's service life. Further, excessive flashing of feed upon entry to the column can cause excessive velocities in the column feed region. These high velocities may damage or erode column internals.

The column is best pressured up with one (or more) of the condensable components of the feed mixture (Fig. 12.2a), or a gas of similar volatility, rather than with an inert gas (i.e., one which is much more volatile than the feed mixtures). Pressuring up with an inert gas will cause colder temperatures and may provide insufficient protection against metal overchilling (Fig. 12.2c). Liquid continues to flash (and chill) until boiling liquid reaches equilibrium with condensing vapor. When the column contains an inert gas, equilibrium will only be reached when the partial pressure of the evaporated liquid builds up sufficiently in the vapor space. Until then, flashing temperatures will be much lower than the feed boiling point at the column pressure (compare Fig. 12.2a and c).

When the column initially contains an inert gas, the flash temperature is considerably higher when the column is pressured than when the column is depressured (compare Fig. 12.2b and c). On the other hand, overchilling at an elevated pressure can at times be more detrimental to the metal than overchilling at atmospheric pressure, because the minimum safe working temperature of the metal increases with column pressure. Both effects should be considered when devising the startup strategy and the only available gas is inert; in most cases, it is still better to pressure up.

If the only available pressuring-up gas in inert, liquid must be introduced extremely slowly. Until the partial pressure of the feed components in the vapor space builds up, the only protection against overchilling is sensible heat transfer from the gas and metal to the flashing liquid. The faster the liquid is introduced, the less effective this mechanism is in evaporating all the feed liquid and preventing overchilling. In addition to overchilling, fast liquid introduction into a column containing an inert gas may quickly raise column pressure and often lift the relief valve. The total pressure that can be reached in the column is the partial pressure of the inert gas (which is equal to the initial pressure at the time of feed introduction), *plus* the vapor pressure of the feed liquid. This total pressure may exceed the relief valve setting. A high column pressure before the column metal has a chance to warm up can be most detrimental to the metal. A sufficiently slow liquid introduction permits adequate venting of the inert gas as the pressure climbs.

Pressuring the column up with a heavy vapor (i.e., one which has a considerably higher boiling point than the feed mixture) will prevent metal overchilling. The heavy gas will condense as feed is added, thus preventing a fall in feed flash temperature (Fig. 12.2d). However, the boiling point of the resulting mixture may be excessive for reboiler startup. This problem is most severe when reboiler and condenser ΔT's are small. Excessive liquid drainage from the column may then be required prior to reboiler startup.

The foregoing emphasizes that the least troublesome startup procedure in pressure columns whose feed is a flashing liquid is to pressure the column up with vapor of one of the key components before introducing feed liquid. A pressuring up line from the vapor space of product storage, or a simple startup vaporizer (which may serve several columns), are minor but often worthwhile investments (Fig. 12.3). The startup vaporizer can be located either on the line from liquid storage or on the liquid feed line. Either technique may prevent equipment damage, material loss, and/or startup delays. The vaporizer techniques are generally faster. If the column is pressured up by storage

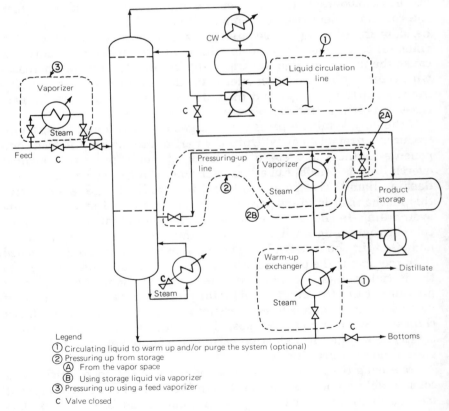

Figure 12.3 Some techniques for preparing pressure columns for liquid introduction.

vapor, a drop in storage temperature should be expected and, if necessary, allowed for.

The above recommendations do not preclude circulating liquid through the column (for warming or flushing purposes) before and while the column is pressured up. This practice is acceptable and often helpful against overchilling upon feed introduction, especially in cold climates. The circulating liquid must be nonflashing (the column feed is therefore unsuitable for this purpose). This practice is described in detail in Sec. 11.10.

Depressuring a column containing a liquid whose atmospheric boiling point is lower than 32°F should be avoided until all of the liquid has been drained. The considerations here are similar to those previously described. In addition, temperatures lower than 32°F can freeze water in the lower part of the column, or freeze steam condensate in heat exchangers.

12.4 Startup/Shutdown Considerations: Vacuum Distillation

In vacuum services, it is best to draw vacuum before introducing the feed. If the feed is introduced before vacuum is fully drawn, the reboiler needs to be started up to avoid liquid accumulation in the column base. Starting the reboiler up before the column is brought to its normal operating vacuum is likely to cause excessive bottom temperatures, accompanied by material degradation, polymerization, and coking. This in turn may lead to plugging, capacity loss, and packed-tower maldistribution. When the chemicals decompose exothermically upon heating, excessive bottom temperature may cause a runaway reaction. In one nitro compound distillation incident (96), failure of a vacuum system raised the bottom temperature of the still to the self-accelerating decomposition level and resulted in an explosion. Drawing vacuum before introducing feed is most important in such services.

By similar considerations, it is best to shut down the reboiler and drain liquid before repressuring the column at shutdowns. This practice is imperative if the column contains hot, flammable liquids and air is likely to enter during repressuring. In one incident (169), violent combustion of liquids upon air ingress into a column at shutdown caused injury and column damage.

It is acceptable (and often desirable) to circulate liquid (even feed) before and while drawing vacuum, as long as its temperature is maintained low enough to avoid decomposition. This practice is detailed in Sec. 11.10.

When a light mixture (e.g., top product, off-spec top product, or a feed stream void of the heavy components) is used for startup, an excessive bottom temperature is unlikely to occur. In this case, feed can be introduced and the reboiler started before vacuum is fully drawn. This practice is often used when startup time is critical.

An alternative technique that prevents excessive bottom temperatures and permits reboiler startup before vacuum is fully drawn is routing the column feed directly into the reboiler inlet (Fig. 12.4). This technique may generate substantial quantities of off-spec bottom product; a provision for recycling bottom (and possibly also top) products into the feed tank may minimize product loss.

As vacuum is pulled on the system, routine frequent tests of air leakage (i.e., oxygen analyses) must be conducted. If an excessive leak is detected and cannot be taken, the unit may require shutdown and repair.

An outage of vacuum-generation equipment (jets, vacuum pump) can lead to pressure surges, which in turn can uplift and damage trays or other internals. The following sequence of events was described by Harrison and France (150a). Consider loss of the vacuum pump. A check valve or seal leg will prevent loss of column vacuum from vapor backflow through the vent system. However, any inerts leaking into or entering the column will no longer be evacuated. They will accumulate in the overhead system and inert-blanket the condenser. Column pressure will rise, possibly causing some of the adverse effects described above.

Once the vacuum pump is restarted, it can almost instantaneously evacuate the few cubic feet of inerts that blanket the condenser (this is the same vacuum pump that may have taken hours to evacuate a column full of inerts at startup). The column will immediately depressure to normal vacuum. This sudden depressurization, compounded by the heat that becomes stored in the column (due to higher temperatures) during the higher-pressure operation, quickly vaporizes liquid. This rapid vaporization generates a rapid and impulsive pressure surge that can damage trays or other internals. This surge can be minimized by (150a) manually opening the vacuum control recycle, inert, or steam loading stream before restarting the vacuum pump or jets. The column is then slowly depressured to the desired vacuum.

12.5 Startup/Shutdown Considerations: Plugging Prevention

During normal operation, liquid downflow sweeps solids across the trays and through downcomers, while vapor upflow prevents solids

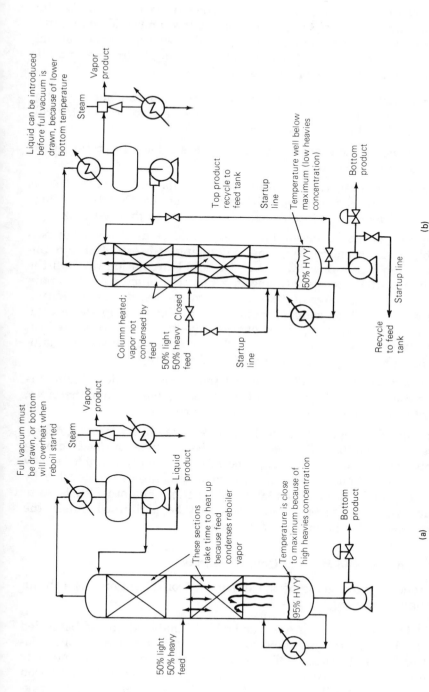

Figure 12.4 Introducing feed at the column base to speed up column startup. (a) Conventional; (b) feed introduced at column base.

Full vacuum must be drawn, or bottom will overheat when reboil started

Steam

Vapor product

Liquid product

These sections take time to heat up because feed condenses reboiler vapor

Temperature is close to maximum because of high heavies concentration

50% light 50% heavy feed

95% HVY

Bottom product

(a)

Liquid can be introduced before full vacuum is drawn, because of lower bottom temperature

Steam

Vapor product

Column heated; vapor not condensed by feed

Closed

50% light 50% heavy feed

Startup line

Top product recycle to feed tank

Startup line

Temperature well below maximum (low heavies concentration)

Bottom product

50% HVY

Recycle to feed tank

Startup line

(b)

from settling in small perforations and in valve units. During startups, liquid and vapor rates tend to be low, and solids may settle in the column, rapidly plugging trays and downcomers and causing "sticking" of valve units (Chap. 6). The following techniques may be useful for minimizing settling of solids during startups:

1. When material solidification is temperature-dependent (e.g., liquids near their freezing points), it is important to preheat the column and feed prior to feed introduction (see Sec. 12.6).

2. Whenever possible, the column should be started up on total reflux, using a solid-free fluid.

3. There is an incentive for minimizing the solid content of the feed at startup. This can be accomplished by temporary filtration, straining, or drawing the feed from a higher point in the feed tank.

4. High vapor and/or liquid flow rates should be used whenever possible. If downcomers are prone to plugging, it is essential to maintain high liquid flow rates (e.g., by liquid circulation through the column) when feed is introduced. When tray plugging is the main concern (e.g., a valve unit sticking), high vapor rates are essential. In one case (294), solids accumulation during startup was a common cause of sticking of peripheral valve units; this restricted capacity in subsequent operation. The problem was solved by maintaining high vapor rates throughout the startup.

5. It is a good practice to desuperheat reboiler or stripping steam during startups in order to minimize hot-spot formation and, therefore, material degradation at the base of the column.

6. In some fouling services, such as chlorinated hydrocarbons, deposits are sometimes "baked" into reboiler tube surfaces during reboiler shutdown (279). This operation thermally degrades sticky deposits and converts them into compounds easily removable by water jets. This operation subjects the tubes and other surfaces to high thermal stresses and should not be performed unless the equipment can handle these stresses.

12.6   Cooling and Heating Considerations

The following are some of the important considerations for column cooling and heating:

1. Columns containing hot, condensable materials must be cooled slowly at shutdown. Rapid cooling may cause excessive condensation, with identical ill effects to those described in guidelines 1 to 4 in Sec. 11.7. In vacuum services, this excessive condensation may also burst the bursting disk and/or induce excessive air leak-

age into the column; this in turn may generate a hot, explosive mixture. In one incident (169), injury and tray damage occurred when hot, flammable liquid contained in a vacuum column caught fire upon opening a manhole.

2. The cooling rate must be slow enough to permit quenching of any trapped heat pockets (e.g., hot liquid which is unable to drain due to plugging). Unquenched heat pockets may catch fire once exposed to air or cause a pressure surge upon contacting wash water. In one incident (23), tray damage occurred when air and wash water were introduced into a column which contained trapped heat pockets.

3. Condenser wind-down is the key to the column cooling rate at shutdown. The condenser cooling should be cut back so that column pressure is kept on control for as long as possible. The cutback may raise the condenser outlet cooling water temperature; this temperature must be kept below the boiling point, even if it means losing column pressure.

4. An inert or process gas can often be admitted into the column to keep pressure up and prevent vacuum while the column is cooled.

5. As much liquid as possible should be drained out of a hot column prior to and during the initial cooling period. This is particularly important if the liquid solidifies or becomes highly viscous as it cools. The author experienced a case where overlooking this guideline caused a hard, solid mass to form at the bottom sump. The mass was extremely difficult to break up.

6. Hot material should be drained from the tower only into transfer and receiving equipment that can handle the hot temperatures. For instance, transportation of hot amine to storage at shutdown via piping that is not mechanically designed for hot alkaline service was proposed as a factor causing or accelerating stress corrosion cracking (337).

7. A number of techniques can speed up column cooling. These include purging the column by a cool inert or process gas (e.g., steam, nitrogen, fuel gas), liquid circulation, and water injection. Water injection should not be applied when column temperature exceeds 200°F, or if any heat pockets are present. Guidelines 1 to 4 and 9 in Sec. 11.7 apply also to water injection for cooling the column.

8. If the column may contain combustible or pyrophoric deposits, it should be cooled to ambient temperature before being open to the atmosphere. This is most important if the column contains internals with a high surface area per unit volume (e.g., struc-

tured packing). Consequences of overlooking this practice are described in guideline 10 in Sec. 11.8. The cooling is best performed using an inert gas.

9. Heating the column prior to startup is essential when the column separates materials which at ambient temperatures are viscous or above their freezing points. Heating is also essential in absorbers and wash columns where condensation of heavier components from the gas is undesirable.

10. Column heating is usually carried out either by steaming (Sec. 11.8), by circulating liquid (Sec. 11.10), or by hot air blowers. In freezing climates, hot air blowers are often used to warm up the column prior to steaming or liquid circulation.

11. If it is decided not to heat up the column prior to startup, it will behave as a major heat sink when reboil is started. Reboil must be slowly applied until the column heats up. Excessive reboil rates during this period may cause premature flooding and even damage to column intervals (105). As shown earlier (Fig. 12.4), it may be desirable to run the column feed directly to the reboiler inlet sump.

12. Reboiler heating upon startup should proceed in slow, incremental steps to avoid excessive thermal stresses. The thermal stress problem is most severe when reboiler tubes are roll-welded to the tubesheet (358).

13. A number of mechanical checks should be performed as the column heats or cools. The effect of thermal expansion or contraction on supports, bellows, and connecting pipelines should be examined (131). One accident was reported (210) where a reflux line, which was rigidly fixed to brackets welded to the column shell, tore one of the brackets from the column because of differential expansion between the hot column and the cold pipe. Sliding supports should adequately respond; if they do not, mechanical damage may occur. Connecting lines should not bow; bowing may overstress pipes and cause them to fail. Bowing may also generate undrainable low points, where undesirable components (e.g., water) may accumulate; an accident resulting from this problem is described in Sec. 13.5. Spring supports on connecting lines should have their pointers between the hot and cold settings (131). Pipe and column joints (e.g., flanges) should be periodically inspected for leakage; a nonleaking joint may spring a leak as temperature is raised or lowered.

12.7 Excess Subcooling or Superheat

Excess feed subcooling steps up condensation and liquid traffic in the column. This is countered by the control system or the operator, who will cut back reflux and condenser duty. Overall, column vapor and liquid loads will change only slightly, and excess subcooling will not be troublesome.

If reflux cannot be cut back (e.g., in an unrefluxed stripper, in azeotropic distillation, or when the packed section above the feed is close to its minimum wetting limit), boilup will need to be raised to compensate for the excess subcooling. Vapor and liquid traffic below the feed and reboiler duty will rise and effectively lower the column feed capacity. Premature flooding may result. If the lower capacity or higher reboiler duty cannot be tolerated, feed preheating (Fig. 12.5a)

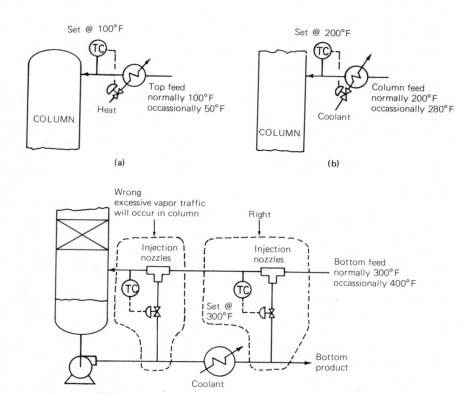

Figure 12.5 Cures for excess subcooling and superheating problems. (a) Preheating a subcooled feed; (b) precooling a superheated feed; (c) injecting a subcooled quench into a superheated bottom feed.

may be required. In three reported cases (19, 76, 239), feed or reflux preheating remedied excess subcooling problems.

The converse applies to excess feed superheat. Excess feed superheat will have only a slight effect on column vapor and liquid loads unless boilup cannot be cut back (e.g., in an unreboiled rectifier, where boilup is provided by the bottom feed). In the latter case, liquid and vapor traffic above the feed and condenser duty will rise and effectively lower the column feed capacity. Premature flooding may result. If the lower capacity or higher condenser duty cannot be tolerated, the excess superheat needs to be trimmed. This is usually accomplished by precooling (Fig. 12.5b) or by injecting subcooled quench liquid into the feed (Fig. 12.5c). The liquid injected must be sufficiently subcooled, or it will vaporize; vaporization will increase the column vapor traffic and will therefore be self-defeating. The injection nozzles must be properly designed to prevent liquid hammer.

In one case (239), column design overlooked the vaporization taking place when a bottom feed with 300°F superheat enters a column. The actual column vapor rate was triple the rate designed for, and premature flooding resulted. The problem was solved by injecting subcooled liquid into the feed line. At a later stage, the subcooled liquid was replaced by a lower-boiling-point liquid which vaporized, and the premature flooding reoccurred.

In another case (150a), a subcooled feed entered the bottom sump of a packed rectifier. Preheating this feed consumed 40 percent of the boilup; the balance entered the rectifier's packed section. When the feed was shut off, the heat sink was eliminated, causing excessive vapor flow and consequent flooding in the packing. The problem was solved by installing an override controller that would cut back boilup upon excessive packing pressure drop.

One designer (76) has argued that in packed columns, it may take several theoretical stage equivalents to heat a subcooled feed to its boiling point; consequently, feed subcooling may incur a substantial efficiency penalty. No experimental evidence was produced to back the argument, and a case history presented in support (76) can also be explained in terms of premature capacity restriction due to subcooling. The author feels that efficiency loss due to subcooling is generally quite small, and is unlikely to have a major impact in other than very short beds. The author's comment is consistent with Strigle's recommendation (386) of adding one theoretical stage (as distinct from several) to allow for reheating a highly subcooled reflux.

The above discussion pertains to subcooling and superheating effects that tend to persist. Additional considerations apply to handling subcooled or superheated feeds during abnormal operation, such as startup, shutdown, or utility failure. Some of these are:

1. When a highly subcooled feed enters a hot column at startup, fast feed introduction or a sudden stepup of feed rate can induce rapid vapor condensation. The ill effects and preventive measures are similar to those discussed in guidelines 1 to 4 in Sec. 11.7.

   Whenever possible, it is best to either preheat the cold feed at startup or to start the column up on total reflux and then slowly admit the feed. These techniques are especially useful when the material distilled is heat-sensitive, of high-viscosity, or when its temperature is close to its freezing point.

2. When the column separates heat-sensitive materials or contains heat-sensitive internals (e.g., plastic packing), it is a good practice to desuperheat direct-injection steam at startups and shutdowns (exception: columns that must remain dehydrated). While during normal operation overheating is prevented by liquid downflow or by a liquid layer covering the steam inlet sparger (Fig. 4.3a), these cannot be relied on during startups and shutdowns. In one incident (440), superheated steam vaporized the liquid layer covering the steam inlet sparger, then melted the plastic packing in the column. Other incidents where plastic and aluminum packings were over-heated and damaged at startup have also been cited (219).

3. In some systems, utility failure may cause hot feed to enter the col-umn. Such systems are best designed with internals that can with-stand the higher temperatures. If the system contains internals that can be damaged by the hot feed (e.g., plastic packings), auto-matic cutoffs or reliable fail-safe instrumentation must be provided to divert the feed away from the column. A typical example is an ammonia plant absorber-regenerator system (Fig. 12.6). If solution circulation fails (e.g., pump failure or power failure), the feed will not be cooled in the regenerator reboiler and will enter the ab-sorber excessively hot. On several occasions, this caused melting of plastic packings in the absorber (349).

12.8 Total Reflux Operation

Columns are frequently operated on total reflux during short (from a few minutes to a few days) feed interruptions. Columns are also fre-quently started up on total reflux. A total reflux startup is one of the most powerful, and often one of the least troublesome, startup tech-niques.

A total reflux startup is best performed with a mixture whose com-position resembles the feed composition. Alternatively, total reflux op-eration can be carried out using one or more of the feed components. The column, reboiler, condenser, and reflux system are stabilized at

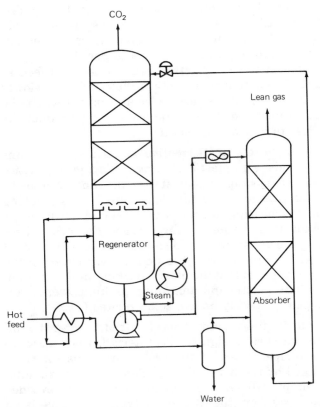

Figure 12.6   A typical simplified ammonia plant CO_2 absorber-regenerator system.

total reflux. Levels are established and placed on control, and instrumentation is debugged. If the mixture is similar to the feed composition, a temperature profile and on-spec products are also established. Upon feed introduction, all that remains to be done is initiating product flows and trimming the column; the products usually stay on-spec.

A total reflux startup is one of the least troublesome startup techniques because

- There is no interference from upstream disturbances.
- On-spec products are easy to achieve at high reflux ratios.
- Column startup is off the critical startup path. This provides more time for establishing good column action.

Additional benefits of total reflux startups are saving in plant startup time (by having the column off the critical path), and a reduced amount of off-spec material.

Total reflux startup, coupled with systematic draining of water from low points, can effectively dehydrate a column. In one case (233), column feed contained strongly acidic components, and these dissolved in small amounts of water inside the column, causing recurring corrosion failures. The problem was rectified by starting the column up on total reflux for 2 to 3 days using one of the pure (acid-free) components. During this period, all low-point drains were intermittently opened until all water was eliminated; the drains in the overhead system were slowest to dry. When dehydrating by total reflux, it is important that all low points are identified and drained. Figure 13.3 illustrates some of the common low points in a column system.

Total reflux startup is most attractive in large superfractionators which use high reflux ratios (e.g., isobutane-normal butane; ethylene-ethane separation) and/or in heat-pumped columns. Such columns take from a few hours to a couple of days to start up and stabilize. Due to the high reflux ratio, they are relatively insensitive to feed variations. These features make them ideal candidates for total reflux startups. Total reflux startup is least attractive when the ratio of reflux to feed is low (in such cases, most of the stabilizing can only be performed after feed is introduced), and when the column is easy and trouble-free to start up.

The two main drawbacks of total reflux operations are the possibility of an adverse chemical reaction and the possibility of concentrating an undesirable component due to leakage. Total reflux operation provides ideal conditions for slow reactions, particularly near the bottom of the column. Bottom product purity, bottom temperature, and bottom residence time are higher compared to normal operation. All these promote reactions or "cooking" of the bottom liquid. Besides yielding undesirable components which may be hard to get rid of, these reactions may cause material degradation, coking, and fouling. The author is familiar with one experience where "cooking" at total reflux generated an undesirable component which was later difficult to remove. Unless such a reaction can be prevented (e.g., by using only one of the components or distilling at a lower temperature), total reflux operation should be avoided.

Leakage converts a total-refluxed column into a batch distillation still. For instance, if leakage occurs near the top of the column, the heavy and medium components will be concentrated. Similarly, any component may become concentrated, depending on the location of the leakage. If concentration of any components may cause an undesirable chemical reaction, fouling, or corrosion, total reflux operation should be avoided. Alternatively, a mixture void of the offending component(s) may be used.

In one case, the author started up a large propylene-propane fractionator on total reflux. Two of the feed components, propadiene and methyl acetylene, are known to detonate in the absence of air if excessively concentrated. To eliminate this possibility, the startup mixture used was a hydrogenated propylene-propane mixture, void of acetylenes and dienes. The above procedure was adopted following a lesson learned from an explosion in a butadiene refining column (180), which demolished the column and caused widespread damage to surrounding units and to a nearby residential area. The column mixture contained vinyl acetylene, which can detonate in the absence of air if excessively concentrated. During a brief shutdown, the overhead product valve leaked while the column was operated under total reflux. The leakage raised the peak vinyl acetylene concentration from 30 to 60 percent, and the highly concentrated vinyl acetylene detonated. Figure 12.7 illustrates the rate of escape of lights from the system and the development of the vinyl acetylene profile. It is important to note that when the explosion occurred, the vinyl acetylene concentration peaked 10–15 trays above where it would peak during normal operation (180). The prime recommendation of the accident investigation was avoiding total reflux operation in this type of column (180).

12.9   Liquid Drainage

Liquid is drained from the column during washing, startup, shutdown, or when the column dumps after being flooded. Liquid draining usually requires no special precautions. However, there are situations where excessively fast drainage can create a vapor gap or induce rapid pressuring or depressuring. In these situations, it is important to restrict drainage rates, as described below.

Drainage can be troublesome when the bottom trays are flooded (i.e., covered with liquid), and the rate of liquid withdrawal from the column exceeds the rate of liquid downflow through the trays and downcomers (61,192,207,356). Under such conditions, a vapor gap may form below the bottom tray. The presence of this gap leaves the bottom tray with the task of supporting all the liquid above it, and may lead to failure of the bottom tray. Once the bottom tray fails and the liquid drains from it, the vapor gap moves to the space below the next higher tray. This tray may fail as well, and a cascade of collapsing trays may ascend the column (Fig. 12.8).

Valve trays are most vulnerable to vapor gap problems, particularly if their downcomers and/or downcomer clearances are small. The

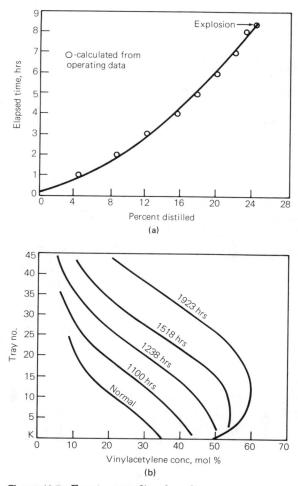

Figure 12.7 Events preceding butadiene column explosion during total reflux operation. (a) Fraction of column contents batch-distilled due to leakage, as a function of time; (b) corresponding rate of vinylacetylene accumulation (*From "Butadiene Explosion at Texas City-2" R.H. Freeman and M.P. McCready, in* Loss Prevention, *vol. 5, p. 65, 1971; reproduced by permission of the American Institute of Chemical Engineers.*)

valves allow only little liquid downflow, and almost all liquid must descend through the downcomer. Thus, the flow area is much lower compared to other tray types (where most of the liquid drains through the tray holes), and the liquid drainage rate is severely restricted.

The author is familiar with one incident where vapor gap formation is believed to have caused severe damage in a 10-ft vacuum column

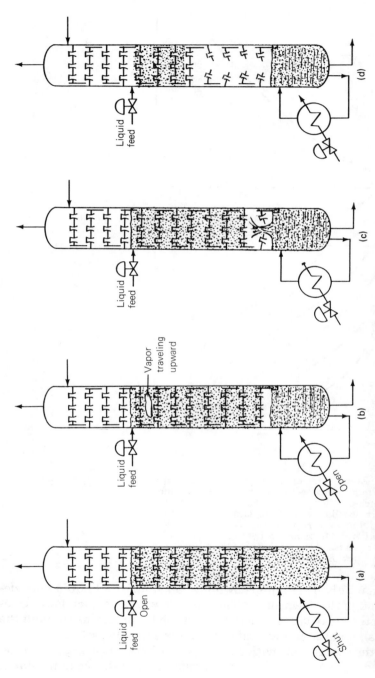

Figure 12.8 A typical vapor gap problem. (a) Lower part of column flooded (full of liquid). (b) Reboiler started. Liquid "drains" into reboiler and vaporizes. Vapor travels up. Liquid travels too slowly into sump. A vapor gap forms under bottom tray. (c) Bottom tray fails when attempting to support liquid column above it. When it fails, vapor gap shifts upward. Second tray now attempts to support liquid column. (d) End result.

containing 30 valve trays. Feed entered 10 trays from the top. The 16 lower trays were distorted downward, with major support beams bent and twisted into V shapes up to 1 in deep (Fig. 12.8*d*). In that incident, the vapor gap was initiated by reboiler startup in a flooded column (Fig. 12.8*b*). Solidification in several downcomers could have contributed to the flow restriction.

Excessively rapid drainage of sieve and bubble-cap trays can cause downcomer damage where downcomer clearances are narrow. Here liquid drains faster through trays than through downcomers. Liquid remaining in downcomers exerts a hydrostatic force that acts to bend downcomers toward the tray. In one refinery stabilizer (206 *a, b*), downcomers in locations of wide tray spacings (e.g., near manholes) pulled out of their braces every shutdown due to rapid drainage. The shutdown sequence involved steaming, filling up with water, then opening the bottom manhole. The last step induced the fast drainage rate that caused the damage.

When a vapor-containing bottom feed enters below the liquid level, fast drainage can induce excessive pressuring or depressuring rates. The liquid above the feed inlet back-pressures the upstream unit. As liquid drains, this back pressure falls, reaching zero when the feed inlet is uncovered. As the back pressure falls, the upstream unit depressures into the column, and the column pressures up. If liquid level drops too fast, the pressure drop between the upstream unit and the column is low, and the vapor space in the upstream unit is much smaller than in the column, then depressuring rates in the upstream unit can be extremely fast. Under similar conditions, column pressuring rates can be very fast if the column's vapor space is much smaller than that of the upstream unit. Potential consequences are described in guidelines 1 and 4 in Sec. 11.2. In one incident (239), rapid draining of liquid from the base of a refinery coker combination tower caused foamover of liquid (due to gas lifting of liquid, a "champagne bottle effect") in the upstream coke drum by the above mechanism.

In situations where either a vapor gap or excessive pressuring/depressuring rates can occur, it is important to monitor the bottom flow and avoid draining the column too rapidly (e.g., by opening the bottom manhole or by widely opening the bottom control valve) after flooding or at shutdowns. The operators should be made well aware of the consequences of too rapid a pumpout, because a natural operator reaction to flooding is to try to withdraw the liquid by pumping out as fast as possible. If column washing is practiced, it is important to restrict column-washing flow rates, and to check the sizes of washing equipment (beware of oversized pumps).

12.10   Sealing Problems

Problems of establishing a liquid seal in the column downcomers are common in low-liquid-rate services during startup. In such services, the downcomer liquid seal may also occasionally be lost during operation. Symptoms of these problems are excessive entrainment and poor separation. As distinct from flooding, column pressure drop may remain low. These problems, and cures that can be implemented at the design stage, are described in Sec. 6.18.

Sealing problems are often overlooked at the design stage, and the operator has to find a solution during startup or operation. A startup-stability diagram can define the range of conditions required for satisfactory sealing. The detailed method has been described elsewhere (196; see also Sec. 6.18). Should the available startup vapor and liquid flow rates fall outside the required range, sealing can only be achieved by judicious trimming (e.g., temporarily lowering liquid levels in upstream units to get a higher liquid flow, or temporarily raising pressure in upstream units to get a lower vapor flow), and/or by changing column pressure. Changing pressure will shift the stability limits on the startup-stability diagram. In one case history (196), the satisfactory startup range (Fig. 12.9) was extended from the area between the dashed lines to the area between the solid lines by changing the column pressure. The pressure change, coupled with judicious trimming, brought the startup vapor rates to within the required range, and permitted the column to start up (196).

In many low-liquid-rate absorbers (e.g., natural gas glycol dehydrators) liquid and gas flows can be varied independently, and a liquid seal is often established by trial and error. Liquid circulation is set at some desired rate, often determined by experience. Gas rate is then set at a desired low value. If this operation does not establish the seal, another gas rate is tried. When the absorber contains bubble cap trays, the lower curve on Fig. 12.9 is extremely low, and sealing can often be established by stopping gas flow altogether (238).

12.11   Reverse Flow Problems

Flow of material into or out of a column in the opposite direction to that intended often takes place during startups, shutdowns, and normal operation. In many cases, reverse flow through a line is of no consequence or is even desirable, but there are also several instances where it is troublesome or hazardous. The consequences of reverse flow through each line connecting to the column should be analyzed under normal operation as well as startup and shutdown conditions. Special attention should be paid to outages of rotating

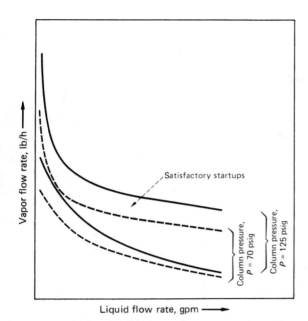

Figure 12.9 Effect of changing column pressure on the startup stability region (*From Henry Z. Kister, excerpted by special permission from* Chemical Engineering, *April 6, 1981; copyright © by McGraw-Hill, Inc., New York, N Y 10020.*)

equipment; under these conditions, reverse flow is most likely to occur.

Figure 12.10 is a simplified example of backflow analysis applied to the main process lines of the separation train in a refinery coker unit. If the compressor trips, gas backflows through the absorber and has the potential of damaging trays, particularly if the absorber contains valve trays (see Sec. 11.1, guideline 2; and Sec. 11.2, guideline 2). If the lean oil pump loses suction, gas from the absorber may flow through the lean oil line back to the stripper and/or to the downstream unit. In one coker plant incident (7), this gas traveled to the downstream (hydrotreater) unit, causing its charge pump to gas up and lose suction. This in turn damaged the reactor catalyst. A check valve in the lean oil line and low-flow trips were added to prevent recurrence (7).

The author is familiar with a case where caustic from a wash tower backflowed into an entirely different section of the plant. The other section was normally at a higher pressure, but not at startup. The caustic attacked an aluminum heat exchanger in that section, resulting in a fire. Piping was later changed to prevent recurrence.

In a third case (125), reverse flow in an absorber-desorber system (sim-

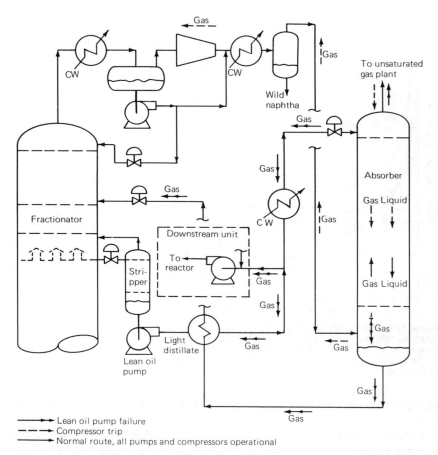

Figure 12.10   Routes of reverse flow, coker plant example.

ilar in principle to the system shown on Fig. 12.6) caused a fire and pump damage in an ammonia plant. The main amine pump was shut down for maintenance and the spare pump started, but it only delivered a third of the normal amine flow to the absorber. The balance was flowing backward through the main pump and its faulty check valve. Poor absorption resulted. This could not be tolerated downstream, and it forced a plant shutdown. During the shutdown, the (operating) spare pump was switched off. This permitted hydrogen-rich gas to backflow from the absorber to the main pump, and to leak through the faulty check valve and the pump seals. The leak fired. To prevent recurrence, motorized isolation valves and a sophisticated switching system were installed. The system enables spare pump isolation, while at the same time permitting autostart of the spare pump. The check valves were also upgraded (125).

In a fourth case (18), ammonia recovery storage tanks were damaged by overpressure due to gas entering their liquid rundown line (Fig. 12.11). The gas backflowed through the upper leg of an absorber pumparound circuit following pump failure at startup. The rundown line branched off the pumparound circuit at an elevated position; the accident could possibly have been prevented had this line branched near grade (Fig. 12.11). The absorber piping and control system were modified to prevent recurrence (18).

Examples of troublesome reverse flow into columns that were shut down are in Sec. 11.4 (also see item 14 in Sec. 11.6). Blinding, double block and bleed, and line disconnection are used to prevent backflow where lines are operated intermittently or when a piece of equipment is shut down. Check valves (double check valves when high reliability is required), good piping and operating practices, trips, and alarms are commonly used for prevention of undesirable backflow in continuously operating lines.

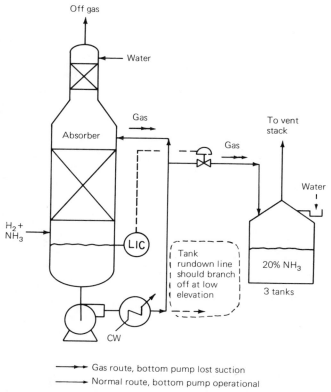

Gas route, bottom pump lost suction
Normal route, bottom pump operational

Figure 12.11 Reverse flow route, ammonia recovery plant.

Chapter

13

Operation Difficulties

There is no sharp demarcation between this and the previous chapter. As an arbitrary demarcation criterion, considerations primarily related to startups and shutdowns were treated in the previous chapter, while this chapter deals with difficulties that frequently affect normal operation as well as startups and shutdowns. Because of this vague distinction, the reader is well-advised to review both chapters when examining either startup/shutdown practices or when reviewing operation difficulties.

Introductory comments made in the previous chapter also extend to this chapter. Failure to recognize and eliminate operation pitfalls or hazards has caused catastrophic accidents, loss of life, injuries, damage to columns and their internals, and lengthy shutdown periods.

This chapter reviews common difficulties in column operation, outlines practices for overcoming them, highlights the consequences of poor practices, and supplies general guidelines for avoiding the hazards and pitfalls. As in the previous chapter, hazards and pitfalls specifically related to the nature of the chemicals processed (e.g., toxicity, flammability) are outside the scope of this distillation text. The safe handling procedures of the relevant chemicals must be followed to cater for the latter hazards and pitfalls.

13.1 Dislodging and Damage of Trays: An Overview

Dislodging ("bumping") and damage of trays is a fairly common occurrence, and as many as 100 trays have been dislodged in a single incident. Packed columns, especially those containing random packings, are prone to similar problems. Causes and preventive guidelines are scattered throughout the book. The "Tray/packing damage" column in

Table 1.2 can serve as a guide to potential culprits. The objective of this overview is to direct the reader to the most frequent causes.

Ellingsen (107) reported a survey of 18 incidents of column damage that occurred in DuPont's columns. Table 13.1 summarizes the results. Over half (55 percent) of the incidents were caused by excessive liquid level in the bottom of the column. In another 17 percent of the incidents, trays ruptured due to a vacuum that was present in some section of the column. Poor installation accounts for another 14 percent of the incidents. Other culprits—including corrosion, high vapor rates, and fatigue failure accounted for the remaining 14 percent of the incidents.

Assuming the sample of incidents is representative, one out of two tray damage incidents can be prevented by avoiding excessive liquid level in the column (Sec. 13.2). Six out of seven incidents can be prevented by also preventing vacuum formation in sections of the column (Sec. 11.2, 11.7, 11.8, 12.6, 12.9, 12.11) and by adequate installation and inspection (Chap. 10).

Ellingsen, as well as other DuPont experts (107), feels that the incidents described probably represent the general distribution of the types of column damage that have occurred throughout the chemical industry. This conclusion matches the author's experience, with three exceptions. First, in services processing hot (>200 to 300°F), water-insoluble materials (e.g., heavy oils), water-induced pressure surges (Sec. 13.5) tend to be the prime cause of tray damage. Secondly, in services processing unstable and exothermically reactive chemicals, runaway reactions and explosions are a common cause of tray (and column) damage (Sec. 13.13). Third, in corrosive applications, corrosion plays a more dominant role in causing tray (and column) damage.

13.2   Liquid Level in Columns

If the liquid level in the bottom sump of the column rises above the reboiler return nozzle (or, alternatively, the bottom vapor inlet nozzle), vapor from the reboiler has to travel upward through a layer of liquid. If this layer is shallow, the vapor can bubble through it or atomize the liquid and carry over liquid as a mist into the first tray from the bottom. This may lead to premature flooding (71, 145, 192, 207, 237, 238) and possibly some wave action that would interfere with level control.

If the liquid layer is deep (i.e., several of the lower trays may be flooded), vapor may travel through the liquid in the form of slugs. These can loosen bolting and other fasteners, damage trays, or lift trays off their supports (61, 150a, 192, 207, 295), collapse packing supports (237), and cause column vibration and structural damage.

TABLE 13.1 Summary of Tray Damage Causes in 17 Columns

Case no.	No. of trays	Diameter of col. in	Operating pressure vacuum	High liquid in bottom	High vapor rate	Fatigue failure	Poor tray installation	Corrosion
1				X				
2				X				
3		30					X	
4		72						X
5	20	96		X				
6		144					X	
7	20	72		X				
8*	30	72					X	
9	20			X		X and		
10	30	120		X				
11	30	120		X				
12*		96	X and	X				
13	20		X					
14	60	90	X					
15	17			X				
16	20				X			
17	20			X				

*Case 12 is counted twice, but 8 once.

SOURCE: W. R. Ellingsen, "Diagnosing and Preventing Tray Damage in Distillation Columns" *DYCORD 86, IFAC Proceedings of the International Symposium on Dynamics and Control of Chemical Reactors and Distillation Columns*, Bournemouth, UK, December 8–10, 1986. Used by Permission.

Several incidents have been reported where trays were lifted off their supports due to such slug (49, 107, 231, 296). The author is familiar with some additional similar experiences. Ellingsen's survey (107) suggests that over half the tray damage incidents in the chemical industry are caused by deep liquid levels in columns (Sec. 13.1).

Vapor slugging is least troublesome when reboiler temperature difference (hot side less cold side) is small, because the deep liquid layer will suppress boiling and, therefore, vapor generation. The column will dump, and once cleared, normal operation can be resumed. On the other hand, vapor slugging is most troublesome when reboiler temperature difference is high or when the vapor originates from a high-pressure source (e.g., live steam). In these cases, the deep liquid layer will do little to suppress vapor generation, and the vapor will force its way through the liquid layer as slugs.

Once a vapor slug passes through a deep liquid layer, the vapor space previously occupied by the slug is left empty and turns into a vapor gap under the bottom tray. Fast drainage of a deep liquid layer can also cause a vapor gap or induce excessive pressuring and depressuring rates. The consequences are described in Sec. 12.9. The description includes an experience where a vapor gap was initiated by vapor slugging.

Flooding initiated by a deep liquid layer can be particularly severe. In one incident (237), a 150-ft absorber filled with liquid. The liquid overflowed into the reflux drum, and then via a vapor vent into the fuel system, spilling out of burners and initiating several fires.

In another accident (16a) failure of a column level controller caused cold liquid propylene to pass out of a relief valve into a flare header. The flashing liquid overchilled the header below its safe working temperature. The header cracked, releasing a vapor cloud that ignited. The accident caused several fatalities and injuries.

A reliable indication of column bottom level is imperative for preventing the liquid level from rising above the reboiler return nozzle. The author recommends refraining from proceeding with a startup when a reliable level indication is absent. It is also important to realize that once the liquid level reaches the top level tapping, level indication will become unreliable and often misleading, and the operator would not know if the liquid level above the nozzle is shallow or deep. Keeping the liquid level well below the upper-level tapping is therefore mandatory. High-level alarms are recommended.

In some instances, installing automatic stripping steam shutoff trips on high bottom level has been highly successful and saved many trays in refinery vacuum columns (296). Unfortunately, this technique is only suitable for the few services where cutting out steam, boilup, or bottom feed is unlikely to greatly upset the column.

In addition, Ellingsen (107) has suggested the following techniques for preventing tray damage due to excessive liquid levels:

1. Pump liquid out rather than attempt to boil it off if liquid level is excessive.
2. Construct the bottom seal pan to be particularly strong.
3. Construct the bottom 25 percent of trays for extra mechanical strength.
4. Provide a liquid level differential pressure measurement for the bottom 25 percent of the column.
5. Provide facilities for easy diversion of bottom liquid to either the feed tank or storage so that liquid level can be readily reduced.
6. Ensure smooth and stable automatic control of boilup to the tower.

The author recommends incorporating at least some of these techniques in services prone to tray damage due to excessive liquid levels.

It is just as important to maintain a good control of the bottom liquid level in columns where the bottom feed, stripping steam, or reboiler return nozzle enters through a submerged sparger (Fig. 4.3a). Here the liquid level serves as a desuperheater, and its loss may overheat the column or its internals. One incident was reported (440) where plastic packings melted because this level was lost.

Low liquid levels can be as troublesome as high liquid levels. When bottom level is lost, vapor can flow out of the column bottom. In one incident (210), such vapor flow ruptured the bottom product storage tank. Low bottom levels can also cause cavitation and overheating of bottom pumps. In some services, a low bottom level can excessively concentrate some chemicals, inducing an undesirable reaction. If these chemicals are unstable (e.g., peroxides, acetylenic compounds), an explosion may result. Some reported accidents (97, 275) were initiated by low liquid levels at the bottom sump.

When a low level can cause such hazards, it is a good practice to install low-level alarms or trips. The trips usually cut the bottom flow rate, or introduce a diluent into the bottom sump.

13.3   Liquid Level in Reflux Accumulators

Liquid level in the reflux accumulator is normally far less troublesome than bottom liquid level. With an all-liquid top product, an overflowing accumulator will usually back up liquid into the condenser, flooding some tubes. This will reduce condensation and prevent further rise in liquid level. Often, column pressure will rise and the relief valve may possibly lift. With a vapor product, excessive accumulator level will cause liquid carryover in the vapor. This will be troublesome when downstream equipment cannot handle liquid. If the vapor feeds a compressor, severe damage may result; in one case (7), compressor internals were destroyed and its turbine driver was damaged by carryover of reflux accumulator liquid.

Low accumulator liquid levels are likely to damage pumps and/or induce flow of vapor to downstream units. One case history of pump damage due to loss of reflux drum level has been described (237).

When the reflux accumulator separates two liquid phases, excessive rise or fall of the interface level can carry over one phase into another, and/or reflux the column with the improper phase. In some situations, such carryover can be hazardous. In one case (7), carryover of hydrofluoric acid into a propane product route from the reflux accumulator of an alkylation depropanizer caused multiple explosions in downstream equipment. The author is familiar with an almost identical incident that overpressured downstream equipment but stopped short of exploding. Refluxing an improper phase into a column can also be troublesome; this is described in detail in Sec. 13.7.

As with column bottom level, when an excessively high or low level can be hazardous, it is a good practice to install high-reliability level instrumentation including alarms, trips, and indicators, and to mount these as per Figs. 5.4 and 15.16. The trips usually switch off rotating machinery downstream of the accumulator and/or cut off product flow.

13.4 Sources and Effects of Water Problems

Water can cause severe operational problems in services which are not meant to handle it. One refiner (296) stated that 99 percent of his fractionator upsets are due to water; other refiners (296) agreed with this evaluation. Most problems occur when the column separates water-insoluble materials such as hydrocarbons. The main adverse effects of water in such services are pressure surges, flooding, cycling, corrosion, hydrates, and off-spec products. Typical sources of water in this type of services are

1. The feed stream. Water may be present in the feed storage tank, may enter the feed from a leaking heat exchanger, or may not be entirely removed in an upstream removal facility.

2. Undrained water in a stripping steam line.

3. Chemical reaction. A condensation reaction forming water may occur between organic chemicals, either in upstream equipment or in the column.

4. A leaking heat exchanger (e.g., reboiler, condenser).
5. A prestartup wash, leak-test, or steam-water operation.
6. The previous campaign run through the column.
7. Condensate formed in previous operations. It could have remained trapped in pipelines or pockets inside the equipment.
8. A water-containing stream that found its way to the column (a typical example is water discharged from the desalter safety valve and ends up in a refinery crude column).

Entry of a slug or pocket of water into a column containing hot (usually >200°F) water-insoluble materials can cause pressure surges. Pressure surges are most damaging and hazardous, and are discussed in Sec. 13.5. Other adverse effects of water are flooding, cycling, severe corrosion, and off-spec products, caused by accumulation of small quantities of water in the column at close to ambient temperatures (usually 0 to 250°F). These are discussed in Sec. 13.7. Accumulation of small quantities of water under cold (less than about 30 to 50°F) conditions may cause hydrates. These are discussed in Sec. 13.10.

In addition, freezing and corrosion due to water accumulation in pipes, dead pockets, and instruments have caused numerous pipe failures. Some of these led to gas releases and fatal accidents. A detailed discussion on preventing water accumulation in column piping is presented elsewhere (210). Extensive discussions of winterization and freeze-prevention techniques are also available (120, 266).

13.5 Water-Induced Pressure Surges

If a slug of cold water is suddenly dumped into a hot column containing water-insoluble materials, or alternatively, if a feed much hotter than about 200 to 250°F suddenly contacts a pocket of water lying in the column, the water will rapidly vaporize. This rapid vaporization is caused by the presence of two liquid phases simultaneously. The water phase will attempt to exert its own vapor pressure and will therefore rapidly vaporize. Due to its low molecular weight, water will expand to about 1000 to 2000 times the liquid volume upon vaporization close to atmospheric pressure; the expansion will be severalfold larger under vacuum. This may generate a rapid and impulsive pressure surge. Such a pressure surge can dislodge or damage trays, packings, and supports.

Figures 13.1 and 13.2 illustrate two refinery case histories of pressure surges resulting from water slugs (4). In Fig. 13.1a, a combination tower (left) was connected by a long horizontal line to coke drum (right). Upon heating, the tower expanded and lifted the tower-end of the line

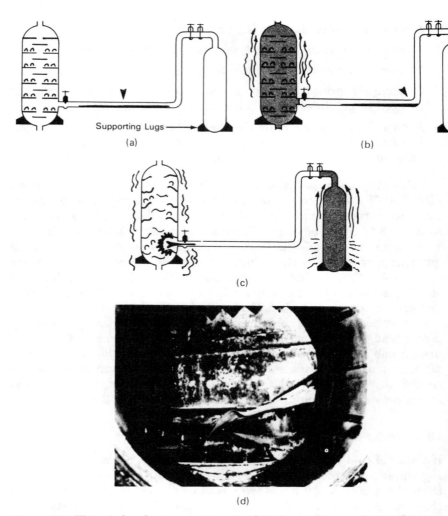

(a)   Supporting Lugs ⟶

(b)

(c)

(d)

Figure 13.1 Water-induced pressure surge incident in a refinery coker combination tower. (*Reprinted with permission of copyright owner from the Booklet "Hazards of Water," copyright 1984, Amoco Oil Company.*)

(Fig. 13.1*b*). Condensate from steam testing, and from water driven off from the coke drum, collected at the low point. When the coke drum was heated, it expanded and lifted the coke drum end of the line. This dumped the water from the low point into the tower (Fig. 13.1*c*). A pressure surge resulted; the relief valve lifted, but was unable to prevent damage to the tower. Figure 13.1*d* is a photograph of the upset trays, taken in tower A following the accident. To prevent recurrence, the startup procedure was altered to require oil flushing before the tower becomes hot.

The accident illustrated in Fig. 13.2 occurred during a catalytic

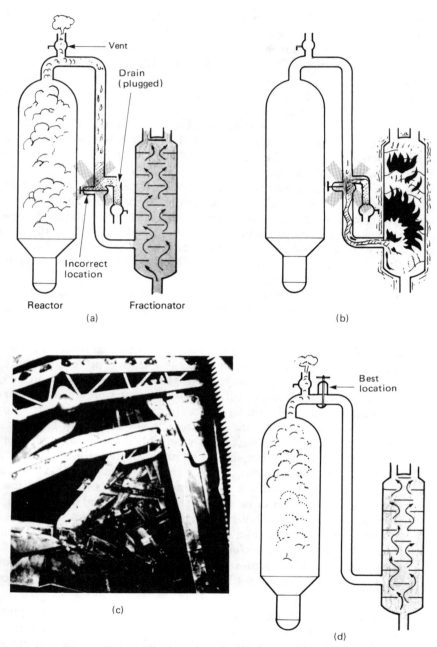

Figure 13.2 Water-induced pressure surge incident in a refinery catalytic cracker fractionator (*Reprinted with permission of copyright owner from the Booklet "Hazards of Water," copyright 1984, Amoco Oil Company.*)

cracker startup. A block valve in the vertical section of the 40-in line isolated the catalytic cracker from the fractionator (Fig. 13.2a). Condensate accumulated above the valve. A ¾-in drain should have drained the condensate, but the drain was plugged. The block valve was opened (Fig. 13.2b), while hot gas oil at 470°F was circulating through the fractionator. The resulting pressure surge severely damaged trays, support beams, and internal piping (Fig. 13.2c). To prevent recurrence, the ¾-in drain was replaced by a 2-in drain. The accident could have been avoided altogether had the block valve been installed in the horizontal leg of the line (Fig. 13.2d), allowing condensate to drain back into the reactor.

The above case histories were described in detail here to illustrate the damage that even small quantities of water can do when they initiate a pressure surge. Several other incidents of water-induced pressure surges in fractionators have been described (4, 7, 23, 145, 146, 237, 296, 358) and the author is familiar with a few others. Amoco's booklet, in particular (4), contains excellent, well-illustrated descriptions of several such incidents.

The following guidelines can help prevent water-induced pressure surges in columns containing hot, water-insoluble materials (Fig. 13.3):

1. Do not feed the column from the bottom of the feed tank. It is best to have the tank outlet nozzle at least 1 ft, and preferably more, above the bottom of the tank. Swing lines ("floating suction") (4, 239) are often used to permit changing the suction elevation during service. Water should be routinely drained from the bottom of the tank to prevent it from reaching the suction point. If routine manual draining is not sufficiently reliable for keeping water out of the feed, consideration should be given to installing automatic water removal equipment, similar to that used in reflux drums (Sec. 13.7). The possibility of layering and emulsification of feed and water in the feed tank should be considered and corrective action taken as needed. When changing over tanks, feed lines should be properly flushed and drained. Amoco's booklet (4) contains an excellent discussion and several additional guidelines. It is important to keep in mind that preventing the damage incurred by one pressure surge is likely to more than pay for the needed water-removal equipment.

2. Install a conductivity probe in the column feed line (237, 239). Water or oil-water emulsions conduct electricity far better than oil. A high-conductivity signal will therefore detect water and can be used for shutting down the feed pump. This technique was found particularly effective (239) for rerunning slop from refinery

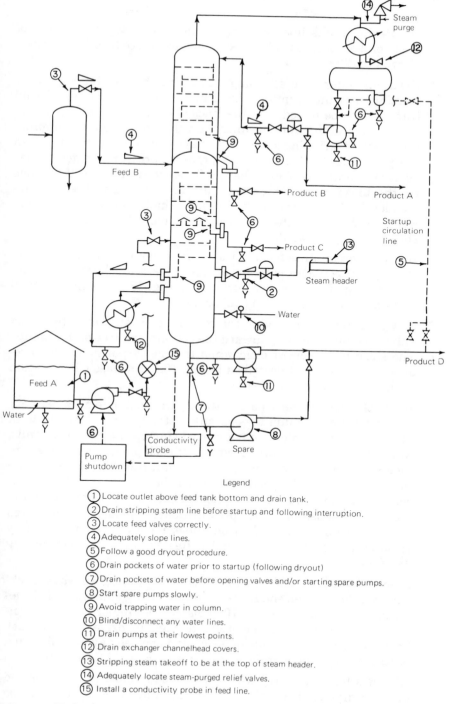

Legend

① Locate outlet above feed tank bottom and drain tank.
② Drain stripping steam line before startup and following interruption.
③ Locate feed valves correctly.
④ Adequately slope lines.
⑤ Follow a good dryout procedure.
⑥ Drain pockets of water prior to startup (following dryout)
⑦ Drain pockets of water before opening valves and/or starting spare pumps.
⑧ Start spare pumps slowly.
⑨ Avoid trapping water in column.
⑩ Blind/disconnect any water lines.
⑪ Drain pumps at their lowest points.
⑫ Drain exchanger channelhead covers.
⑬ Stripping steam takeoff to be at the top of steam header.
⑭ Adequately locate steam-purged relief valves.
⑮ Install a conductivity probe in feed line.

Figure 13.3 Techniques for eliminating pockets of water.

dewatering tanks into refinery fractionators. Regular testing of these probes is essential; a recommended procedure is described elsewhere (239).

3. Drain all water from stripping steam lines before starting (or restarting after interruption) stripping steam to the column. Steam traps should be tested, and the lines should be blown free of any remaining water before opening the valve to the column. Improperly drained stripping steam lines have caused pressure surges in the past (231, 296). In services prone to pressure surges, it has been recommended (296) to have a flange valve right at the flange on the tower with a blowdown drain. It has also been recommended (231) that the tie-in for the stripping steam line should come off the top of the steam header pipe. The stripping steam should be superheated.

4. Block valves on lines feeding the column should be located so that the piping sections upstream of the valve drain back to the upstream unit. The incident described in Fig. 13.2 is a classic illustration of the importance of this rule.

5. Any feed lines connected to the column, downstream of the block valve, should be sloping toward the column and free of low points. The slope must persist even when thermal expansion acts to reduce it. The incident described in Fig. 13.1 is a classic illustration of the importance of this rule.

6. When the column feed passes through a heater (e.g., a refinery fractionator or vacuum tower), any water lying at low points in the coils must be blown out prior to startup. In multipass coils, water must be separately blown out of each pass; block valves are sometimes installed on each pass to permit this (7). If blowing into the tower, it must be performed when the tower can still tolerate water. The coils must be kept hot and/or purged from then on to prevent condensation. One pressure surge incident (7) occurred when water accumulated in one heater pass entered a refinery vacuum tower which was under full vacuum and circulating 280°F oil.

7. A good dryout procedure, which includes draining all water pockets from the system, should be followed during startup. This is described in detail in Sec. 11.10. This is most important if the column was steamed out beforehand (e.g., to drive the air out). Several pressure surge incidents could have been avoided if such a procedure was adequately carried out (4, 5, 7, 23, 237, 296).

8. All lines likely to carry water should be blinded or disconnected prior to dryout. The only water-containing lines which may be reconnected following dryout are those leading to the cooler parts

of the system (e.g., the reflux drum boot), where removal facilities are available.

9. It is best to always assume that water is present on the other side of any valve which is about to be open during operation (e.g., cleaned exchanger, line to a spare pump, additional feed line, drain, vent, instrument, or sample line). Such a line must be thoroughly drained before the valve connecting it to the column is opened. Failure to do this has been a common cause of pressure surges (4, 145, 296).

10. Opening of any valve to a vacuum column is likely to suck in the contents of that line, and possibly materials from other equipment connected to it. These must be thoroughly dried prior to opening the valve. A pressure surge incident resulting from sucking water from a drain line into a vacuum column has been described (4).

11. When a steam purge is applied at the base of the column relief valve (to prevent the valve and its inlet line from plugging), the relief valve should be located so that any condensed water drains into sections of the system that can tolerate it (e.g., the reflux drum). In one case (239), this condensed steam dripped back into the column during short outages, repeatedly causing pressure surges upon restart.

12. A common troublespot in refinery crude units is the desalter relief valve, which often relieves into the crude column. Lifting of this relief valve has caused pressure surges. Guidelines 4 and 5 are important with regard to the location of this valve (296). It is best to have this relief valve discharge elsewhere.

13. Spare pumps should be started slowly (296), in case a small residual amount of light material or water had not been properly drained. Two pressure surge incidents resulting from water or flushing oil in spare pumps entering a hot vacuum column have been described (7).

14. Any unnecessary pockets where water can collect should be eliminated from the column and piping. Common troublespots are drawpans, internal heads, and seal pans. Incidents of pressure surges initiating from water trapping in such locations have been reported (4, 23, 237, 296). Further discussion is in Sec. 4.10, guidelines 11 and 13; in Sec. 6.21, guideline 7; and in Sec. 7.11.

15. Sometimes, the bottom liquid outlet is elevated 6 to 12 in above the bottom of the sump to prevent solids or polymer leaving with the bottom product. In such columns, water is likely to accumulate in the dead space below the outlet. Attention should be paid to adequately draining water from this area during dryout. Such

designs are best avoided in services prone to water-induced pressure surges.

16. In services using water-cooled condensers, a tube leak can lead to water being refluxed into the column. In one case (358), this caused a pressure surge and severe tray damage. The effect of a condenser tube leak should be examined; if a pressure surge may result, water removal facilities should be provided at the reflux drum.

17. Attention should be paid to draining exchanger channelheads and the lowest points on pumps. Experiences of water collecting at these points have been reported (5, 233).

18. Operators and supervising personnel should always be on the lookout for low points that may be caused by expansion, sagging, or misalignment. This is most important during startups or before opening valves.

19. At shutdown, any hot liquid lines (e.g., bottom line from another column) must be blinded and any hot liquid drained prior to steaming out. In one incident (3), hot liquid leaked into a column and contacted condensate that formed when an upstream unit was steamed into the column. This caused a pressure surge and extensive tray damage.

20. Water-induced pressure surges are often experienced during short unit outages (231, 239). At such times, steam or water may back into the column and form water pockets. Special caution is required during such periods. Following a pressure surge accident in one refinery vacuum tower (7), it was recommended to drain all liquids and follow the full startup procedure when tower temperature drops below 300°F during an outage.

21. Where pressure surges are expected, support beams are frequently designed for a high uplift resistance (Sec. 7.4). The use of shear clips, which nonrigidly fix the support lip of a tray to the column wall, has been proposed (15) for improving tray resistance to uplift. The use of heavier-gage trays and clips, adding washers, tack welding sections together, and other means of attachment are often of marginal value for that purpose (15).

13.6   Lights-Induced Pressure Surges

A pressure surge similar to that described in Sec. 13.5 may occur when a slug of cold light material is suddenly dumped into a hot column. In practice, sources of light material are usually less abundant than water, and the molecular weight of such a material is likely to be considerably higher than water. Pressure surges induced by light material are therefore less frequent, and usually less potent, than those caused by water. Such pressure surges are seldom troublesome in atmospheric and pressure columns, but they can upset hot vacuum columns. One pressure surge incident resulting from light flushing oil in the piping of a spare pump entering a hot vacuum tower has been described (7).

A lights-induced pressure surge problem is usually tackled by keeping the lights out. The techniques are similar to those discussed in Section 13.5; the specific techniques that are likely to be most effective depend on the sources of the lights.

13.7 Intermediate Impurity Accumulation

Column feed often contains components whose boiling points are intermediate between the light and heavy key components. In some cases, column top temperature is too cold, and column bottom temperature is too hot to allow those components to leave the column at a rate sufficiently high to match their feed rate into the column. Having nowhere to go, these components accumulate in the column causing flooding, cycling, and slugging. If the intermediate component is water or acidic, it may also cause accelerated corrosion; in refrigerated columns it may cause hydrates. A high temperature difference between column top and bottom, a large number of components, and high tendencies to form azeotropes or two liquid phases are conducive to intermediate component accumulation.

A typical symptom of this problem is cyclic slugging, which tends to be self-correcting. The intermediate component builds up in the column over a period of time, typically hours or days. Eventually, the column floods, or a slug rich in the offending component exists either from the top or from the bottom. The column end from which the slug leaves varies unpredictably. Once the slug leaves, column operation returns to normal over a relatively short period of time, often without operator intervention. The cycle will then repeat itself. A number of typical experiences have been reported (61, 203, 206b, 362); the author experienced two more.

Intermediate component accumulation may interfere with the control system. For instance, a component trapped in the upper part of the column may warm up the control tray. The controller will increase reflux, which pushes the component down. As the component continues accumulating, the control tray will warm up again, and reflux will increase again; eventually, the column will flood. One case describing a similar sequence has been reported (352).

If the intermediate component is water, and the bulk of the material distilled is water-insoluble, the water will form a second phase that will tend to dissolve small quantities of acidic impurities. This circulating acidic water can severely corrode column internals. Experiences illustrating this problem have been reported (232, 233, 237, 239).

Cures that can overcome intermediate impurity accumulation problems include:

1. *Avoiding refluxing of the component back into the column:* A typical example is the reflux drum boot technique, commonly applied in hydrocarbon separations where water is the impurity. The boot may be an integral part of the drum (Fig. 13.4), or a separate drum located at ground level. Some guidelines for boot sizing are in Sec. 15.14.

   Plugging may be a problem in the water outlet line from the boot because of low flow rates and because solids and corrosion products tend to entrap in the boot and the water stream. The converse problem is leakage rates across the water outlet control valve exceeding the rate of water inflow into the boot. This makes maintaining boot level difficult and causes loss of product in the water stream.

   Both the plugging and leakage problems are most troublesome when there is a high pressure difference across the water outlet control valve. A high pressure difference promotes valve leakage; it also tends to keep the valve opening narrow, which promotes plugging. Both problems can be overcome by adding an external water stream (which may be a circulating stream) to the boot outlet (Fig. 13.4). This stream boosts velocity (232, 237) and safeguards against a loss of liquid level. The external water flow rate should be low enough to prevent excessive water backup from overflowing the boot during fluctuations.

2. *Reducing column temperature difference—either by raising top temperature, or lowering bottom temperature, or both:* This enables the accumulating component to escape with one of the product streams.

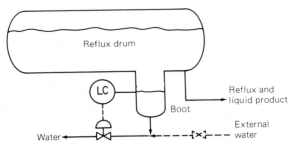

Figure 13.4   Reflux drum boot arrangement.

The effectiveness of this technique may be limited, and it can also cause excessive product losses.

This technique was successfully applied in one of the experiences reported (203), and in one other case the author experienced. In the former case (203), absorption oil was injected to prevent excessive product loss at the higher temperature. In the latter case, which occurred in an azeotrope column, the additional product loss due to the higher top temperature was negligible. Others (61) also advocated this technique.

3. *Changing column feed temperature:* Lowering feed temperature may prevent accumulation of the component in the top section; raising feed temperature may prevent accumulation of the component in the bottom section. Similarly, a feed point change may encourage the component to leave the column at one end or another. The effectiveness of these techniques may also be limited. In one reported case (203), changing feed temperature was effective in curing an intermediate accumulation problem.

4. *Providing component removal facilities from inside the column:* Usually, this technique involves drawing a small liquid or vapor side stream from the column, removing the intermediate component from the side stream externally, and returning the purified side stream to the column. The side stream drawn should be large enough to remove the amount of the component entering in the feed. Since at the drawoff location the component is normally far more concentrated than in the feed, the stream drawn is usually small. In one case (352), it was sufficient to remove a purge side stream from the column without providing purification and return facilities.

A typical example of this technique is using an external boot for removing water from inside a hydrocarbon distillation column (Fig. 13.5). Proper design of piping to and from the boot is essential for avoiding syphoning, choked flow, and excessive downcomer backup. A slightly different arrangement is described elsewhere (233). A pumped system, or a more sophisticated draw pan (e.g., a chimney tray), can also be used, and the boot may be located at grade. As with the reflux drum boot (1 above), an external water supply may be desirable if the water-line plugs or if the water control valve leaks excessively.

Another typical example is removal of higher-boiling alcohols ("fusel oil") from ethanol-water columns. Unless removed, they will concentrate in the column, and upon reaching their solubility limit, form a second phase and cause cyclic flooding (362) as described earlier. Fusel oil is commonly removed by a similar scheme to Fig. 13.5, except that the side stream is usually cooled prior to phase

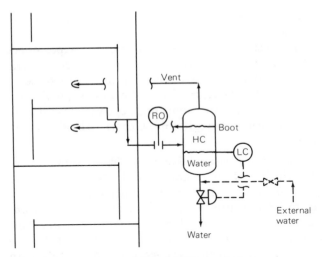

Figure 13.5 Water removal from inside a column.

separation, and the aqueous phase (rather than the organic phase) is returned to the column.

Another application of this technique is in separation of ethyl ether from aqueous ethanol, where benzene tends to build up in the bottom section. Removal of a small benzene-rich side stream out of the bottom section was found effective in increasing column overall capacity (98).

The component-removal technique need not be confined to gravity settling. Other separation techniques such as stripping, adsorption, and others may also be employed.

5. *Keeping the top tray temperature at an adequate margin above the dew point of the component:* A typical example is a refinery crude distillation column. Should top tray temperature fall below the dew point of water, water will condense on the top trays. The result will be equivalent to refluxing water back into the column.

Top impurity accumulation. In some columns, overhead is totally condensed and then decanted to form an aqueous stream and an organic product. The organic product is then sent to a stripper to remove traces of the aqueous phase. The stripper overhead is then recycled to the column condenser inlet. A typical example is a refinery HF alkylation unit depropanizer/HF stripper system.

When a light condensable organic enters the column, it will end up in the organic phase. In the stripper, it will be stripped and return to the condenser. Thus, it will entrap in the system, traveling back and forth between the condenser and the stripper. The stripper temperature controller will act to keep the light in the system, because the presence of the light will reduce the control temperature, which in turn will increase stripping heat input. The trapped light will raise column pressure, as well as the heat loads on the column condenser and stripper reboiler.

To provide an outlet to the light, either venting from the reflux drum or reducing the stripper heat input (thus allowing the light out of the stripper bottom) is necessary. Each has a disadvantage; venting may incur excessive product losses, while reducing the stripper heat input may lead to a breakthrough of aqueous phase into the product. Generally, venting is the more effective technique when the light is far more volatile than the organic product. Reducing stripper heat input is more effective when the light is only slightly more volatile than the organic product. One case has been reported (239a) where reducing stripper heat input effectively provided outlet for ethane trapped in the overhead of an alkylation unit depropanizer; in that case, venting was relatively ineffective.

13.8 Absence of an Expected Component

A distillation system may perform unsatisfactorily if one of the feed components is missing, present in insufficient quantity (or concentration), or is vaporized too early upon entering the column. This component may be required for the formation of a volatile azeotrope, or for preventing other materials from precipitating out of solution, or for keeping certain components in the bottom liquid.

Cures which may overcome problems of absence of an expected component are:

1. *Modifying the system to live with the problem:* In one instance (98), removal of a volatile solvent from an aqueous solution caused precipitation of dissolved solids in the section below the feed, thus plugging the column. The problem was solved by replacing packings by trays to prevent plugging.

2. *Injecting the absent component:* In one instance (13), water was essential for converting some heavy components into volatile azeotropes, which would leave with the top product. When water was absent, the azeotropes would not form, and the heavy components would leave with the bottom product. During periods when the quantity of water in the column feed was insufficient for azeotrope formation, or when excess reboil boiled off the water in the feed, steam was injected below the feed to assure azeotrope formation.

3. *Modifying control system:* In the instance described above (13), the main problem was excessive reboil causing water to boil off and dry

out upon entry to the column. Judicious modification of the control system to prevent excessive reboil was sufficient for assuring azeotrope formation without steam injection.

13.9 Cooking

When column feed consists mainly of lights and a small amount of residue, the column base will contain the residue and a small amount of lights. These lights will tend to preferentially vaporize out of the column base, and the material remaining in the column base and the reboiler loop will thicken over a period of time. Bottom temperature and viscosity will slowly creep up, and eventually reach a point where viscosity tends to "run away." Poor material balance control, which is a common problem under those circumstances, as well as polymerization and condensation reactions at the high temperature, will aggravate the problem. The author is familiar with a case where cooking severely plugged a bed of packing, and experienced cases in which cooking almost solidified the contents of a large bottom sump.

The following techniques have been effective for tackling cooking problems:

1. Draining a large portion of the inventory so that it is replaced by lighter liquid from the column.
2. Injecting a low-viscosity heavy solvent into the column to dilute the bottom. The author is familiar with cases where this is done routinely.
3. Adding heavier components into the column feed. These will replace the small quantity of lights in the column base, and suppress the preferential vaporization. The author had one experience in which replacing a reflux stream by a much heavier one in a short column receiving a vapor feed completely eliminated a cooking problem.

13.10 Hydrates

Hydrates are loosely bonded chemical compounds of hydrocarbons or similar substances and water. Hydrates are solids which resemble packed snow or chipped ice. Low temperatures, high pressures, and turbulence promote hydrate formation. In process equipment hydrates behave like ice and can cause plugging. Hydrates commonly form and accumulate in columns operating below 30 to 50°F, eventually flooding the column prematurely in a sudden and unpredictable manner. Hydrates may also restrict flow in heat exchangers and product control valves.

Flashing of a product across a control valve, or chilling it in a heat exchanger, can generate sufficiently cold temperatures to form hydrates even when the column is warm enough to prevent them.

Proper drying of column feeds is essential when column temperature or column product temperature is less than 30 to 50°F. Adequate column dehydration at startup is also essential.

Drying cannot be entirely relied on to prevent hydrate formation. Minor quantities of water (even at a level of a few parts per million or less) can accumulate in a column over a period of time, and eventually cause a hydrate problem. Facilities for injecting a suitable antihydrate such as methanol or propanol (which act like antifreeze) must be provided. The author found that appropriate differential pressure recording can provide an excellent tool for detecting and diagnosing hydrate problems.

13.11 Freezing

Freezing can occur whenever the solidification point of components in the column fluids is higher than the cooling medium temperature. Freezing is most frequently experienced at startups, low-rate operation, and during cold weather spells. Freezing usually occurs near the top of the column or in the overhead system. It can cause condenser tube plugging, premature flooding, or erratic control action. If water is present, freezing may also damage equipment. In one instance (150), freezing of one of the components in the top mixture of an air-cooled vacuum column caused erratic pressure control and frequent loss of reflux flow and accumulator level. In another case (421), water would freeze on the tubes of a reflux coil of a glycol still at low rates. When rates would increase, the ball of ice melted and the water would flow into the still, causing the still to "puke" over.

When the column contains waxy components (e.g., natural gas absorbers processing wax-contaminated feeds, or when leaking oil enters a refrigerated column), congealing rather than solidification is likely to occur. The congealed wax may not adhere to exchanger tubes; instead, it tends to settle in the column and plug trays or downcomers. In one cryogenic plant incident (251), turboexpander lube oil leaked into a stripper feed stream and congealed at the top of the stripper, blocking all flow.

Attention to low-rate operation is the key to avoiding freezing problems. The following techniques can be used.

1. Providing facilities for warming the cooling medium during low-rate operation or cold periods (e.g., recycling warm cooling water into the inlet of the condenser, or providing a warmup exchanger).

This technique was recommended for the glycol still problem above (421).

2. Whenever feasible, components that lower the freezing point of the top mixture can be introduced or injected. In the vacuum column above (150), the freezing problem was successfully tackled by deliberately running the bottoms of an upstream column "off-spec" for lights. The presence of these lights prevented freezing in the next column. Similarly, in natural gas absorbers, pressure or reflux is often raised and bottom temperature is lowered to induce more lights into the absorption oil when a wax problem is detected (241).

3. Injecting a solvent for dissolving the solid or wax. For instance, wash oil or naphtha with a low freezing point is often used for removing congealed oil (e.g., from pump seal leaks) from refrigerated columns. In many cryogenic services, warming the column may be necessary before solvent injection can be effective. Warming up followed by solvent injection was successful in one reported case (251); the author is familiar with several more.

4. Heating the overhead line upstream of the condenser (e.g., using a steam jacket or simply steam lances). The effectiveness of this method is usually limited.

13.12 Precipitation

Dissolved solids often enter with the column feed. As the component that retains the solids in solution is evaporated up or down the column, the dissolved solids may be precipitated out of solution, causing plugging. The problem is most common in aqueous distillation systems where the water contains dissolved solids. Incidents where dissolved solids precipitated in columns and caused plugging were described (71, 98, 236, 237, 239). In another incident (107), precipitation caused foaming. In one nitro-service incident (96), precipitation led to a detonation.

A precipitation problem can often be cured by an on-line wash. This may require operating with off-spec products during the wash. In one methanol-water column (236, 237) experiencing caustic precipitation, bottom rate was reduced and heat input increased, forcing water to travel to the upper trays. The water washed the hydroxide deposits on these trays. The column was then allowed to dump, the deposits disappeared, and column capacity returned to normal. Other techniques for preventing precipitation include keeping the solids out of solution upstream of the column and injecting solvent.

13.13 Chemical Reaction

Undesirable chemical reactions are frequently encountered in distillation and absorption columns, especially in distillation of organic materials. An undesirable reaction may cause an explosion, produce compounds that contaminate column product(s), interfere with the distillation process (e.g., by forming azeotropes), degrade column chemicals, and plug or corrode column internals. Generalizing on means of diagnosing and preventing undesirable reactions is difficult because reactions depend on the chemical nature of the components. Guidelines which may be useful for some common situations follow below:

1. Many organic reactions are temperature sensitive, sometimes to the extreme, and tend to be favored by high temperatures. Lowering the bottom temperature by about 30 to 40°F is often a useful test to confirm the occurrence of a reaction. If this stops the reaction, it may be possible to operate the column permanently in that mode. Product lost to the column bottoms can be recovered in a downstream separation facility (e.g., another column), and recycled to the column feed (Fig. 13.6). The author is familiar with one experience where this technique not only prevented the reaction but also improved plant product recovery (194).

   When the column chemicals are thermally unstable and decompose exothermically, an excessive bottom temperature can cause a "runaway" reaction and sometimes lead to an explosion. Some experiences with such explosions have been reported in distillation of peroxide, nitro, hydrocarbon oxide, and acetylenic compounds (16a, 96, 97, 209a, 275).

2. Many organic reactions are catalyzed by the presence of even small quantities of certain oxidizing or reducing agents. Compounds such as corrosion or polymerization inhibitors often fall into this category. If an adverse reaction is suspected, using a more inert inhibitor should be investigated and plant-tested. In one case (180), a reducing polymerization inhibitor contributed to an undesirable reaction. In another incident (131), excessive air leakage into a vacuum column caused polymerization, which in turn led to tray plugging and tray damage.

3. Many organic reactions are influenced by the materials of construction of the column or its internals. The author is familiar with one experience where use of any packing material other than ceramic induced the formation of a compound that contaminated the bottom product. In water peroxide distillation, certain materials of construction are known to catalyze decomposition (97). In

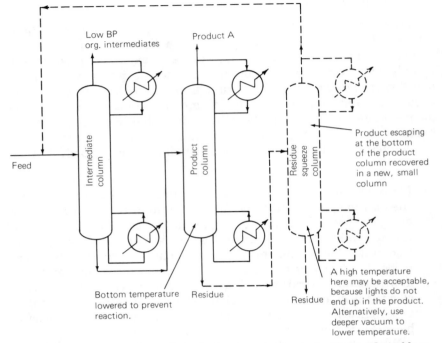

Figure 13.6   Reducing product column bottom temperature to avoid an undesirable reaction.

hot activated potassium carbonate (hot pot) service, tests revealed (349) that ceramic packings from different suppliers had widely different reactivities toward the solution. Laboratory tests under simulated column conditions can often detect the interaction between the materials of construction and the chemicals.

4. Long residence times, often associated with large recycles, can aggregate small amounts of reaction products in the system. Recirculation systems such as solvent recovery or absorber-regenerator system are particularly prone to this problem. In one solvent recovery system (98), the solvent slowly hydrolyzed and reaction products built up to the point of interrupting the action of the solvent recovery column. The cure was periodically distilling the reaction products from the solvent. The author is familiar with a similar experience where agglomerated products of solvent hydrolysis extensively corroded a solvent recovery distillation unit. In glycol dehydration systems, slow oxidation of glycol into organic acids tends to occur over an extended period of time, causing severe corrosion problems (84).

Facilities to either inhibit these reactions remove the reaction products, or render them harmless should be incorporated. In the glycol example above, the products are usually rendered harmless by neutralization (84).

5. Excessive concentration of an unstable component is a common cause of reactions in columns. Several explosions resulting from inadvertent concentration of peroxides in a column or reboiler have been reported (97, 275); similar incidents may occur with other unstable chemicals (e.g., 96, 180, 209a).

6. Feedstock impurities, or chemicals left over in the system from a previous operation, are often the cause of undesirable reactions in columns. Should an undesirable reaction be suspected, it may pay to look into the nature and effects of feedstock impurities, and to attempt deriving plant feedstock from an alternative source for a trial period. If the impurity is chemically unstable, or reacts with the column chemicals explosively, a detonation may result; some examples were reported (96, 275). In absorber-regenerator systems, contamination of the solvent charge can lead to undesirable reactions or poor stripping; one example has been reported (14a).

7. Impurities introduced in the feed in even small quantities may sometimes build up in some section of a column and eventually cause an undesirable reaction. For instance, small quantities of some hydrocarbon impurities, ingressed in the air intake to air-separation plants, tend to aggregate as solids in the liquid oxygen at the reboiler of the air-separation column (225, 427). When built up sufficiently, they react explosively with the oxygen. An air separation column explosion caused by this mechanism has been reported (96). The hazard is greatest during column shutdown, because the more volatile liquid oxygen vaporizes preferentially, thus concentrating the remaining impurities.

In some organic processes, small amounts of oxygen dissolved in the feedstock or leaking into vacuum equipment may become concentrated in the light-end vapor stream leaving a distillation column. This is particularly troublesome if the light ends constitute only a small portion of the plant products.

A persistent impurity problem can be tackled at the source or in the column, often in both. In the air separation example, hydrocarbon impurities are often separated out at plant inlet. In addition, pockets where impurity-contaminated liquid oxygen may aggregate are avoided in the column; an impurity adsorber may be installed on the oxygen liquid stream to the reboiler; the oxygen side of the reboiler may be run flooded and liquid oxygen is continuously blown down from the reboiler (225). Special attention is paid to cleaning and shutdown practices, and to keeping the heating side of the air separation column reboiler well vented. Several additional techniques are spelled out elsewhere (427).

8. A feedstock may contain an unexpected chemical which shows up as one of the normal components in the routine laboratory analysis. Similarly, one of the feed components may be present in an unexpected molecular form. In either case, the volatility and solubility characteristics of the unexpected compound may cause it to distill toward the wrong end of the column. The routine lab analysis will indicate poor column separation. In one case (338), hydrogen fluoride (HF) was present mainly as carbonyl fluoride in a feed to an aqueous wash column. While HF can readily be absorbed, carbonyl fluoride could not; the apparent symptom was poor absorption of HF. If an adverse reaction is suspected, routine lab analyses should be checked against comprehensive chemical assays, and attention should be paid to concisely defining the components entering and leaving the column.

9. Most chemical reactions are sensitive to the pH of the system. Altering the pH in the column may alter the course of reaction. Maintaining tighter pH control in the unit supplying column feed may also affect the reaction.

10. The presence and the quantity of water in the system often affect the course of reactions, and more important, the azeotroping of compounds.

11. Catalyst carried over from a reactor may find its way into the column, converting it into a "secondary reactor." Carryover of solid (e.g., metal) fines from upstream equipment may have an identical effect. One experience has been reported (131) where carried-over catalyst, together with excessive air leakage into a vacuum column, polymerized the bottom product and caused plugging and damage to trays. In another incident (209a), iron carried over from an upstream column is believed to have catalyzed a polymerization reaction in an ethylene oxide column. This in turn caused overheating and a decomposition explosion.

   When suspecting an undesirable reaction, it is worth checking column products and deposits for the presence of reactor catalyst and/or metal fines.

12. An undesirable reaction may only occur at turned-down conditions, either because of excessive reagent, or because of longer residence time. In one instance (150), an additive was continually injected into a column feed to eliminate small quantities of aldheydes. At turned-down conditions, the same additive also re-

acted with ketones, causing excessive product losses. The problem was overcome by running the column always at high rates and shutting it down when product stocks built up. Possible alternative solutions are devising an effective control system for injecting additives or injecting the additives at a location that minimizes their residence time at low rates.

13. Several reactions are influenced by column pressure. In one case (96), a chemical which was nonreactive when heated at atmospheric pressure detonated when heated at column pressure.

14. Material precipitated out of solution or emulsion may cause an undesirable reaction. In one case (96), explosive-grade nitrocellulose precipitated out of emulsion in a solvent recovery column and detonated.

13.14 Leaking Heat Exchangers

Tube leaks may occur in the reboiler, condenser, preheater, precooler, or any other heat exchanger linked with the column. Several experiences where leakage caused poor performance have been reported (71, 238, 239). The effect of a leaking heat exchanger tube depends on the magnitude, the direction, and the location of the leak.

Effects of heat exchanger tube leaks into the column include off-spec products and/or undesirable chemical reactions. In some cases, this reaction may lead to rapid corrosion or plugging. It is important to realize that material leaking at the condenser or at an intermediate exchanger may travel to the column base and decompose there. In one vacuum column, water leaking at the condenser reached the reboiler and caused a pressure surge and tray damage (358).

Leaks are usually best detected by analyzing column products for materials likely to be present in the heating or cooling medium. Alternative techniques include isolating and draining one side of the exchanger (if feasible), shutting down and pressure-testing the exchanger, or using chemical or radioactive tracers.

Effects of heat exchanger tube leaks out of the column include inability to maintain column pressure, erratic performance of reboiler or condenser, and erratic performance in other units because of contamination of the heat transfer medium. One classic experience (239) of an extremely sick column performance caused by a leaking reboiler was described in Sec. 1.2.

Techniques for detecting leaks out of the column are similar to those for leaks into the column, except that the heat transfer medium, rather that the products, should be analyzed. Often, testing a heat exchanger drain or vent can reveal the presence of a leak. Specific symptoms of hydrocarbons or organic compound leakage into the cooling water are discussed in detail elsewhere (239). These include accelerated biological growth and an increase in the need for chlorination; gas rising from the cooling tower (either an observed haze, or can be determined by analysis), and shaking of the cooling water return line due to expanding vapor.

13.15   Heat Integration "Spins"

The response of a column to a disturbance can interact with the column's heat integration equipment to amplify the disturbance. This aggravates the column's response, which further amplifies the disturbance. An unstable cycle or "spin" develops. Factors conducive to spins are a high degree of heat integration (e.g., heat pumps, feed-product interchangers), latent heat transfer, and shortage of flywheels (e.g., exchangers heated or cooled by outside sources, spare capacity, or bypasses) which can dampen the cycle. Spins are most troublesome during startups or column upsets.

A well-known example is the "cold spin" (16, 394, 421), often experienced in cryogenic gas plant demethanizers operated close to their capacity limits. These are highly heat-integrated columns. Their feed is usually chilled by the column reboilers and by a feed-product interchanger (Fig. 13.7a), and they exchange little heat (if any) with outside sources. If an upset causes liquid carryover from the top of the column, the entrained liquid will step up condensation and chilling of the feed in the top product/feed interchanger. Since the feed to the packed sections is the liquid condensed either in the drum or at the expander (the expander gas bypasses the packed section; see Fig. 13.7a), the load on the packed section increases. This in turn causes more liquid carryover, and often flooding, which further steps up feed chilling. A cycle is initiated; frequently, it can only be broken by a column shutdown (394), or a major rate reduction. Increasing column pressure, or utilizing any available flywheel to reduce chilling (e.g., bypasses around exchangers; stepping down refrigeration chilling if used in the feed chilling train) can also help break the cycle. Avoiding operation close to the capacity limit, or increasing column capacity (e.g., by using larger packings) can also help prevent cold spins.

The above is a typical example of a spin in a heat-integrated column. Similar principles and corrective actions are frequently used for tackling other heat-integrated spin problems.

Figure 13.7b illustrates another example of a highly heat-integrated column that was reported to be troublesome at startup (357). The problem occurs when the column is taken out of total reflux and forward feed is started. A rapid rise or fall in product flow rate

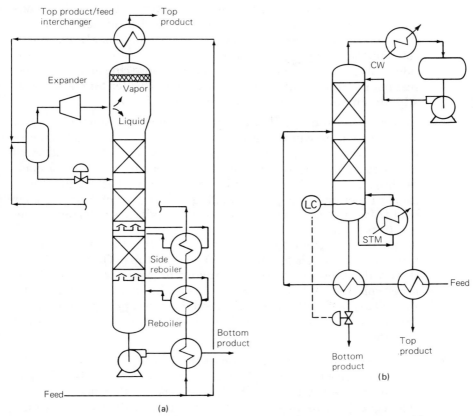

Figure 13.7   Examples of heat-integrated columns that can experience startup problems. (*a*) A typical cryogenic gas plant demethanizer that can experience cold spins; (*b*) a heat-integrated column that can experience surging.

(e.g., due to rapid level changes) can effect a sudden change in the vapor content of the feed. This, in turn, can cause pressure surges in the column (357). A cycle similar to the cold spin can also develop with this type of system. Increasing feed rate by small increments during startup (357) or utilizing any available flywheel (e.g., exchanger bypasses) can help overcome this problem.

13.16   Instrument Problems

Few column startups are completed without instrument problems. Instrument problems are not unique to startups; they also occur during normal operation. Below are some of the most common instrument problems encountered during column startup and operation:

1. *False signals:* When an instrument indicates an unexpected reading, it should be adequately checked rather than disbelieved, as it may indicate an unexpected process condition. It is always worth verifying this reading by looking at other instruments. In one incident (210), column instrument readings were overlooked for several hours after the bottom temperature controller failed. Eventually, flammable oil spilled from the reflux drum.

2. *Incorrect level indication:* Level indication is of prime importance; a faulty reading can be hazardous (Sec. 13.2). Level instruments should always be suspected, especially during startups, and their readings frequently verified (e.g., by comparing to the level glass readings). Impulse lines should be routinely checked to ensure proper functioning (e.g., no plugging, no oil accumulation, proper heat tracing, etc.). Level measurement pitfalls are described in detail in Sec. 5.3.

   Special attention should be paid to level float devices (e.g., controllers, switches, indicators) at startups. Occasionally, a float will collapse or drop off during pressure testing (421) or other commissioning operation. One technique for checking the float (421) is by lifting it with a rod, then wriggling the rod; one can then usually hear the float hitting the side of the float chamber.

3. *Incorrect bottom (or reboiler) temperature indication:* This can be due to a faulty thermocouple, but a more likely cause is fouling of the thermowell. Ironically, bottom or reboiler thermocouple fouling tends to occur in services that are most vulnerable to a malfunction of this instrument, i.e., when heat-sensitive materials are distilled. A fouled thermowell will read low; this in turn will enhance heat input into the column bottom, either automatically or by operator action. The greater heat input will accelerate thermal degradation; in one peroxide service incident (97), it caused overheating and an explosion.

   A bottom temperature indicator measuring liquid temperature may read vapor temperature (which may be considerably lower) when the level drops. The consequences are similar to those of a fouled thermocouple. A malfunctioning level indicator can thus lead to a dangerously misleading, but apparently consistent, indication of both level and bottom temperature (Fig. 13.8). In one case (275), this led to overheating and an exothermic reaction at the column base, which in turn caused residue discharge from a column vent.

   In heat-sensitive services, it is best to provide an additional temperature indication point near (but a sufficient distance above) the bottom of the column, and to use it as a cross check

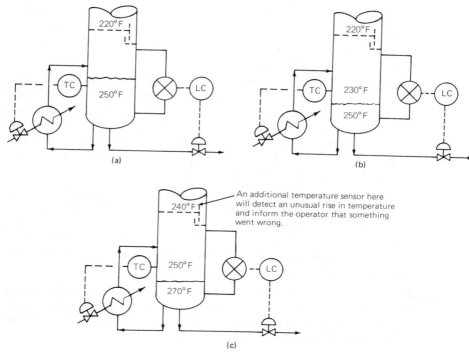

Figure 13.8 A faulty level sensor can cause overheating. (*a*) A malfunctioning level sensor mistakenly senses a high level. It opens bottom valve. (*b*) Level drops and uncovers temperature sensor. If vapor temperature is lower, the controller will call for more heat. (*c*) Level is low, and bottom sump liquid is overheating, but instruments indicate normal level and normal temperature.

(Fig. 13.8). Thermowells should be regularly inspected and, when possible, cleaned. An independent high-temperature alarm may also be helpful. Any suspect temperature readings near the column bottom should be immediately investigated.

4. *Controller cycling:* This is usually caused by inadequate tuning. Switching the controller to manual should stop the cycling if the controller is the culprit. Special caution is required with cycling of reboiler controllers; these can cause surging (98).

5. *Faulty instrument action:* This should be picked up during loop testing prior to startup but can easily be missed.

6. *Troublesome impulse lines:* This is one of the most common instrument problems. Impulse lines are discussed at length in Chap. 5.

7. *Faulty control valve action:* Common causes include a faulty air or power supply, a leaky or ruptured actuator diaphragm, a malfunc-

tioning positioner, a sticking control valve stem, or a faulty current-to-pressure (I/P) transducer. Control valves should be fully stroked prior to startup. Control valve action should be locally inspected at startups.

8. *Mislocated temperature control point:* Often, the temperature control point is located on an incorrect tray, especially if the system distilled is uncommon. It is worthwhile to try hooking the controller to other temperature-measurement points and checking if this improves performance. Having a reasonable estimate of the expected temperatures at these other points is essential for the success of this technique. Further discussion is in Sec. 18.2.

9. *Wet or dirty instrument air supply:* This may cause erratic controller action and/or incorrect signals, and may affect one or several instruments. The author witnessed one entire small distillation unit operated for months without a single reliable flow measurement because of this problem. This was a major factor restraining good performance.

10. *Interface controllers:* These, especially the displacer types (357), are notoriously troublesome during startups when levels change rapidly and frequently. Special attention is required to lining these controllers out during startups.

13.17   Other Typical Operating Problems

Previous chapters described poor column performance caused by troublesome column internals and their installation. Earlier sections in this chapter described several other sources of operation difficulties. A number of the simpler sources of problems remained unmentioned so far. Although these sources appear obvious, they must not be overlooked when diagnosing column problems. Some of these sources are

1. Column upsets and operation difficulties are often initiated in an upstream unit. For instance, poor column performance may be due to an upstream column experiencing control problems, a malfunctioning pH regulation or additive injection system upstream, or a reactor performing in a different manner than design.

   Several incidents of column problems initiated in an upstream unit have been reported. In one incident (98), solids suddenly appeared in the feed and plugged a column when the feed tank was emptied. This occurred because settled solids from the bottom of the tank were stirred up and entered the feed pipe. In another incident (391), oil from an overflowing inlet separator entered an absorber system; it took two days to clean this oil out.

2. Poor performance can be caused by leaks (e.g., block valves on connecting pipes, vent valves) into or out of the system. The effects are similar to those described in Sec. 13.14. In one case (206b), a column separating propane from butadiene experienced excessive propane in the bottom. The cause was a seal failure on the reboiler pump; the pump used propane as seal gas.

3. Poor performance is often suspected due to misleading laboratory analysis, even when the column performs well. Some classic distillation (206b, 268) and nondistillation (131) experiences illustrating this have reported. The author recalls several such experiences. Malfunctioning instruments can lead to similar suspicions. A thorough check of the analytical procedure and analytical calculations, and complication of mass, component, and energy balances can determine whether performance is really poor. This is discussed further in Sec. 14.3.

4. Poor performance may be caused by control valves sticking or clogged or block valves being inadvertently partially open or partially throttled. A few experiences (206b, 231) illustrating this have been reported.

5. Poor performance can be caused by inadequate winterization of pipes, dead legs, and instruments. There have been cases of equipment damage, and even explosions, initiated by freezing of water or process fluids in dead legs of a distillation system. Careful design, inspection, and testing is required in cold climates. Detailed discussion is available elsewhere (120, 266).

6. Poor performance may be caused by an error or oversight in the primary (or process) design stage. [Emphasize the word "primary," as distinct from the secondary (or internals) design stage.] Poor performance due to an oversight in the primary design stage is uncommon. (On the other hand, poor performance due to an oversight in the secondary design is very common. Guides for preventing oversights in the secondary design stage have been the subject matter of Chaps. 2–8). Poor performance due to a primary design oversight is most likely to occur

   - In a first-of-a-kind process, or
   - When the engineering effort is meager, or
   - When the engineering quality is poor, or
   - When the designer has little experience with the service, or
   - When the column is designed very tightly, or
   - When turndown of internals is low

A number of illustrative cases are described elsewhere (85, 203, 278, 338). Good design practices to prevent this are described in most texts dealing with the primary distillation design stage (38, 49, 73, 123, 193, 194, 243, 257, 319, 338, 371, 386, 409). Generally, this type of poor performance rarely results from obvious errors (98), but rather from subtle design pitfalls and unpredictable plant conditions. Typical examples of such pitfalls are (19, 98, 268, 338) uncertainties in equilibrium data, column efficiency calculations, scaleup, and feed point location; unsuspected trace components in the feed; and differences in feed compositions and feed rates compared to design values.

Column Field Testing:
Flooding, Foaming, Efficiency

Capacity and efficiency are the major column performance criteria. Field tests are by far the best means for evaluating these criteria. Field test data are used as a basis for column performance evaluation, troubleshooting, optimization (e.g., reducing energy consumption), and debottlenecking. This basis, however, is only as good as the test data, and these, in turn, are only as good as the testing techniques. Many troubleshooting, optimization, and debottlenecking efforts fell short of their expectations because of poor test data. It is therefore essential to recognize and avoid potential pitfalls when testing column performance in the field.

Column capacity is usually restricted by the onset of flooding. This chapter describes the common flooding mechanisms; outlines techniques for identifying, testing, and curing flooding in the field; and provides guidelines for avoiding common traps inherent in these techniques. One type of premature flood which persistently occurs in certain applications is foaming. This chapter describes the nature of foaming; outlines techniques for identifying, testing, and curing foaming problems; highlights pitfalls in dealing with a foaming problem; and provides guidelines for avoiding these pitfalls.

Efficiency testing requires elaborate preparations if test data are to be meaningful. There is a multitude of considerations, and lack of attention to what may appear a simple detail can lead to an erroneous and misleading test result. This chapter discusses the strategy, preparation, and execution of efficiency tests; outlines techniques for processing their results; highlights common traps in planning and executing efficiency tests; and provides guidelines for avoiding these traps.

Field tests for detecting pinching, packing maldistribution, and column minimum throughput limit are reviewed in this chapter. Finally, the application of radioactive techniques to column testing and troubleshooting is described.

14.1 Flooding

14.1.1 Tray flooding mechanisms

Flooding is excessive accumulation of liquid inside the column. This accumulation is generally caused by one of the following mechanisms.

Spray entrainment flooding (Fig. 14.1a). At low liquid flow rates, trays operate in the spray regime, where most of the liquid on the tray is in the form of liquid drops (Fig. 6.1b). As vapor velocity is raised, a condition is reached where the bulk of these drops is entrained into the tray above. The liquid accumulates on the tray above instead of flowing to the tray below.

Froth entrainment flooding (Fig. 14.1b). At higher liquid rates, the dispersion on the tray is in the form of a froth (Fig. 6.1a). When vapor flow rate is raised, froth height increases. When tray spacing is small, the froth envelope approaches the tray above. As this surface approaches the tray above, entrainment rapidly increases, causing liquid accumulation on the tray above.

When the tray spacing is large (> 18 to 24 in), the froth envelope seldom approaches the tray above. As vapor velocity is raised, a condition is reached when some of the froth inverts into spray. Flooding will then take place by the previously described spray entrainment mechanism.

At high liquid rates (>6 gpm/inch of outlet weir), high ratio (>3) of flowpath length to tray spacing, and a high-fractional hole area (>11 percent), cross flow of vapor in opposite direction to the liquid can build up froth near tray inlet. This channels more vapor to the tray outlet region, thus accelerating the cross flow. The inlet froth keeps rising until it reaches the tray above.

Downcomer backup flooding (Fig. 14.1c). Aerated liquid is backed up into the downcomer because of tray pressure drop, liquid height on the tray, and frictional losses in the downcomer apron. All of these increase when liquid flow rate is raised, while tray pressure drop also increases when vapor flow rate is raised. When the backup of aerated liquid in the downcomer exceeds the tray spacing, liquid accumulates on the tray above, causing downcomer backup flooding.

Downcomer choke flooding (Fig. 14.1d). As liquid flow rate increases, so does the velocity of aerated liquid in the downcomer. When this veloc-

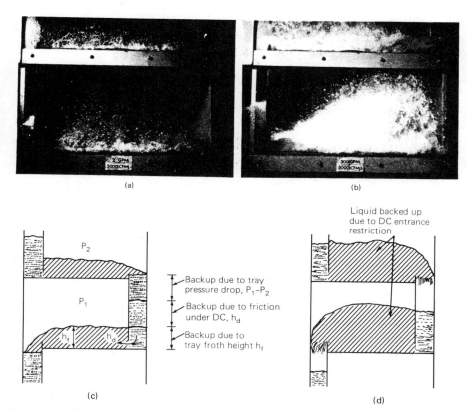

Figure 14.1 Common flooding mechanisms in tray columns. (a) Spray entrainment flood; (b) froth entrainment flood; (c) downcomer backup flood; (d) downcomer choke flood. (*Parts a and b reproduced from Dr. D. C. Hausch, "Discussion of Papers Presented in the Fifth Session," Proceedings of the International Symposium on Distillation, the Institution of Chemical Engineers, London, 1960; reprinted courtesy of the Institution of Chemical Engineers, UK.*)

ity exceeds a certain limit, friction losses in the downcomer and downcomer entrance become excessive, and the frothy mixture cannot be transported to the tray below. This causes liquid accumulation on the tray above.

Effect of pressure and L/V. Figure 14.2 is a rough application chart showing the effect of pressure and liquid-to-vapor (*L/V*) ratios on the mechanism of flooding. This chart does not take into account the tray and downcomer geometry, type of system, and operating conditions, all of which strongly influence the flooding mechanism. For this reason, the chart is suitable for defining general guidelines only.

Low pressures favor high vapor velocities and low liquid flow rates

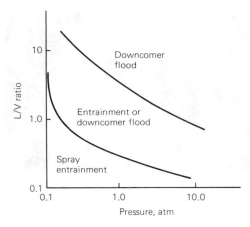

Figure 14.2 A rough flooding mechanism application chart.

and, therefore, spray regime dispersions. Flooding in vacuum columns operating at a low L/V ratio is usually caused by the spray entrainment mechanism.

At high pressures, the difference between vapor and liquid density becomes smaller, and separation of vapor from liquid in the downcomer becomes difficult. Because of the more difficult separation, downcomer aeration increases, raising both downcomer frictional losses and froth backup in the downcomer. High liquid flow rates also increase tray pressure drop, tray liquid level, and frictional losses in the downcomer. For this reason, downcomer flooding is favored at high pressures and high liquid rates.

At moderate pressures and L/V ratios, the dispersion on the tray tends to be in the froth regime, as vapor velocities are lower compared to vacuum and because higher liquid rates favor froth formation. Also, the density difference between liquid and vapor is sufficiently high to ensure good vapor separation in the downcomer. Under these conditions, any of the above mechanisms can dominate. Generally, at low tray spacing (< 12 to 15 in), froth entrainment flooding is favored. At higher tray spacing, and when conditions do not favor vapor cross flow (see above), the froth regime will turn into a spray as vapor velocity increases, and spray entrainment flooding is favored. Finally, when downcomers are small or downcomer backups are high, downcomer flooding is favored.

Effect of design parameters. A number of design parameters have a far greater effect on one flooding mechanism than on others. These parameters are listed in Table 14.1.

Low bubbling areas or low fractional hole areas enhance the flooding tendency of all types of flooding except downcomer choke. Low

TABLE 14.1   Effect of Tray Geometry on Various Types of Flooding

Design parameters that lower the flooding point	Spray entrainment flooding	Froth entrainment flooding	Downcomer backup flooding	Downcomer choke flooding
Low bubbling area	X	X	X	
Low fractional hole area (<8%)	X	X	X	
Low tray spacing	X	X	X	
High weirs (> 4 in)		X	X	
Small weir length		X	X	
Small clearance under downcomer			X	
Small downcomer top area				X

bubbling areas and low fractional hole areas generate high vapor velocities, thus enhancing entrainment, pressure drop, and downcomer backup. These parameters have little effect on downcomer liquid velocity or downcomer froth density, and, therefore, on downcomer choke flooding.

Low tray spacing also enhances the tendency of all types of flooding other than downcomer choke flooding. As tray spacing diminishes, drops have to travel a shorter distance to be entrained (spray entrainment flooding), the froth envelope becomes closer to the tray above (froth entrainment flooding), and a lower downcomer backup is sufficient to cause flooding. Tray spacing has little effect on either downcomer liquid velocity or downcomer froth density, and, therefore, on downcomer choke flooding.

High weirs and small weir lengths slightly reduce spray action and therefore decrease the tendency for spray entrainment flooding, but they increase the height of the froth envelope and therefore increase the tendency for froth entrainment flooding. They also increase liquid height on the tray and tray pressure drop, and, therefore, downcomer backup. Weir height and length have little effect on either downcomer liquid velocity or downcomer froth density, and, therefore, on downcomer choke flooding.

The other two parameters, small clearance under the downcomer and small downcomer top area, have little effect on entrainment flooding, as they are associated with the downcomer only. Downcomer clearance affects downcomer backup, but not downcomer liquid velocity, while downcomer area affects the velocity but has little effect on downcomer backup.

Prediction.   Prediction methods for spray entrainment, froth entrainment, and downcomer backup are available in distillation texts and articles (e.g., 73, 193, 201a, 243, 257, 319, 371, 404, 409). Criteria for predicting downcomer choke flooding are discussed in Sec. 6.16.

14.1.2. Packed tower flooding

Flooding in packed columns can be discussed with reference to a typical pressure drop characteristics diagram (Fig. 14.3).

At very low liquid flow rates (region A-B), the open cross-sectional area of the packing is about the same as in a dry bed. The pressure drop is almost entirely caused by frictional losses through a series of openings and, therefore, is approximately proportional to the square of the vapor flow rate. At higher liquid rates, the effective cross-sectional area open for vapor flow is smaller because of the presence of liquid (region A'-B'). Pressure drop is higher, and may be proportional to the vapor rate raised to a power other than 2.

As vapor flow rate is raised, a point is reached when vapor begins to interfere with liquid downflow and holds back liquid in the column. At this point, pressure drop starts to rise steeply because the rapid accumulation of liquid in the column reduces the cross-sectional area available for vapor flow. In this region (B-C and B'-C' in Fig. 14.3), the slope of the curve increases to a power distinctly above 2. This region is termed the *loading region*.

As liquid accumulation increases, a condition is reached when the liquid surface becomes continuous across the top of the packing. The slope of the curve increases further, until it becomes very steep, where

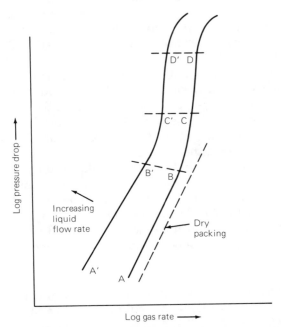

Figure 14.3 Typical pressure drop characteristics of packed columns.

small increases in vapor flow significantly raise the pressure drop. When this occurs (points C and C' in Fig. 14.3), the column is flooded.

A stable operating condition beyond flooding (above points D and D') was observed (230). Because of the limited information available on this region, it is presently of little significance in industrial practice.

The exact definition of the flood point in packed towers is uncertain. Silvey and Keller (367) list ten different flood point definitions which were used by different investigators. Perhaps the most popular definition is (51, 319) "the vapor velocity above which liquid accumulates uncontrollably in the packed bed and continued operation becomes impossible."

Effect of geometry and pressure. Generally, the larger the openings available to vapor flow, the higher the vapor rate at which loading and flooding will occur. For this reason, smaller packings flood at lower vapor flow rates than larger packings. With high-capacity packings (e.g., 2-in pall rings or packings of equivalent or higher capacity), and/or under vacuum, capacity is usually restricted by excessive entrainment rather than by flooding.

At low and moderate pressures and liquid flow rates, structured packings flood at higher vapor flow rates than random packings of the same open area. Pressure drop through a randomly packed bed is mainly due to expansion and contraction friction losses, while pressure drop through a structured packing is mainly due to friction losses through bends in a uniform channel. The expansion and contraction mechanism incurs higher friction losses when the flow-through bends in a uniform channel, resulting in a lower flood point.

At high liquid loads (> 15 gpm/ft^2 of bed cross section) and high pressures, vapor entrainment in the liquid may restrict structured packings capacity well before flooding is approached. This phenomenon causes efficiency to rapidly diminish as throughout is raised. Random packings also experience this type of limitation (386), but to a lesser extent, because of the unrestricted lateral movement of vapor and liquid.

Prediction. Prediction methods for packed tower flooding are available in most standard distillation texts (e.g., 51, 193, 257, 319, 386, 404, 409) and in packing manufacturers' literature.

14.1.3 Flood point determination in the field

Flooding is characterized by the accumulation of liquid in the column. This accumulation propagates from the first flooded tray (or from the lowest flooded packed section) upward. Accumulating liquid backs up into the tray (or packed section) above, and so on, until the whole column is full of liquid or until an abrupt change in tray design or flow conditions (e.g., feed point) is reached. Flooding may or may not propagate above that point.

Flooding can be recognized by one or more of the following symptoms:

1. Excessive column differential pressure
2. Sharp rise in column differential pressure
3. Loss of bottoms
4. Rapid rise of entrainment from column top tray
5. Loss of separation (as can be detected by temperature profile or product analysis)

Pressure drop measurements across various column sections are the primary tool for flood point determination, particularly in superatmospheric and high-liquid-load columns.

Excessive column pressure drop.   In general, a pressure drop per tray greater than 50 to 60 percent of the tray spacing in the relevant column section indicates flooding (2, 186, 231). In packed columns, the following rules of thumb are useful.

1. Pressure drop lower than 1 in of water per foot of bed indicates no flooding (74, 235).
2. Pressure drop greater than 3 in of water per foot of bed indicates fully developed flooding (235, 327).
3. Pressure drop between 1 and 3 in of water per foot of bed generally indicates a high probability of flooding (74, 235).

Sharp rate of rise of pressure drop.   A sharp rate of rise of pressure drop with vapor rate may be an even more sensitive flooding indicator than the magnitude of pressure drop. The flood point can be inferred from a plot of pressure drop against vapor or liquid flow rate, and is the point where the slope of the curve changes significantly (Figs. 14.3, 14.4). In tray columns, the slope change can be relatively mild (curve 1 in Fig. 14.4), which is generally characteristic of entrainment flooding or a small number of flooded trays, or relatively steep (curve 2 in Fig. 14.4), which is generally characteristic of downcomer flooding or flooding which propagates throughout several trays. It is not unusual to find a vertically rising pressure drop curve once the flood point is reached (132, 182).

In packed columns, defining the flood point by use of a pressure drop

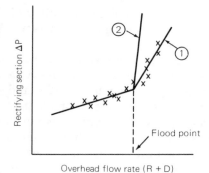

Figure 14.4 Plots of pressure drop versus vapor flow rate for tray columns.

versus load curve (Fig. 14.3) is generally less satisfactory, because the slope begins changing at the loading point (point B), and the change may be continuous (curve BCD), rather than abrupt, in the vicinity of the flood point. Further, in many packed columns a rapid drop in efficiency occurs well before the hydraulic flood point. Here throughput is limited by loss of separation, and the hydraulic flood point is of little practical value.

The author, as well as several others (2, 186, 268), recommends that for best results, differential pressure recorders should be installed at least across each section of the column prior to flood testing. Since only part of the column may be flooded (e.g., only the section above the feed), the overall column pressure drop may not increase markedly, but the pressure drop in the affected section will.

It is important that the differential pressure recorders be installed and read adequately, and do not suffer from problems of condensation or refluxing in the lines to the transmitter. Several application guidelines are discussed in Sec. 5.4.

Manometers are often suitable for temporary differential pressure measurements at low pressures with nonhazardous systems. Manometers are portable, easily installed, and have a wide range of operation. Caution is required to ensure that the manometer used is suitable for the service. A comprehensive set of application guidelines for manometers in flood testing is available elsewhere (2).

Since differential pressure recorders and their installation are expensive, there is often a tendency to minimize their use. One alternative proposed is the installation of static pressure gages. This may be satisfactory if the slope of the pressure drop versus load curve is known to be very steep at the flooding point (Fig. 14.4, curve 2), but is very poor otherwise. Even with steep slopes, this alternative will lower the quality of the data and is not recommended (268).

Another alternative often suggested is using a single differential pressure instrument and switching by valves from one location to another in the column. This technique, which was applied by Kelley et al. (186), is a more satisfactory option than pressure gages, but is not suitable for more than a preliminary analysis (268).

If it can be afforded, a multichannel strip-chart recorder with a chart speed sufficiently high to trace the sequence of events is ideal for flooding tests. This type of instrument has been highly recommended (268) and successfully applied (132, 268). The record of this instrument can identify the exact location where flooding starts, the exact conditions when it takes place, how the flooding propagates, and which remedial action is working. This device may be particularly useful if flooding is caused by plugging of the top tray or downcomer (e.g., by corrosion products). In one case (238), conventional differential pressure devices failed to indicate this condition. Figure 14.5 (268) shows the type of information the multichannel strip-chart recorder can convey.

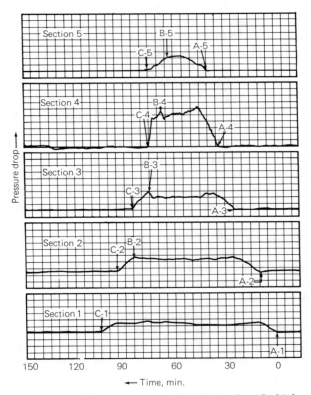

Figure 14.5 Pressure drop profile obtained with high-speed, multichannel strip-chart recorder *(D.B. McLaren and J.C. Upchurch, excerpted by special permission from* Chemical Engineering, *June 1, 1970; copyright © by McGraw-Hill, Inc., New York, NY 10020.)*

Loss of bottoms. Loss (or a shop decline) of bottom flow is a common indicator of flooding, and is one of the main criteria to determine the flood point (2, 367). There are two problems with using this as the main indicator of flooding. First, the fractionator may flood without losing the bottom flow. For instance, if flooding occurs in the rectifying section, while most of the feed is liquid, the bottom section may continue to operate normally without a loss of bottoms. Secondly, if the flood point is well above the bottom, there may be a significant delay from the onset of flooding to the time the bottom flow is lost, which makes accurate measurements of the flooding conditions difficult.

Loss of bottoms is therefore a good indicator of flooding in columns which are relatively short, particularly if flooding occurs between the feed point and the bottom. These conditions are typical of the fractionator in Ref. 367. For columns containing a large number of trays, loss of bottoms may not be a good primary indicator of flooding.

Rapid rise in entrainment. A rapid rise in entrainment is another common flooding indicator (2, 91, 367). This is often recognized as a large rise in reflux or product rate for a small or no increase in reboil. Bleeders (see below) have also been used to detect this rise in entrainment (186).

The main disadvantage of this indicator is that it may fail to indicate a stripping section flood that does not propagate to the rectifying section. On the other hand, this indicator is particularly useful when the pressure drop rise is not sharp (Fig. 14.4, curve 1). In some instances (91) it may even be a better indicator than column pressure drop.

Loss of separation. Loss of separation is apparent when the column floods. As flooding is approached in either tray or packed columns, the rate of liquid entrained by the vapor sharply rises. At high pressures and/or high liquid rates, the quantity of vapor entrained in the downflowing liquid also rises. As either type of entrainment accelerates in the vicinity of the flood point, efficiency and separation plunge (Fig. 14.6). The drop in efficiency tends to occur closer to the flood point with downcomer flooding than with entrainment flooding; in Fig. 14.6, it is believed (429) that flooding was caused by a downcomer limitation in the butane system, and by excessive entrainment in the cyclohexane-heptane system. In packed columns, the drop in efficiency occurs closer to the flood point with smaller packings. With large packings (e.g., 2-in pall rings or packing of equivalent or higher capacity), and/or under vacuum or high pressure, the drop in efficiency may begin at rates well below the hydraulic flood point.

Since loss of separation begins before the column is fully flooded, us-

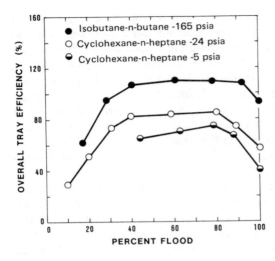

Figure 14.6 Overall tray efficiency, 4-foot-ID tower, at total reflux, illustrating drop of efficiency near the flood point (*Reprinted with permission from T. Yanagi and M. Sakata, Ind. Eng. Chem. Proc. Des. Dev., vol. 21, p. 712, copyright (1982) American Chemical Society.*)

ing it as a flooding indicator can suggest a lower flood point than other indicators. This is hardly a disadvantage, because the exact location of the hydraulic flood point is of lesser practical significance than the point where column efficiency starts to rapidly deteriorate. The latter point is often referred to as "the maximum operational capacity," and usually occurs at flow rates about 0 to 20 percent below the hydraulic flood point in tray columns, and about 10 to 30 percent below the flood point in packed columns. In most atmospheric and superatmospheric tray columns, this point occurs at flow rates of 5 percent or less below the hydraulic flood point.

The loss of separation is best recognized from laboratory analyses of column products. A plot of column efficiency (Fig. 14.6) or separation ratio (ratio of light to heavy key for the top product times ratio of heavy to light key in the bottom product) against flow rate at a constant reflux ratio is commonly used to identify the point where loss of separation occurs.

Another good indicator of separation loss is the column temperature profile. Hausch (152) describes the use of temperature profiles for investigating flooding problems. The application of this method requires a good knowledge of the normal and flooded temperature profiles under similar feed conditions.

Figure 14.7 gives some examples of temperature profiles under normal and flooded conditions. When the entire column is flooded, little separation (and therefore, temperature change) is achieved throughout most of the column. Note that some separation (and therefore temperature change) still occurs near the top and bottom of the column. Another example (Fig. 14.7) is bottom section flooding. Here the section below the feed accomplishes little separation, but the rectifying section operates normally.

Caution must be exercised when curves of this type are interpreted, because they may also indicate a "pinch" condition (i.e., poor separation due to insufficient reflux or reboil). In order to tell the difference, reflux and reboil can be raised. If separation improves, pinching is indicated. If it deteriorates or stays the same, flooding is indicated.

In order to reliably establish the column temperature profile, a large number of temperature measurement points is required. A multipoint temperature recorder with a short recording cycle is particularly suitable for obtaining a time record of temperature profiles. Alternatively, a vertical temperature survey can be conducted. The survey involves (238) cutting holes (diameters of about 1 in each) in the insulation covering the centers of several preselected downcomers, and measuring their external temperatures with a surface pyrometer. The pyrometer readings should be cross-checked against the available column temperature indicators. An infrared camera may in principle substitute for the surface pyrometer, but the author is not aware of any reported column temperature surveys using such a camera.

Temperature gradients are an effective, low-cost method of determining the flood point, but the method's success depends on the existence of a sufficiently large temperature gradient under normal operating conditions. If the normal tray-to-tray temperature difference is small, as in close separations, the flooded temperature profile will not vary a great deal from the normal profile, and temperature profiles will be poor indicators of flooding.

Bleeders. One technique found effective for flood testing is the use of vapor bleeders (186). Each bleeder is located in the vapor space above

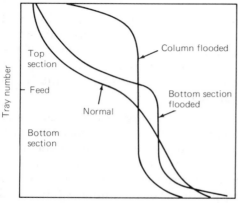

Figure 14.7 Normal and flooded temperature profiles.

a tray and/or in the overhead line upstream of the condenser. If the bleeder is opened during normal operation, vapor will come out; if it is opened while the tray is flooded, liquid will spray out. When the tray liquid is above its atmospheric boiling point, it will flash and chill upon bleeding. The presence of liquid can thus be detected by a temperature recorder as a sharp temperature drop of gas issuing from the bleeder (186).

The bleeder technique is not very popular and can be hazardous or environmentally unacceptable. Other disadvantages are a need to know which trays are most likely to flood, and lack of indication of an approaching flood condition.

Gamma ray scanning.   Gamma ray scanning is one technique particularly suitable for flooding detection. Although expensive, it is powerful in diagnosing a flooding problem, identifying the flooded regions, and often also providing insight into the nature of the flood. Detailed discussion is in Sec. 14.5.

Sight glasses.   These have been used to give visual indication of flooding (186). This technique is not commonly used in commercial columns (Sec. 5.7).

14.1.4 Flood testing

In order to determine the flood point, either vapor or liquid flow rate or both are raised. Most commonly, both are raised, because otherwise column material balance is affected and one product will have poor purity before flooding conditions are reached. The following techniques are commonly used for raising vapor and liquid rates during flood testing:

1. Raising feed rate, while simultaneously increasing reflux and reboil in proportion or in a manner that keeps product composition constant. This technique gives the most direct measurement of the maximum feed rate that can be processed through the column, but can only be applied when upstream and downstream units have sufficient capacity to handle the additional feed.

2. Raising reflux and reboil rate while keeping feed rate constant. This is probably the most popular technique used. Only two variables (instead of three) need to be changed, and it is independent of the capacities available in other units, making it simpler and easier to implement. In most cases, data provided by this technique can be easily extrapolated to predict the maximum column feed rate.

3. Varying preheater or precooler duty while adjusting reflux and

reboil. This method, which can only be used if the feed is preheated or precooled, is often restricted by the exchanger capacity and is least popular. In some multicomponent distillations, it can give misleading results because it may induce accumulation of an intermediate impurity in one section of the column (Sec. 13.7).

Using any of the above techniques, reflux and reboil rates are varied. The procedure of varying these rates is important, and must take the column control system into account.

Most column control schemes (see Sec. 16.5) use the composition (or temperature) controller to manipulate either the reflux or reboil, directly or indirectly. The stream which is not controlled is commonly "free," i.e., on flow control. This "free" stream is usually manipulated during flood testing, while the stream on temperature control will be automatically adjusted to maintain product composition. For instance, if reflux rate is on temperature control and reboil rate is on flow control, flood testing is performed by raising the reboil rate. This warms up the control tray and increases condensation. The temperature controller will call for more reflux, and the column will reach new stable conditions with both reboil and reflux increased.

This procedure may induce the problem of "overshooting" the flood point. In the above example, reflux rate will increase shortly after a change in reboil. The column may look stable for quite some time following the change, even if the reflux and reboil rates which cause flooding have been exceeded, as it may take the liquid some time to reach the tray where flooding initiates. This is particularly true for tray columns containing a large number of trays. In the meantime, vapor and liquid rates are raised further as the test progresses. When overshooting occurs, the flood point determined will be higher than the actual one.

The problem of "undershooting" the flood point is just as common. Its occurrence depends on the dynamics of the column. For instance, in the above example, increasing reboil can displace some liquid from the trays. This liquid increases the internal liquid flow rate on the tray below, and it may flood prematurely. When this occurs, the flood point determined will be lower than the actual one.

In order to avoid overshooting and undershooting, little can be offered as a substitute to raising vapor (or liquid) rates in extremely small steps, and allowing long stabilization periods between steps. This is most important in columns containing a large number of trays. It may pay to carry out a preliminary flooding test, in which the steps are relatively large and fast. Typically, vapor (or liquid) rates are raised by 5 to 10 percent increments at 15- to 30-minute intervals during the preliminary test (2). Increments as small as 1 to 2 percent are preferable, even at the preliminary test. It was found (186) that frequent small increases in vapor (or liquid) rates are less likely to prematurely upset the column, and generally require shorter stabilization periods (186).

Although results of the preliminary test may suffer from overshooting or undershooting, they are likely to determine the flood point within ±10 percent, and often within ±5 percent. The results of this test are used to determine a good starting point for the slow test. The preliminary test technique was found effective both for improving accuracy and reducing time requirements for flood point determination (186).

With columns containing a large number of trays, a test lasting several days is sometimes best when accurate determination of the flood point is required. In most plant situations, weekend tests are ideal, as changes to the fluctuations in upstream units are minimized.

Accurate material and energy balances are important for flood point determination, and these should close within 3 and 5 percent, respectively (2, 186) and be checked prior to the test and during the test. There is generally no need for accurate component balances. Several of the key guidelines described in Sec. 14.3, particularly those pertaining to material and energy balances, are also useful for flood testing. Note, however, that flood tests are far less sensitive to analytical errors than efficiency tests, and therefore require a much lower level of effort.

14.1.5 Field cures for flooding

Overcoming a flooding limitation usually involves changes to the hardware of the column. Such changes are discussed elsewhere (193). At other times, they involve column shutdown in order to clean the column or to remove blockage. On-line cleaning, antifoam injection, and solvent injection to dissolve frozen particles are methods of overcoming specific problems which cause premature flooding.

Other techniques are useful for increasing column capacity during operation and when flooding occurs at "normal" conditions (as distinct from "premature" flooding). These techniques are described below.

Unloading. This is by far the most common technique and requires little discussion. Column unloading can be achieved either by reducing plant rates, by changing feed composition, or by diverting one of the feed streams or a portion of it away from the column. It often pays to closely examine each stream that ends up in the column and test the effect of taking it out of the feed. In one example (203), a column bottleneck was eliminated with minimum adverse effects by taking a major stream out of a column feed.

Preheating/precooling changes. If the column feed is preheated or precooled, feed temperature can be varied in order to unload the section above or below the feed.

When the flooding limitation occurs below the feed, a hotter feed can reduce the reboiler heat load and the vapor and liquid traffic in the section below the feed at the expense of higher vapor and liquid traffic above the feed. Conversely, reducing feed temperature unloads the section above the feed at the expense of higher loads in the section below the feed.

Pressure changes. Distillation column pressure variation can increase column capacity (80, 93, 152, 197–199). Capacity gains can be achieved either from raising or lowering the pressure. Raising the pressure reduces gas density, thus allowing a greater vapor flow rate through the column, but it also reduces relative volatility, causing a higher reflux and reboil requirement for the same separation. Either factor may predominate, and the dominant factor dictates the direction in which pressure should be changed. In several cases, both factors are balanced (80, 197) and changes in pressure have little effect on column capacity.

A number of guidelines have been proposed (197–199) for pressure variation when entrainment flooding limits capacity (Fig. 14.8; Table 14.2 provides a key to the curves in Fig. 14.8). The rate of reduction of relative volatility with an increase in pressure is characterized by the exponent m in the following equation.

$$\alpha_1 = \alpha_0(P_1/P_0)^{-m}$$

where α is relative volatility and P is absolute pressure; 0 and 1 denote the initial pressure and final pressure, respectively. A low value of the exponent m implies a low rate of reduction of relative volatility with pressure.

Figure 14.8 shows that capacity gains achieved from increasing column pressure prevail near atmospheric pressure conditions. These gains are greatest at low pressures, high relative volatilities, and at low rates of reduction of relative volatility with pressure. Capacity gains from lowering the pressure become significant at elevated pressures. These gains are greatest at high pressures, low relative volatilities, and high rates of reduction of relative volatility with pressure.

Similar guidelines apply when column capacity restriction is primarily related to column vapor load [e.g., a downcomer backup restriction when backup is primarily due to dry tray pressure drop (199); or in packed columns in which flooding is induced by excessive vapor loads (195)]. The guidelines do not apply when column capacity restriction is primarily caused by excessive liquid load (e.g., downcomer choke or downcomer backup restriction when the backup is primarily

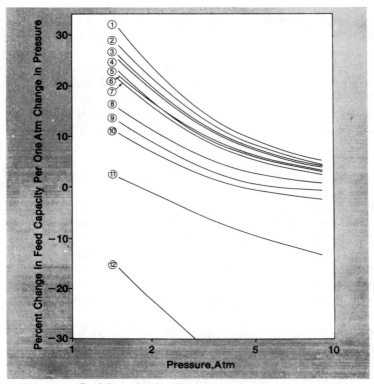

Figure 14.8 Guidelines for the effect of pressure on column feed capacity when limited by entrainment flooding (*From H. Z. Kister and I. D. Doig,* Hydrocarbon Processing, *July 1977. Reprinted courtesy of* Hydrocarbon Processing.)

TABLE 14.2 Key for Fig. 14.8

Curve	α_o	m	α_o/m
1	Any	0	∞
2	3.0	0.05	60
3	2.0	0.05	40
4	3.0	0.10	30
5	3.0	0.15	20
6	1.5	0.05	30
7	2.0	0.10	20
8	2.0	0.15	13
9	1.5	0.10	15
10	1.2	0.05	24
11	1.5	0.15	10
12	1.2	0.10	12

caused by wet pressure drop). In the latter situations, the enhancement of gas density with pressure does little to raise capacity, the relative volatility effect dominates, and capacity gains are normally achieved by reducing column pressure.

Improved stability. When columns operate close to their capacity limits, even small disturbances can carry the column beyond the flood point (152). Once flooding sets in, it may be difficult to return the column to normal operation unless vapor and liquid traffic are largely reduced. In order to operate the column at a point close to maximum capacity, stable operation and reduction of the magnitude and frequency of outside disturbances are essential. The column control system and instrument tuning play a major role in achieving maximum column capacity. It has been the author's experience that the improved stability can usually enhance capacity by 2 to 5 percent, and sometimes by up to 10 percent.

14.2 Foaming

14.2.1 Nature of foaming

Foaming in fractionation and absorption columns can drastically lower capacity and lead to premature flooding, liquid carryover, and solvent losses. In packed columns, foaming can also lead to poor distributor and redistributor action.

The principles of foam formation, stabilization, structure, dynamics, and destruction are presented in detail elsewhere (36, 157, 216, 243, 319, 339). A summary is presented below.

Mechanism. A foam forms when bubbles rise to the surface of a liquid and persist without coalescence with one another, or without rupture into the vapor space. The life of foams varies over many orders of magnitude—from seconds to years, but in general is finite. A quick-breaking foam may reach a life of 5 s; a moderately stable foam can persist for 2 to 3 minutes (319). Maintenance of a foam, is therefore, a dynamic phenomenon.

The stability of a foam can be described in terms of the forces acting on a liquid film surrounding a bubble in a foam structure. The following forces are important:

1. *Gravity:* Drains liquid from the film between two adjacent bubbles.
2. *Interfacial tension:* Favors the coalescence and ultimate disappearance of the bubble.
3. *Capillary forces:* Tend to stabilize the bubble surface.
4. *Viscosity:* Opposes the drainage of liquid from the film.

The balance of the above forces alone is insufficient to stabilize a foam. Liquid drains from the film surrounding a bubble by gravity, and the interfacial tension eventually causes rupture. Four common mechanisms cause foam stabilization:

1. *The Marangoni effect:* A surface-active solute tends to concentrate at the liquid surface. When liquid drains from a film surrounding a bubble, it thins a part of the film. In the thin part, surface area increases (Fig. 14.9a). The additional surface is supplied by liquid from the bulk, which is leaner in the surface-active solute. Surface tension at the thinned surface therefore rises. This causes a surface flow from the nonthinned (low-surface-tension) surface to the thinned (high-surface-tension) surface, which counteracts film drainage and restores the film.

2. *The mass-transfer-induced Marangoni effect:* This effect stabilizes the film when liquid surface tension increases due to mass transfer (i.e., the less-volatile component has a higher surface tension than the more-volatile component). The liquid film just between two adjacent vapor bubbles is closer to equilibrium than the liquid at some distance from these vapor bubbles, and therefore has a higher surface tension than the bulk of the liquid (Fig. 14.9b). The surface-tension gradient sucks liquid into the film between the two bubbles, thus counteracting drainage.

3. *Ross-type foaming (340):* A weak solvent-solute interaction may cause surface activity. This activity, and therefore the foaminess of the solution, increases with the tendency for separation of the solution into two liquid phases. As solute concentration increases toward the critical solution point, or the plait point, foaminess increases. Once this point is reached and two phases are formed, one phase may act as a foam inhibitor and destroy the foam.

4. *The formation of a gelatinous surface layer:* Such a layer immobilizes the liquid inside the film between two adjacent bubbles so that gravitational and capillary stresses are insufficient to cause flow. The gelatinous surface layer is formed by chemical or intermolecular interactions occurring in the film.

Foam structure and dynamics.   Surface layers surrounding the bubbles in a foam act as a membrane or skin that can stretch and relax in response to the lateral forces acting on it. At first, drainage of liquid taking place at the surface layer is entirely hydrodynamic, but once spherical bubbles are in contact, flat walls develop between them, and polyhedral cells appear in the foam (Fig. 14.9c). Capillary forces be-

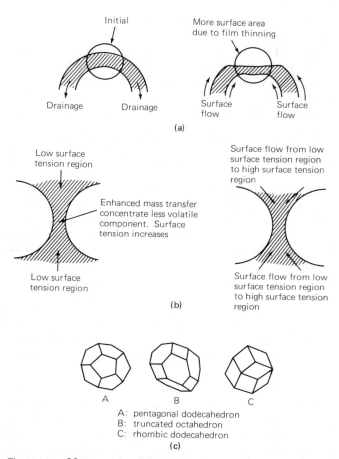

A: pentagonal dodecahedron
B: truncated octahedron
C: rhombic dodecahedron
(c)

Figure 14.9   Marangoni stabilization of foams. (a) The Marangoni effect; (b) The mass-transfer-induced Marangoni effect; (c) examples of polyhedral foam cells. (*Part c from G. E. Ho, R. L. Muller, and R.G.H. Prince, I. Chem. E. Symp. Ser. 32, p. 2:10, 1969. Reprinted courtesy of the Institution of Chemical Engineers, UK.*)

come important, and these tend to redistribute the liquid along the cell walls. Marangoni forces act to supply liquid to the cell surface from the central liquid; at the same time, liquid is lost by the cells due to drainage. At the top of the foam, the effect of drainage is strongest, because cells do not receive liquid draining from the cells above, and the structure is least stable there. After a period of time, the liquid film becomes so thin (200 to 300 Å) that the central liquid itself is affected by the surface forces, and the Marangoni effect is lost. This causes the resilience of the liquid film to be replaced by brittleness, and it is soon ruptured by relatively slight mechanical shocks that it had previously been able to withstand. The foam, therefore, begins to collapse from the top down.

Foams resulting from the Marangoni effect alone, often referred to as *physical foams,* are relatively unstable (e.g., champagne foam); they drain rapidly to the critical thickness of 200 to 300 Å, at which point they become brittle and soon rupture. Foams produced by the mass-transfer-induced Marangoni effects are more stable (407), and often lead to severe foaming in columns. However, it has been argued (340) that the severe foaming sometimes attributed to the mass-transfer-induced Marangoni effect is actually caused by Ross-type foaming.

The very stable foams (e.g., whipped cream, firefighting foam, shaving cream, beer foam) initially start the same way, being stabilized by Marangoni forces. This effect is then replaced by the growth of gelatinous surface layers, which immobilize the fluid in the film so that gravitational and capillary forces are both insufficient to cause flow, leading to the formation of stable films.

Gelatinous surface layers are formed due to the presence of some impurity in the liquid that interacts with the liquid to form a hydrous gel structure, extending from the surface to a depth of about 900 Å. The liquid makes up the main portion of the surface film (at least 97 percent by weight). The liquid is oriented and partly immobilized in an icelike configuration which acts as the linkage between solute molecules.

An important characteristic of the gelatinous surface film is its transition to a freely flowing film over very narrow range of temperatures. The transition is sharp and reversible. A practical example occurs with certain fuel oils which foam excessively in winter when pumped into oil tanks but show no signs of foaming during summer. Chemical characteristics of gelatinous surface layers are discussed in detail elsewhere (36, 339).

Occasionally, when imperfectly wetted solid particles are present, stabilizing of the foam may be caused by the solid particles holding entrapped gas between their hydrophobic surfaces and hindering coalescence of bubbles.

14.2.2 Foam inhibitors

Foam inhibitors are always insoluble materials, usually liquids, but some solids such as waxes, gums, or soaps are used. Their foam-inhibiting action arises from their ability to spread spontaneously over the surface of the foamy liquid. A single drop of the spreading liquid, once it has arrived at the surface of the liquid film, acts effectively as a venturi pump, ejecting on every side all the liquid lying beneath it and thinning itself out to the point of rupture. The film of foamy liquid is replaced by film of spreading liquid, which cannot support an extended liquid film, causing rupture.

The major requirement of a foam inhibitor is cost-effectiveness; accordingly, some useful characteristics are low volatility (to prevent stripping from the system before it is dispersed and does its work), ease of dispersion and strong spreading power, and surface attraction origination. Also important are effects on product quality, downstream units, and on the environment and health. Several common types are discussed elsewhere (36, 216, 319, 339). Often, a foam inhibitor does not use a single compound, but a combination also including a carrier (usually a hydrocarbon oil or water, which supports the release and spread of the primary defoamer); a secondary defoamer; an emulsifier, which enhances the speed of dispersion; and a stabilizer, which enhances the inhibitor's stability.

One of the most popular groups of foam inhibitors commonly used in distillation and absorption is silicones, the most common compounds being dimethyl silicone and trialkyl and tetraalkyl silanes (319). These are basically finely divided solid silica particles dispersed in a hydrocarbon or silicone oil, which serves as a spreading vehicle. These inhibitors combine several desired properties such as involatility, low surface tension, chemical inertness, and insolubility in water and in lubricating oil. They are usually effective in concentration of less than 10 ppm (339), while other organic compounds are often used in the range of 100 to 1000 ppm. On the other hand, silicones cost per pound is about 10 to 20 times higher than the cost of common organic defoamers.

14.2.3 Identifying foaming in fractionators and absorbers

Foaming commonly occurs in some systems, while it seldom occurs in others. Some guidelines which may assist in identifying the potential of foam formation are presented below.

1. Foaming is a common problem in absorbers and regenerators using aqueous solutions of high-molecular-weight organic solvents. Typical examples are absorbers and regenerators using ethanol amine, glycol, and potassium carbonate solutions.

2. In the above type of application, the presence of compounds such as liquid hydrocarbons and organic acids often promotes or induces foaming.

3. Corrosion inhibitors are often surface-active agents and generally severe foamers. For example, corrosion inhibitors injected into

natural gas–gathering systems are known to have caused severe foaming problems in gas plant amine absorbers (368, 373). The source of corrosion inhibitors is commonly an upstream process, but they may also be present in the water used to make up a lean absorption solution. In one reported case (373) a severe foaming problem in an amine absorber occurred because steam condensate, which contained the steam system corrosion inhibitors, was used to make up the amine solution. In another reported case (203), a corrosion inhibitor injected into a natural gas well ended up causing severe foaming in the gas plant absorber.

4. Solvents may dissolve or react with the material used in filters or similar equipment, and the reaction products may cause foaming. Elements containing paper, fabric, glue, or thermally unstable materials are often troublesome, especially when improperly pretreated (22). Pretreatment of filter elements, and/or good choice of element materials can eliminate this type of problem. One case has been reported (373) where foaming in an ethanolamine absorber was caused by the elements of a filter which was packed with cotton linter. The ethanolamine saponified the cotton oil in the element, and this led to a foaming problem.

5. Foaming is commonly experienced in absorbers using certain hydrocarbon solvents such as light cycle oil (299) or crude oil (135) to absorb hydrocarbons in the C_3 to gasoline range. Experience has indicated (299) that replacing the solvent by a lighter solvent such as heavy naphtha is often the most effective means of overcoming this problem. Use of antifoam is also effective (71, 135, 299).

6. Foaming is often experienced in some extractive distillations (243, 407, 408). Examples are extractive distillations using acetonitrile solvent (407), such as those used in butadiene plants and sulpholane extractive stripping (407, 408). This foaming is attributed to mass-transfer-induced Marangoni forces (407, 408), is more likely to occur under stripping conditions (407, 408), is sensitive to concentration of high-surface-tension components (e.g., in some cases, increase in water concentration promoted foaming while adding minor quantities of heavy homologues suppressed foaming (407, 408)), and is sensitive to the fraction of feed which is vaporized (in some cases, a smaller fraction enhanced foaming (407, 408)).

7. Foaming is commonly experienced in the "cold" (90 to 100°F) hydrogen sulfide–water contactors of the GS (girdler-sulfide) heavy water process (132, 182). This foaming is promoted by a high content of suspended solids in the feed water, is sensitive to this solids content,

and is usually suppressed by antifoam injection. Certain tray designs appear to reduce the foaming tendency (182).

8. Foaming is a common problem in the stripping section of refinery crude atmospheric columns (24, 85). This is often related to the long residence time of the residue at high temperature (85) and to the presence of trace impurities (24). The problem may occur only while processing some crudes, but not others.

9. Foaming is a common problem in refinery preflash crude towers (298; the author is familiar with other experiences). The problem may occur only while processing certain crudes.

10. Foaming is sometime experienced in moderate-pressure (100 to 200 psi) strippers that strip light from heavy hydrocarbons (**209**). The presence of small quantities of water may promote this foaming.

11. Some guidelines to the tendency of foaming in different systems are given in Tables 6.1 and 6.2.

12. Suspended solids and polymers generally tend to stabilize foams. One example is iron sulfide particles in amine solutions (26, 239, 318); also see item 7 above. In one case (107), precipitation of sodium chloride in a solvent-water separation column caused foaming.

13. Wall-stabilized foaming (cellular foam) may occur in small and pilot-size columns (Fig. 14.10). This type of foaming differs from the foaming experienced in industrial-scale columns by being wall-stabilized. A comprehensive description of this foaming phenomenon is given elsewhere (159).

14. A foaming problem may occur in one installation, but not in another, even though both handle the same materials. Often, a foaming problem may remain "hidden" because a column is overdesigned (26, 50), or may not occur in one installation because a surface-active impurity is absent. One experience has been reported (50) where a new absorption unit experienced a foaming problem, while a similar but older unit did not. This occurred because the older unit had oversized downcomers, which were capable of handling the foam.

15. Foaming is unlikely (107) in distillation columns where the bottom product has a lower surface tension than the top product ("negative-surface-tension systems"). Here the mass transfer counteracts the Marangoni effect. Foaming is also unlikely when two liquid phases are present throughout, because one of these phases will tend to act as a defoamer (107).

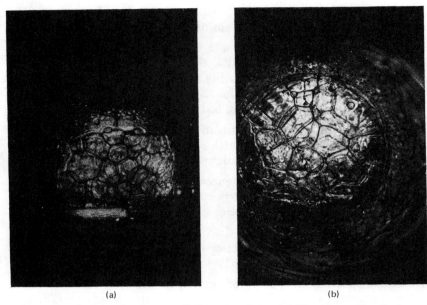

(a)   (b)

Figure 14.10 Cellular foam on small-diameter plates. (a) View from side of column (b) Looking on the bottom of the plate (*From G. E. Ho, R. L. Muller, and R.G.H. Prince, I. Chem. E. Symp. Ser. 32, p. 2:10, 1969. Reprinted courtesy of the Institution of Chemical Engineers, UK.*)

14.2.4 Symptoms of foaming in fractionators and absorbers

Numerous fractionator or absorber foaming case histories have been reported in the literature (24, 33, 50, 85, 135, 203, 209, 239, 318, 373, 408). These suggest several symptoms which are typical of a foaming problem. These symptoms are not universal; a foaming problem may exist even if some of these symptoms are not observed. The following symptoms have been observed:

1. Premature flooding and massive entrainment. Flood symptoms often suggest downcomer flooding (e.g., sharp rise in pressure drop accompanied by massive entrainment) (50, 85, 135, 209, 373). Continuous monitoring of column differential pressure is one of the best tools for diagnosing foaming problems (22, 135, 239, 349, 373) and should always be practiced in foaming services. A foaming condition, or aggravation of a foaming condition, is suggested by one or more of the following:

   ■ A sudden increase in differential pressure (Fig. 14.11a)
   ■ A differential pressure exceeding 40 to 50 percent of tray spacing in the

relevant section (239); alternatively, a pressure drop exceeding 1 in of liquid per foot of packed bed

- An erratic differential pressure (Fig. 14.11a)

2. Flood data are nonreproducible. Figure 14.11b is a typical example of this type of symptom (50).

3. Flood often initiating under perfectly steady-state conditions for no apparent reason (50).

4. Flood problem very sensitive to temperature (Fig. 14.11c). At higher temperatures, the problem disappears (50).

5. An abnormal temperature profile. For instance, in amine absorbers, foaming will cause the reaction to occur higher up in the column (22). The rich solution will become cooler, while the lean gas will become warmer. A significant decrease in the temperature difference between the inlet and outlet gas, accompanied by a drop in the temperature difference between the rich and lean solution, will suggest foaming.

6. An additional point of inflection (Fig. 14.11d) on a downcomer gamma ray scan (33). On a tray gamma ray scan, foaming resembles flooding, but may appear somewhat less dense.

7. Throughput effectively increases by antifoam addition (Fig. 14.11c) (33, 50, 71, 85, 203, 328, 373, 408).

14.2.5   Testing for foam

A number of tests for foaming are reported in the literature (50, 85, 135, 368, 373, 408), and those have been applied with various degrees of success. When choosing a test method, one must keep in mind that tests for foaming may be inconclusive, or even misleading, if they are not done at plant operating conditions (50, 85, 203). Generally, only the last method below is capable of achieving these conditions, and it is often relatively simple to set up. This method is therefore strongly recommended. The common test methods are

1. The "bottle shake" test. A clean bottle is filled with solution from the system. The bottle is left open for some time to allow all the dissolved gas to escape. A cap is put on, and the bottle is vigorously shaken up and down, then set on a table. The foam height and the time taken for the foam to settle back to the liquid level are recorded. If it takes more than 5 s to settle, foaming is indicated. This test was successfully used in amine absorbers (22). This method is simple, but may often fail to detect foaming.

2. Dispersing air or nitrogen through a solution sample in a small

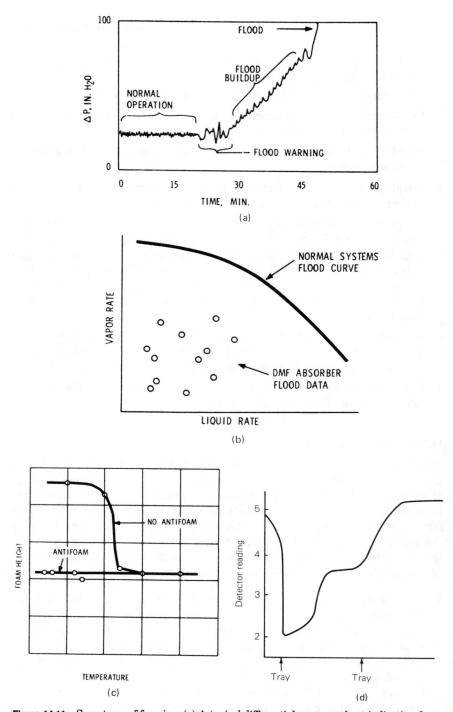

Figure 14.11 Symptoms of foaming. (*a*) A typical differential pressure chart indicating foam-induced flooding; (*b*) nonreproducible flood data, indicating foaming; (*c*) effect of temperature on foam height; (*d*) gamma scan across the downcomer, parallel to the weir, suggesting foaming. (*Parts a, b, and c from "The Solution of a Foam Problem," W. L. Bolles*; Chemical Engineering Progress, *vol. 63, no. 9, p.48 (September 1967). Reproduced by permission of the American Institute of Chemical Engineers; part d based on B.W. Betts and H. N. Rose, Joint Symposium on Distillation, The University of Sydney/The University of NSW, Australia, 1974.*

vessel. Foam heights and/or foam decay times can be used for measuring the tendency to foam. This is a simple method, but cases have been reported (50, 203, 407, 408) where this method indicated no foaming when the actual column foamed. If this method is used, it is desirable to carry out the test at the process temperature, but sometime the temperature makes little difference (373). A standardized procedure is described elsewhere (373).

3. An apparatus in which gas is dispersed through a solution sample (Fig. 14.12*a,b*). The gas is usually dispersed via a fritted filter to give good dispersion. Details are described elsewhere (135, 373). This test, at least in principle, may suffer from the same problem as (1) above, but the experience reported with it has been favorable (135, 368, 373).

4. An Oldershaw column and a pilot test column were successfully used for detecting foaming (407, 408). This technique is most suitable for atmospheric and vacuum columns, but may have difficulty reproducing all conditions such as pressure and solids content. It is best to use a sample of the actual column mixture in this test.

5. An armored level glass technique (Fig. 14.12*c*), which is capable of duplicating plant conditions of temperature, pressure, and composition. The level glass can be mounted either in the laboratory (50) or in the field (203). This technique is more reliable than any of the above, as it reproduces actual plant conditions. Both case histories above reported success with this technique where other techniques failed.

14.2.6 Curing a foaming problem

The simplest solution to a foaming problem is to inject a foam inhibitor. A number of practical guidelines are listed below:

1. A correct choice of foam inhibitor is important. This is often achieved by trial and error. Experiences were reported where injecting the wrong inhibitor aggravated a foaming problem (26, 239). Some inhibitor selection guidelines are available elsewhere (319, 339, 373).

2. It is important to correctly inject the inhibitor. Injecting the inhibitor upstream of a point of high turbulence such as a pump suction or ahead of a pump letdown valve has been recommended (373). An injection point a long distance upstream of the column should be avoided whenever possible (328). It is also important to disperse the inhibitor correctly. One case has been reported (50)

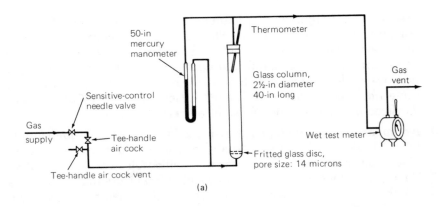

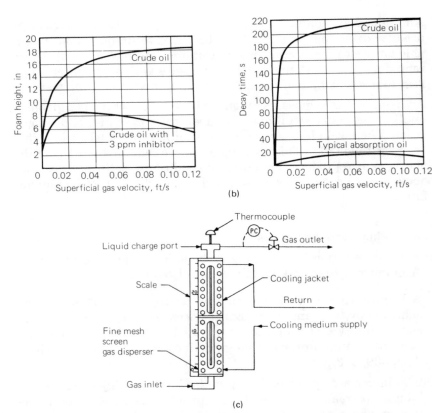

Figure 14.12 Testing for foam. (*a*) An apparatus for dispersing gas through a solution sample; (*b*) a typical foam height and decay time record from this dispersing apparatus; (*c*) an armored level glass technique. (*Parts a and b from "Foaming in a Natural Gasoline Absorber"; W. E. Glausser; Chemical Engineering Progress, vol. 60, no. 10, p. 67 (Ocotber 1964). Reproduced by permission of the American Institute of Chemical Engineers. Part c from "The Solution of a Foam Problem," W. L. Bolles; Chemical Engineering Progress, vol. 63, no. 9, p.48 (September 1967). Reproduced by permission of the American Institute of Chemical Engineers.*)

where antifoam injection was ineffective because it was applied by massive injection without effective dispersal; success was eventually achieved by dissolving the inhibitor in a solvent prior to injection. In another absorber (85), foam inhibitor injection into the lean solution drum was ineffective, presumably because of insufficient mixing with the lean solution.

A fairly recent technique recommended by one source (328) is to inject the inhibitor into the gas as an aerosol; however, little experience has been reported on the effectiveness of this technique.

3. It is essential to inject the amount of inhibitor recommended by the supplier or found optimum in laboratory tests. Excessive inhibitor injection can be harmful. One experience has been reported (85) where application of three times the recommended amount of antifoam inhibited tray action, causing product purity to drop. In other circumstances, excessive inhibitor addition was reported to aggravate a foaming problem (84, 238, 239).

4. Continuous addition of inhibitor is usually preferred to batchwise addition (85, 328). Batchwise addition may suffer from "fatigue," i.e., a loss of inhibitor effectiveness with time (328); it is often less economical when foaming is a persistent problem (373), it may lead to excessive antifoam injection (85), and it is at times unreliable. Two cases have been reported (85) where batchwise addition of antifoam was unreliable; in both cases, continuous and controlled injection solved the problem. In other cases, however, batchwise addition was effective (26, 135).

5. Foam inhibitors are often expensive and may adversely affect products or downstream equipment. Downstream filtration is often needed to remove them (373).

6. Activated carbon adsorption is effective in adsorbing high-molecular-weight foaming constituents. One technique often used in absorption-regenerator systems is to run a small fraction (about 0.25 to 0.5 percent) of the lean aqueous solution through a carbon bed (22, 26, 84, 239, 373). This can reduce inhibitor consumption; however, the activated carbon may also adsorb a defoamer (318). Careful design and operation of the activated carbon vessel and bed are essential for the success of this technique. Excellent detailed guidelines are discussed elsewhere (22).

7. Reclaiming solvent in an absorber-regenerator system is effective for removing degradation products and solids which promote foaming (84, 238, 239). The reclaimer (Fig. 14.13) is a reboiler, often equipped with a packed wash section, which receives a small portion of the lean solution. It boils off water and solvent and returns them to the process. The liquid portion, containing the sol-

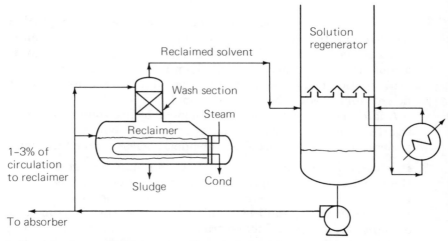

Figure 14.13 A typical reclaimer installation in an absorber-regenerator system.

ids and degradation products, is discarded. A reclaimer can be used continuously or batchwise. Reclaimer operation in amine systems is described in detail elsewhere (84, 238, 239).

8. Increasing downcomer size and reducing the potential for downcomer flooding often eliminates foaming problems. Several cases in which large downcomers were effective in eliminating foaming problems have been reported (33, 50, 209).

9. Heavy hydrocarbons may condense out of warm saturated gases upon contact with a cold aqueous solution; this may promote foaming. Upstream removal of the heavy constituents can reduce foaming. It is best to have the entering lean solution at least 10°F warmer than the dew point of the gas (238, 239).

10. Whenever possible, an effort should be made to identify the cause of the foaming problem and to minimize its effects upstream of the column (26, 373). This can drastically reduce the cost of the inhibitor and the adverse effects of the inhibitor on the product or downstream units.

11. Extensive filtration, both of the feed and the lean solution (in absorbers), is often practiced to remove solid particles which may stabilize foam (22, 26, 318, 373). Attention should be paid to correct selection of the materials and micron rating of the filter elements. Several useful guidelines are discussed elsewhere (22, 318, 319). Filters capable of removing extremely fine particles, such as diatomaceous earth filters, were found most effective (84). In one monoethanolamine absorption system (318), a foaming problem which was treated with only limited success by antifoam injection was cured by improving lean amine filtration.

12. When the liquid flows through the column in a once-through manner (e.g., absorption with no regeneration, such as caustic scrubbers), it is generally more economical to use an oversized column than to inject an inhibitor (135).

13. In some applications, it may be feasible to replace an absorption solvent by one that is less likely to foam. For instance, replacing light cycle oil by heavy naphtha was found the most effective means of preventing foaming in refinery LPG absorbers (299).

14. In extractive distillations, foaming caused by stripping off of light components that lower the surface tension can often be suppressed by adding a heavy homologue (407, 408). In one case (407, 408), adding a minor quantity of kerosene to the feed of a sulfolane extractive stripper effectively suppressed foaming.

15. In column piping, it is best to avoid oils or greases (e.g., plug valve greases) that have a soap or detergent basis.

16. A poor startup or shutdown washing technique is a frequent cause of foaming problems. Several guidelines are discussed in Sec. 11.6.

17. Foaming is a common problem in the initial run period following the installation of new plastic packing in some systems (e.g., 349). The foam is caused by leached-off additives from the plastic. It is best to request the manufacturer to guarantee that the supplied packing is nonfoaming.

18. After a new catalyst charge is installed in an upstream unit, catalyst dust may be carried over into the column and induce foaming. Measures for preventing this should be considered.

14.3 Efficiency and Performance Testing

14.3.1 Purpose of performance testing

The incentive for collecting performance data on a malfunctioning column is obvious. On the other hand, when the unit performs well, there appears to be little incentive to carry out a performance test, because other problems are more pressing. The latter reasoning is usually fallacious. A performance test is essential even for a well-performing unit, for the following reasons:

1. When the unit malfunctions at a future date, availability of performance test data tremendously reduces the troubleshooting effort.

2. A unit may appear to perform well, even while it is running at nonoptimum conditions. For instance, energy usage may be excessive; a pinch condition may reduce the apparent efficiency. Because of overdesign, these factors may be hidden. This reduces the profit

margin for the unit. Yarborough et al. (429a) presented several case histories (not all distillation-related) where performance testing directly led to major improvements in profitability.

3. Computer simulations are often used to determine best running conditions and to assess the effectiveness of proposed modifications. These computer simulations may be misleading unless tested against reliable plant data.

4. Many companies have an internal research organization which is responsible for providing the company with proprietary procedures. Improving these procedures leads to significant long-term savings. Reliable plant performance data is the most important tool available to the research engineer for improving these procedures. This factor is often overlooked by plant engineers, as the benefits may not occur until several years later.

5. In a new plant, the designers are often interested in the performance test, partly as a means of ensuring smooth operation, partly as customer service, and partly as a means of updating their own design data. Working with the designers on a performance test provides a forum for exchange of ideas, which may lead to further improvements. It also reduces the work load on plant personnel.

Acceptance test. An acceptance test checks whether the column achieves guaranteed or designed performance. Criteria often checked are product purity, utility consumption, maximum capacity, and sometimes also turndown. The strategy and planning of an acceptance test are identical to those of a normal performance test. In addition, in an acceptance test, every effort should be made to run the column as close as possible to the guaranteed or design operating conditions. Any deviations require extrapolation when comparing actual performance to guaranteed or designed performance. Such extrapolation may lead to uncertainties and inconclusive results.

14.3.2 Rigorous versus shortcut performance tests

The procedure recommended below for performance testing is rigorous, effort consuming, and time consuming. Concerns are often expressed about the cost effectiveness of the rigorous procedure, and it is often argued that a shortcut version should suffice. A suitable shortcut procedure can be derived from the rigorous procedure described here by skipping over guidelines which are considered less important.

The best procedure to adopt depends on the objective of the test. A shortcut test is best suited for detecting gross abnormalities, and is often performed as a part of a troubleshooting effort. When investigating a gross malfunction, rigorous testing is seldom justified, because it is likely to seriously delay identification and rectification of the fault. When a column appears to perform well, a shortcut test can provide a useful, albeit somewhat unreliable, set of data for future reference.

A shortcut test is unsuitable and often misleading for detecting subtle abnormalities, for determining column efficiency, for checking the design, for optimization, and as a basis for performance improvement or debottlenecking modifications. The author has experienced several cases and is familiar with many others where shortcut tests applied for these purposes needed repeating several times, yielded conflicting data, provided inconclusive results, and led to ineffective modifications. In most of these cases, the test objectives were not met even though the total time, effort, and expense spent were severalfold those that would have been spent had a single rigorous test been performed. The author therefore strongly warns against applying a shortcut procedure for these purposes. This recommendation is shared by others (429a).

Shortcut tests range from those that do little more than take a set of readings and samples from the column to those that incorporate checks of material, component, and energy balances and some key instrumentation. Even for shortcut tests, the author recommends incorporating the above checks and spending some time on preparations to ensure that key indicators are working. These checks will permit at least identification of major problem areas and enable a rough assessment of data reliability. These key items can be extracted from the list of preparations and checks recommended below for rigorous tests.

14.3.3 Strategy of performance testing

General. It is best to carry out a performance test engulfing the whole unit. Individual testing of columns, one at a time, increases the total effort and time consumed and reduces the reliability of measurements. Testing the entire unit provides several material balance cross checks, and enables better identification of erroneous meters and lab analyses. For instance, if the column feed analysis is off, the column component balance may not be sufficient to point out which analysis is suspect; but if data from a component balance on upstream and downstream equipment is also available, the incorrect analysis can easily be identified.

A shortcut to the above recommendation may be acceptable when the plant inlet rate and compositions have not significantly changed since the last plant test and the problem areas are well-known. In such cases, it may suffice to test only the specific column area (429a).

Another shortcut is often acceptable when the column is near the end of a processing train which yields reasonably pure products. Product analysis and metering tend to be far more reliable than intermediate stream measurements, and there is usually more to be gained from cross checks with downstream measurements than with upstream measurements. In such cases, it may suffice to restrict the testing to the column area and downstream equipment.

Duration. It is best to carry out a performance test over a 2- to 3-day period (2) and add another 1 to 2 days for column pinch testing. If shorter periods are used, variations in plant conditions may introduce serious errors. Over a period of 2 days, errors are averaged out. Further, column control problems may make it difficult to obtain a sufficiently long period of stable operation if the test is short. Over a 2-day period, the column should be running under stable conditions at least for some of the time.

Timing. The best time to carry out a performance test is when the plant is stable. In most plant situations, weekends are ideal, as changes due to fluctuations in upstream units are minimized.

Organization. Shaw et al. (359) present a comprehensive description of techniques that can be applied to performance test organization. In general, organizing and documenting the type of performance test considered here to anywhere near the extent described by Shaw et al. is seldom practiced. Nevertheless, the paper contains a good overview of the factors that should be thought of while organizing the test, and many of the techniques described can be adopted in specific situations.

Safety and environment. Test procedures must conform to all statutory and company safety and environmental regulations. The test plan should be reviewed with persons familiar with safety and environmental requirements and amended to fully conform to these.

14.3.4 Early preparations for performance testing

This is the most important phase of a performance test. A malfunctioning meter, a leaking block valve, or a poor laboratory analysis during the performance test can dramatically reduce the reliability of the results and defeat the purpose of the test. This is the time to sort out all potential bottlenecks. The following considerations are important and are based on the author's experience supplemented by Refs. 2, 61, 133, 239, 268 and 429a.

1. *Prepare a brief tentative outline:* This outline must give a good indication of the efforts required from the various departments including production, laboratory, instrumentation, and mechanical. Check with these departments to find out how long it takes to carry out the preparations for the test. Add a "Murphy's law" factor and determine the earliest date at which the test can be carried out. A good Murphy's law factor is half the longest preparation period, or 1 week (whichever is longer). For instance, if the instruments department advises that it will take 3 weeks to calibrate all the instruments, allow another 1½ weeks for delays. Better Murphy's law factors can be determined by experience in each situation.

2. *Prepare a needs list for work required from other departments:* This should include a detailed list of instruments to be tested and required laboratory analyses. When preparing the needs list, consider the following:

   ■ Review the instrument list to ensure that no important meters are left out, and that the meters included are sufficient to entirely define and check unit performance. Add temporary instrumentation to fill any gap in the information.

   ■ Prepare a list of all hand-held monitoring devices (e.g., timers, contact pyrometers) required. Check their availability and have them ordered as necessary.

   ■ Survey all meters to ensure that they are operational and look for oddities in field instrument installation. Check that no indicator is reading off-scale. Inspect flow meter installation. A flow meter will read inaccurately if an insufficient pipe run separates it from a pipe fitting or an obstruction, when a pipe contains two-phase flow, when an orifice plate is incorrectly sized or damaged, or when a tapping is plugged. Check all orifice plates to ensure they are the proper size; if practicable, inspect them to ensure they are in good condition. Failure to check orifice plates has ruined the results of many test runs (2). All orifice tappings should be blown. Inspect pressure indicator installations and check their tappings for blockage. A pressure gage located near an elbow or a tee will read the impact pressure rather than the fluid pressure. Inspect thermocouples and thermowells. A thermocouple or a thermowell may read incorrectly if it is too short or in the presence of deposits. Pay attention to impulse lines, heat tracing equipment, and vibrations near transmitters. Plan to have any oddities rectified prior to the test.

   ■ Ensure all flow and differential pressure indicators are zeroed and calibrated. All control room pressure indicators need to be checked with a calibrated gage. All dial thermometers and outside pressure gages should be removed and checked for accuracy. The unsatisfactory ones should either be replaced or removed prior to the test and read with calibrated thermocouples, thermometers, or gages during the test. These calibrated devices should adequately fit in the existing fittings.

   ■ Determine the location of sample points to be used in the test. Identify locations where cooling coils are required, and add these to the needs list. Samples should be taken where a stream is either all vapor or all liquid. Liquid samples (rather than vapor) normally give better accuracy. Special care is needed in vapor lines, where atmospheric condensation may occur, and in liquid lines which may contain vapor bubbles. If any samples are to be withdrawn from the column, a phase separation technique may be required, because some vapor is likely to be present even at the bottom of the downcomer, while some liquid is likely to be present in the vapor phase between the trays.

- Ensure all sampling lines are operable, free of blockage, and comply with the relevant safety requirements. Valuable and thorough sets of application guidelines for sampling are available elsewhere (2, 268).

- Prepare a list of all samples required for the test and their frequency. Discuss those samples with the lab supervisors to determine the sample containers suitable for each analysis (e.g., bombs, glass bottles, plastic bottles) and the best way of sealing the containers (plugs, corks, screwed tops). Determine which samples need special handling (e.g., refrigeration, stabilization). Attempt to use as few types of sample containers as possible to minimize errors.

- Check with lab supervisors, safety supervisors, production supervisors, and the environmental officer what measures and equipment (e.g., protective clothing) are needed for sampling and for other tests duties. Order any unavailable items.

- Any unavailable items should be marked on the needs list and ordered. The list should clearly identify the type and quantity of items required and who is responsible for ordering them.

- Contact the research department for input. Decide whether the designer should be contacted. Check with all the various departments whether the needs list has changed their timing requirements.

3. *Set a tentative date:* Attempt to set the time most convenient to the plant production people. Stable operation during the test is essential, and the production personnel can anticipate periods of potential instability best. Contact the laboratory supervisor. A performance test usually means increasing the lab workload, and the lab supervisor must schedule additional personnel. Obtain a commitment from the lab supervisor to complete the analysis by a reasonable date. The longer a sample sits on the shelf, the smaller is the reliability of the analysis because of possible leakage or chemical reaction, and the greater is its chance of being lost.

   Inform all parties involved on the proposed dates and duration of the test. Do not neglect to inform upstream and downstream units and check for potential bottlenecks.

4. *Become familiar with the control room computer and data logger:* Collect data on the reliability of these devices, and check to what extent you can use them without interfering with normal op-

eration. Also find out which instruments provide the input to the computerized data, and compare these to control board readings. Investigate any discrepancies.

5. *Prepare data collection sheets for the test:* Attempt to use modified versions of the log sheets used by the shift personnel. This is particularly important if the shift team is requested to take readings over night shift. A new data sheet may confuse them and reduce their cooperation.

   Readings such as levels, control valve positions, and times at which readings are taken may not appear important but often turn out to be extremely valuable in analyzing test results. For instance, a reboiler control valve setting can provide useful information for determining if the steam side is flooded with condensate. These should be included on the data collection sheets.

   Attempt to split the sheets into control room and outside data. Usually, such a split will coincide with the log sheet used by the shift personnel. Aim to include a many sets of data as possible on each sheet to reduce the volume of paper. A computer-typed log sheet is easiest to edit. When using a personal computer, use a high-quality printer for the final version. An attractive flow sheet enhances cooperation from those taking readings.

14.3.5 Checks prior to performance testing

Good preparation alone is not sufficient to eliminate bottlenecks or unreliable data during the test. It is just as important to check and troubleshoot for bottlenecks and sources of unreliable data. The following considerations are important and are based on the author's experience supplemented by Refs. 2, 61, 133, 239, and 268.

1. *Carry out a preliminary test run:* Use this to test your data sheet and see what measurements were left out. It is too late to find these out during the test. Eliminate any readings which are unnecessary, but exercise special caution in doing this. For instance, a pressure gage reading which appears unnecessary may turn out to be the one used as a judge between two conflicting readings.

2. *Using the preliminary test run, perform material balances on the unit and each column:* Ensure that these close within ±5 percent. Do not proceed with the performance test if the closure is worse than ±10 percent (118). Leaking valves and exchanger tubes will affect the material balance, and these leaks need to be identified prior to the final test. Meters have happened to read incorrectly even at times when the instrument people were sure they were

right. A material and component balance closure within ±3 percent and an energy balance closure within ±5 percent are recommended for the final test (2, 186).

Focus attention on flow measurements of streams close to their bubble point or dew point. Flashing of liquid or condensation of vapor can lead to major errors. Investigate any "noisy" or heavily dampened flow meter signals. This may indicate unexpected flashing or condensation.

Whenever practicable, cross check the material balance against direct volume or weight measurements (e.g., from product tank levels). Also, ensure that density corrections are applied for all flow meter readings; correction equations are described elsewhere (2).

Component balances can often be checked by plotting the ratio of distillate to bottom concentration for each component against its relative volatility at average column conditions (2). Except for columns that are primarily rectifying or stripping, the plot on log-log paper should give a straight line or a smooth curve approximating a straight line (2).

Table 14.3 is a checklist for troubleshooting nonclosure of material, component, and energy balances.

3. *Carry out an energy balance:* An incorrect reflux meter will not show up in the material balance. Ensure closure is similar to the material balance.

Focus attention on exchanger duties which are calculated from small temperature differences (e.g., a condenser duty calculated from the inlet and outlet temperatures and cooling water flow, where the temperatures are less than 10°F apart). Often, the flow can be throttled to increase the temperature difference; if this is impractical, high-accuracy temperature indicators may be required.

Pay attention to any streams entering the column as a vapor-liquid mixture; determining their enthalpy from the feed temperature is often inaccurate, particularly if their boiling range is narrow. A better estimate can often be obtained from measurements at the last point where these exist as a single phase.

If the temperature of a vapor-liquid mixture needs to be measured (e.g., feed line or reboiler return line), an effort should be made to measure pressure as close to that point as possible. Alternatively, the temperature measurement should be performed as close to the column as possible, where the pressure is known.

In some high-temperature vacuum columns or in uninsulated columns, radiation and convection heat losses may be significant

TABLE 14.3 Checklist for Troubleshooting Nonclosure of Material, Component, and Energy Balances

1. Are any valves leaking? (check vents, drains, crossovers, relief valves)
2. Are the orifice plates the correct size? Is the flow factor correct?
3. Are the orifice plates in good condition?
4. Are any orifice tappings plugged?
5. Are any flow meters measuring two-phase flow?
6. Can any condensation occur in measured vapor streams?
7. Can any vaporization occur in measured liquid streams?
8. Is there sufficient pipe run free of obstruction upstream and downstream of all flow meters?
9. Are any spiral wound (seamed) pipes used? Are they discontinued in the vicinity of all meters?
10. Are the appropriate density corrections applied to adjust flowmeter reading?
11. Are all meters zeroed and calibrated?
12. Do flow measurements match volume or mass measurements?
13. Are the lab analyses correct?*
14. Are the sampling procedures adequate?*
15. Can there be any liquid in vapor samples or vapor in liquid samples?*
16. Are the correct sample containers used? Are they adequately purged?*
17. Do the sample containers leak?*
18. Are the analyzers functioning properly? Are the sample handling loops adequately purged?*
19. Does a log-log plot of distillate/bottom components concentration vs. relative volatility produce a straight line or a smooth curve?*
20. Does a reaction or accumulation occur in the column?*
21. Is any exchanger duty calculated using a small temperature difference (10°F) between inlet and outlet stream?†
22. Are temperature gages accurate?†
23. Do thermocouples adequately fit in the thermowells? Are there deposits in the thermowells?†
24. Is feed enthalpy based on temperature measurement of a 2-phase flow?†
25. Does the steam trap leak?†
26. Do heat losses from the column need to be allowed for?†

---

*Check for nonclosure of component balances only
†Check for nonclosure of energy balances only
  All unmarked items affect mass, component, and energy balances.

(2). In these cases, these need to be estimated or measured and accounted for. In most cases, these losses are negligible (2).

Table 14.3 provides a checklist for troubleshooting nonclosure of material and energy balances. Reference 2 contains several useful suggestions for carrying out material and energy balances on plant columns.

4. *Monitor the reliability of laboratory analyses:* One useful trick is to take several samples of the same stream one after the other. If possible, take them yourself rather than involve the lab tester. Then give the lab tester one and ask for an analysis. Plug the

other sample containers and ensure no leakage. Give the same lab tester the second sample container several hours later, and ask for another analysis (don't forget to remove the plugs first—otherwise the tester may get suspicious!). Give the other containers to the lab testers on afternoon and night shifts, and leave one for the day tester the next day. Compare the analyses. You may surprise yourself with the difference in results! Check with the lab supervisor if necessary, but avoid getting personal. Don't forget, your objective is to obtain reliable data, and bad personal feelings will not help.

Another useful check is to obtain a number of the test containers early, fill them with samples, then close and plug them to prevent leakage, and record their weight. Reweigh them a number of days later; a loss of weight will indicate leakage despite all the necessary precautions. Alert the laboratory supervisor as necessary.

5. *Check lab analyses against design data, previous performance test results, or by component balances:* Look for anything that does not make sense. Also check the sampling technique used by the lab tester for each sample. Volatile samples need to be taken in bombs, and these must be adequately purged.

6. *Explain the importance of the performance test to the lab people and to the shift personnel by informal discussions:* Their cooperation is essential for obtaining reliable results.

7. *Keep a checklist and a schedule of the test preparations:* Use this to monitor progress.

8. *Tag and label instruments and sample points with weatherproof tags:* Remember, several of the people that will be taking readings are only partially familiar with some sections of the unit. Assign areas of reading for each person. Ask each of the people taking readings to perform a "dry run" and to point out any instruments they cannot locate. Ensure that each person is familiar with his or her function.

9. *Prepare a "flag sheet" of the unit:* This is a process flow diagram showing items of equipment, temperature, pressure, and key component compositions on it. Use this to take key readings during the test and to troubleshoot for gross deviations. Perform a "dry run" using this sheet before the test.

10. *When reboiler steam flow is unmetered, it can often be measured by running a hose to a 45-gallon drum and timing it:* Check whether this technique can be used. Remember that steam condensate flashes unless it is cooled, giving rise to incorrect readings. It may be necessary to cool the condensate or run it into a drum partially

filled with cold water. Check for special safety precautions associated with such measurements.

11. *Keep an eye on filters, coalescers and storage tank levels:* Ensure that sufficient operation margin is available in these units so that they do not disturb the performance test.

12. *If the material and component balances are to be carried out using a computer, preparing the program and simulation at this stage may prove very beneficial:* Once the program is ready, data can be quickly processed and inconsistencies easily detected before or during the test, before it is too late.

14.3.6 Last-minute preparations

Last-minute preparations are critical for a successful performance test. Problems which crop up at the last minute are most likely to persist during the test and may lead to meaningless test results. The guidelines below are based on the author's experience supplemented by Refs. 2, 61, 133, 239, and 268.

1. Check that all test supplies ordered were received and are adequately located. Ensure that the required number and type of sample containers and plugs is ready at the lab. Tag them with weatherproof tags. Ensure that these tags are different from those used for normal analysis. Ensure that all the required safety clothing has been received by those who will use them.

2. Check that all sample points are unplugged and operable.

3. Check that the computer or data logger is operational and that the data it supplies are consistent with other data. When such a device is counted on, a computer bug can ruin the entire test.

4. Check that the shift supervisors and laboratory personnel are aware of the test. They will tend to carry out their duties more carefully if they appreciate the importance of the test. Check for any last-minute safety requirements.

5. Ensure that a sufficient number of blank data sheets is available. Photocopying machines have a habit of breaking down when extra copies are needed. Also prepare nylon sheets to cover the data sheets while carrying these in the plant. Rain tends to occur more often on test days than at other times.

6. Difficulties encountered during data recording may lower the quality of the data and the cooperation of those taking readings. Make use of tricks that make data recording easier. These include (133):

- Supply dark ballpoint pens for recording data. Inked readings run when they become wet; penciled readings are hard to record when the paper is wet and are difficult to reproduce.
- Ensure that each person has a clipboard for holding the data sheets and taking readings.
- Supply light, sturdy ropes for tying the clipboard or hand-measuring devices and draping them over the shoulders while climbing ladders.
- Supply pocket notebooks to eliminate the need for carrying clipboards while climbing up tall towers.

7. Ensure the required instrument, mechanical, and lab personnel are available. Ensure all the people assigned are aware of their duties and of any special safety requirements.

8. Check safety valve bypasses and drains for leakage. Place your hand on the downstream side of each to find if these are warm. Leakage will affect the material balance. If it cannot be prevented, check if it can be estimated from flare or vent header analysis for certain components. Adjust the nitrogen flow to the flare or vent header accordingly.

9. Check steam traps for leakage. A leaking steam trap may cause errors in column heat balance.

10. Perform a last-minute check using the flag sheet. Look for any glaring inconsistencies.

11. Deoil refrigerated heat exchangers as necessary. Inject methanol into columns susceptible to hydrate formation. Backflush strainers.

12. Check that there is a sufficient safety margin between the plant operating rates and its maximum limit. A popping relief valve is the last thing you want during a performance test.

13. Make a last-minute check with upstream units, preferably with the shift supervisors as well as the plant superintendent, and ensure that they are aware of the test.

14. Postpone the test if an extreme weather condition is expected. Plant personnel may be too busy keeping the plant on-line, instability is likely, and the weather may adversely affect meters.

14.3.7   The test day(s)

If preparations have been adequately carried out, the test should proceed smoothly. The main considerations in this period are

1. Ensure that all protective clothing is worn and that safety procedures are adhered to. Unsafe practices are not only hazardous, but

can also generate friction with operating supervisors and may lead to discontinuation of the test.

2. Mark all strip charts. Markings should be made at the beginning and the conclusion of the tests and several times during the test. Operators often "move charts along" when a chart sticks or when inking recorder pens. Check if charts can be removed following the conclusion of the test.

3. Keep an eye on the flag sheet. Look for any major inconsistencies.

4. Keep in touch closely with the plant superintendent, shift supervisor, panel operator, and upstream units control rooms. Quickly question any decisions that may affect the test.

5. Keep an eye on the sampling technique. Attend personally when key samples are being taken. A common problem is the sampler not allowing sufficient purge time through the sampling line and sample bomb, especially on rainy days and when sample lines are long. For key samples, one useful technique is to obtain these in duplicate, and keep one in case the other is lost. Check that sample labeling is correct.

6. It is best to have readings taken around the clock. If test duties are delegated to the shift team overnight, be sure to check with them to see if they know exactly what you want. A phone call to the plant will help sort out problems and will affirm the importance of the test to night-shift personnel.

14.3.8   The day(s) after

The day after is best for carrying out "pinch" tests. These are simply performed by raising the reflux ratios in the columns. If the apparent efficiency significantly increases, chances are that pinching reduces the apparent efficiency of the column. Reflux ratios for each column should be increased in three to four increments, and samples taken for each. Use a minimum of four reflux ratios (including the one used during the performance test), in case one reading is off. The desired reflux ratios should be determined ahead of the test and checked to ensure that column capacity is not exceeded. If the column operates close to its capacity limit, feed rate may need to be reduced during the pinch test.

During the pinch test try to get samples analyzed as soon as possible. Continuously plot the Fenske parameter for the key components $x_{LK,D} \, x_{HK,B}/ \, (x_{LK,B} \, x_{HK,D})$ against the reflux ratio and ensure a smooth curve. In this group, x indicates product composition; B, bottoms; D, distillate;, LK, light key; and HK, heavy key.

In some cases, it may be desirable to determine the variation of efficiency with throughput. This test may be useful where excessive entrainment lowers efficiency well below the flood point (a common situation in many packed columns and/or vacuum distillation) or where it is desired to determine the column turndown. In such cases, the day after (or the second day after if a pinch test is included), feed rates can be increased or lowered in three to four intervals, keeping reflux and boilup ratios fairly constant, and samples taken for each interval. Use a minimum of four feed rates (including the one used during the performance test), in case one reading is off.

Tidyup.   No test is complete until all special equipment (e.g., instruments and labels) is properly removed and the plant is returned to its initial status. Any temporary equipment remaining may become a safety hazard or a nuisance; any temporary instruments may become damaged or lost. Labels may confuse the operators. Poor tidyup is one of the most frequent sources of complaints against test organizers.

It is best to first obtain input from plant and shift supervisors as to which items need removing (and how soon) and which items (e.g., some labeling) constitute a beneficial addition to the plant. Any items to be retained (e.g., instruments) should be discussed with the department supplying them. The rest should be immediately removed.

14.3.9   Processing the results

The first step is compiling material, energy, and component balances for the unit. A good way to tackle this is to fill a blank process flow diagram with the test data. The performance of each piece of equipment can then be determined. Check laboratory analyses using dew point and bubble point calculations. Some flows and compositions may need readjustment to satisfy the balance equations. Any inconsistencies must be resolved before proceeding with result processing.

Column Efficiency Determination.   To determine column efficiency, follow the steps below:

1. Make an initial guess of the column efficiency and assume it is uniform throughout the column. This initial guess need not be accurate. From knowledge of the actual number of trays and the guessed tray efficiency, estimate the number of theoretical stages.

2. Using the test material balance and the estimated theoretical number of stages, run a computer simulation for the test conditions. Adjust the number of stages in the column to give the measured product purities.

3. Check that simulated exchanger duties match measured values. If they do not, carefully investigate the cause. Pay attention to latent heat data used in the simulation.

4. Compare the simulated temperature profile with the profile measured during the test. If significant discrepancies exist, alter the number of stages in the relevant section to improve the match. Similarly, if internal column samples were obtained, check them against stage compositions predicted by the simulation and adjust the number of stages accordingly.

5. If the number of stages in one column section was significantly altered in step 4 above, the total number of stages in the column may also need adjustment. Repeat steps 2 and 4 until the simulation matches both the component balance and the temperature or internal composition profile.

6. Apply the simulation to examine the sensitivity of product purity to changes in the number of stages. This is perhaps the most critical step in processing test data; overlooking it has been a prime source of grossly misinterpreted test data. It is not uncommon to find that column efficiency was overestimated by a factor of 2, and even more, in columns where product purity is insensitive to the number of stages. Scale-up of such misinterpreted data has proved disastrous on many occasions.

   When a column operates near minimum reflux, contains an "excess" of trays, or operates under other pinched conditions, product purity is often insensitive to the number of stages. The author is familiar with one deethanizer operating close to minimum reflux where product purity remained practically unchanged when the number of stages was halved. Another designer (145) experienced a similar problem in a refinery FCC unit sponge oil absorber. When the column was simulated with less than three stages, product purity was sensitive to the number of stages. In the same column, changing the number of stages between 5 and 15 yielded product purity changes far smaller than those that could be detected by the lab analysis.

   Pinching (either due to a mislocated feed, proximity to minimum reflux, or a tangent pinch) is commonly implicated by the above insensitivity. A McCabe-Thiele diagram and a key ratio plot can help identify the cause; application of these techniques for this purpose is described elsewhere (193). Any scale-up of such efficiency data must be conservatively performed.

7. Similar to step 6 above, check the sensitivity of the number of stages to errors in reflux rate. Near minimum reflux or a pinched condition, minor changes (equivalent to typical flow meter errors)

can have a greater effect on product purity than doubling (or halving) the number of stages in the column.

8. Check the sensitivity of the simulation to a reduction in the number of stages in each section. In multicomponent distillation, examine the effect of reducing the number of stages on the key component ratios (ratios of light key to heavy key concentration) in the top and bottom products. The author experienced one case where using an efficiency ranging from 40 to 80 percent in the stripping section matched test data quite well; in that case, the key component ratios gave a closer estimate.

9. Allow for stages contributed by reboiler, condenser, interreboiler, and intercondenser. Common rules of thumb are

   - A single stage is allowed for a once-through reboiler, a kettle reboiler, or when the bottoms draw compartment is separated by a preferential baffle from the reboiler compartment (Sec. 4.5). Half or zero theoretical stages are allowed for an unbaffled recirculating reboiler arrangement.
   - A single stage is allowed for a partial condenser, and none for a total condenser. Note, however, that most computer simulations count a total condenser as a stage.
   - Determine whether an interreboiler or an intercondenser approximates a theoretical stage. If so, allow for it.

   Subtract the total number of stages contributed by these devices from the number of stages calculated by the simulation. The difference is the number of stages in the column.

10. Repeat the above steps for data obtained during the pinch test. If the efficiency significantly improves with reflux ratio, pinching is indicated. Explore this by examining the composition profile, by varying the feed point in the computer simulation, and by plotting simulated compositions on a McCabe-Thiele diagram.

11. Compare test run efficiency with the design efficiency. This can be done by comparing the designer's efficiency value with the calculated test efficiency if relative volatility is high (> 2). This procedure, however, is likely to be misleading if relative volatility is low (< 1.5), unless the designer used identical vapor-liquid-equilibrium (VLE) data to those used in the test run simulation. Differences in VLE values will be reflected as differences in column efficiency when relative volatility is low. In one case experienced by the author, a 2 percent difference between the relative volatility used by a designer and an operator in a low-volatility ($\cong 1.1$) system was sufficient to account for a difference of 50 percent in the efficiency value. With low-volatility systems, it is best to simulate the design conditions using identical VLE data to those used for test run simulation.

12. Unless the design or operating conditions of the top and bottom section are widely different, column efficiency should be reasonably uniform throughout the column. If the simulation indicates wide variation from top to bottom, it may suggest an error in the simulation or an actual performance problem. Check it carefully.

14.4 Other Column Tests

14.4.1 Field tests for minimum acceptable throughput

Minimum acceptable throughput tests are often conducted when the column is anticipated to operate at low throughput for lengthy periods. The tests supply information on whether column internals should be modified or replaced to improve performance (e.g., reduce energy consumption) during low-rate operation.

Minimum throughput tests are usually carried out by reducing the feed rate to the column while at the same time reducing reflux and reboil rates in proportion until the separation falls off sharply or until an instability is observed. The procedure for reducing the reflux and reboil rates should take the control system into account, and should be carried out in a manner similar to that used for flood testing (see Sec. 14.1.4).

Consideration must be given to the thermal state of the feed; this state should either be kept constant or be adjusted in the manner anticipated under low-rate operation conditions. Often, spare capacity is available in a preheater or a precooler when the column operates at low rates, and this can be utilized.

For best results in a minimum throughput test, it has been recommended (2) to carry out a series of tests at diminishing feed rates (and reflux and reboil in proportion). An efficiency versus throughput curve can then be plotted from which the minimum throughput for acceptable separation can be determined.

14.4.2 Detection of packed column maldistribution in the field

Maldistribution in packed columns can be detected in the field by the following techniques

1. *Water testing prior to startup:* When possible, water can be run through the packed bed before startup, and the quantity reaching the bottom of the bed can be measured at different points.

2. *Temperature measurements around the circumference of the column at different vertical heights:* Maldistribution causes a variation in

composition along the column cross section, and this causes the temperature to change from one point to another.

A surface pyrometer, often with an extension probe to penetrate through the insulation, can be used. Success with this technique was demonstrated in a couple of cases (34, 121). Temperature differences greater than 15°F are usually indicative of a maldistributed pattern; temperature differences smaller then 5°F suggest no evidence for a maldistributed pattern.

It is important to note that small temperature differences along the column circumference do not necessarily imply the absence of maldistribution. In column sections where temperature does not vary greatly along the column height (e.g., in sections where there is almost a pure product in the column, or in superfractionators where the boiling points of both products are close), temperature may be insensitive to composition, and large composition differences may exist and still indicate the same temperature along the circumference. Also, when the maldistribution is between the side and center (e.g., excessive wall flow or excessive center flow), circumferential temperature measurements will be unable to detect maldistribution.

3. *Gamma-ray scanning techniques:* These can be used to detect areas in the column where packing damage or blockage occurs. In either case, maldistribution may result. This technique, however, is often ineffective for detecting a poor irrigation or vapor maldistribution. Gamma-ray scanning may also be ineffective with structured packings. This technique is discussed in detail below.

14.5 Radioisotope Troubleshooting Techniques

14.5.1 General

Radioisotope troubleshooting techniques can be classified into two types:

1. *Sealed-source techniques:* In these techniques the radioactive source remains within a sealed container. Radiation penetrates the column, and variations in the amount transmitted or scattered radiation provide information on the inside of the column. This is analogous to the use of X rays in medicine.

2. *Radiotracer techniques:* In these techniques the radioactive source is injected into a process stream. Subsequent movement of the source is monitored using radiation detectors located outside the column.

The sealed-source techniques are the most common radioisotope techniques used in distillation. The most popular of these is the gamma-ray absorption technique. Other techniques used are neutron absorption, X rays, and gamma-ray backscatter. Of the radiotracer techniques, the pulse technique is the most popular.

When applying any radioactive technique, one must bear in mind that radioactive materials are hazardous and must be handled properly. Therefore, prior to application of any of the techniques outlined below, the user must be familiar with all the relevant safety precautions, and adequately follow these precautions.

These techniques and their distillation/absorption applications are discussed below. A comprehensive description of these techniques and their applications in process plants is available elsewhere (71).

14.5.2 Gamma-ray absorption techniques

Gamma scanning of distillation columns employs radioactive sources in the 500- to 2500-keV range (71). Compton scattering is the chief process responsible for the attenuation of these rays. The radioactive sources used are normally cobalt-60 and cesium-137. A summary of the principles of this technique follows; a detailed description is available elsewhere (71).

When a gamma ray passes through a medium from a radioactive source to a detector, some of its radiation is absorbed by the medium. The amount of radiation that is not absorbed is given by (71, 72, 136, 352)

$$I = I_0 e^{-\mu\rho x}$$

where I is the radiation intensity in counts per second, as seen by the detector; I_0 is the radiation intensity of the source, in counts per second; ρ is the density of the medium; x is the thickness of the medium; and μ is the absorption coefficient, which depends on the γ-ray source and the medium material.

When the energy of the gamma ray exceeds 200 keV, μ becomes independent of the chemical composition of the medium, and the absorption becomes a function of the product of the density and thickness of the medium (71, 72). In column gamma scanning, the thickness of the medium in the line of light is constant. For instance, when a gamma ray passes through the center of the column in Fig. 14.14, the thickness equals the column diameter. For a source emitting gamma rays toward a detector, the intensity of radiation received at the detector is therefore a function of the density of the medium. If the ray passes through metal (very high density) or liquid (high density), the intensity received by the detector is relatively low, but if the ray passes through a vapor space (low density), the detector reading is be high.

Figure 14.14 illustrates the principle of column gamma-ray scanning. The source and detector are lined up on the same horizontal plane, and a reading is taken either across the tray or across the

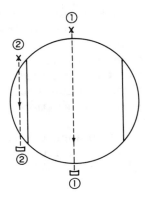

X = Source

🖳 = Detector

① Scan across the tray

② Scan across the downcomer

Figure 14.14 Gamma ray scanning.

downcomer (as desired). Both source and detector are then simultaneously moved to the next higher (or lower) vertical position, often a few inches above or below the previous height, where the next reading is taken. The source and detector are thus simultaneously moved up or down the column, and radiation intensity reaching the detector is recorded at each vertical position. High recorded intensities are indicative of vapor spaces along the ray's path; low recorded intensities are interpreted as passage through liquid or solid. Analysis of the intensity profile thus obtained is used to provide information on column behavior and to identify the location and nature of column irregularities.

Figure 14.15a (71, 72, 136, 229, 424) and 14.15b (229, 424) are fault-condition diagrams illustrating how different types of irregularities show up on a gamma scan. For normal operation, a tray scan will show a high-density (or low-detector-reading) region just above the floor of each tray (due to the presence of liquid) followed by a low-density (or high-detector-reading) region in the vapor space between the trays. A high-density region between trays implies flooding; a region of uniform intermediate density between trays implies foaming; while a low-density region where a tray is expected implies a missing or damaged tray. In a randomly packed column, a packed area will show up as a high-density region (low detector reading), a collapsed bed as a low-density region (high detector reading). If the density is extremely high, plugging or flooding is implicated. No experiences have been reported for structured packings. The author is familiar with one case

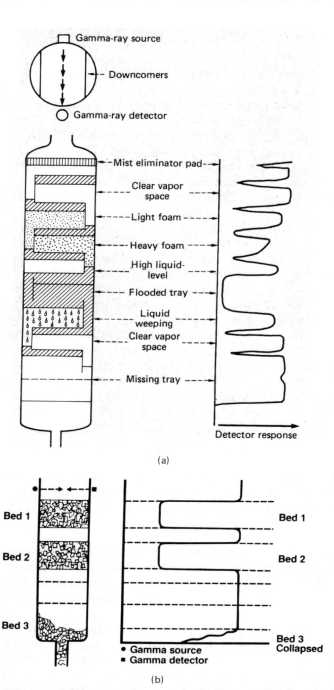

Gamma-ray source

Downcomers

Gamma-ray detector

Mist eliminator pad

Clear vapor space

Light foam

Heavy foam

High liquid-level

Flooded tray

Liquid weeping

Clear vapor space

Missing tray

Detector response

(a)

Bed 1

Bed 2

Bed 3

Bed 1

Bed 2

Bed 3 Collapsed

● Gamma source
■ Gamma detector

(b)

Figure 14.15 Illustrative gamma scans depicting various types of column irregularities. (a) Plate columns; (b) packed columns. (*Part a from J.S. Charlton and M. Polarski, excerpted by special permission from* Chemical Engineering, *January 24, 1983; copyright © by McGraw-Hill, Inc., New York, NY 10020; part b from "Radioisotope Techniques for Solving Ammonia Plant Problems," V.J. Leslie and D. Ferguson,* Plant/Operations Progress, *vol. 4, No. 3, p. 145 (July 1985). Reproduced by permission of the American Institute of Chemical Engineers.*)

where gamma scans were unsuccessful for detecting plugging in a structured (not a grid) packing.

Gamma-ray scanning can readily detect gross abnormalities such as missing trays, collapsed trays, flooding, or heavy foaming. This technique can also detect more subtle abnormalities such as high or low tray loadings, foaming, excessive entrainment (71), excessive weeping (72), blockage, and multipass liquid maldistribution. Gamma scans performed on a routine basis can also be used to monitor deterioration in column performance due to fouling, corrosion, and other factors.

To detect these more subtle abnormalities, it is important to have a reference gamma scan, either of the empty column or when the column operates satisfactorily, or, preferably, both. The reference scan enables distinguishing the subtle abnormalities from interferences from column internals, lagging, flanges, piping, platforms, and the like. It is also important to ensure that the measurement geometry approximates to narrow beam conditions. This is achieved by collimation or by electronic discrimination (71). Accurate positioning of the source and detector may also be required, particularly in smaller-diameter columns. Metal guides attached to the column, and sometimes mechanized systems which move the detector and source synchronously up and down the guides, are used for this purpose. On large columns, these are often unnecessary, as small lateral movements of source or detector have a small effect on the transmission count rate (71).

Some pitfalls. Column gamma scans can be of little value, even misleading, unless scanning pitfalls are avoided. The author had experience with the following pitfalls that adversely affect the quality of information supplied by gamma scans:

1. Interpretation of gamma scans may at times be difficult and requires good knowledge of both scanning technology and column operation. For best results, it is important to have a person or persons capable of interpreting the scans on the spot, so that one can plan the next move then.

   In many cases, a gamma scan contractor is brought into the plant, takes a scan, and then writes a report. The author has seen many such reports containing interpretations that were way off the mark. To avoid this, it is important to have a person or persons who are familiar with the tower design and operating history closely communicate with the contractor both when the column is being scanned and when the scans are being interpreted.

2. The metal guides along which the source and detector descend are usually suspended from the top of the tower. Strong cross winds may transversely sway the source, affecting the amount of radiation measured at the detector. Each guide should be adequately anchored at the bottom and along its length to minimize sway, especially under windy conditions. It is best not to scan columns in strong winds.

3. Some commercial scanners prefer to shoot downcomer scans perpendicular to the outlet weir (instead of parallel to it as shown in Figure 14.14). The author is yet to have a satisfactory experience with downcomer scanning perpendicular to the weirs. The appearance of these perpendicular scans will vary from column to column, depending primarily on downcomer width. This makes interpretation extremely difficult. The problem is aggravated when the downcomer is sloped. Further, the gamma rays used in scanning the top few inches of every downcomer travel through support beams, which may be quite thick; their presence shows up in the same manner as extra liquid. For these reasons, the author recommends always performing downcomer scans parallel to the weir (Figure 14.14) and not perpendicular to it.

   Scanning downcomers parallel to the weir (Figure 14.14) also has pitfalls. If the scan chord is shallow, scattering and reflection may lower its accuracy. When the column shell is thick (e.g., high-pressure services), the metal can absorb a larger portion of the radiation than the fluid. Consider a scan along a 12-in chord in a downcomer containing liquid of 0.5 specific gravity at 50 percent aeration. If the column shell is 1 in thick, it will absorb approximately five times more of the radiation than the process fluid. For this reason, in high-pressure columns it is best to scan center rather than side downcomers.

4. It is important to supply the people scanning the column with good drawings of the column and with good data on tray location and column shell thickness. This information is essential for good planning and interpretation of the scans.

5. Quantitative analysis can provide a valuable tool for analyzing a gamma scan. It is most useful when seeking the location of a capacity bottleneck. Unfortunately, most quantitative techniques tend to focus only on selected features shown by the scans, and should therefore be only used as a single step in the ladder to scan interpretation. Some of the common quantitative techniques are:

   a. A *peak-valley ratio analysis*. For each tray, the tray's peak transmission count (in the vapor space) is divided by the least transmission count (near the tray floor). The ratio for each tray is then plotted against the tray number. Low ratios signify high loadings (i.e., high entrainment and/or high aeration at the tray floor). This technique is useful for interpreting scans with shifting base lines. The author had several favorable experiences with ths technique.

b. A *peak analysis* (182). For each tray, the peak transmission (vapor phase) is plotted against tray number. Low values signify high entrainment and proximity to flooding. This technique was used successfully by Jones and Jones (182).

c. A *spray height ratio analysis* (150b). This technique extrapolates a tray's absorption peak descent to an established clear vapor baseline at each side of the peak. The interval on the clear vapor baseline is interpreted as the spray height. The ratio of this spray height to the tray spacing is termed the spray height ratio, and is plotted against the tray number. High ratios signify high entrainment, and a ratio of 1.2 signifies entrainment flooding. This technique was developed and successfully used by Harrison (150b).

Applications. Several case histories involving gamma-ray scanning have been reported (33, 57, 72, 102, 126, 136, 150b, 229, 264, 352, 353, 375, 424). These sources illustrate, with the aid of detector intensity diagrams, the application of gamma-ray scanning to diagnose the following abnormalities:

1. Damaged trays (126, 352, 353)

2. Missing trays (72)

3. Collapsed trays (72)

4. Flooded trays (33, 57, 102, 150b, 352, 353)

5. Rate of flooding a tray (57, 102, 150b)

6. Foaming (33, 71)

7. Monitoring tray plugging (71, 126)

8. Indication of tray froth height (33, 126, 150b)

9. Dry tray panels (353)

10. Collapsed packed beds (229)

11. Bottom liquid level (71, 229, 375)

12. No abnormality where one was expected (33, 352, 353)

13. Two different types of abnormalities in one column (57, 72)

14. Plugging in a sidestream line (424)

14.5.3 Other radioisotope techniques

Neutron backscatter. Neutrons are high-energy particles which are capable of penetrating a substantial thickness of metal. These fast neutrons, however, are slowed down by collision with hydrogen nuclei. These collisions transfer energy to the hydrogen atoms, and slow neutrons are reflected back toward the source. This is analogous to rebound of balls on a pool table. The intensity of the rebounded neutrons is proportional to the concentration of hydrogen atoms in the medium adjacent to the source and can be measured by a detector. A more detailed description of this technique is provided elsewhere (17, 71, 424).

Neutron backscatter techniques are suitable for locating the interface between two materials that have different hydrogen atom concentrations. Figure 14.16 shows a typical example of such an application. The probe is positioned near the wall of the vessel and is moved up and down along the surface. The signals show where the interface exists.

The most common applications of this technique in distillation and absorption columns is for liquid level and liquid level interface detection, especially when normal level-measuring techniques suffer from plugging. Neutron backscatter techniques have also been used for froth height measurements on trays and downcomers, and for measuring the top and bottom of packed beds. One case history has been described (71) where downcomer froth height measurements using the neutron backscatter technique led to a detection of downcomer deposits which caused premature flooding of the column. The author is familiar with one case where this technique successfully detected overflow of a packed tower distributor.

Neutron backscatter techniques can detect a liquid interface far

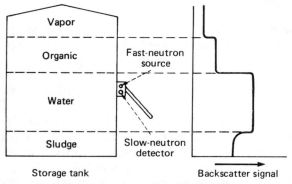

Figure 14.16 Level and interface measurement using a neutron backscatter technique (*From J. S. Charlton and M. Polarski excerpted by special permission from* Chemical Engineering, *February 21, 1983; copyright © by McGraw-Hill, Inc., New York, NY 10020.*)

more effectively than the gamma-ray absorption techniques, particularly in large-diameter columns and when the densities of the two liquids are similar. It also enables level measurements from one side of the vessel only, and is easier to apply in portable devices. On the other hand, neutron backscatter level indicators are more expensive than those using gamma rays and are difficult to apply through wall thicknesses exceeding 1½ inches or through insulation. They are only useful when the liquid contains hydrogen atoms and give a noisy signal when liquid level fluctuates. Because of the above, the use of neutron backscatter level indicators is limited compared to gamma-ray absorption level indicators (71).

Gamma-ray backscatter.   When gamma rays collide with electrons, the rays scatter in many directions at a lower energy level. Some of the rays are scattered back toward the source. The amount of backscatter, at a given angle of incidence, is proportional to the density of the medium and can therefore provide a measure of the average density over the path of the radiation. The radiation only needs to penetrate a few inches into the column. Compared to the gamma-ray absorption technique, gamma-ray backscatter requires a low-intensity radioactive source, minimizes alignment problems, and is less expensive (264). On the other hand, its low penetrating power is restrictive, it cannot be applied through thick walls or insulation, and it may be difficult to apply when the liquid-gas boundary is nonuniform or poorly defined (e.g., on a distillation tray). The main application of gamma-ray backscatter in distillation and absorption columns is for level measurement. Even in this application, gamma-ray backscatter is often considered inferior to neutron backscatter (71). A detailed description of gamma-ray backscatter and some applications is available elsewhere (71, 264).

X rays.   X rays have limited process troubleshooting applications in distillation columns. They are sometimes used to detect blockages, up-ended trays (231), and some types of mechanical damage.

Radiotracer techniques.   These involve injection of a radioactive tracer into sections of the plant and monitoring its movement with the aid of radiation detectors. Depending on the application, the tracer may be injected either as a pulse or at a constant rate. These techniques are often applied for leak detection and for flow measurement. For instance, a tracer can be injected into the reboiler steam line, and a detector on the process side will determine whether any of it found its way into the process fluid. A case where this technique successfully diagnosed a reboiler leak and measured the rate of leakage has been reported (71). Radiotracer techniques are discussed in detail elsewhere (71, 72, 229, 424).

15

Reboiler and Condenser Operation and Troubleshooting

It is difficult, perhaps altogether impossible, to do justice to the wide topic of reboilers and condensers in a modest amount of space. There are texts entirely devoted to this topic, or even only to a fraction of it (e.g., 41). Several published texts and reviews (e.g., 41, 69, 78, 81, 82, 114, 115, 124, 187, 267, 282, 291, 310, 311, 381) can testify to the large volume of information available.

Nevertheless, reboilers and condensers are an integral part of a distillation system. As shown in Chap. 1, problems with reboilers and condensers account for a sizable fraction of distillation malfunctions. Distillation supervisors or troubleshooters cannot perform their duties effectively unless acquainted with the pitfalls in the operation of the column heat exchangers. There have been many instances in which the cause of an apparent poor column performance problem was traced back to the reboiler or condenser.

For this reason, a chapter on these heat exchangers is included in this text. This chapter omits information pertaining to the design and rating of these exchangers; it does not contain a single equation for predicting heat transfer coefficients or pressure drop. This is left to most heat transfer texts and the references cited above. Instead, this chapter provides an in-depth coverage of the common causes and cures of poor performance in these exchangers, and of interactions between a malfunctioning exchanger and the column. Principles of boiling and condensation are overviewed only to the extent of providing the necessary background for understanding how these exchangers operate and the causes and cures of malfunctions.

A brief review is also given of the types of reboilers and condensers commonly encountered and their selection criteria. Poor match of the exchanger type to the service is a common source of operation problems. Insight into the pros and cons of the different reboiler and condenser types is needed for assessing whether switching exchanger type can solve an operation problem, and, if so, whether the switch is likely to create new problems.

Finally, the function of the reflux drum is reviewed. Although reflux drums are seldom troublesome, they influence the ability of the column to dampen upsets and the extent to which upsets are transferred to downstream units.

15.1   Types of Reboilers

The common types of reboilers (Fig. 15.1) are vertical thermosiphon, horizontal thermosiphon, forced-circulation, kettle, and internal. The advantages and disadvantages of each type are compared in Table 15.1 (68, 82, 114, 124, 178, 253, 430). This table and the following discussion are intended to serve as a general guide. In a specific application, reboiler selection may also be influenced by other factors. The application of the common reboiler types is described below.

Vertical thermosiphon reboilers. This is the most common type of reboiler in distillation practice. It achieves high heat transfer rates, has a low fouling tendency, and a low residence time of boiling material in the heated zone; it is compact, requiring little plot space and simple piping configuration; and it has a relatively low capital cost and no operating cost (e.g., pumping). Unless one of the alternative reboiler types can offer some distinct advantages, a vertical thermosiphon reboiler is usually preferred. Some services where other reboiler types are often preferred are:

- Where the heating medium is fouling and is best kept in the tubes. This cannot be achieved with a vertical thermosiphon reboiler.

- Where sufficient liquid head is unavailable, or additional head is expensive to provide. Alternative reboiler types generally have a lower liquid head requirement.

- Where a large surface area is required. Alternative reboiler types can generally offer more area per shell and more shells per column.

- Where a positive method of circulating liquid is required. The forced circulation reboiler can achieve this better.

- In vacuum services, where thermosiphon reboilers can be troublesome. Forced circulation and kettle reboilers are often preferred.

- Where high reliability is a primary consideration. Forced circulation and kettle reboilers are often more reliable.

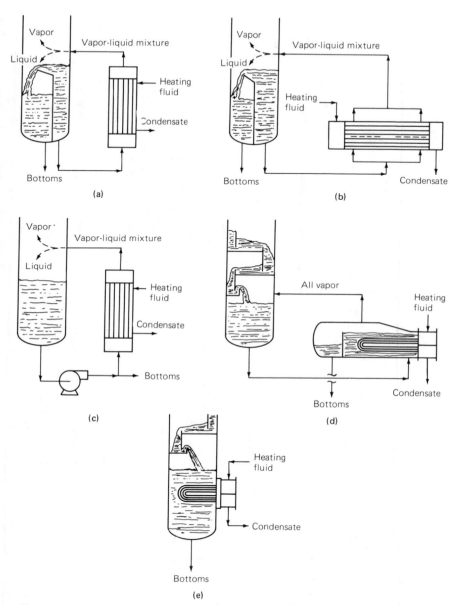

Figure 15.1 Reboiler types. (*a*) Vertical thermosiphon; (*b*) horizontal thermosiphon; (*c*) forced circulation; (*d*) kettle; (*e*) internal.

TABLE 15.1 Comparison of Reboiler Types

	Vertical thermosiphon (Fig. 15.1a)	Horizontal thermosiphon (Fig. 15.1b)	Kettle (Fig. 15.1d)	Forced circulation (Fig. 15.1c)	Internal reboiler (Fig. 15.1e)
Boiling side	Tube	Shell	Shell	Tube	Shell
Heat transfer rate	High	Moderately high	Low to moderate	High	Low to Moderate
Space requirement (Horizontal)	Small	Extra space required	Extra space required	Small (if vertical)	None
Process piping	Little amount of piping of simple configuration required	Extra piping required	Extra piping required	Extra piping required	None
Pump requirement	None	None	None	Yes	None
Additional column skirt requirement (If bottoms is not pumped)	Additional column skirt height required	Additional column skirt height required but less than in vertical thermosiphon	Small	Small, but can be large if NPSH requirement is high or if residence time in column sump is insufficient for vapor disengagement	Small
Residence time in heated zone	Low	Low	High	Low	High
Fouling tendency (process side)	Low	Moderate	High	Very Low	Moderate
Performance with high-viscosity liquids (> 25 cP)	Poor	Poor (but better than vertical)	Poor	Good	Poor

Characteristic					
Handling large surface area	More than 1 shell required	Easier than vertical thermosiphon but not as easy as kettle	Large surface can be accommodated in a single shell	More than 1 shell required	Limited by column size. Generally very difficult
Maintenance and cleaning	Can be difficult, depending on congestion	Relatively easy	Relatively easy	Can be difficult depending on congestion (if vertical)	Difficult on-line. Easy at shutdowns
Process side controllability	Good	Good	Poor	Excellent	Poor
Susceptibility to instability	High	High	Low	Low	Low
Use of fin tubing on boiling side	—	Possible	Possible	—	Possible
Design data	Readily available	Some available	Readily available	Readily available	Readily available
Vapor disengagement area	None	None	Built in	None	—
Capital cost	Low	Moderate	High	Moderate	Very low
Operating cost (not counting the heating medium)	None	None	None	Pumping, possibility of pump gland leakage	None G
Reliability	Good	Medium	Excellent	Good to excellent	Medium
ΔT required	High	Moderate	Low	High	Moderate to high

Horizontal thermosiphon reboilers. Compared to the vertical thermo-siphon reboiler, the horizontal thermosiphon reboiler generally requires more piping and plot space, has a higher fouling tendency (the boiling fluid is in the shell) and a lower reliability, and usually costs more. Its advantages often outweigh the above limitations in the following situations:

- When the heating fluid is fouling, and preferably kept in the tubes.
- Where the liquid head required either for a vertical thermosiphon reboiler or for the pump in a forced circulation reboiler is not available or is expensive to provide.
- Where a large surface area is required and is difficult to achieve with a vertical thermosiphon reboiler.
- Where it is desired to maximize circulation rate or minimize boiling point elevation due to liquid head (e.g., heat-sensitive materials distilled under vacuum).
- When low-finned and enhanced boiling tubes on the boiling sides are beneficial.

Note that the second and third considerations make horizontal thermo-siphon reboilers attractive in many superfractionators (e.g., propane-propylene splitters), where high heat transfer areas are required and where increasing column height to supply additional reboiler head is costly.

Forced-circulation reboilers. These are usually avoided because of the cost of pumping and the possibility of leakage at the pump. In the following situations, however, they are preferred:

- In highly fouling or solid-containing systems. Forced-circulation reboilers can achieve higher velocities and can operate at lower vaporization rates per pass than others.
- In fired heaters, where continuous circulation at an adequate and controlled rate is required to avoid tube overheating.
- In highly viscous systems (> 25 cP), where liquid must be "pushed" through the reboiler.
- Where the reboiler is located a fair distance from the column.
- In vacuum systems (< 4 psia). Thermosiphon reboilers are often troublesome in such systems.
- Where good control is required.

Kettle reboilers.   Kettle reboilers have low heat transfer rates, a high fouling tendency, a high plot space consumption, and are expensive. They are therefore not very popular in distillation practice. Situations in which they are preferred are:

- Insufficient vertical head in the column.
- Where the heat transfer area required is large.
- Where instabilities are anticipated.
- Where frequent cleaning is anticipated, particularly in vacuum columns.
- Where it is desired to minimize liquid in the reboiler outlet.
- Operation near the critical pressure, where reliability is a primary consideration.

Internal reboilers.   These are usually avoided when fitting them into the column requires a significant increase in column diameter or height, and also where reboiler cleaning is expected while the column is on-line.

Since the boiling liquid forms froth, which may vary in density, controlling bottom level is difficult. This makes these reboilers even less attractive, particularly in foaming and vacuum services. Applications where these reboilers are sometimes used are

- Batch distillation, where the tube bundle can easily be fitted into the batch drum, and periodic cleaning can be easily accommodated.
- Very low heat duty clean services, where column diameter is large due to other considerations, and the reboiler tube bundle required is small.

15.2   Principles of Tube-Side Boiling: A Brief Overview

Principles and design of tube-side reboilers are discussed in several texts and papers (e.g., 41, 81, 113–114, 124, 174, 267, 282, 311) and are outside the scope of this distillation text. A brief overview of some of the main principles is presented below.

Circulation.   Natural circulation occurs because of the density difference between the liquid in the bottom of the column and the two-phase mixture in the heated tubes.

Liquid flows from the bottom sump of the column to the reboiler base, where it is distributed to the tubes. The feed leg contains resistances to flow, such as valves, orifices, expansion and contraction pieces, and, in the case of a forced-circulation system, also a pump.

When the liquid arrives at the reboiler base, it is usually subcooled because of the effect of static pressure and heat losses from the line. When the liquid enters the tubes, heat is applied to the liquid. Initially, the subcooled liquid is heated to its boiling point by sensible heat transfer only. After the boiling point is reached, vaporization begins and two-phase flow regimes are established.

Circulation rate through the reboiler is fixed by the driving force and the resistance to flow. In a fixed piping system, it is a function of the liquid level in the reboiler sump in the case of natural circulation, and of pump design and operation in the case of forced circulation.

Boiling.   Tube-side boiling generally takes place in one out of two mechanisms: nucleate and convective. Nucleate pool boiling occurs when the heated surface is surrounded by a large volume of liquid, with nucleation taking place at the tube wall. In order for nucleate boiling to occur, the temperature of the heated wall must exceed the saturation temperature of the boiling liquid. In order to maintain nucleate boiling, the vapor bubbles on the surface must be surrounded by a layer of superheated liquid.

If the liquid in contact with the wall is not sufficiently superheated to sustain bubble nucleation, then heat removal is by convection through the liquid film, with evaporation occurring at the liquid-vapor interface of the vapor core. This is often the dominant mechanism in thermosiphon reboilers, particularly when they operate in the annular flow regime (discussed below).

Flow regimes.   Five main flow regimes are important in reboiler operation (Fig. 15.2). In order of increased vaporization these are

1. *Single-phase flow:* Where the entering subcooled liquid is preheated. Heat transfer in this zone is by sensible heat transfer only. Heat transfer rate in this regime is low compared to boiling heat transfer.

2. *Bubble flow:* Where the gas is dispersed as fine bubbles in the liquid. This flow regime is subdivided into two regions. In the "subcooled boiling" region, bubbles are formed on the heating surface, although liquid in the middle of the tube is still subcooled. The bubbles formed either condense directly on the wall or become detached and condense in the subcooled bulk liquid. The second, "nucleate boiling," region, forms once the bulk liquid reaches saturation, and the bubbles produced at the wall are maintained.

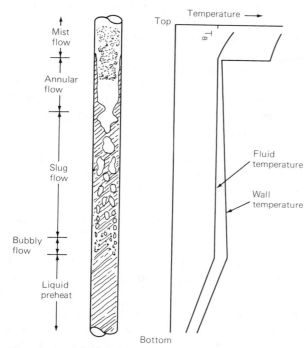

Figure 15.2   Reboiler flow regimes.

3. *Slug flow:* Where the bubbles coalesce to form slugs of vapor, which may occupy the cross section of the tube. Slug flow is similar to the flow pattern in a percolating coffee pot and is the predominant flow pattern in vertical thermosiphon reboilers.

  Slug flow in reboiler tubes was described in detail by Fair (114). A summary of this description follows. Bubble coalescence into slugs is rapid once boiling begins. The slugs accelerate faster than the liquid, and "slippage" between the phases occurs. Some liquid tumbles backward due to gravity, leading to fluctuating flow and pressure drop. Although each tube undergoes such transient operation, the combined effect of many tubes is a steady reboiler outlet condition.

4. *Annular flow:* Where the liquid flows as an annulus and the vapor flows inside as a core. The vapor shear drags the liquid film continuously upward. Heat is removed from the wall by convection, with evaporation taking place at the liquid-vapor interface.

5. *Mist flow:* Where the liquid is dispersed as fine drops in a continuous vapor phase. The liquid film at the wall disappears, and heat is transferred to the dispersed drops through the vapor. Since heat

transfer through a vapor film is low, heat transfer rates sharply drop. Metal temperature approaches the heating medium temperature (Fig. 15.2). This may lead to fouling, metal overheating, and failed tubesheet joints (312).

For the above reasons, mist flow should be avoided. Mist flow tends to occur when fractional vaporization across the reboiler is high. The transition to mist flow is termed "dryout" or "burnout" (81, 113, 282, 312). A detailed mechanistic description of this regime is available elsewhere (29).

Quantitative criteria for determining the flow regimes existing in a specific reboiler are presented elsewhere (e.g., 113, 114). An extensive mechanistic review of boiling is also available (282).

Some design considerations.   Some design considerations are listed below:

1. *Mist flow and film boiling:* Avoid film boiling (Sec. 15.4) and mist flow by restricting maximum heat flux and ensuring sufficient circulation.

2. *Tube length:* Shorter tubes reduce pressure drop, thus increasing circulation rate and reducing outlet vapor fraction and preheating zone length, but they are more expensive; 8- to 10-ft tubes are common (124, 178, 311).

3. *Tube diameter:* Larger tube diameters reduce pressure drop, thereby increasing circulation rate, reducing outlet vapor fraction, and shortening the preheat zone; 1-in-OD tubes are common (124, 311), but larger ones are often used in vacuum, with high-viscosity systems, and where a temperature pinch of the Fig. 15.5c type may be expected (372).

4. *Inlet line:* This line is usually sized for a cross-sectional area of 25 to 50 percent of the total tube cross-sectional area of the reboiler (124, 253, 358). A relatively high pressure drop in the inlet line is important for preventing excessive circulation, dampening oscillations, and for minimizing preheat zone length.

5. *Outlet line:* This line is usually sized so that its cross-sectional area is at least the same as the total cross-sectional area of the reboiler tubes (253, 358, 360). Excessive pressure drop in this line promotes oscillations and elongates the preheat zone. Excessive velocity in this line may also be undesirable at the column inlet (see Sec. 4.1, guideline 5). On the other hand, if the outlet line rises vertically before bending toward the column, velocities should be kept above 15 ft/s (237), or slug flow may result.

6. *Top head and reboiler return nozzle:* An improperly designed top head and reboiler return nozzle can lower reboiler capacity (253, 254). Transition from shell diameter to nozzle diameter should be smooth in order to avoid eddies and internal recirculation.

15.3 Effect of Liquid Level on Vertical Thermosiphon Reboiler Operation

Figure 15.3 illustrates the effect of liquid level in the reboiler sump on vertical thermosiphon reboiler performance. The diagram describes typical trends based on some published commercial-scale data (114, 181, 357, 372). The analysis assumes a narrow boiling range bottom product (e.g., a pure product); extension to wide-boiling mixtures is addressed following the analysis.

When the reboiler sump level is very low, there is little driving head to force liquid into the reboiler, and circulation rate is low. The little liquid entering the reboiler is essentially totally vaporized. Since there is little liquid head to suppress the boiling point, the liquid preheat zone is small and nucleate boiling begins almost immediately. A mist flow zone is formed above the nucleate boiling zone, where the

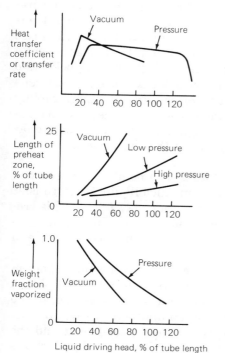

Figure 15.3 Effect of liquid level on vertical thermosiphon reboiler performance.

heat transfer mechanism is gas superheating. In this zone, the heat transfer coefficient is low, the temperature difference becomes smaller as the gas heats up, sometimes causing a temperature pinch, and the fouling tendency increases.

As the reboiler sump level rises, so does the circulation rate. This lengthens the nucleate boiling zone and shrinks the mist flow zone. Heat transfer rate continually increases until it reaches a maximum where the mist flow zone is entirely eliminated. This maximum usually occurs when the liquid driving head is about one-third of the tube length (68, 124).

Once the mist flow zone is eliminated, a further rise in reboiler sump level will lessen the heat transfer rate. The greater liquid head increases the boiling point suppression and the reboiler (process side) pressure. The enhanced boiling point suppression raises the preheating requirement, which in turn elongates the preheat zone. Since the heat transfer coefficient in the preheat zone is relatively low, the heat transfer rate declines. The higher reboiler pressure elevates the boiling point, which in turn narrows the temperature difference between the heating medium and the boiling liquid. This effects a further decline in heat transfer rate.

For atmospheric and superatmospheric systems, the rise in liquid head has a relatively small impact on the boiling point and on the preheat zone length, and the preheat zone occupies a relatively small portion of the tube length. Heat transfer rate dwindles slowly with a rise in level, giving a wide region of stable operation. In contrast, even a small rise in liquid head substantially elevates the boiling point and lengthens the preheat zone in vacuum systems, and the preheat zone is quite long. In these systems, heat transfer rate diminishes rapidly as the level rises, and the region of stable operation is narrow. In summary, the heat transfer rate and the weight fraction vaporized are far more sensitive to level in vacuum systems than in atmospheric and pressure systems.

If the reboiler sump level is raised further, a point will be reached where the reboiler outlet is flooded with unboiled liquid. Heat transfer slumps and, in some cases, hydraulic hammer may follow. This scenario is not uncommon during column dumping (356). This point is depicted by the sharp drop in heat transfer at the high end of the scale in Fig. 15.3.

While the above analysis applies to a narrow boiling range bottom product, it can readily be extended to wide-boiling mixtures. With a narrow-boiling mixture, reboiler ΔT is practically independent of the fraction vaporized. For this reason, the top diagram in Fig. 15.3 depicts the effect of level on both the heat transfer coefficient and the heat transfer rate. With wide-boiling mixtures, the top diagram shows only the variation in the heat transfer coefficient with level. The heat transfer rate is proportional to this coefficient times the reboiler ΔT. The reboiler ΔT increases as the fractional vaporization declines, and therefore as the level increases. This will mitigate the rate of decline in heat transfer as level increases.

In order to design and operate at optimum conditions, the following recommendations have been made:

1. Reboiler sump liquid level is best maintained at the top of the tube bundle for atmospheric and superatmospheric systems (68, 124, 150a, 181, 253, 254, 356–358). For these systems, a level lower than 40 percent of the tube length should be avoided (357, 358). For vacuum columns, the level is typically about 50 percent to 70 percent of the tube length (150a, 181). It will tend to be higher for wide-boiling mixtures and lower for pure bottom products.

2. Atmospheric and superatmospheric systems are usually designed for 5 to 25 percent vaporization (68, 356, 357, 372), with 20 to 25 percent being typical. For nonfouling vacuum systems, 50 percent vaporization has been recommended (181).

3. Keeping the liquid level constant is important. Installing a preferential baffle in the column bottom compartment (Sec. 4.5) has been recommended (178, 357). This baffle separates the bottom and reboiler sumps and ensures a constant head to the reboiler.

4. Since vertical thermosiphon reboilers are sensitive to level variations in vacuum systems, and because of the high-viscosity liquids often processed, forced-circulation reboilers are frequently advantageous in such systems (114, 124, 358).

5. Controlling heat input by a control valve in the reboiler condensate outlet line may adversely affect reboiler stability, especially in vacuum reboilers. This valve varies condensate level in the heating side of the reboiler. It therefore varies the point at which heating is first applied to the process side. This is equivalent to operating the process side at a fluctuating reboiler sump level. Further details of this and alternative control schemes are in Sec. 17.1.2.

15.4 Vertical Thermosiphon Reboilers: Common Operating Problems, Process Side

Excessive circulation. Excessive circulation occurs when reboiler sump level is too high and cannot be lowered (e.g., when the level is set by the top of a preferential baffle in the bottom sump). This may restrict heat transfer rate as described in the previous section. This problem is uncommon in pressure systems, but more widespread in vacuum and atmospheric reboilers.

In wide-boiling mixtures, this problem can be recognized by a low temperature difference between reboiler outlet and reboiler inlet on the process side. Alternatively, if reboiler heat transfer rate improves when the reboiler inlet valve is throttled, an excessive circulation problem is implied. If a valved dump line (Fig. 15.4a) exists, the valve can be opened, the level in the reboiler and bottom sumps equalized, and the effect of level (and therefore, circulation) on reboiler heat transfer examined.

An excessive circulation problem can be cured by adding a restriction to the reboiler inlet line. In one case (237), this almost tripled the heat transfer coefficient. Extensive experimental investigations of plant-scale reboilers (313, 360) also reported large enhancements in heat transfer upon restricting reboiler inlet. Implementation is usually either by installing a restriction orifice, which takes 25 to 40 percent of the total pressure drop in the reboiler loop (124), or by installing a throttling valve in the inlet line to the reboiler (66, 178, 311, 313, 356, 360). The latter technique (Fig. 15.4a), while more expensive, is also more flexible and permits on-line maximization of reboiler heat transfer, as was demonstrated by plant-scale tests (313, 360). The throttling valve or restriction orifice should be located as close to the reboiler as possible to prevent flashing in the reboiler inlet line. Either technique may also help cure other reboiler operating problems such as oscillations and some temperature pinch problems, as described below. Note, however, that reducing circulation rate may promote reboiler fouling.

Insufficient circulation. Insufficient circulation forms a mist flow zone in the upper portions of the reboiler tubes. This gives rise to poor heat transfer, accelerated fouling rates, and possible tube overheating as described earlier.

In wide-boiling mixtures, insufficient circulation can be recognized by a high temperature difference between reboiler outlet and reboiler inlet on the process side. Alternatively, if heat transfer coefficient improves when reboiler sump level is raised or when the reboiler inlet valve is opened, insufficient circulation is implied. Very low heat transfer coefficients, excessive fouling, and low pressure drop across some sections may also suggest insufficient circulation (360).

Insufficient circulation is usually caused by plugging, a leaking reboiler preferential baffle or draw pan, or by insufficient liquid head (alternatively, excessive pressure drop in the reboiler loop). Leakage across the preferential baffle is implied when the bottom sump level influences reboiler heat transfer rate despite the presence of a baffle.

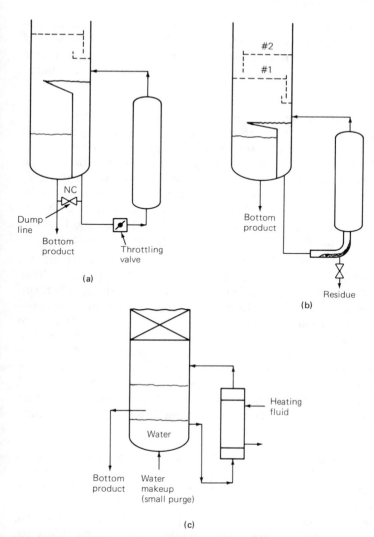

Figure 15.4 Techniques for overcoming thermosiphon reboiler problems. (*a*) The dump line technique and the throttling valve technique; (*b*) residue removal from reboiler base; (*c*) water accumulation to solve a water-induced reboiler surging problem.

Plugging can be recognized by a pressure drop well in excess of the calculated pressure drop in a section of the reboiler loop.

Remedies for a leakage problem are discussed in Secs. 4.7 and 4.8. Remedies for a deficient liquid head are raising liquid level or cutting down pressure drop in the reboiler loop. If impractical, a forced-circulation reboiler should be considered.

Temperature pinch.   A temperature pinch is process temperature approaching the heating medium temperature. It therefore occurs only when the reboiler ΔT is small. A temperature pinch lessens reboiler heat transfer. Two mechanisms are usually responsible for temperature pinches:

1. Boiling point elevation due to excessive pressure drop in the reboiler tubes and outlet line (Fig. 15.5a,b). This pinching occurs in vacuum reboilers, where a rise in reboiler pressure substantially elevates the boiling point. The pinch point can be anywhere along the tube. Reducing the pressure, and therefore, temperature at which boiling takes place alleviates this pinch. The pressure can be reduced by decreasing liquid head in the reboiler sump, or by enhancing the restriction (e.g., throttling a valve) in the reboiler inlet line (Fig. 15.5a,b). These techniques were discussed earlier.

2. Boiling point elevation due to vaporization of light components, leaving the less volatile components in the liquid (Fig. 15.5c). This pinching occurs with wide-boiling mixtures, particularly when the reboiler operates at high fractional vaporization, and may be experienced under pressure or vacuum. This mechanism is the more common of the two, and the pinch point is at the reboiler outlet. Lowering liquid level in the reboiler sump or adding a restriction to the reboiler inlet line aggravates this pinching, because they reduce circulation, thereby increasing the fractional vaporization and further raising the boiling point (Fig. 15.5c). This type of pinching can be alleviated by increasing the circulation rate, which in turn reduces the outlet temperature.

Either type of pinching can also be alleviated by increasing the heating medium temperature or by injecting sparge gas such as nitrogen at the reboiler bottom. Sparge gas injection, however, may adversely affect overhead condenser action or top product purity.

Pinching is sometimes difficult to distinguish from fouling, because most symptoms are similar. Recording reboiler temperature difference (outlet minus inlet, process side), while experimenting with a temporary increase of the lights content in the column bottom or with circulation rates, can help identify a pinch. Alternatively, a computer simulation of the reboiler can often detect a temperature pinch (372).

Accumulation of heavy residue at the reboiler base may cause pinching. This was demonstrated by one case history (134); the author has had a similar experience. A symptom of this problem is heat transfer rate declining steadily with time. In such cases, residue drawoff facilities should be provided at the reboiler base (134, 421) (Fig. 15.4b). Residue drawoff facilities at the column base are often ineffective (134).

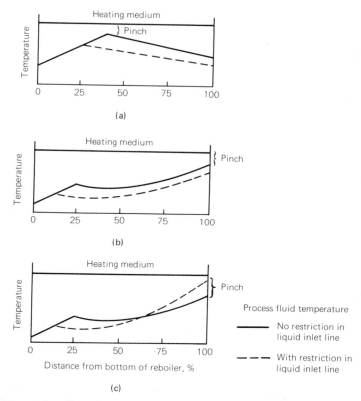

Figure 15.5 Temperature pinches in a reboiler. (*a*) Induced by pressure drop, single-component; (*b*) induced by pressure drop, wide boiling range; (*c*) induced by composition, wide boiling range.

Surging. Surging (or "burping") is an instability caused by depletion of lights and consequent drop in heat transfer and boilup rate. Columns with bottom liquid consisting mainly of high boilers, together with a small fraction of lights, are prone to surging. The problem is most potent when the boiling point of the high boilers approaches the heating medium temperature, and when liquid viscosity rapidly rises as lights are depleted (358). Surging is known to have damaged trays in some installations (358).

Figure 15.6 illustrates the sequence of events of surging. Once lights are vaporized, the mixture temperature rises, and heat transfer is reduced. This retards thermosiphon action, increases *fractional* vaporization, and further depletes the lights. As vaporization rate drops, or even stops, the column dumps, giving a high liquid level and excessive lights in the bottom. Vaporization is resumed, and the cycle repeats itself.

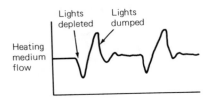

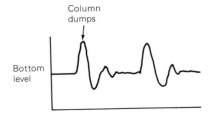

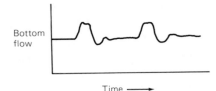

Figure 15.6   Thermosiphon reboiler surging.

To avoid reboiler surging, the following have been recommended (358):

1. Avoid subjecting heavy components to repeated thermal contact—polymerization contributes to the surging problem.
2. Periodically flush the reboiler with the light component. One source (358) reports favorable experience with this approach.
3. Consider a forced-circulation reboiler.
4. Consider a hotter heating medium.

Surging may also occur when the column bottom contains water-insoluble components along with a small quantity of water. Because of steam distillation, the water acts as a light component. A case where severe tray damage was caused by this mechanism has been reported (358). The author is familiar with another experience where small quantities of water caused reboiler surging. The problem was eliminated by elevating the reboiler liquid offtake about a foot (Fig. 15.4c), and converting the section below the offtake into a reservoir which constantly supplied a small amount of water to the reboiler.

Surging may also occur when reboiler ΔT is small and either column pressure is poorly controlled or column vapor products are improperly vented. A column pressure rise, uncountered by control or venting, raises the bottom temperature. Boiling drops or stops because of the lower reboiler ΔT. This causes the column to dump and the pressure to fall; both of these increase reboiler ΔT and the pressure takes off again. In one case known to the author, a similar sequence of events in a low-ΔT reboiler was triggered whenever steam pressure suddenly dipped, even though the column was properly vented.

Oscillations. Oscillations cause reboiler instability and often prematurely restrict reboiler heat input. Oscillations usually imply a pressure drop limitation in the reboiler outlet or outlet piping. The generated vapor cannot find its way out in sufficient quantity, and some accumulates as a pocket near the top of the reboiler. Expansion of the vapor pocket produces a recoil and a momentarily reversed process flow (360). Upon expansion, the pressure in the vapor pocket drops, and liquid rushes back in and compresses it. The process then repeats itself. This phenomenon is analogous to U-tube oscillations (67, 68, 357, 358) with one end of the U tube restricted. An extensive experimental investigation of reboiler oscillations in a commercial-scale reboiler has been reported (360). High heat fluxes (67, 68, 313, 360) and low operating pressure (313, 358) are conducive to oscillations. Monitoring reboiler pressure drop was found useful for diagnosing this type of problem; a high, oscillating reboiler pressure drop is indicative of this limitation (Fig. 15.7).

Reboiler oscillations can often be dampened by increasing flow resistance in the reboiler inlet lines or reducing flow resistance in the reboiler outlet lines. This shifts flow resistance from the outlet to the inlet lines. This is analogous to dampening U-tube oscillations by increasing friction in the bottom of the tube while easing the restriction at the open end of the tube. Design practices for reboiler inlet and outlet lines were discussed in Sec. 15.2.

In an existing installation, or in a service where oscillations are anticipated, an oscillation problem can be overcome by installing a restriction orifice or a throttling valve (see earlier in this section; Fig. 15.4a) in the inlet line to the reboiler (150a, 313, 358, 360). Figure 15.7 (360) demonstrates the effectiveness of a throttling valve in dampening reboiler oscillations.

Failure to thermosiphon. At low heat fluxes, reboiler liquid flow is sometimes difficult to initiate. If the flow is not started, the reboiler may vaporize some of the liquid, leaving a residue too heavy to lift or too heavy to boil. The problem is common at startups, especially when

Point A — Steam flow rate increased
Point B — Butterfly valve closed to 66% of full open
Point C — Butterfly valve closed to 36% of full open
Point D — Butterfly valve closed to 24% of full open
Point E — Butterfly valve closed to 14% of full open

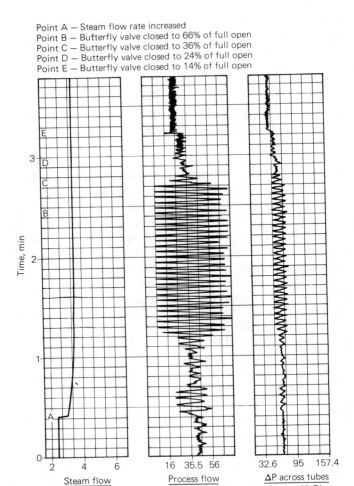

2 4 6	16 35.5 56	32.6 95 157.4
Steam flow	Process flow	ΔP across tubes
(lb/hr × 10⁻³)	(GPM)	(Inches H₂O)

Figure 15.7 Thermosiphon reboiler oscillations and their damp-ening by throttling reboiler inlet valve (*"Experimental Study of a Vertical Thermosiphon Reboiler,"* K. R. Shellene, C. V. Sternling, D.M. Church, and N. H. Snyder, Chemical Engineering Progress Symposium Series, *vol. 64, no. 82, p. 102, 1968. Reproduced by permission of the American Institute of Chemical Engineers.*)

the reboiler heats a multicomponent mixture at a small ΔT. Reboilers operating in vacuum, where preheat zones are large, and reboilers lo-cated well below the reboiler return nozzle (e.g., interreboilers located at ground level; a typical example is gas plant cryogenic demethanizer side reboilers), where the difference between reboiler inlet and outlet densities is relatively small, are particularly prone to this problem.

The following techniques are commonly used to start the thermo-siphon:

1. Installing a throttling valve in the reboiler inlet line (Fig. 15.4a). Throttling this valve reduces the length of the preheat zone. This technique is mainly effective in vacuum reboilers.

2. Reducing the liquid level in the column. This technique can only be used if no baffle is present in the bottom compartment. This technique is most effective in vacuum reboilers, but it is also frequently applied in pressure reboilers.

3. Installing a valved dump line connecting the column bottom outlet line with the reboiler inlet line (Fig. 15.4a). This technique is only needed when the column reboiler sump is separated from the column bottom sump by a baffle or when the reboiler liquid comes from a trapout pan. The valve remains shut during normal operation, but is opened during startup to lower the level and inspire thermosiphon action during startup. One case where this technique was successfully used has been described (237); the author has had several similar experiences.

4. Blowing down outlet reboiler lines.

5. Injecting lift gas into the outlet liquid line. This reduces the density in this line and initiates flow. This technique is commonly used in pressure fractionators and when the reboiler is located well below the reboiler return nozzle (e.g., in cryogenic gas plant demethanizer side reboilers) (170, 394, 421). The injection nozzle should be downstream of the reboiler outlet temperature indicator (421).

Film boiling. As the heat flux from the heating fluid to the boiling fluid increases, vapor becomes more predominant at the tube wall. When the heat flux reaches a critical value, a vapor film forms between the bulk liquid and the wall, a condition termed *film boiling*. Since this vapor film has a high resistance to heat transfer, the transition into film boiling is accompanied by a drop in heat transfer rate and a sudden rise in wall temperature. Note that film boiling refers to a vapor film at the wall, with liquid being the continuous phase in the core, in contrast to mist flow, where vapor is the continuous phase throughout. Film boiling therefore usually occurs at high liquid circulation rates (282, 313).

Film boiling should be avoided in thermosiphon reboilers. A common rule of thumb (123, 124, 253, 254, 358) suggests that the temperature difference across the wall should not exceed 90 to 100°F. An alternative rule of thumb (187) suggests that maximum heat flux in thermosiphon reboilers should not exceed 12,000 and 30,000 Btu/h/ft^2 for organics and aqueous solutions, respectively. Palen et al. (313) demonstrated that such rules of thumb are grossly oversimplified and fail to take physical properties and tube geometry into account. Experimental investigations of commercial-scale reboilers (313, 360) have demonstrated that the above maximum heat flux criterion may be grossly conservative. It was also shown (313) that as the critical pressure of the boiling fluid is approached, the temperature difference at which film boiling begins rapidly diminishes. Palen et al. (311, 313) proposed more rigorous criteria for predicting film boiling.

One experience of film boiling in a high-ΔT reboiler has been reported (290). The heating and process fluids were arranged cocurrently, giving a ΔT of almost 300°F at reboiler inlet. Reversing the passes lowered the maximum temperature difference to less than 200°F, and solved the problem.

Change in tube vapor volume with change in heat input. Stepping up heat input generates more vapor and expands the volume of fluids in the tubes. Fluid swell following an abrupt increase in heat input may displace liquid back into the column base. If the liquid to the reboiler comes from a trapout pan, this can flood the bottom tray. This problem is most pronounced at low heat loads (67, 68).

Fouling. Reboiler fouling is common in services handling unstable or corrosive chemicals. Foulants either form scale on heat transfer surfaces, or plug tubes, or both. Increased fouling is often accompanied by increased degradation of material.

Fouling is usually hindered by proper selection of materials of construction and the use of antifoulants and corrosion inhibitors. These mainly depend on the chemicals processed and are outside the scope of this text. Nonetheless, several design and operation features also help moderate reboiler fouling; these are described below.

Important variables affecting fouling rates are the reboiler ΔT (256) and the fraction vaporized. The fraction vaporized should be minimized and should not exceed 25 percent in fouling services (372). If a reboiler which utilizes a condensing heating fluid has some spare capacity (reboilers designed for fouling services often have spare capacities during initial or clean operation), fouling may be retarded by installing a condensate pump and a seal pot at the condensate outlet. This system (see later, Fig. 17.1e) is discussed in detail in Sec. 17.1.2.

In fouling services where on-line operation must be continuously maintained, the reboiler is frequently spared to facilitate cleaning while the column is on-line. An extensive discussion on decommissioning, cleaning, and recommissioning a reboiler while the column is on-line is described elsewhere (279).

Fouling caused by corrosion products, dirt, and migrating pieces of packings can be retarded by installing a suction screen at the reboiler inlet. This technique was effectively applied in demethanizer reboilers in natural gas processing plants (394). The screen is designed for easy bypassing and isolation to permit on-line cleaning.

Liquid distribution. Liquid distribution among the tubes of a vertical thermosiphon reboiler is seldom a problem, although at least one author (358) feels that a liquid distributor is desirable for reboilers greater than 4 ft in diameter.

Liquid distribution can be a severe problem and a source of instability when more than one reboiler is used, and the reboilers share common inlet and outlet lines (237). It is best to provide separate liquid draw and vapor return nozzles on the column for each reboiler.

15.5 Troubleshooting Horizontal Thermosiphon Reboilers

In these reboilers, the mechanism of heat transfer and circulation is similar to that of vertical thermosiphon reboilers. Considerations affecting liquid level are similar to those described for vertical thermosiphon reboilers, and it has been recommended (358) to keep this level at the top of the tube bundle.

Problems described for vertical thermosiphon reboilers (see previous section) are shared by their horizontal counterparts. Some may even be more potent in horizontal thermosiphon reboilers. Operating problems experienced with horizontal thermosiphon reboilers and not experienced in vertical thermosiphon reboilers are:

Liquid distribution. Liquid needs to be uniformly distributed to the shell, particularly when boiling a multicomponent mixture. Uneven distribution may locally deplete the lighter component and result in localized pinching and loss of heat transfer. Uneven distribution can also promote uneven heating, resulting in further loss of heat transfer. In extreme cases, maldistribution can lead to stratified flow, local mist flow, excessive thermal stresses, and accelerated corrosion (430).

The following techniques are commonly used to prevent liquid maldistribution is horizontal thermosiphon reboilers:

1. Two or more inlet liquid and outlet vapor lines (Fig. 15.1b) are often used (82, 178, 358) in long shells (> 12 ft; 358).

2. Horizontal baffles may be used to force the liquid to travel back and forth in the shell (115, 178). Longer units with multiple inlets and outlets use split baffles to permit upward flow (dashed lines in Fig. 15.1b). The baffles must be adequately vented. In one case (87), vapor generated beneath a horizontal baffle could not be satisfactorily

released, causing a buildup of pressure under the baffle. When enough pressure built up, the vapor was released as a "puff"; the process would then repeat itself a few minutes later. The puffs interfered with tray hydraulics and reduced column separation efficiency. Drilling vent holes in the baffle alleviated the problem.

3. If more than one reboiler is used, separate feed and return lines to each were recommended (358).

Impingement. If the inlet liquid enters at excessive velocities, it may damage tubes near the inlet. Impingement baffles are a common preventive measure. The inlet arrangement should not incur excessive pressure drop.

Flow-induced vibrations. With high shell velocities perpendicular to tube spans, tubes may fail from flow-induced vibrations (430). The impingement baffles mentioned above also offer some protection against flow-induced vibrations. Full or partial support baffles are another preventive measure (430).

15.6 Forced-Circulation Reboilers: Common Operating Problems, Process Side

Forced-circulation reboilers are similar to vertical thermosiphon reboilers, but do not depend on the natural thermosiphon action, and commonly operate at high circulation rates. The pump replaces the level in the reboiler sump as the driving force that sets reboiler circulation. Heat transfer is mainly by sensible heat, often followed by nucleate boiling. Additional boiling occurs when the heated mixture is flashed at the reboiler outlet. Forced-circulation reboilers may be installed horizontally or vertically; horizontal units are often easier to clean, but consume more plot space.

A major consideration with these reboilers is pump-system compatibility. Since the liquid is near its boiling point, and liquid head is costly, NPSH (net positive suction head) is critical. Oversized pumps can be detrimental to NPSH and should be avoided (134, 253, 358). One case has been reported (134) of poor reboiler performance due to pump oversizing; reboiler performance was the same whether power to the pump was on or off.

Figure 15.8 illustrates the pitfalls of pump oversizing. The desirable capacity-head operating point is point A. The pump in Fig. 15.8a can deliver much more head than the system dissipates, and the system will operate at point B. At that point, the required NPSH will exceed

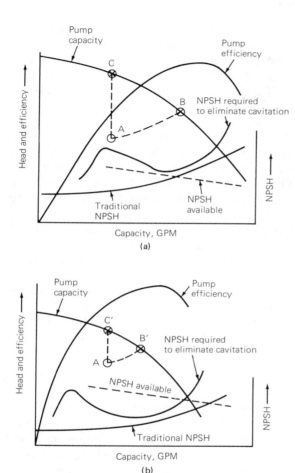

Figure 15.8 Pump and forced-circulation reboiler compatibility. (a) Oversized pump, expect NPSH problems; (b) well-sized pump.

the available NPSH, and the pump will cavitate. It may appear that installing a control valve will solve the problem by shifting the operating point to point C. Although the available NPSH at that point exceeds the required "traditional" NPSH (NPSH from pump curves), operation at point C is conducted in the lower-efficiency portion of the pump characteristic curve. Recent literature (e.g., 395) shows that in this region, suction recirculation, vortex-type cavitation, and low-frequency pulsation and vibrations can occur, which raises the effective NPSH requirement for eliminating cavitation. In Fig. 15.8a, a modified NPSH curve taking these effects into account (based on a similar curve in Ref. 395) is above the available NPSH, suggesting that cavitation may still occur at point C.

Point A and the available NPSH curve were duplicated in Fig. 15.8b, but fitted with a well-sized, smaller pump with a lower NPSH requirement. Either point B (no control valve) or point C (with a control valve) are suitable for operation.

It has been recommended to design a forced-circulation reboiler for a high pressure drop (134). In some cases (68), a restriction is placed in the vapor line downstream of the reboiler and sized to prevent vaporization in the reboiler. This restriction is often placed at the column inlet nozzle (68), but this may generate an undesirable high-velocity jet at the column inlet (see Sec. 4.1, guideline 5). Further, a restriction downstream of the reboiler may interfere with the action of a control valve located in the liquid line to the reboiler. The author is also familiar with a case where such a restriction experienced erosion at an intolerable rate.

To obtain good heat transfer coefficients, it has been recommended to maintain liquid velocities of 10 to 15 ft/s in the reboiler (253, 254, 358). Low velocities promote fouling and overheating, while high velocities erode tubes and add little to the heat transfer rate (253, 358). Film boiling should be avoided, in a similar manner to thermosiphon reboilers.

Fired heaters.   When the process temperature and the heat duty are high, a fired heater is occasionally used as a forced-circulation reboiler. Fired heater technology is outside the scope of this distillation book, and the reader is referred to review articles (30, 319) as well as some troubleshooting guidelines (237, 239) on this subject. A few of the process-side-related considerations pertaining to fired reboilers are

- Considerations regarding pump selection for forced-circulation reboilers also apply to fired heaters.

- Compared to a forced-circulation reboiler, a fired heater often operates at a higher pressure drop, higher velocities, and a larger fractional vaporization. It is common to have fired reboilers operate at 30 to 50 percent vaporization.

- Maintaining even distribution of liquid among heater passes is essential. Uneven distribution can lead to fouling, metal overheating, and poor heat transfer.

15.7   Kettle Reboilers: Common Operating Problems, Process Side

With kettle reboilers, the liquid level at the column base is set by the reboiler liquid level plus the head for overcoming reboiler circuit friction. The boiling mechanism is pool boiling with some convective effects. Kettle reboilers normally operate with high ($\cong$ 80 percent) fractional vaporization and are therefore prone to fouling.

The following operating problems are commonly experienced with kettle reboilers:

Fouling.   The high fractional vaporization rates, low velocities, and high retention time in the heated zone are conducive to fouling. Buildup of degradation products in the reboiler increases the boiling point and aggravates the problem. Adequate blowdown is frequently needed to purge out degradation products and avoid residue accumulation.

Film boiling.   When boiling inside a tube, the fluid velocity helps sweep the vapor bubbles from the tube wall, thus retarding vapor film formation. In kettle reboilers, the effect of fluid velocity is much smaller, and film boiling may be a severe problem. It is therefore important to ensure that vapor can escape faster than it is generated. Some quantitative criteria are presented elsewhere (187, 253, 314).

Liquid level.   When liquid level falls below the top of the bundle, the unflooded tubes heat vapor rather than perform nucleate boiling. Since vapor heating has a much lower heat transfer coefficient than nucleate boiling, reboiler heat transfer suffers. Further, the tube wall temperature in the unflooded portion approaches the heating fluid temperature, which may overheat tube metal. For these reasons, it is important to maintain the desirable liquid level in the reboiler.

If tube metal can be overheated, the tubes must always be flooded. This is usually achieved by an overflow weir (Fig. 15.1d). The overflow liquid constitutes the tower bottoms stream. Even with an overflow weir, it has been recommended (68) to monitor the liquid level in the tube chamber in order to protect the tubes in the event that boilup temporarily exceeds reboiler feed. A low level can be alarmed or used to cut back on heat input until the level is reestablished.

If overheating is not a concern (e.g., in refrigerated services), a reboiler level controller can replace the overflow weir, and used to control the boiling rate by flooding or unflooding tubes. In some cases, bottom product is drawn from the bottom of the column, and

neither an overflow weir nor a level controller is used in the kettle
(Fig. 15.9a). This arrangement is less expensive, but it loses a theoretical separation stage and is more sensitive to fouling. It is also
more troublesome, because liquid level in the reboiler rides on the
column bottom level and also becomes dependent on pressure drop
in the lines to and from the reboiler. In one instance (134), excessive pressure drop in the reboiler outlet line caused the reboiler
level to fall below the column bottom level, unflooding tubes and

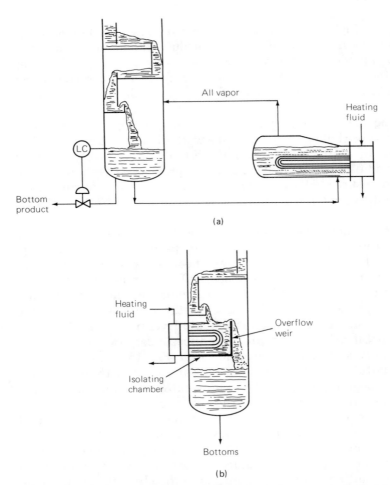

Figure 15.9 Variations on kettle and internal reboiler arrangements. (a)
Kettle reboiler with liquid drawn from column sump; (b) internal reboiler
with an isolating chamber.

reducing heat transfer. If the arrangement in Fig. 15.9a is used, a level indication on the reboiler and sizing the vapor outlet line for low pressure drop are mandatory.

Disengagement. Sufficient disengagement space needs to be provided above the bundle to disentrain liquid droplets. The disengagement space begins at the top of the liquid level and must allow for the liquid head over the weir. A minimum height of 12 in, or 1.3 to 1.6 times the bundle diameter, whichever is more, has been recommended for vapor disengagement (253). An alternative rule of thumb suggests that the top tube row should not be higher than 60 percent of the shell diameter (187, 314, 356, 357). A more rigorous criterion for setting the disengagement space is also available (393). In long units, it was recommended (358) to provide more than one vapor outlet nozzle. Sometimes demisters are used to improve disengagement.

Liquid distribution. The liquid distribution requirement of kettle reboilers is similar to that of horizontal thermosiphon reboilers. The techniques for overcoming liquid maldistribution in horizontal thermosiphon reboilers (Sec. 15.5) also apply to kettle reboilers. The author is familiar with one case in which channeling in a kettle reboiler was evidenced by the shell surface being much warmer in the center (above the inlet) than near the shell ends. The problem was eliminated, and heat transfer largely improved, after a horizontal baffle (which directed liquid toward the sides) was installed above the inlet nozzle.. The baffle arrangement must not restrict inlet liquid flow, as this may back liquid up in the column. In one reported case (237), such liquid backup caused column flooding.

Impingement. As per horizontal thermosiphon reboilers (Sec. 15.5).

Bottom product surge. The liquid draw compartment of kettle reboilers is much smaller than most column bottom sumps, and usually provides less liquid residence time and product surge. It is often impractical to incorporate the desired residence time (Sec. 4.4) in this draw compartment, and one needs to either live with the lower residence time or add a surge drum downstream of the reboiler.

15.8 Internal Reboilers

Internal reboilers are similar in principle to kettle reboilers. The main operating difficulties experienced with internal reboilers are:

Liquid level.   Measurement and control of liquid level is a major problem in columns equipped with internal reboilers. Because of vapor bubbling through the liquid, froth rather than pure liquid exists at and above the bundle, as was demonstrated by field tests (156). The froth aeration tends to increase with reboil rate, and is difficult to predict. This makes it difficult to relate apparent to actual liquid level.

The consequences of unflooding the top of the tube bundle are similar to those described for kettle reboiler. On the other hand, a liquid level too high above the bundle may initiate column flooding. The following can be helpful for minimizing problems of level measurement and control in columns equipped with internal reboilers:

1. It is best to perform field tests to determine the apparent liquid levels at the point where column flooding initiates and at the point where reboiler tubes become unflooded (as indicated by a loss in reboiler performance). The tests should be performed at various reboil rates and will define the satisfactory operating range as a function of the boilup rate. Hepp's article (156) provides an excellent illustration of how to conduct these tests.

2. Sufficient height should be incorporated above the top of the bundle for avoiding froth carryover into the column or bottom seal pan. Typically, this height is at least 110 to 150 percent of the bundle diameter (156).

3. A good elevation for the lower level tap is 25 percent of the bundle diameter below the bottom of the bundle (156). A good elevation for the upper level tap is 210 to 250 percent of the bundle diameter above the bottom of the bundle (156). The relationship determined in (1) above can be used to determine whether a measured level is within the desired operating range.

4. In wide-boiling mixtures, temperature points located at different elevations were successfully used for monitoring liquid level and checking the action of the level controller (99). Since the vapor temperatures are lower than the liquid, this technique can indicate the vapor-liquid boundary (99). This technique, however, can suffer from "geysering" of liquid from the bundle and may not be always successful.

   In order to overcome the level measurement problem, an isolating chamber arrangement or "bathtub" (68) is sometimes used (Fig. 15.9b). The chamber overflow weir keeps the bundle submerged, and the column liquid level can be separately controlled. This arrangement may increase column height requirement. If the column level is too close to the isolating chamber, it may overflow, flood the reboiler, back up liquid into the column, and possibly also flood the column. When boilup exceeds liquid downflow, or when the isolating chamber excessively leaks, tubes may become unflooded. The isolating chamber arrangement is also more sensitive to fouling (below).

Distribution. With tray towers, liquid distribution to the reboiler is uneven. With wide-boiling mixtures, this may prompt composition pinches. Further, light components can be depleted near the inlet or the top of the bundle, causing temperature pinches. In most cases, however, distribution is not a major problem because the bundle is small and internal reboiler ΔT is large.

Fouling. The downflow movement of liquid makes fouling a less severe problem with conventional internal circulation reboilers (Fig. 15.1e) than with kettle reboilers. Fouling, however, can be severe if tubes are unflooded. The bathtub arrangement (Fig. 15.9b) does not share the liquid downflow benefit, and fouling can be as much as a problem as with kettle reboilers. The isolating chamber forms a dirt trap, and must be blown down (e.g., by perforating the chamber floor).

15.9 Common Operating Problems—All Reboilers, Heating Side

Inerts. Accumulation of inerts can drastically reduce heat transfer, particularly in steam reboilers. Accumulation of acidic or oxidizing inerts such as CO_2 is also known to have caused severe corrosion (234, 253, 254). Numerous troublesome case histories of inert accumulation in the heating side of reboilers have been reported (28, 96, 232, 239). Inert venting facilities must be adequate. The following guidelines have recommended for venting inerts (28, 377, 381):

1. There should be a well-defined flow path to positively guide noncondensables from the reboiler inlet to the vent outlet. If the flow path is defined by baffles, baffle design should ensure adequate sweeping of inerts.

2. The vents must be located at the end of the vapor flow path, regardless of where this is located. Figure 15.10a, based on Bell's case history (28), shows a reboiler that performed poorly due to inert accumulation, and where a vent should have been located.

3. The flow path should preferably be tapered. This maintains high velocities and avoids collection of noncondensables in stagnant pockets.

4. Possible variations in heat transfer conditions on the cooling side

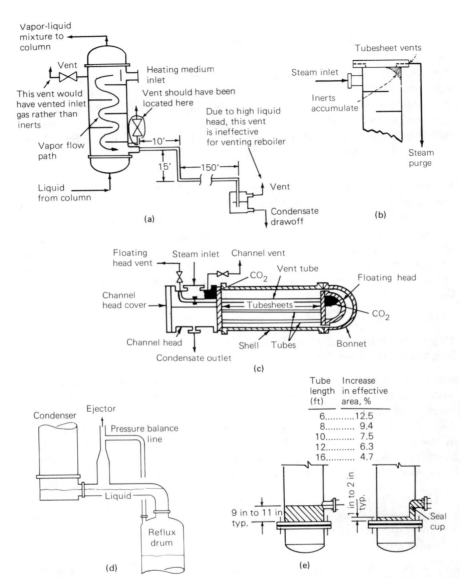

Figure 15.10 Techniques for preventing heating-side problems in reboilers. (a) A reboiler that performed poorly due to inerts accumulation, and where a vent should have been located; (b) a tubesheet vent in a vertical steam-heated reboiler; (c) a vent tube arrangement for venting a horizontal reboiler floating head; (d) a pressure balance line to prevent vent choking; (e) effect of condensate seal cup on effective heat transfer area; (*Part a, based on "Coping with an Improperly Vented Condenser," K. J. Bell, Chemical Engineering Progress, vol. 79, no. 7, p. 54 (July 1983). Reproduced by permission of the American Institute of Chemical Engineers. Parts b and g from R. C. Lord, P. E. Minton, and R. P. Slusser excerpted by special permission from Chemical Engineering, June 1, 1970; copyright © by McGraw-Hill, Inc., New York, N Y 10020. Part c from N. P. Lieberman, Hydrocarbon Processing, January 1979, reprinted courtesy of Hydrocarbon Processing. Part e from "Improving the Performance of Vertical Thermosiphon Reboiler," J. V. Smith, Chemical Engineering Progress, vol. 70, no. 7, p. 68 (July 1974). Reproduced by permission of the American Institute of Chemical Engineers.*)

should be considered. These may alter the condensing rates in different locations, and, therefore, the regions where noncondensables accumulate.

5. All vapors to be condensed contain noncondensables. A vapor containing as little as 100 ppm noncondensables can fill a reboiler up with noncondensable gas within less than 10 h if there is no venting (28).

6. It is good practice to provide tubesheet vents (134, 254, 358; Fig. 15.10b) opposite the top vapor inlet nozzle in a vapor-heated vertical reboiler (condensing on the shell side). This is essential in steam reboilers, where a small steam purge should be continuously vented to atmosphere. Shell corrosion due to CO_2 accumulation is known to have occurred in steam reboilers that had no such vent (234, 254).

7. Venting inerts from the floating head end of a horizontal reboiler can be difficult. A novel method which successfully accomplished this (232) was installing a 1-in internal pipe that extended a top tube from the channelhead tubesheet to a vent in the channel head (Fig. 15.10c). The top tube was thus converted into a "vent tube." The internal pipe was coupled to the vent from inside the channel head, to permit removal. This technique cured a CO_2 corrosion problem attributed to poor venting at the floating head end.

8. When inerts are continuously vented (e.g., vacuum condenser), it is important to prevent the vent line from choking with liquid. In one condenser, the arrangement shown in Fig. 15.10d experienced choking and instability before the pressure balance line was installed. The pressure balance line eliminated the problem.

9. Adequate venting is most important during startup, when air or nitrogen are likely to be present.

Condensate removal. A submerged surface transfers sensible heat only and attains a heat transfer coefficient far lower than condensing. Adequate removal of condensate is essential to prevent flooding of the tube surface. In steam reboilers equipped with steam traps for condensate collection, a condensate removal problem can often be diagnosed by drawing condensate via the trap bypass and comparing reboiler performance before and after (239). The following guidelines are useful for avoiding condensate accumulation problems in reboilers:

1. Undersized steam traps, dirty traps, and incorrectly installed traps should be avoided. Oversized traps can cause water hammer.

2. Lines leading to the trap should be sloped.

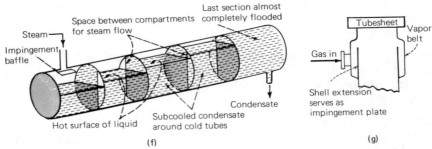

Figure 15.10 *(Continued)* Techniques for preventing heating-side problems in reboilers. *(f)* an arrangement which restricted vapor flow passages and caused hammering and baffle erosion when surface was partially flooded; *(g)* a vapor belt to minimize entry impingement problems. *(Part f, from N. H. Wild excerpted by special permission from Chemical Engineering, April 21, 1969. Copyright © by McGraw-Hill, Inc., New York, NY 10020.)*

3. When more than one reboiler is installed, each should be provided with its own trap (358).

4. When a condensate pot is shared by a number of reboilers, each reboiler should have a separate condensate line leading to the pot (358).

5. Adequate sizing of condensate outlet lines is important (134, 290; see also Sec. 15.12).

6. In short reboilers (< 10 ft), seal cups (Fig. 15.10e) have been successfully used (372) to reduce the area covered by condensate. Seal cups must be properly engineered and avoid obstructing condensate passage.

7. Many reboilers are designed to operate with a portion of the tube surface continuously flooded, for instance, when a control valve is located at the condensate outlet line (see Sec. 17.1.2). With these designs, partial flooding may restrict vapor flow passage when liquid level is high (e.g., during low-rate operation). In one case (426), severe liquid hammering and baffle erosion resulted from a restrictive arrangement (Fig. 15.10f). Expanding the baffle windows near the top and injecting noncondensable gas (e.g., nitrogen) into the steam entering the exchanger eliminated the hammering. The noncondensable gas lowered the heat transfer coefficient, thereby raising the demand for exposed condensing surface. The control system obliged by reducing liquid height, thus negating the restriction.

Control.   See Section 17.1.

Blown condensate seal. When this occurs, uncondensed vapor blows and channels right through the reboiler and out the condensate drain line. Heat transfer slumps and water hammer may follow. Experience shows that as much as half the reboiler duty is lost by a small amount of vapor blowing (239). Throttling the reboiler outlet reestablishes the seal. Installation of a condensate seal drum can cure this problem. Additional discussion is in Sec. 17.1.2.

Impingement. When the heating fluid enters the shell at a high velocity, an impingement plate should be provided to protect tubes from damage. Vapor belts (Fig. 15.10g) are sometimes used to reduce shellside inlet velocities.

Tube leakage. This subject is discussed in Sec. 13.14.

15.10 Principles of Condensation: A Brief, Selective Overview

This overview selectively describes only those principles of condensation that directly pertain to operation and troubleshooting of distillation condensers. This overview omits several considerations foremost for optimizing condenser pressure drop and heat transfer, and leaves their coverage to most standard heat transfer texts and review articles (e.g., 69, 187, 310, 319). Even the principles covered are discussed rather briefly; the reader is referred to the cited references for in-depth treatment.

Modes of condensation. Figure 15.11 illustrates the principal mecha-

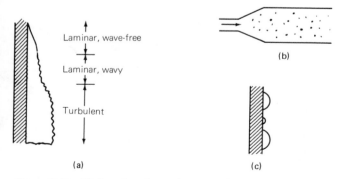

Figure 15.11 Modes of condensation. (*a*) Filmwise; (*b*) fog formation; (*c*) dropwise.

nisms of condensate formation. These are:

1. *Filmwise condensation* (Fig. 15.11a): Condensate forms a continuous liquid film on the cooling surface. This is the most important mode in industrial equipment.
2. *Fog formation* (Fig. 15.11b): Vapor condenses as droplets suspended in the gas phase. This is similar to atmospheric fog formation and requires vapor below its saturation temperature.
3. *Dropwise condensation* (Fig. 15.11c): Condensate forms as droplets on the cooled surface instead of a continuous film. This mode features a high heat transfer rate, but is difficult to maintain continuously in condensers.

Film condensation is the prime condensation mode in distillation condensers. Dropwise condensation can help improve heat transfer, but it is seldom relied on. Fog formation constitutes an operation problem rather than a desirable condensation mode and is further discussed in Sec. 15.13. The following discussion focuses on film condensation.

Single-component condensation. Figure 15.12 shows the resistances to heat transfer in filmwise condensation. When a single component is condensed (Fig. 15.12a), the entire heat transfer resistance on the condensing side is in the liquid phase. This heat transfer resistance is usually small, and the heat transfer resistance on the coolant side almost always controls the heat transfer rate.

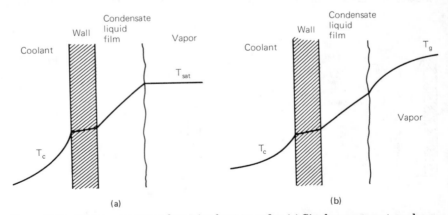

Figure 15.12 Resistances to condensation heat transfer. (a) Single-component condensation; (b) condensation in the presence of a noncondensable gas.

Multicomponent condensation. In multicomponent condensation, the heavier components are preferentially condensed first. As these are removed from the vapor, the dew point and condensation temperature of the remaining vapor decrease. Equilibrium occurs at the vapor-liquid interface, and the remaining vapor must be cooled down to stay in equilibrium at the interface. Sensible heat removal from the vapor becomes important and has a low heat transfer coefficient. In such cases, there is a resistance in the vapor phase as well as in the condensate film (Fig. 15.12b).

When the condensing components have similar vapor pressures, or the quantity of inerts is small, the vapor-phase resistance is low. If the mixture contains both light and heavy components, or a significant concentration of inerts, the vapor-phase resistance is high and may be controlling. Even more important is the diffusional resistance to mass transfer. For film condensation to occur, the heavier components must diffuse to the interface. Heavy components diffuse slowly, and a high concentration of inerts retards their diffusion. As heavy components condense out, the vapor becomes leaner in heavy components and richer in lights or inerts, which further retards diffusion. The above mechanisms are responsible for the phenomenon of "inert blanketing," which constitutes perhaps the most common operating problem in distillation condensers.

The mode of liquid removal is important. If liquid and vapor flowing through the condenser are thoroughly mixed, the condensing temperature would change from the initial dew point to the bubble point (path T_1 to T_2 on Fig. 15.13). However, if the liquid is removed as soon

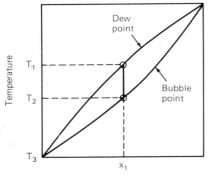

x, fraction of MVC in liquid

Figure 15.13 A temperature-composition diagram for a condensing vapor.

as it is formed, or if the diffusional resistances are controlling, the condenser will operate like a Rayleigh fractionator, and the last remaining vapor is essentially the pure light component. In this case, the temperature across the condenser will follow the temperature path T_1–T_3 on Fig. 15.13, and the effective condensation temperature will be much lower.

Condenser design determines the extent of Rayleigh fractionation. Vertical upflow condensers ("vent" or "inerts" condensers (e.g., Fig. 15.14c,e; see Sec. 15.11) are intended to operate as Rayleigh fractionators. Other condensers which exhibit this fractionation behavior are unbaffled vertical downflow shell-side condensers (if baffles are included, they would interrupt the free fall of liquid), and multipass horizontal in-tube condensers where liquid is segregated at the end of each pass. One case was reported (381) where this behavior caused severe inerts blanketing in an unbaffled, downflow shell-side condenser.

Desuperheating. Vapor entering distillation condensers is usually saturated, unless the column is heat pumped. In the latter case, the condenser first desuperheats the vapor, then condenses it. In the desuperheating zone, heat is removed from the vapor by sensible heat exchange.

Although sensible heat transfer coefficients are considerably lower than condensing coefficients, heat transfer rates are quite high in the desuperheating zones of distillation condensers. The low heat transfer coefficient in the desuperheating zone is compensated for by the higher temperature difference between the superheated vapor and the coolants (compared to the temperature difference between the saturated vapor and the coolant). In most cases, moderate variation in superheat has little effect on condenser performance and is seldom troublesome in distillation operation.

Subcooling. Distillation condensers are often designed to perform a limited amount of subcooling in order to provide NPSH for a pump, to suit a control system, to reduce storage temperature, or to recover additional heat. When the condenser has spare capacity, it often performs some additional subcooling. If a large amount of subcooling is desirable, a separate exchanger is usually used.

The heat transfer mode for subcooling liquid is that of sensible heat exchange. Subcooling heat transfer rates are much lower than condensation heat transfer rates for two reasons. First, sensible heat transfer coefficients are much lower than condensation coefficients; and second, as the liquid subcools, the temperature difference between the concoolant and the subcooled liquid becomes smaller compared to the temperature difference between the saturated vapor and the coolant.

If liquid accumulates (e.g., due to poor drainage) in the condensing side of the condenser, the flooded surface subcools condensate. This robs the condenser of condensation area and lowers the overall heat transfer rate. This mechanism is responsible for "liquid removal" problems in condensers.

When the condenser contains excess heat transfer area, the fraction of condenser area that is flooded can be manipulated to vary the rate of condensation. This principle is used in "flooded condenser" control schemes (Sec. 17.2.2).

15.11 Types of Condensers

Horizontal in-shell condensers. This is the most common type used for water-cooled condensers. TEMA E-type shells (Fig. 15.14a) with vertical baffle cuts are most popular (206, 291). These shells give good heat transfer and permit true countercurrent flow, but also have a high pressure drop (206). For applications where pressure drop is important, particularly vacuum systems, J-type shells (Fig. 15.14b), or unbaffled shell condensers (291) are often used.

Vertical in-shell condensers. This is a common condensation mode in reboilers but not in overhead condensers. Vapor flows downward, (Figs. 15.1a, c, and 15.10a). The shell can be baffled (Fig. 15.10a), but often it is not for a single-component vapor.

Horizontal in-tube condensers. Typical applications of this condensation mode are air condensers and horizontal condenser-reboilers. These condensers are often made in single-pass or U-tube arrangements, but multipass arrangements are also used. With multipass arrangements, successive passes should be below one another (291). Multipass condensation may be troublesome with multicomponent mixtures as described in the preceding section.

Vertical in-tube condensers—upflow. This type of condenser is often termed *inerts condenser, vent condenser,* or *knockback condenser.* It is used in some partial condensers when small quantities of vapor flow upward, while most of the condensate drains backward by gravity. Some fractionation takes place, and the mode of flow is similar to a wetted wall column. The most common distillation application of vent condensers is in internal condensers (below), or as a secondary

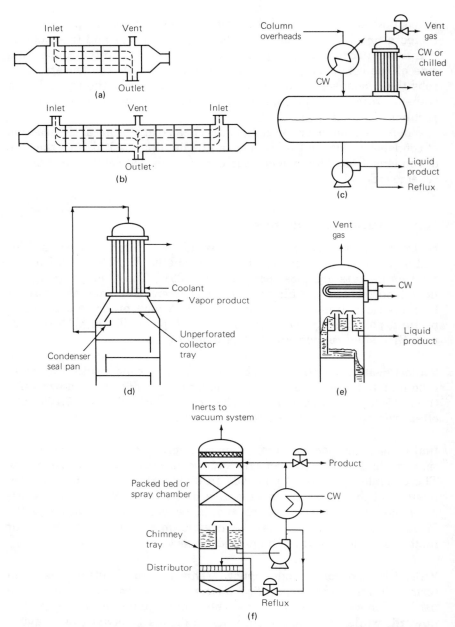

Figure 15.14   Common types of condensers. (*a*) Horizontal in-shell condenser, TEMA E-type shell; (*b*) horizontal in-shell condenser, TEMA J-type shell; (*c*) a vent condenser mounted on top of a reflux drum; (*d*) an internal downflow in-tube condenser; (*e*) an internal horizontal condenser; (*f*) a direct contact condenser.

condenser. In the latter application (Fig. 15.14c), it usually recovers heavy components from an inerts-rich stream leaving the reflux drum.

The major problem with these condensers is the possibility of flooding and excessive entrainment. A sufficiently high vapor velocity will carry the liquid film upward. Flooding velocities can be estimated using the Diehl and Koppany correlation (92), as recommended (319, 381, 392). Vapor velocities in these exchangers are kept low, and are highest near the entrance. To reduce vapor velocities, vent condensers are usually short with large diameters.

Vertical in-tube condensers—downflow. This configuration is popular in the chemical industry (68). It can lead to heat transfer coefficients even higher than shell-side condensation (319), and has distinct advantages for multicomponent mixtures if significant pressure drop can be tolerated. It also minimizes condenser costs when highly corrosive materials are handled (68). An application of this condenser type as an internal condenser is illustrated in Fig. 15.14d.

Internal condensers. These are used to a modest extent in distillation. They can be designed for upflow or downflow through the tubes or shell. Common configurations are shown in Fig. 15.14d,e.

An internal condenser eliminates the reflux drum, reflux pump, and associated piping. It saves both capital and operating costs, and simplifies column operation. On the debit side, an internal condenser increases column height, which somewhat offsets the savings. Further, in an attempt to prevent flooding and to minimize the added height, these condensers are usually designed with short vertical tubes. This practice leads to low vapor velocities and, therefore, to low heat transfer coefficients and higher exchanger costs. Indication and control of the reflux flow are usually not available. At startups, establishing liquid seal in the condenser seal pan (Fig. 15.14d) may be difficult.

Direct-contact condensers (Fig. 15.14f). These are used for minimizing pressure drop in vacuum condensation. To accomplish this, the direct-contact zone contains low-pressure-drop internals such as packings, or is a spray chamber. Another common application is intermediate heat removal ("pumparounds") in refinery fractionators. Here the main purpose is to maximize heat recovery at the highest possible temperature levels. A third common application is intermediate heat removal from absorbers or reactive distillation columns in which an exothermic reaction takes places. In all these applications, condensation is effected by contact with an externally subcooled circulating liquid stream. The main operating problems are the performance of the liquid collecting devices (this is discussed in Chap. 4) and the internals (e.g., packings and distributors).

15.12   Common Operating Problems with Condensers

Inerts accumulation, condensate removal, and condenser fouling are by far the most common problems that adversely affect condenser operation (381). One author (28) states that "more than half of all condenser problems are due to poor venting." Fouling in condenser usually occurs on the coolant side and will not be dealt with here. Other problems occasionally also affect condenser performance.

Inerts.   Accumulation of even a small fraction of noncondensables can impair condensation heat transfer. The mechanism by which noncondensables accumulation reduces heat transfer is described in Sec. 15.10. Inerts problems are most common in shell-side condensation, where gases can segregate in pockets, and are difficult to remove unless sufficient pressure drop is used to force them to the vent outlet (291, 381). Cross-flow condensers are particularly prone to inerts accumulation. Numerous cases where inerts venting was a problem have been reported (28, 134, 239, 381). Adequate venting facilities are needed at all locations where noncondensables are likely to accumulate. Guidelines for venting are presented in Sec. 15.9.

Figure 15.14a,b shows preferred vent locations on the common types of in-shell horizontal condensers. Gilmour (134) described a case history of inert blanketing in the condenser shown in Fig. 15.15a before the vents were added; a similar problem has been reported by others (381). Figure 15.15b shows a good vent location on a two-pass, in-tube condenser.

An inert blanketing problem in a condenser shell can often be diagnosed by measuring surface temperature. Areas blanketed by inerts tend to be considerably cooler. If safety regulations permit, feeling the shell by hand at different spots can sometimes be an effective surface thermometer.

Condensate removal.   If condensate is removed at an insufficient rate, or if the condenser traps condensate, heat transfer area will become flooded. This will lower condenser heat transfer rates, as described in Sec. 15.10. Numerous case histories of troublesome condensate removal have been reported (134, 290, 381).

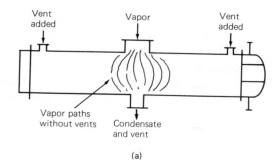

(a)

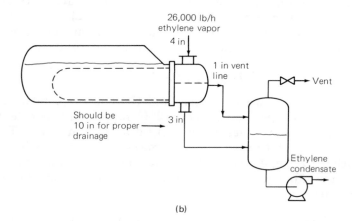

(b)

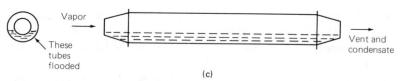

(c)

Figure 15.15 Techniques for preventing operating problems in distillation condensers (a) Adding vents in a horizontal, in-shell condenser (b) Incorrect and correct sizing of gravity outlet line. Note also correct vent location (c) Tubes flooded because condensate outlet was not at the bottom of the condenser *(Parts a and c from Charles H. Gilmour, excerpted by special permission from* Chemical Engineering, *June 19, 1967; copyright © by McGraw-Hill, Inc., New York, NY 10020. Part b, contributed courtesy of John A. Moore, Consulting Engineer, Galveston, Texas. Part d from G. C. Shah, excerpted by special permission from* Chemical Engineering, *July 31, 1978; copyright © by McGraw-Hill, Inc., Neu York, NY 10020.)*

A condensate removal problem in shell-side condensers can sometimes be detected by surface temperature measurements. Where safety permits, feeling the exchanger by hand can be effective. Regions where condensate builds up tend to be considerably cooler.

Some guidelines for condensate removal have been discussed in Sec. 15.9, but those specifically address reboilers. For condensers, the following guidelines are proposed:

1. Adequate sizing of outlet condensate lines is essential.

2. If no liquid level is maintained in the condenser, and the liquid is not substantially subcooled, vapor bubbles are likely to be entrained in the liquid condensate. The outlet line should therefore be sized for self-venting flow (Fig. 4.5). Moore (290) has suggested a slightly more conservative rule of thumb for gravity drainage of condensers (assuming no liquid accumulation)

$$D = 0.1\sqrt{GPH}$$

where D is the line size in inches, and GPH is the condensate flow rate in gallons per hour. Moore (290) supported this rule by a case history of a troublesome gravity-drained condenser (Fig. 15.15b). The outlet line was much smaller than would be required either by Fig. 4.5 or by the above equation. Gilmour (134) described a similar (possibly identical) case history, but did not supply quantitative data.

3. If no liquid level is maintained in the condenser, the liquid outlet line should enter the vapor space of the reflux drum and should not be submerged.

4. Unless it is intended to maintain a liquid level in the condenser, all liquid outlet piping should leave the condenser at the bottom and slope toward the reflux drum (or condensate pot) with no high points. This practice is recommended in most cases, and is essential for gravity draining. The condenser in Fig. 15.15c experienced a liquid-removal problem (134) because its outlet was not at the bottom of the condenser. A similar problem has been reported by others (381).

5. When a condensate pot is shared by a number of condensers, separate lines should lead from each condenser to the pot.

6. It has been recommended to have air condenser tubes slope downward (381) at a small angle. One troublesome experience in which tubes sloped upward has been reported (254).

Blown condensate seal. This is the same problem as experienced with reboilers (Sec. 15.9). Condensate pots (Fig. 15.15b) are effective for overcoming this problem. These are discussed in Sec. 17.1.2.

Fog formation. This is discussed in the next section.

Flooding. Flooding and liquid carryover are often experienced in vertical upflow, tube or shell side. This limit can be predicted for tubeside upflow (see in Sec. 15.11).

Some troublesome experiences with liquid carryover from these condensers have been reported (381). The author had an experience with liquid carryover from a vent condenser mounted on top of a C_3 splitter reflux drum (similar to Fig. 15.14c). Carryover occurred whenever the vent control valve opened excessively, and was recognized by "watering" or icing up of the line downstream of the valve (due to liquid flashing). The author is familiar with other similar experiences. Installing a valve limiter was sufficient to prevent carryover in the above case.

Condensing side fouling. Fouling on the condensing side is seldom troublesome. In cases where problems have been experienced, they have often been caused by sticky or viscous materials which condense near the inlet. Flushing tubes near the inlet with lighter material (Fig. 15.15d) can minimize this type of fouling (357). A proper external solvent can also be used as the flushing liquid.

Slug flow. When partial condensers are located below the reflux drum (Fig. 15.15e), and the velocity in the riser is too slow, vapor and liquid segregate in the riser. A head of liquid builds up and exerts back pressure against the column. Periodically, a slug of liquid breaks through and releases the back pressure. The riser then gradually fills up with liquid, and the cycle repeats itself. This causes fluctuations in column pressure and accumulator level. One troublesome case history of slug flow has been reported (70).

If raising the flow rate or temperature in the riser improves column stability, slug flow should be suspected. A rough rule of thumb (239) suggests that slug flow is likely at velocities lower than 15 ft/s, and unlikely at velocities greater than 25 ft/s. Alternative criteria are discussed elsewhere (70, 188).

To overcome slug flow, dual risers are often installed (188, 239). The small-diameter riser is operated at low throughputs, while the larger is operated at higher throughputs.

Control problems. These are discussed in Secs. 17.2 and 17.3.

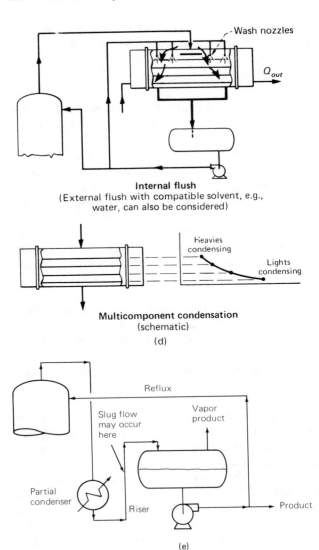

Internal flush
(External flush with compatible solvent, e.g.,
water, can also be considered)

Multicomponent condensation
(schematic)
(d)

(e)

Figure 15.15 (*Continued*) Techniques for preventing operating problems in distillation condensers (*d*) Flushing tubes near condenser inlet to prevent process-side fouling (*e*) Slug flow may occur in risers of partial condensers located below the reflux drum.

Distribution problems. Uniform distribution is important when vapor entering the condenser splits into parallel streams (e.g., tube-side condensers, cross or split-flow shell-side condensers) (188, 381). High entrance resistances and multiple inlets are often used to gain good distribution. The use of two or more inlets has been recommended for distributing vapor to the tubes of large air coolers (381). Such inlets should be symmetrical (188). When a horizontal internal condenser is used (Fig. 15.14d), achieving good vapor distribution and preventing vapor bypassing can be difficult.

Vapor maldistribution is most troublesome when column overhead vapor is split between two or more total condensers in parallel. In a newly designed plant, maldistribution is usually prevented by designing, piping, and locating the condensers absolutely symmetrically. When a condenser is added at a later stage, absolute symmetry can seldom be achieved. Vapor maldistribution may result and channel the bulk of the vapor via a lower-pressure-drop flow path, thus impairing heat transfer. In one case (237), a new set of condensers was added in order to increase condensing capacity. Instead of increasing, condensation capacity actually decreased because of maldistribution.

One technique recommended for avoiding maldistribution is to add total condensers in series, not in parallel (237). Adding condensers in series need not add pressure drop to the overhead circuit. The new condenser can be designed for a low pressure drop and located upstream of the existing one. When the overhead stream reaches the second (existing) condenser, it is partially condensed, and both its velocity and pressure drop are lower. In one case (237), this technique was shown to actually lower overhead circuit pressure drop. A possible drawback of adding condensers in series is the difficulty in evenly distributing two-phase flow to the second condenser. This may be troublesome with a tube-side condenser, but it is of little concern for a single-entry, shell-side condenser. Some piping arrangements that minimize maldistribution at the entrance to a tube-side condenser receiving two-phase flow are discussed elsewhere (188).

With partial condensers, vapor maldistribution among parallel condensers is less troublesome and can usually be avoided by careful exchanger design (237).

Two liquid phases. When a single liquid phase condenses, the liquid exerts it own vapor pressure. When two liquid phases are present, each exerts its own vapor pressure. The net effect will be a higher total vapor pressure at the same temperature; conversely, at the column pressure, both components will condense at a lower temperature. If the condenser is designed to handle only a single liquid phase, an unexpected appearance of the second liquid phase will lower the heat transfer. The author is familiar with one experience where as a result of an upstream energy-saving revamp, an aromatics separation column received a lot more water in the feed. The water distilled up and formed a second liquid phase upon condensation. The condenser was not revamped and was unable to achieve its duty when the two phases separated. Excessive product loss resulted. A similar occurrence is known to happen when a column handling water-insoluble materials suddenly experiences more water in the feed (e.g., upon a feed tank change). The remedy is either to supply ample condensing capacity or to bar an excessive quantity of the second liquid from entering the column.

15.13   Fog Formation

Fog can form when a vapor mixture cools below its saturation temperature, in a similar manner to atmospheric fog formation. In condensers, it takes place when the heat transfer rate from the cold surface exceeds the mass transfer rate to the surface. Fog droplets range in size from 0.1 to 40 μm (78, 380). Below about 3 to 5 μm, the droplets tend not to collect on the cool surface but remain with the vapor and pass down the vent line in partial condensers. Reported measurements (381) suggest that in some cases, 20 to 40 percent of the mixture can be in the form of fog droplets.

Fog is difficult to separate from the vapor. Fog droplets containing corrosive, reactive, or noxious materials may settle and create problems downstream (242). If vented to the atmosphere, they may create an emission problem. Other problems are reduced condensate yield and possibly a somewhat reduced heat transfer rate (78).

Fog formation is favored by conditions which slow the mass transfer rates and enhance the heat transfer rate. Factors which favor fog formation by slowing the mass transfer rates are a high ratio of noncondensables to condensable vapor, and a high molecular weight (low diffusivity). Factors which favor fog formation by speeding the heat transfer are a high temperature difference between the vapor and interface and low initial superheat.

In order for fog to form, a nuclei must first be formed. Homogeneous nucleation is difficult because of the energy barrier associated with the creation of an interface. The rate of nucleation is a very steep function of the supersaturation ratio. Generally, below a certain critical supersaturation ratio nucleation is slow enough to be ignored, and above this, it will be substantial. The critical supersaturation ratio can be predicted from Amelin's equation (78, 242, 380). Procedures for fog formation prediction are outlined in the cited references.

The following procedures have been recommended (78, 242, 380) for minimizing fog formation:

1. Reducing temperature difference in the condenser. This is the most common cure used (242), but it may increase the exchanger surface.

2. Using mist eliminators. These can be expensive and unreliable for fog elimination purposes (242, 381). Specialized mist eliminators have been recommended (380, 381). Bench scale testing prior to use has been recommended (242).

3. Seeding the gas stream with condensation nuclei in order to produce drops that can be captured by conventional mist eliminators or a settler. Testing has been recommended before using this approach (242).

4. Reducing presence of dust, ions, and entrained drops prior to condensation.

5. Entering the condensation step with a high superheat.

6. Reheating the gas. However, this may be self-defeating.

15.14   Reflux Drums

Reflux drums perform one or more of the following functions:

1. Providing surge volume to protect the column, the reflux pump, and downstream equipment in case of upsets or failure.

2. Facilitating control and smoothing out fluctuations of product going to a downstream unit and of tower reflux.

3. Providing knockout capacity to separate a vapor product from a liquid product.

4. Providing settling time for separating two liquid phases.

A typical reflux drum (Fig. 15.16) consists of four portions, which perform different functions:

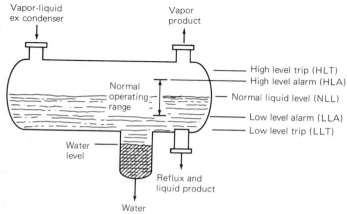

Figure 15.16   A typical reflux drum with water-removal facilities.

1. *The portion between the high- and low-level alarms:* This portion provides surge for smoothing out product fluctuations and for taking action should product or reflux flow fail. Recommended residence times for this portion (237, 365) are in Table 15.2. The recommendations made by the two sources (237, 365) are not always in agreement. For product going to storage, the author feels that 2 minutes is usually adequate if overflowing the drum is of no major consequence, but 5 minutes is more appropriate if vapor from the drum goes to a unit that cannot tolerate liquid (e.g., a compressor). For tower reflux, the author feels that 2 minutes (based on reflux plus product) is more appropriate.

In applying the guidelines in Table 15.2, a surge volume is calculated separately for each product or reflux stream based on the destination of that stream. The largest surge volume thus calculated sets the desired surge volume between the low- and high-level alarms. Note that the vertical height of this portion must be sufficient to accommodate the level transmitter span.

Watkins (417) has proposed a more rigorous procedure for sizing reflux accumulators. The procedure takes into account the availability of instrumentation, the quality of control, and the expected speed of operator response. Drum size calculated by Watkins' procedure may differ widely from that calculated from Table 15.2. The author feels that it is best to estimate drum size using these two alternative procedures, and to use judgment in deciding which estimate suits the service better.

2. *The portion below the low-level alarm:* This portion provides surge for protecting the reflux pump against losing suction. The residence time

TABLE 15.2   Reflux Drum Residence Time (Low Liquid Level to High Liquid Level)

Operating condition	Basis for residence time definition		Recommended residence time, min	References
	Volume	Flow		
Product goes to storage	HLL to LLL	Product	2	365
			5	237
Product goes to another tower	HLL To LLL	Product	10–12	237
Product goes to heater	HLL TO LLL	Product	10	365
Product goes to heat exchanger along with other process streams	HLL to LLL	Product	5	365
Tower reflux	HLL to LLL	Product	5	365
	HLL to LLL	Reflux+product	2	237

in this portion should give the operator enough time to reach the pump before it loses suction, and therefore, it depends on drum location. A residence time of 2 minutes based on the product flow rate has been recommended (237) when the drum is adjacent to a fully staffed control room, with greater residence times for more remote locations. An alternative criterion for this portion (365) is to make its height 10 percent of the drum diameter, but at least 5 in. The author prefers the residence time criterion to the height criterion.

3. *The portion between the high-level alarm and the high-level trip:* This portion provides surge for protecting downstream equipment from liquid overflow in case of a malfunction in the column system (e.g., reflux pump trip). If the downstream unit cannot tolerate liquid (e.g., a compressor), it has been recommended (237) to make the residence time in this portion at least 3 minutes. Presumably, this residence time is based on the liquid flow through the reflex pump. A greater residence time is required (237) if the distance between the vessel and the nearest operator is large, or if keeping the downstream unit on line is vital.

   If the downstream unit can tolerate liquid (e.g., no vapor product; liquid backs up to the condenser in case product stops), a small residence time in this portion is usually satisfactory.

4. *The portion above the high–level trip:* This portion performs vapor-liquid separation and is sized using a knockout drum-sizing criterion (e.g., 237, 365). If all product is liquid, this portion is not needed.

   When the drum is horizontal and the incoming stream is all liquid, it is best to enter the feed near one dished end, and use an internal pipe terminating with a bend or a baffle that points the entering liquid jet at the dished end. This breaks the incoming liquid jet and minimizes turbulence and disturbance to any liquid-liquid separation.

Larger residence times may be desirable when the column occasionally experiences slugs of light ends or inerts in the feed (68). The condenser is essentially inert-blanketed during the period required for venting the lights. If reflux is lost, recovery from the upset will be appreciably lengthened. Residence times of up to 10 to 30 minutes based on the total drum volume and the reflux plus product flow are sometimes used (68). Upgrading vents to cope with these upsets, or adding override controls, are more satisfactory ways of dealing with the problem and should be practiced to keep the extra residence time required to a minimum.

Unnecessarily large reflux drum residence times should be avoided. These do not only increase drum cost but also enhance liquid inventories (a distinct drawback when the liquid is hazardous) and incur an excessive composition lag that interferes with column control.

Separation of two liquid phases. If two liquid phases are to be separated, the reflux drum must provide sufficient settling time. The rate of settling can usually be calculated from Stoke's law. A more detailed discussion on settling and decanting is available in Ref. 319 and its cited literature.

Separation of small quantities of water from water-insoluble organics. This is usually achieved using a reflux drum boot. A boot can be integral (Fig. 15.16) or external; in the latter case, it can be located at ground level and piped to the reflux drum.

A general rule of thumb for settling rates of water in hydrocarbons varying in density and viscosity from butane to diesel oil is (237) 20 ft/h for substantially complete settling; 30 to 50 ft/h for reasonably good settling, and 90 ft/h or more for incomplete settling. When the boot is drained on level control, a residence time of 7½ minutes half full has been recommended (237); if drained manually, it should be sized to hold the volume of water expected to accumulate over 4 h (237). Other recommendations (365) are to size the boot to a maximum velocity of 0.5 ft/s, minimum length of 3 ft, minimum diameter of 16 in for small reflux drums (diameter <8 ft), and minimum diameter of 24 in for large reflux drums (diameter >8 ft).

Flooded reflux drums. These are used as part of flooded condenser control schemes (Sec. 17.2.2). A flooded drum is only suitable when substantial fluctuations in product rate can be tolerated (e.g., product goes to storage). A flooded drum provides no surge for controlling and smoothing out product fluctuations, but it maintains surge for the reflux pump and reflux circuit.

No reflux drum. The consequences of eliminating the reflux drum are poorer controllability, fluctuating feed to downstream units, and a higher probability of losing suction to the reflux pump during upsets. If the consequences of these are operationally acceptable, the reflux drum can be entirely eliminated. Further, if the condenser can be mounted at a higher elevation than the column, the reflux pump can also be eliminated. In small columns (2 to 4 ft in diameter, less than 50 ft tall), it often pays to overdesign the column to accommodate for fluctuations in order to save the cost of a drum and pump. In larger columns, the use of an internal condenser (see Sec. 15.11) or an elevated condenser is also occasionally practiced, thereby eliminating the pump and drum.

16

Basic Distillation
Control Philosophy

A malfunctioning control system causes instability. The instability can adversely affect product purity, column capacity, economy, and ease of operation. Instabilities are often transmitted to downstream or upstream units, or can amplify small disturbances. In extreme cases, an instability can also lead to column damage or safety hazards.

As in the case of reboilers and condensers, distillation control is too wide a topic to be adequately covered in a handful of chapters. Entire texts (68, 89, 301, 332, 362) deal exclusively with distillation control. Most of these strike a balance between theory, practice, controls design, and controls optimization. In contrast, the coverage here emphasizes operational aspects: what various control schemes can and cannot do, how to put together a control system (not necessarily optimum, but one that works), how to recognize and avoid a troublesome system, what are the ill effects of various poor control schemes, and what corrective action can restore trouble-free operation.

Computer controls and advanced controls are outside the scope of this text. Although these controls are widespread and can be of primary importance for column optimization, their role in setting stable operation is usually secondary. Advanced controls often enhance column stability, but they seldom assure that the primary stability objectives are met. A troublesome computer control loop can usually be taken off control, and stable (albeit nonoptimum) operation can be restored. On the other hand, an unstable basic control loop usually means an unstable column, even when computer control is used.

The next four chapters deal with column control practices. This chapter addresses the overall philosophy of column control, while the next three examine the performance of various individual control loops.

A column control philosophy that is either defective or unsuitable for the service is frequently responsible for column instability. Devising an appropriate control philosophy is seldom easy. Prediction of column dynamic behavior during the design is extremely difficult, and designers resort to their previous experience in similar columns to guide their control philosophy. Although relative gain analysis (362) has proved invaluable for control system synthesis, it is often helpless when data on column dynamic responses are lacking.

Nevertheless, a lot is known about the overall philosophy of column control. Although this knowledge may fall short of predicting the optimum philosophy, it is usually sufficient for detecting poor and troublesome practices and distinguishing them from good practices. This aspect is of greatest interest to the distillation troubleshooter, superintendent, and operator, and is promoted here.

This chapter reviews the philosophy of overall column control system synthesis, examines common practices, outlines their pitfalls, supplies guidelines for avoiding these pitfalls, and discusses the pros and cons of the common control philosophies and their effects on column operation and stability.

16.1 Conventions Adopted for Diagrams and Discussions

Certain conventions were adopted in the diagrams and discussions in this chapter. These are

1. For convenience only, the diagrams in this chapter depict manipulation of condensation rate by varying coolant flow rate. Alternative techniques for manipulating condensation rate are described in Chap. 17. The discussions in the present chapter generally apply regardless of the manipulative technique.

2. For convenience only, the diagrams in this chapter (except Fig. 16.1) depict composition control as temperature control. Alternative variables (e.g., product analysis) used for composition control are discussed in Chap. 18. The discussions in the present chapter generally apply regardless of how composition is controlled.

3. The discussions and diagrams in this chapter exclude manipulation of column feed. This manipulation seldom belongs to the column control system. Feed is usually manipulated as part of an upstream control system (e.g., to regulate bottom level of an upstream column), or is maintained constant by a flow controller. One exception

is briefly discussed (Fig. 16.6c). Several of the concepts discussed in this chapter can be judiciously extended to exceptional situations where feed manipulation is incorporated into the column control system.

4. For simplicity, the discussions and diagrams assume that the column produces a bottom product, a top product, and no side product. The concepts discussed are extended to columns with side products in Chap. 19.

5. For simplicity, most diagrams assume a total condenser. The discussions do not. Extension to two-phase and to vapor top product is discussed in this chapter and in Chap. 17.

16.2 Controlled Variables and Manipulated Streams

A column control system has three main objectives:

- To set stable conditions for column operation
- To regulate conditions in the column so that the products always meet the required specifications.
- To achieve the above objectives most efficiently. This could mean maximizing product recovery, minimizing energy consumption, and often both.

Variables. Figure 16.1 shows the variables typically controlled in a column. Excluding flows, these include pressure, bottom level, accumulator level, top product composition, and bottom product composition. These can be classified into two groups:

1. *Single-loop variables:* These include pressure and levels. They are controlled in order to achieve the first objective, i.e., setting stable conditions for column operation. The set points at which these are controlled are established by stability considerations alone, regardless of product specifications.

   Controlling pressures and levels regulates material accumulation in the column. Keeping the levels constant prevents liquid accumulation, while keeping the pressure constant prevents vapor accumulation. Unless accumulation (positive or negative) is prevented, a continuous system will not operate at steady state and will not be stable.

2. *Unit objective variables:* These include top and bottom compositions. They are regulated to achieve the second objective, i.e., meet-

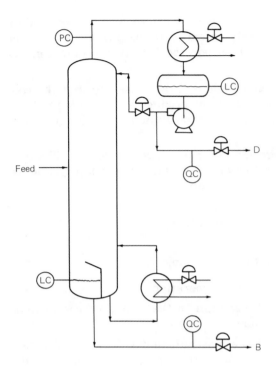

Figure 16.1 Controlled variables and manipulated streams in a typical column.

ing product specifications. The set points at which these are controlled are determined by product purity considerations alone.

Composition controls can be direct, i.e., using composition measurements of the product streams, or indirect, using a physical property representative of product composition. Typical physical properties used are refractive index, density, vapor pressure, freezing point, and, most commonly, tray (equilibrium) temperature.

In summary, beside flows, five variables are controlled in a typical column. Three of these (pressure and two levels) are controlled to set stable conditions, and two (compositions) are controlled to achieve the desired product purities.

Manipulated streams. A stream is manipulated by varying the opening of its control valve. The stream flow rate is thereby varied to control a desired variable. Figure 16.1 shows positions of control valves in a typical distillation system. There are five manipulated streams: top and bottom product flow rates, condensation rate, boilup rate, and reflux flow rate.

Some ground rules for control system synthesis . The essence of control system synthesis is suitable pairing of the five control variables with the five manipulated streams. Theoretically, 5! (= 120) pairing combinations are possible, but only a handful of these are commonly used. The following ground rules can be applied for initial screening out of undesirable combinations:

1. Figure 16.1 shows a column producing a single top product stream. Many columns produce a top vapor product and a top liquid product. This adds a manipulated stream (the second product) and two controlled variables (lights in the liquid product and liquid components in the vapor product). Since only one additional manipulated stream is available, only one of these compositions is controlled, while the other is allowed to float. Overall, one manipulated stream and one control variable are added, and these are usually paired together. The rest of the pairing is the same as for a column generating a single top product stream. The discussion below, therefore, applies to this situation.

2. In some systems, the reflux drum is omitted or run flooded (e.g., Fig. 17.5b). This eliminates one variable (accumulator level). This also eliminates a manipulated stream by making the condensation rate a slave to the top product rate. Similarly, when an internal condenser is employed, the accumulator level is eliminated, together with one manipulated stream (reflux becomes a slave to the condensation rate). In either situation, the rest of the variables are paired in the normal manner. The discussions below therefore apply to that situation too.

3. Ideally, both top and bottom product compositions should be controlled to maintain each within its specifications. In practice, simultaneous composition control of both products suffers from serious coupling between the two composition controllers (301, 332). Unless this interaction is decoupled, instability will result.

   This interaction can be illustrated by examining the action of the control scheme in Fig. 16.4a and supposing that a top-section temperature controller manipulates reflux instead of the shown flow controller. Suppose the concentration of lights rises in the column feed. This will be sensed by a drop in both control temperatures, and the temperature controllers will increase boilup and decrease reflux. If the actions of the two controllers are perfectly matched and instantaneous, both control temperatures will return to their set points without interaction. However, the two actions are rarely perfectly matched, and their dynamics are dissimilar. Suppose the boilup response is faster, as is usually the case. The additional boilup will return the bottom section temperature to its set point, but it will also raise the top section temperature. The top section temperature controller will now call for more reflux. In the meantime, the previous reduction in reflux will be felt by the bottom temperature controller, that will now call for less boilup. Each subsequent change will affect both reflux and reboil, and the two will cycle. The author and others (170, 309, 361) experienced column instability because of this interaction. A relative gains analysis (362) can often predict this interaction (361, 400).

In sophisticated control systems, this interaction is eliminated by decoupling. In basic control systems, it is avoided by controlling only one of the two product compositions. On some occasions, the second composition may be controlled with a detuned controller (309) which gives a very slow response. This is essentially the same as eliminating one composition control.

This leaves four variables to be controlled by manipulating five streams. The fifth stream, often termed the "free" stream, is usually flow controlled. If the free stream is the boilup rate, it is sometimes controlled by differential pressure. This technique is discussed in Chap. 19. At other times, the free stream is controlled by ratio control (often reflux to feed) or by a detuned composition controller (above).

4. Pressure is often considered the prime distillation control variable. Pressure affects condensation, vaporization, temperatures, compositions, volatilities, and almost any process that takes place in the column. An unsatisfactory pressure control often implies poor column control. Pressure is therefore paired with a manipulated stream that is most effective for providing tight pressure control. When the top product is liquid, this stream is almost always the condensation rate; when the top product is vapor, this stream is almost always the top product rate (see Sec. 17.2.).

Figure 16.2 illustrates the impact of the above ground rules. For a column producing a top liquid product (Fig. 16.2a), three variables (one composition and two levels) remain to be paired with three out of four stream (two product streams, reflux, and boilup). The unpaired stream is "free," i.e., on flow control. For a top vapor product (Fig. 16.2b), the condensation rate replaces the top product as one of the streams to be paired. In a column generating both a top vapor product and a top liquid product (Fig. 16.2c), the variables and streams remaining to be paired are the same as in a column producing a liquid top product only (Fig. 16.2a). For clarity, the second composition con-

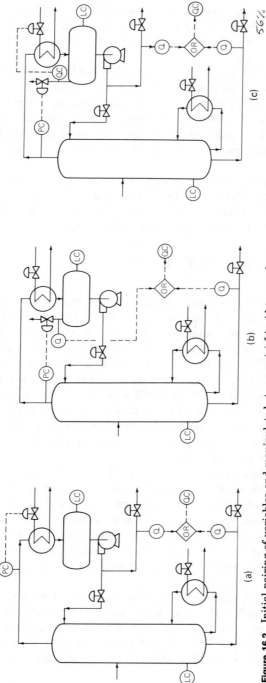

56%

Figure 16.2 Initial pairing of variables and manipulated streams. (a) Liquid top product; (b) vapor top product; (c) liquid and vapor top products.

491

trol was shown on the top vapor product, but it can also be positioned to control the concentration of lights in the top liquid product. Also, the pressure and second composition control may sometimes be interchanged (Chap. 17).

If the condenser is flooded (Fig. 16.2d), the unpaired variables and streams are the same as with a top liquid product (Fig. 16.2a), except that the reflux drum level and the top product stream are out of the pairing considerations. In the case of the internal condenser in Fig. 16.2e, the pairing is the same as for a vapor top product (Fig. 16.2b), except that reflux drum level and the reflux rate are out of the pairing considerations.

To prevent repetition, control system philosophy from this point on is discussed mainly with reference to the common case of a single liquid top product. Similar considerations apply to the other cases and can be inferred from the discussion, keeping in mind the above ground rules.

16.3 The Material Balance (MB) Control Concept

In an MB control scheme, product composition is controlled by manipulating the flow of material into and out of the column. This concept can be illustrated by examining the action of one of the common MB control schemes (Fig. 16.4a). Suppose the concentration of lights rises in the column feed. This will be sensed by a temperature drop, and the temperature controller will increase boilup. This will raise column pressure, and the pressure controller will step up condensation. Accumulator level will rise, and the level controller will increase distillate rate. Meanwhile, the increased boilup mentioned above will reduce the amount of liquid reaching the bottom sump, and the level controller will lower bottom product rate.

The net result of the control action is a shift in the material balance, so that more of the feed leaves with the distillate and less with the bottom. The shift transports the light component up the column and keeps bottom (and therefore, top) composition constant.

The vast majority of distillation columns use material balance control schemes. Only in circumstances where a satisfactory MB control scheme cannot be devised should an alternative control scheme be considered.

Perhaps the greatest obstacle to appreciating the need for MB control is the common belief that to obtain on-spec product, the reflux needs to be controlled. This reasoning is fueled by a column design practice: in order to reach product purity with a desired number of stages, reflux is raised until the product attains specification. While

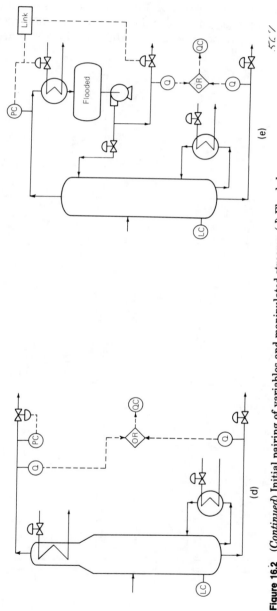

Figure 16.2 (*Continued*) Initial pairing of variables and manipulated streams. (*d*) Flooded condenser; (*e*) internal condenser, vapor product.

this is a common column design procedure, this is seldom the way to control a column.

Figure 16.3, based on the work of McCune and Gallier (265), shows the effects of varying the material balance split (characterized by the distillate to feed, *D/F*, ratio) and of varying the reflux (characterized

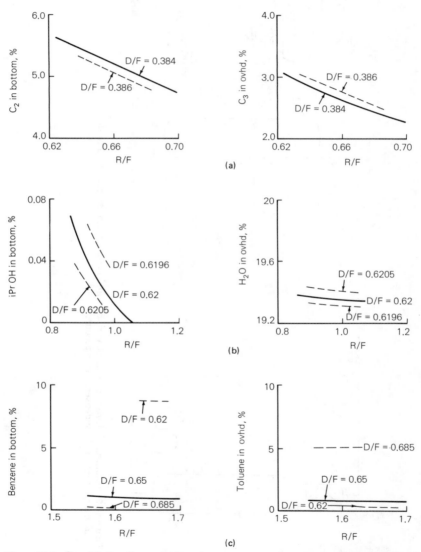

Figure 16.3 Sensitivity of composition to material balance and to reflux (data from *McCune and Gallier*, ISA Transactions, *12(3), 193, 1973.) (a)* Deethanzier, 14 trays, 400 psia; *(b)* ethanol/isopropanol/water, 10 trays, 14.7 psia; *(c)* benzene/toulene, 16 trays, 14.7 psia.

by the reflux to feed, R/F, ratio) on top and bottom compositions in three columns. It is clear that in all cases, the sensitivity of product composition to shifting the material balance split is severalfold greater than its sensitivity to reflux changes. This is why material balance control is almost always used in distillation columns.

One implication of MB control is that a product stream cannot be flow-controlled (or be the free stream). If flow rates of both the feed and one product are fixed, then the flow rate of the other product must be the difference between them, or accumulation will occur. This fixes the material balance and precludes shifting it for product composition control.

Running a product stream on flow control therefore leads to poor product purity control. In one case (268), a column failed to meet its bottom product spec even at high reflux rates. Setting the top product flow at too low a rate was the culprit; the remainder of the light component was forced out of the bottom. Unless feed composition and feed flow are always uniform, or there is a lot of room for increasing reflux, the above type of problem should be anticipated. One source of trouble is using a cascade controller of some column variable to drive a product flow controller. This works fine as long as the cascade mode is sustained. The moment the controller comes off cascade, the system will no longer be a material balance control, and the above problem will crop up.

A non-MB control scheme, in which a product stream is the free stream, can work if feed flow rate and feed composition are extremely steady, although it requires some operator intervention. In one case (151a, 151b) a column performed poorly with the MB control scheme in Fig. 16.4a. Performance was sharply improved by changing to a non-MB control scheme similar to that of Fig. 16.4e, but with the bottom stream flow-controlled (instead of temperature-controlled). In that case (151a, 151b) the improvements resulted from the superior response of scheme 16.4e (compared to scheme 16.4a) to the large ambient disturbances that were experienced in the column (items 7 and 8, Sec. 16.6) and from the elimination of an insensitive temperature controller (Sec. 18.1). For further improvements, the plant is planning to add a product analyzer for manipulating the bottom stream, thus completing the conversion to scheme 16.4e (151a, 151b) and restoring MB balance control. With the present non-MB controls, bottoms flow is operator-adjusted according to charts based on feed assays (151b). This works well because the column feed is steady on flow control and the feed composition is steady per assay.

Another situation where a product stream can be satisfactorily flow-controlled is when column feed is incorporated into the column control system. In that case, one product flow can be fixed by a flow control, while the feed and other product are manipulated to maintain the material balance. This case is briefly discussed later (e.g., Fig. 16.6c).

16.4   Pairing for MB Control

Application of MB control, as well as the constraints discussed in the previous section, considerably narrows the options for pairing the remaining variables and streams. For the system shown in Fig. 16.2a, the free stream can only be either the reflux flow or the boilup rate. The other remaining variables are two levels and a composition.

Additional pairs can be eliminated by recognizing that accumulator level is insensitive to bottom flow rate and bottom level is insensitive to distillate flow rate. This leaves six possible control schemes (Table 16.1). The first three, and to a lesser degree, also the fourth, are the most common schemes and are discussed in detail below. The others are uncommon. An additional uncommon scheme, which defies one of the previously stated ground rules (it does not use pressure to control condensation), but is reported to have worked well in some installations (234), is also included in Table 16.1.

The author recommends caution if the control system used does not conform to one of the schemes in Table 16.1. Unless different parameters are used or unique circumstances exist, such a method is likely to violate one of the ground rules mentioned earlier, and may be troublesome. The first four schemes are generally the least troublesome and should be used whenever possible.

16.5   Direct and Indirect MB Control

Indirect MB control.   Here, the composition (or temperature) controller does not directly regulate a material balance (i.e., a product) stream. Instead, it regulates reflux, boilup, or condensation. The product streams are controlled by level or pressure. Adjustments to the material balance are thus performed indirectly, i.e., by working through the pressure or levels. This becomes apparent when the stepwise action of the scheme is considered. The stepwise action of the indirect scheme in Fig. 16.4a is described in Sec. 16.3.

Two indirect MB control schemes are common (Fig. 16.4a,b). For columns producing two liquid products, both products are on level control. In the Fig. 16.4a scheme, the composition (or temperature) controller regulates boilup, with reflux being the free stream. In the Fig. 16.4b scheme, these controls are interchanged.

When the top product is vapor, it is on pressure control, with the accumulator level controlling the condensation rate. The Fig. 16.4c

TABLE 16.1 MB Control Configurations

Scheme (Fig.) no.	Type	Accumula- tor level	Sump level	Composi- tion	Free	Pressure	
16.4a	Ind	D	B	H	R	C	
16.4b	Ind	D	B	R	H	C	} Most
16.4d	Dir	R	B	D	H	C	common
16.4e	Dir	D	H	B	R	C	
16.6a	Dir	D	R	B	H	C	
16.6b	Dir	H	B	D	R	C	} Uncommon
17.7e	Dir	C	B	D	R	H	

LEGEND: B = Bottom   Dir = Direct material balance
         C = Condensation   H = Heating medium
         D = Distillate   Ind = Indirect material balance
                             R = Reflux

scheme is analogous to the Fig. 16.4a scheme, and has similar advantages and disadvantages. Similarly, a scheme analogous to that in Fig. 16.4b, but with the pressure and accumulator level controls interchanged, can be used with a vapor top product.

Direct MB control. Here the composition (or temperature) controller directly regulates a material balance (i.e., product) stream. The other product is regulated by a level or by pressure. The stepwise action of the Fig. 16.4d scheme is as follows. Suppose the concentration of lights rises in the column feed. This will be sensed by a drop in column temperature, and the temperature controller will increase distillate flow. Accumulator level will fall, and the level controller will reduce reflux. This will lower the bottom level, and the level controller will reduce bottom flow.

One direct MB control scheme (Fig. 16.4d) is common. This scheme is similar to the Fig. 16.4b scheme, but with the reflux and product controls interchanged. The Fig. 16.4d scheme is suitable for two liquid products. An analogous scheme for a vapor top product requires that the composition controller regulate top product flow, a function better performed by the pressure controller. For this reason, a scheme analogous to the Fig. 16.4d scheme is seldom used with a vapor top product.

A direct MB control scheme with the composition controller regulating the bottom product flow (Fig. 16.4e) is also used. This scheme is similar to Fig. 16.4a, but with the boilup and bottom level controls interchanged. This scheme is less popular than the other common scheme, mainly because of its high sensitivity to inverse response (65, 66). This is described in the next section.

Richardson's Law - Never control level w/ a small strm

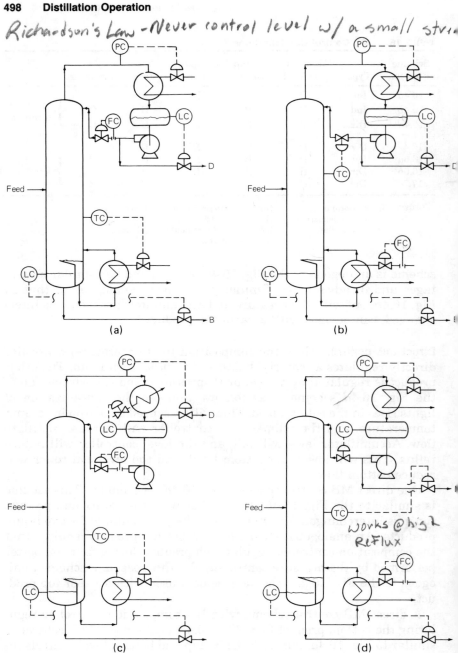

works @ high Reflux

Figure 16.4 Common MB control schemes. (*a*) Indirect control, composition regulates boilup; (*b*) indirect control, composition regulates reflux; (*c*) as *a*, but with a vapor product; (*d*) direct control, composition regulates distillate flow.

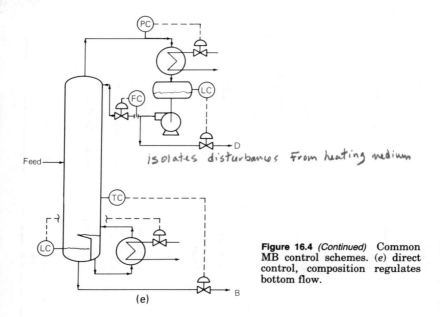

isolates disturbances from heating medium

Figure 16.4 *(Continued)* Common MB control schemes. *(e)* direct control, composition regulates bottom flow.

Uncommon schemes. Three additional direct MB control schemes are shown in Table 16.1. Two of these (Fig. 16.6a and b) are briefly discussed in the next section, and one (Fig. 17.7e) is described in Sec. 17.2.4. All three schemes have serious shortcomings that restrict their use to fairly narrow ranges of applications and make them uncommon. Nevertheless, successful application of each one of these has been reported (65, 66, 234, 258).

16.6 Pros and Cons of the Common Column Control Schemes

Schemes in Fig 16.4a, b, d, and, to a lesser degree, also e, are the common control schemes. Other control schemes are relatively uncommon, are generally only used in special circumstances, and can be troublesome. Unless experience shows that an uncommon scheme works well in a particular service, or that none of the common schemes can be made to work in this situation, uncommon control schemes are best avoided.

The following considerations apply for the selection, design, and operation of the common control schemes. These are also summarized in Table 16.2.

1. Schemes in Fig 16.4a,b, and e are suitable when column top product is either liquid or vapor. Scheme 16.4d is not suitable when

TABLE 16.2 Pros and Cons of the Common Control Schemes

Scheme (Fig.) No.	16.4a	16.4b	16.4d	16.4e
Type	Indirect	Indirect	Direct	Direct
Stream used for composition control	Reboil	Reflux	Top product	Bottom product
Top product for which scheme works	Liquid or vapor	Liquid or vapor	Liquid only	Liquid or vapor
Favored when more important product purity is...	Bottom	Top	Top	Bottom
Favored when smaller product flow is...	Bottom	Top	Top	Bottom
Favored when better control tray temperature is in the...	Bottom	Top	Top	Bottom
Speed of response in large tray columns	Fast	Slow	Slow	Reasonably fast
Suitability for essentially stripping columns	Excellent	Least satisfactory	Least satisfactory	Good
Response to disturbances in coolant system	Least satisfactory	Reasonable	Excellent	Least satisfactory
Response to disturbances in heating medium system	Reasonable	Least satisfactory	Least satisfactory	Excellent
Response to a variety of disturbances in small columns (129, 265)	—	Inferior	Superior	—
Performance in the presence of inverse response	Affected, but stable	Unaffected	Unaffected	Unstable
Smoothness of bottom flow when bottom flow and surge volume are small	Fluctuating	Fluctuating	Fluctuating	Smoothest
Effect of small distillate flow	Accumulator level drifts; mass balance poor	As 16.4a	Loose composition control	As 16.4a
Suitability to handle low reflux flows	Excellent	Good	Poor; reflux fluctuations likely	Excellent

the only top product is vapor. This was discussed in the previous section.

2. Schemes 16.4a and e give a closer control of the bottom composition and are therefore favored when bottom product purity is more important. Similarly, schemes 16.4b, d are favored when top product purity is more important.

3. The sensitivity of the material balance to changes in product flows influences the favored scheme when product flows are grossly unequal. Composition control by manipulating the larger product stream can cause major swings in the composition of the smaller stream. Conversely, composition control by manipulating the smaller stream may give slow and sluggish variation of the composition of the larger stream. In most cases, the swings are less desirable, and it has been recommended (363) that the smaller of the two product streams be manipulated by the composition controller. Schemes 16.4a and e are therefore favored when the bottom flow is smaller, while schemes 16.4b and d are favored when the distillate flow is smaller.

4. Speed of response to a composition change is best with scheme 16.4a, good with scheme 16.4e, and rather slow with schemes 16.4b and d.

   Schemes 16.4a and e respond to a composition disturbance by changing a vapor flow rate (reboil). This change propagates almost instantaneously, even through a large column, giving a fast, effective response. On the other hand, schemes 16.4b and d respond to a composition disturbance by changing tray liquid flow. This change must travel with the liquid from tray to tray, and it will be delayed by the combined hydraulic lags of all the trays above the control point. Pilot-scale measurements (27, 118, 362) suggest that the delay may be of the order of 0.1 minute per tray. Further, while a change in vapor affects all tray compositions simultaneously, a change in liquid will affect the top trays' compositions several minutes before it reaches the bottom trays. For these reasons, a change in liquid rate may incur a slow and sluggish response in large columns, particularly when the temperature control point is located several trays below the top. A far more responsive composition control with scheme 16.4a compared to scheme 16.4d has been demonstrated in the field (400, 422).

   The response of a column to a composition change can be significantly slowed down when the temperature difference between the control point and the bottom tray (schemes 16.4a and e) or top tray (schemes 16.4b and d) is large. Considerable sensible heat may be stored in the temperature gradient across the column, and it will act to delay and reduce the effectiveness of the changes (362). The response of a column to a composition change can also be considerably slowed down when the residence time in the sump feeding the reboiler is large (400), e.g., in a small column heated by a large kettle reboiler.

The slow speed of response of schemes 16.4b and d is particularly troublesome in trayed superfractionators. The author has experienced situations where the slow and sluggish response of scheme 16.4d made it unsatisfactory for controlling an ethylene-ethane splitter and a propylene-propane splitter. In the ethylene-ethane splitter, response was immensely improved when the control scheme was changed to scheme 16.4a, even though top product purity was far more important than bottom product purity. In the propylene-propane splitter, the energy balance control method (Fig. 16.7) was a significant improvement in performance compared to scheme 16.4d; again, the top product was far more important.

In packed columns, response to liquid changes appears faster, and the speed of response of the schemes 16.4b and d is less troublesome. In one case (388), considerable improvement in the speed of response of a butene splitter was experienced after trays were replaced by packings. A feed-forward loop, used for speeding the response of trays, was found unneeded and undesirable following the revamp.

5. Generally, schemes 16.4a and e are favored when the better control tray is below the feed, while schemes 16.4b and d are favored when the better control is above the feed. This is especially important when the feed flow rate is high compared to column internal flows, or when a large temperature gradient exists near the feed point. The sensible heat stored in the feed and in the temperature gradient near the feed point will dampen a vapor or liquid change. This reduces the control point sensitivity to vapor or liquid changes traveling across the feed point.

   Dynamic considerations spelled out in item 4 above also affect this guideline. Since vapor travels quickly through the column, a top section control tray will respond rapidly in schemes 16.4a and e. On the other hand, since liquid travels slowly through the column, a control tray located in the bottom section will give a slow and sluggish response in schemes 16.4b and d unless the column is short or packed.

6. Many columns are essentially "stripping" columns, with the bulk of the product leaving in the bottom. Here the feed effectively provides most of the column reflux, and the purity of the bottom product is more important than the distillate purity. This type of col-

umn is an ideal candidate for scheme 16.4a, as was demonstrated by field experience (52).

7. A major strength of direct MB control schemes is their ability to minimize the impact of disturbances in the cooling medium (scheme 16.4d), heating medium (scheme 16.4e) or in the heat losses from the column (both schemes 16.4d and 16.4e) on column operation. Consider the reaction of scheme 16.4d to a sudden rise in reflux subcooling (e.g., due to a rainstorm cooling the reflux drum). Column pressure will drop, and the pressure controller will reduce condensation rate. The accumulator level will fall, and the level controller will reduce reflux. Reducing reflux is the desired action, and it maintains the internal reflux in the column constant. The disturbance does not propagate beyond the overhead loop.

   In contrast, consider the response of schemes 16.4a,b, or e to a similar change. The drop in accumulator level will reduce distillate flow, while reflux flow rate will remain unaltered. The same quantity of reflux will enter the column, but at a lower temperature. It will be reheated upon entry by vapor condensation, and this will increase liquid flow down the column. This is not the desired response. Eventually, the appropriate response will be established, but not until the control tray temperature drops and the temperature controller takes corrective action. In scheme 16.4b, a subcooling disturbance disturbs at least the top part of the column. With the Fig. 16.4a and e schemes, it will disturb the entire column.

   The above is a major advantage of scheme 16.4d when coolant temperature experiences sudden or frequent changes. The author has experienced a case where cooling water temperature periodically rose or fell by as much as 20°F within a short time interval. The column, which was controlled by scheme 16.4d, was barely affected. Nisenfeld (300) reports another case where a column controlled by scheme 16.4b experienced high sensitivity to changes in ambient conditions. Switching to scheme 16.4d desensitized the column and considerably reduced the need for overrefluxing.

   In tall and thin and/or poorly insulated columns, sudden rainstorms can substantially increase cooling at the tower wall and condensation inside the tower. Column pressure will rapidly drop and bottom level will rapidly rise. The composition (and, therefore, temperature) changes are generally slower and will lag behind. Scheme 16.4d will give the best and desired response because the pressure controller will immediately act to reduce condensation, which in turn will reduce drum level, and the level controller will in turn cut reflux. This action will be augmented by the temperature controller, which will increase product flow, thus effecting a further reduction in drum level and reflux rate. Scheme 16.4e will give a good response when the column operates well below the capacity limit. Here the rise in bottom level will raise boilup, which will rapidly counteract the increased condensation rate. Again, the action will be augmented by the temperature controller, which will also act to raise boilup. On the other hand, the response of schemes 16.4e and b will be far less satisfactory. In either case, the scheme will initially attempt to raise product flow rates. This may cause the products to go off-spec while doing little to counteract the disturbance. With these schemes, the appropriate response will be established only after the slower temperature (or composition) controller takes action.

Hatfield (151a) has reported one column experiencing major disturbances from sudden rainstorms, where control problems with scheme 16.4a were severe enough to annul the benefits from a column revamp with high-efficiency packings. Reflux reduction due to the higher packing efficiency rendered the column more sensitive to the rainstorms. The problem was solved by implementing a non-MB control scheme similar to scheme 16.4e (see Sec. 16.3 for further description).

8. In a similar manner to item 7 above, scheme 16.4e is best, while schemes 16.4b and d are least satisfactory for minimizing the impact of disturbances in column heating medium on column operation. Reported plant tests (259) demonstrate this conclusion. Some authors (258, 259, 309) recommend the use of scheme 16.4e because of this advantage. However, note items 10, 11 below.

9. The responses of scheme 16.4d to a variety of material and energy balance disturbances were demonstrated by Gallier and McCune (129, 265) to be superior to the responses of scheme 16.4b. The study was based on simulation of columns containing 14 to 17 trays. Earlier statements by Nisenfeld (300) support this conclusion.

One case history contributed by J. A. Hatfield (151b) describes a methanol-water column, reboiled by live steam injection, which used a non-MB control scheme. The scheme was the same as those in Fig. 16.4a or b, but there was no automatic temperature control and both steam injection and reflux were flow-controlled. The operators manually adjusted the set point of these flow controllers to suit a tray temperature. The system performed poorly and the column expereienced upsets and off-spec products. Changing the system to scheme 16.4d eliminated the problems and gave excellent control.

10. Scheme 16.4e becomes unstable or unsatisfactory when the column and/or reboiler experience inverse response. This is a major shortcoming and is responsible for the relatively low popularity of this scheme.

In the froth regime, an increase in vapor flow reduces tray froth density. Froth height above the weir rises, and some of the tray liquid inventory spills over the weir into the downcomers. The expelled liquid ends up in the bottom of the column, and bottom level initially rises (Fig. 16.5). This is opposed to the expected response, and is termed *inverse response*.

The reboiler may also experience inverse response, often referred to as *reboiler surge* or *reboiler swell*. An increase in heat input may increase the volume of vapor in the reboiler or the pressure drop in the reboiler and its outlet piping. This will temporarily back up liquid into the column bottom, causing liquid level to rise.

Inverse response does not affect the operation of schemes 16.4b, d (except when boilup fluctuates), because boilup is maintained constant. Inverse response of the column may have some effect on scheme 16.4a. In principle, it can temporarily raise the lights concentration in the bottom product, and momentarily lower the control tray temperature, following a heat input increase. However, the author is not aware of any troublesome experiences, even though scheme 16.4a is widely used. It appears that controller tuning is sufficient to prevent any significant ill effects of inverse response on scheme 16.4a, and that inverse response is seldom, if at all, troublesome with this scheme.

Inverse response has a destabilizing and detrimental effect on

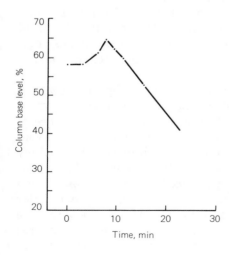

Figure 16.5 Open-loop response of bottom level to an 8 percent steam rate increase, featuring inverse response. (*From "Inverse Response in Distillation Column,"* P. S. Buckley, R.K. Cox, and D.L. Rollins, Chemical Engineering Progress, *vol. 71, no. 6, p. 83 (June 1975). Reproduced by permission of the American Institute of Chemical Engineers.*)

scheme 16.4e (65, 66). Suppose the concentration of lights rises in the column feed. The control temperature will fall and reduce the bottom flow. Bottom level will rise, and the level controller will increase the heat input and vapor flow. When inverse response is significant either in the column or in the reboiler, increasing the heat input and vapor flow will raise the bottom level. This in turn will enhance heat input and vapor flow, and so on, and an unstable response will develop. Figure 16.5 illustrates the open-loop response of bottom level to an increase in reboil in a column experiencing inverse response (65, 66).

It is difficult to predict whether inverse response is likely to occur and whether it will constitute a severe problem. A predictive model was developed by Buckley et al. (66, 68), and it gave good predictions in at least two case histories (52, 66). Field determination of inverse response is far easier. If stepping up heat input raises bottom level, an inverse response is implicated.

The incidence of inverse response depends on column internals and vapor rates. It has been suggested (362) that inverse response may occur with valve trays at all vapor rates, while sieve trays may give inverse response at low rates, direct response at high rates, and effectively dead time at intermediate rates. It is unknown whether columns operated in the spray regime experience inverse response; in principle they should not, since tray liquid holdup is unaffected by vapor flow rates within this regime (204, 244).

Packed columns are not likely to be affected by inverse response, but may in principle experience a related problem. In the loading regime, a rise in vapor flow increases liquid holdup in the column. With scheme 16.4e, raising boilup will therefore induce a drop in bottom level, which in turn will reduce boilup. If sufficiently vigorous, this phenomenon can unsettle column boilup. The expectation of this behavior is based on first principles only; the author is not aware of any troublesome reports of this behavior.

Even though some favorable experiences were reported (52, 258, 259, 309) with scheme 16.4e, its susceptibility to instability when inverse response occurs makes it undesirable. Unless absence of inverse response can be ascertained, it is best (68, 301) to avoid this scheme whenever possible (but see also items 8, 11). Buckley et al. (65, 66) could not get this scheme to work in a column containing 100 valve trays which experienced inverse response. Even when the column experiences no inverse response, the reboiler may, and this will be as detrimental to the performance of scheme 16.4e. Bojnowski et al. (52) reported a case where reboiler inverse response caused sluggish control with scheme 16.4e; the remedy was switching to scheme 16.4*a*. In another column described by these authors (52), and in a case reported by others (259), no inverse response existed and scheme 16.4*e* worked well.

11. With schemes 16.4*a*, *b*, and d, fluctuations in bottom level cause swings in bottom flow. This may be troublesome if the bottoms product flows into a unit that requires a steady feed, such as a furnace, a reactor, and sometimes even another column. An unsteady bottom flow may also be troublesome when the bottoms product preheats a stream going into such a unit, or even if it preheats the column feed, and preheat control is slow or ineffective.

Providing a large residence time at the column bottom, or surge capacity between the column and the downstream unit, can normally smooth out these fluctuations, but it may be costly. When the chemicals at the column base are thermally unstable or hazardous, a large surge volume at the hot temperature is also undesirable.

Bottom flow swings are particularly vigorous when the bottom stream is only a small fraction of the liquid arriving at the column base. Level fluctuations are most likely to originate from disturbances to column liquid flow, and those will be amplified when transmitted to the bottom product flow. For instance, a 5 percent disturbance to column liquid flow will be reflected as a 50 percent disturbance in bottom flow when column liquid flow is 10 times greater than the bottom flow. Under these conditions, not only would the swings be large, but bottom flow changes would be quite ineffective for controlling bottom level.

When the bottoms stream provides the bulk of the column preheat, bottom flow swings may cause fluctuations in feed enthalpy. Unless the feed temperature controller can suppress these rapidly and effectively, the disturbances will reenter the column and interact with the composition controller. In one column controlled with scheme 16.4*a*, this resulted in severe oscillations of the composition controller.

When a steady bottom flow is desired, scheme 16.4*e* is at a major advantage and is usually preferred (258, 259, 309, 332, 362) unless inverse response is likely. When inverse flow is troublesome, an unconventional direct MB control scheme (Fig. 16.6*a*) can provide the cure. Buckley et al. (65, 66) reported satisfactory performance of this type of scheme in a column where scheme 16.4*e* failed due to inverse response. In that column, bottom flow was too small to effectively control bottom level, and column base holdup was to be kept at a minimum to avoid material degradation (i.e., schemes 16.4*a*, *b*, and *d* were unsuitable).

Scheme 16.6*a* suffers from similar drawbacks to scheme 16.4*b*,

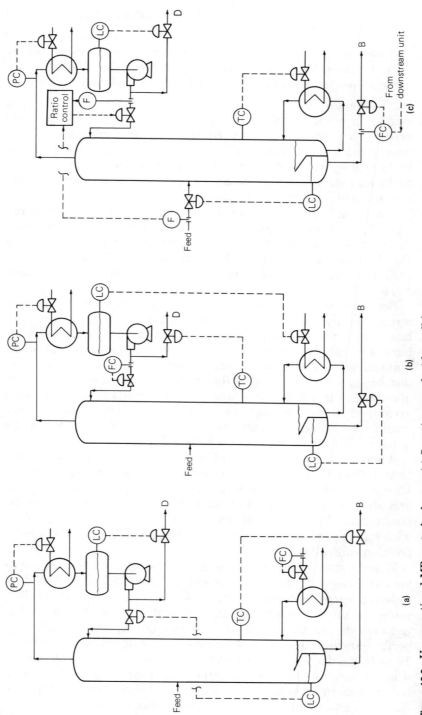

Figure 16.6 Unconventional MB control schemes. (a) Sometimes used with small bottom flows; (b) sometimes used with small distillate flows; (c) bottom product demand, analogous to Fig. 16.4a.

except for the problem of bottom flow swings mentioned above. It also suffers from a delay from the time bottom level begins to change to the time the added reflux arrives at the bottom. In the 100-tray column described by Buckley et al. (65, 66), this delay was of the order of 3 minutes. Although this was undesirable, the column could live with it. The bottom level controller must be loosely tuned to prevent reflux-bottom level instability from developing.

Scheme 16.6a can be troublesome when column feed is mostly liquid and its flow is large compared to reflux flow. Bottom level may then rapidly rise or fall in response to feed flow fluctuations. A large surge volume at the column base can abate the level variations, but providing this surge is self-defeating. The level variations will be aggravated by the temperature controller action. A drop in feed rate will raise the control temperature, which in turn will open the bottom valve and further drop the bottom level. For these reasons, scheme 16.6a is best avoided where column feed flow is large compared to reflux. In this case, a bottom product demand scheme (item 14 below) may be worth considering.

12. Control of the column material balance may be difficult when the distillate stream is small relative to reflux flow. Unless large swings in distillate flow are acceptable, changes in the small distillate flow will have little impact on accumulator level, and schemes 16.4a, b, and e will not maintain a steady accumulator level. In most cases, the level in the accumulator will be allowed to drift and will be periodically adjusted manually by changing reflux, boilup, or condensation rate. However, this mode of control does little to maintain the column material balance, and over a period of time may cause accumulation or depletion of the light components. Scheme 16.4d does not suffer from the above problem and is often favored with small distillate flows (300).

If the main source of disturbance in the column is the heating medium, the disturbances may interact with the accumulator level control, and scheme 16.4d may lead to reflux flow fluctuations. The unconventional Fig. 16.6b scheme may be better for smoothing these disturbances. Favorable experiences have been reported with this scheme, and its use has been recommended when distillate flows are small (258). Like its unconventional counterpart (Fig. 16.6a), scheme 16.6b may not be suitable if feed flow is large compared to the column boilup. It may also suffer from interaction between the level and pressure controls (301).

Both schemes 16.4d and 16.6b have the drawback of providing a fairly loose composition control when distillate flow is extremely small. If a tighter composition control is desired, or if the main source of disturbances is the feed stream, it may be better to use scheme 16.4a, b, or e, and to live with the component accumulation problem. Alternatively, it may be worth considering a sophisticated control scheme (e.g., using feed forward or computer control). In one extreme case (269), when reflux to distillate ratio was 70:1, the composition control of scheme 16.4d was successfully tightened by applying on-off control of the distillate.

13. When the reflux flow is small compared to distillate flow, scheme 16.4a (and in the absence of inverse response, also scheme 16.4e) is most satisfactory, while scheme 16.4d is least satisfactory. With scheme 16.4d, the accumulator level controller will amplify small disturbances in column vapor flow into large swings in the reflux flow. For instance, consider a 5 percent disturbance in the vapor rate in a column operating at a reflux to distillate ratio of 1 to 4. This will generate a 5 percent disturbance to the total liquid condensed. With scheme 16.4d, all this disturbance will be transmitted to the reflux flow, since the distillate rate is unaffected by drum level. Since reflux flow is only one-fifth of the total liquid flow, the disturbance will be reflected as a 25 percent change in reflux flow. In one troublesome case (52), reflux flow swung widely in response to small changes in column heat input, and at times fell below the minimum required for tray wetting. Reflux was controlled by accumulator level, and reflux-to-distillate ratio was 0.43.

14. Until now, it was assumed that feed is controlled to satisfy some control criterion upstream of the column or is on flow control. While this is usually the case, there are instances when it is beneficial to incorporate the feed into the column control system. A typical example is when it is desired to maintain a constant bottom flow rate to a downstream unit (e.g., a reactor or a furnace), and it is impractical to smooth out bottom fluctuations by one of the common control schemes or by adding surge capacity at the hot bottom temperature.

Scheme 16.6c is a bottom product demand scheme analogous to scheme 16.4a. Similarly, bottom product demand schemes analogous to schemes 16.4b and d, and distillate demand schemes analogous to schemes 16.4a,b, and e can be devised. Often, a ratio control is used on the free stream (e.g., reflux-to-feed ratio in Fig. 16.6c) to minimize the impact of feed fluctuations on the column. Several product demand schemes with more sophisticated ratio controls are described by Buckley et al. (68).

Product demand schemes are generally either inferior to or more complex and expensive than the conventional schemes, and are best avoided if possible. Feed fluctuations interact with other control variables, and introducing them on purpose does not improve column stability. Nevertheless, if supplying a stable product to a downstream unit is a prime consideration and cannot be economically achieved otherwise, they may be necessary. Feed surge capacity is often required to minimize upset to upstream units and to permit fluctuation of feed to suit the downstream unit demand. Feed surge is far less prone to thermal degradation problems than column base surge because of the cooler temperature, and may therefore be more desirable than bottom surge.

16.7 Energy Balance Control

In energy balance control schemes (e.g., Fig. 16.7), energy balance variations control product composition, and the free variable is one of the product flows.

The main shortcoming of an energy balance control scheme is that material balance variations interact with the controls. Further, the controls are more sensitive to these variations than to changes in the energy balance, as described in Sec. 16.3. For these reasons, energy

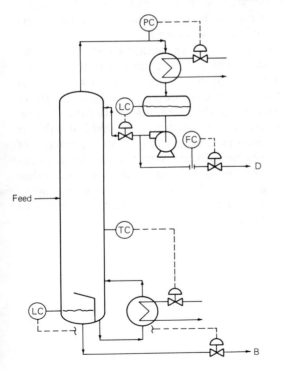

Figure 16.7 An energy balance control scheme.

balance controls are usually only used if a satisfactory material balance control scheme cannot be implemented, or in conjunction with advanced or computer control.

An energy balance scheme as in Fig. 16.7 usually requires continuous adjustment of a product rate (in Fig. 16.7, the distillate rate) by the operator, and a very slow level control action (in Fig. 16.7, accumulator level). Consider a rise in concentration of the light component in the feed. The bottom section temperature will drop, and the temperature controller will raise boilup. Column pressure will rise, and the pressure controller will increase the condensation rate. The accumulator level will rise, and the level controller will pour more reflux into the column. This in turn will reduce control tray temperature, and the temperature controller will raise boilup again. This will continue until reflux and boilup sufficiently rise to keep the bottom section temperature up. In the meantime, the light component accumulates in the system, and this will cause further increase of reflux and boilup.

Eventually reflux and boilup rates will exceed their desirable flow rate. At this time, the operator will let out more product, which stops reflux and reboil from rising. The product is thus operated in a "semibatch" manner. If the operator increases product rate too much, reflux and reboil will start falling. With this scheme, slow continuous cycles of reflux and reboil are often experienced.

Due to the above limitations, an energy balance control scheme such as scheme 16.7 is not recommended, but some situations exist where this scheme can offer better product composition control than many other alternatives. Superfractionators with a reflux to distillate ratio of 10 to 1 or more are one example. Here, distillate flow may be too small to satisfactorily control either accumulator level or column temperature. The author has experienced a satisfactory operation of scheme 16.7 in a propylene-propane splitter, with intermittent operator intervention to adjust the material balance. The cycles in reflux and reboil (see above) could be tolerated, as the column was not operating close to its limits. In this column, scheme 16.7 gave tighter composition control than scheme 16.4d.

Reboiler, Condenser, and Pressure Controls

Reboiler and condenser controls regulate the energy inflow and out-flow in a distillation column. These controls must adequately respond to column changes, minimize transmission of disturbances into the column, and be energy efficient. Failure to achieve the first two functions will lead to column instabilities; failure to achieve the third wastes energy.

Column pressure control is frequently integrated with the condenser control system. This control is often regarded as the most important control in the column. It has been the author's experience that a column will not achieve stable operation unless steady pressure can be maintained.

This chapter surveys the multitude of control schemes available for reboiler, condenser, and pressure controls; critically examines the benefits and shortcomings of each scheme; outlines common pitfalls in their implementation; and supplies guidelines for avoiding these pitfalls.

17.1 Reboiler Control

Reboiler control should not only provide good response to column disturbances, but should also isolate the column from disturbances occurring in the heating medium. The best variable for regulating column boilup is selected when the column overall control philosophy is devised (Chap. 16). In the majority of cases, boilup is regulated either to achieve a desired product purity, normally in the bottom section, or at a constant rate. When boilup is kept at a constant rate, the reboiler control valve is usually manipulated by a heating medium flow controller. When boilup is regulated to achieve a desired product purity, the reboiler control valve is directly or indirectly manipulated by a tray temperature, a product analyzer, or the base level. Indirect manipulation is performed by a cascade controller that varies the set point of the heating medium flow controller. The flow controller, in turn, manipulates the reboiler control valve.

Temperature, analyzer, or level regulation of column boilup is generally far smoother and gives superior response if performed by a cascade controller rather than by direct manipulation. Nevertheless, direct manipulation is often satisfactory.

Sluggishness and oscillations in boilup regulation are far more potent and interactive when boilup is manipulated to achieve a desired product purity than when boilup is kept constant. This is because the temperature or composition controller feeds back any fluctuations in boilup manipulation as delayed signals calling for further manipulative actions. The author experienced a case where this feedback action rendered a sluggish boilup control system inoperable during even mild upsets.

The following discussion generally applies regardless of the variable that regulates boilup and the directness of control. Perhaps the only exception is that the need to avoid sluggish and oscillating boilup manipulation is emphasized when boilup is manipulated to achieve a desired product purity. For simplicity only, most of the examples discussed assume that the heating medium flow rate manipulates the reboiler control valve.

17.1.1 Reboiling with a condensing fluid

Typical examples are steam reboilers and refrigeration vapor reboilers. The control valve may be located either in the reboiler inlet line (Fig. 17.1a) or in the reboiler condensate outlet line (Fig. 17.1b).

When the control valve is located in the reboiler inlet line (Fig. 17.1a), heat transfer rate is regulated by varying the reboiler condensing pressure and therefore the reboiler condensing temperature. When additional boilup is needed, the valve opens and raises the reboiler pressure, which enhances the reboiler temperature difference, which in turn increases the boilup rate.

Instead of controlling flow to the reboiler (in Fig. 17.1a), one could use the pressure at the reboiler as the control parameter. Controlling reboiler pressure is not recommended because the relationship between pressure and condensing temperature, and therefore between boilup and pressure, is nonlinear. Further, the relationship between boilup and pressure changes as the reboiler fouls and when the heat transfer coefficient varies.

When condensate flow rate is manipulated (Fig. 17.1b), vapor al-

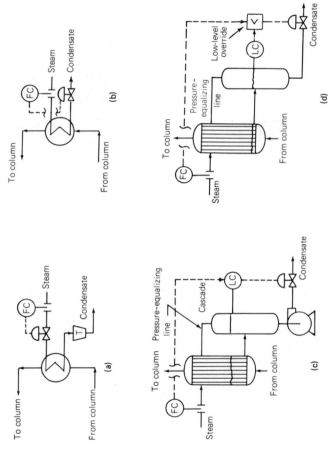

Figure 17.1 Reboiler steam control. (*a*) Inlet control, no condensate pot; (*b*) outlet control, no condensate pot; (*c*) outlet control via condensate pot level; (*d*) outlet control, with condensate pot level override.

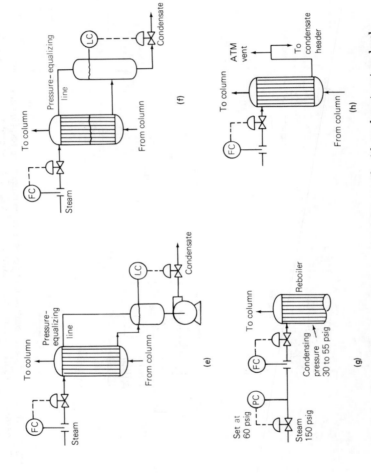

Figure 17.1 *(Continued)* Reboiler steam control. (*e*) inlet control, with condensate pot, no level in reboiler; (*f*) inlet control, with condensate pot, level kept in reboiler; (*g*) inlet control with a pressure regulator; (*h*) inlet control with a seal loop.

ways condenses at essentially the supply header pressure. Heat transfer rate is varied by partially flooding the exchanger with condensate, thereby varying the surface area available for condensation.

17.1.2   Vapor inlet valve versus condensate outlet valve

The location of the reboiler control valve often has a major bearing on the performance, operation, and heat efficiency of the entire column. In an excellent paper, Mathur (280) reviewed pros and cons and several control variations for each location. This review is supplemented and expanded below. The following considerations are important:

1. The dynamic response of the vapor inlet control scheme is far superior to that of the condensate outlet scheme. Manipulating the inlet valve immediately changes the vapor flow. Reboiler pressure and heat transfer rate only lag a few seconds behind. On the other hand, the condensate outlet valve has no direct effect on vapor flow. Condensate flow determines condensate level, and this level changes slowly. Reboiler vapor rate and heat transfer tread on the heels of the level change. Because of this slow response, manipulating vapor flow is a far better means of control than manipulating condensate flow.

   With the condensate outlet scheme, the response time varies with the actual condensate level in the exchanger. For instance, a considerable volume of condensate will need to be displaced to achieve a 10 percent increase in flow area if the level is near the exchanger bottom. A far smaller volume of condensate will need displacing to achieve a 10 percent increase in exposed area when the level is near the top. The response to a condensate flow change therefore speeds up as the condensate level becomes progressively higher. Where the response is overly fast, it may be unstable (280); where it is relatively slow, it may be sluggish.

2. The condensate outlet scheme is often troublesome. If the condensate valve cannot handle the amount of condensate that the reboiler can generate, a maximum vapor flow rate may be reached with condensate still covering a portion of the tubes. The remedy depends on the cause of the problem. If there is ample pressure difference between the reboiler and the condensate system downstream of the valve, it may be sufficient to resize the valve or condensate line. If the pressure difference is small, a condensate pot with a pump (Fig. 17.1c) may be needed to overcome the problem.

   The converse of this phenomenon can be just as troublesome with this control scheme. When the reboiler cannot condense vapor as fast as the condensate valve removes liquid, the liquid seal in the reboiler may be lost and vapor will pass into the condensate system. This is accompanied by a dramatic loss of heat transfer, and in the case of steam, also hammering in the condensate system. One seal loss incident in a reboiler condensing refrigerant vapor using the control system of Fig. 17.1b has been described (203). This problem can be overcome by the Fig. 17.1c scheme without the pump. Figure 17.1d shows an alternative solution; here the flow controller normally controls the condensate valve, with the level override cutting in whenever the level falls too low. The level override should not be set at a vertical height greater than the bottom of the reboiler; otherwise, some of the tubes will always be covered and the effective reboiler capacity will be reduced.

Sometimes, monitoring the liquid level in the reboiler shell and using it for level or level override control is satisfactory; in this case, the drum can be deleted.

3. The condensate outlet scheme can destabilize thermosiphon reboiler operation. The condensate-side level fluctuations inherent in this scheme are just as detrimental to thermosiphon reboiler stability as process-side level fluctuations (Section 15.3). This is most troublesome in vacuum services.

4. The condensate outlet scheme permits reboiler (condensate-side) operation at a higher pressure because it eliminates the pressure drop in the inlet control valve.

This is a major advantage when refrigerant vapor is the heating medium. Refrigeration compressor interstage pressures are often set to "ride" on the reboiler condensing pressure, or to ensure satisfactory condensation of refrigerant vapor in the reboiler. The higher this pressure is, the lower is the refrigerant compressor power consumption. When the reboiler heat load is large, lowering the interstage pressure by even a few psi can significantly raise compressor power consumption. Because of this, the condensate outlet control scheme is usually preferred in refrigerated reboilers, despite the disadvantages discussed in items 1 to 3 above.

The same argument may also apply to steam-heated reboilers, but here the steam supply pressures are usually set to suit the boilers or turbines rather than the reboilers, and the pressure differences between adjacent supply pressures are generally high. For this reason, sufficient pressure difference is usually available between the steam pressure required at a given reboiler and that available in the next higher steam supply pressure, and the difference is utilized for regulating the inlet valve. Energy can be saved by going to the condensate outlet scheme only when eliminating the inlet valve pressure drop will permit stepping down of the reboiler steam supply pressure. Since this is infrequently the case, the vapor inlet scheme is usually preferred in steam-heated reboilers.

5. In steam-heated reboilers, the vapor inlet scheme minimizes the reboiler tube wall temperatures. This suppresses reboiler fouling (process side), and lowers thermal stresses at the reboiler heads. These thermal stresses often cause leakage at the channelhead to tubesheet gasket (234).

   If fouling or leakage is a serious problem, it is often desirable to keep the reboiler wall temperature as low as possible. The arrangement in Fig. 17.1e fully utilizes the reboiler area to automatically minimize the condensation (and, therefore, tube wall) temperature. The pump permits the reboiler pressure at the lowest desired condensation temperature to approach, or even fall below, the condensate header pressure. Both the reboiler and pump may need to be designed for vacuum.

6. Corrosion due to the condensate level maintained in the reboiler often occurs with the condensate outlet scheme. In one case (239), a rust layer on the steam side of the reboiler showed the level at which steam condensate usually ran.

7. A smaller control valve is required with the condensate outlet scheme.

8. The vapor inlet scheme may be troublesome when excess surface is available at the reboiler (e.g., when the reboiler is over- designed, or during initial operation when the tubes are clean). In either case, the inlet control valve closes in an attempt to reduce condensing temperature ($Q = UA \, \Delta T_{lm}$; UA is large; where Q is heat duty, Btu/h; U is the overall heat transfer coefficient, Btu/(h · ft^2 · °F); A is reboiler area, ft^2; ΔT_{lm} is log mean temperature difference, °F). By closing, it also lowers the condensing pressure. If the condensing pressure falls below the condensate header pressure, it will be impossible for the condensate to be removed. The unremoved condensate will back into the reboiler and flood some of the tube surface until a new equilibrium is reached. The point at which condensate will start backing up can be calculated using a procedure similar to that in Ref. 315a.

   With condensate tubes partially flooded, any further variations in steam flow to the reboiler will affect both the ΔT in the reboiler and the fraction of tube surface covered by condensate. These two often interact, giving rise to a sluggish, sometimes erratic response. The steam trap will offer little to assist with the control of the condensate level. Further, if the reboiler load changes are sudden, the above-mentioned equilibrium will be difficult to establish or sustain. Instead, cycling may develop, with the control valve hunting as exchanger surface is covered and uncovered (20, 234, 280). One incident of such cycling has been described (239). This cycling causes swings in the reboiler duty and column vapor rate, as well as backflow from the condensate header, noise, hammering, and mechanical damage (280).

To overcome this problem, a submerged condensate pot is often installed instead of the steam trap (Fig. 17.1e) as described earlier (item 5 above). An alternative remedy is replacing the steam trap by a level condensate pot (Fig. 17.1f). By varying the level control set point, the surface in the reboiler can be adjusted so that the reboiler operates at a pressure high enough to ensure condensate removal at all times without a pump. Note that the bottom of this drum is located below the bottom of the condensing side of the reboiler (189); otherwise, "dry" reboiler operation at high rates will not be possible, and reboiler capacity will be reduced.

It is important to properly design and operate the condensate pot. In one case history (351a), a column preheater equipped with a steam inlet control scheme and with a condensate pot (no pump) experienced condensate removal problems upon turndown. It was not stated whether the Fig. 17.1e or f arrangement was used. Arrangement 17.1e needs the pump for avoiding this type of problem. Arrangement 17.1f needs a sufficiently tall condensate pot (Sec. 17.1.3) and adequate operation of the level controller at turned-down rates in order to avoid this problem. The author suspects that in this case (351a), one of these needs was not fulfilled.

Successful application of the condensate pot technique above has been reported (155, 280). Sometimes, self-priming condensate pumps without condensate pots (155) or pumping traps (20) are installed in the Fig. 17.1e arrangement. A cheaper alternative to both, but one which suffers the disadvantages described in items 1 to 3, 5, 6, 12, and 13 is using the condensate outlet scheme.

9. The vapor inlet scheme may be troublesome when there is a small pressure difference between the reboiler heating medium and the condensate header (e.g., with steam reboilers using 15 to 35-psig steam). The problem is identical to that described in item 8 above, but it is caused by insufficient ΔP rather than oversizing or clean surface. In this case, the condensate outlet scheme of Fig. 17.1d is often preferred (234). The vapor inlet scheme shown in Fig. 17.1e can also be used and will provide better control, but at the expense of the additional pump.

10. Steam traps are considered generally troublesome because they are prone to plug or stick wide open. The use of a steam trap isconsidered a disadvantage of the Fig. 17.1a arrangement. The vapor inlet

schemes in Fig. 17.1*e* (without the pump) or 17.1*f* overcome this problem.

11. Tuning can be troublesome with the vapor inlet scheme if flow across the valve changes from noncritical to critical upon reboiler turndown (67, 68, 362). As boilup falls, so does the absolute pressure downstream of the valve. When the ratio of upstream to downstream pressure exceeds a critical value, critical flow is established through the valve, and the downstream pressure ceases to affect the vapor flow rate. The controller dynamics differ under critical and noncritical flow. A loop tuned for noncritical flow tends to be unstable when flow becomes critical, while a loop tuned for critical flow tends to be sluggish when flow becomes noncritical (67, 68).

To prevent this problem, it is best to design the system to operate over its normal range in one flow regime or another. A level condensate seal pot (Fig. 17.1*f*) can keep up the downstream pressure during turned-down conditions, thereby avoiding this problem. Alternatively, installing a pressure regulator upstream of the flow controller (Fig. 17.1*g*) can lower the pressure upstream of the inlet valve. Both techniques are also useful for minimizing valve erosion at high pressure drops. The author experienced one installation where the latter technique worked well.

12. With the condensate outlet scheme, condensate accumulation in horizontal shells at turned-down conditions can flood most of the exchanger baffle windows and restrict vapor passage through the window. This may result in liquid hammering (426); Sec. 15.9.

13. The vapor inlet scheme is more effective in reducing disturbances in the steam supply than the condensate outlet scheme (332).

It has been argued (234) that a level indication of the condensate in the reboiler shell is useful for the operator. The arrangements shown in Fig. 17.1*a* to *f* can be designed to provide this.

17.1.3 Condensate pots

Figure 17.1*c* to *f* features different condensate pot arrangements. Some of the main considerations in their design and operation are

1. A pressure-equalizing line must be provided. This is a small line (often 1 in diameter) connecting the top of the reboiler with the top of the condensate pot. Without this line, it will be impossible to maintain a steady pressure and level in the condensate pot.

2. Condensate drums may be vertical or horizontal.

3. Arrangements 17.1*c, d,* and *f* require a large surge volume in the condensate pot in order to prevent reboiler level variations from flooding or draining the pot. In arrangement 17.1*e* (with or with-

out a pump), a lower volume is adequate; here the main consideration is providing sufficient seal height to avoid vapor breakthrough. In this arrangement, liquid level is kept below the reboiler bottom.

4. Bertram (32) has discussed sizing condensate pots. The paper presents a sizing chart and a few examples, but no firm guidelines. Based on the author's interpretation of the information presented by Bertram, the following guidelines can be inferred:

   a. All arrangements (Fig. 17.1*c* to *f*)

   ■ The maximum recommended liquid velocity through a vertical drum is generally a function of the condensate valve size, as follows:

Condensate control valve size, in	Maximum liquid velocity, gpm/ft^2
1	25
1½–3	37
>4	50

   ■ Allow at least 6 ins above the upper-level nozzle and below the lower-level nozzle to mechanically accommodate the nozzles.

   ■ Allow a vapor space of 12 to 24 in between the highest liquid level expected and the upper level nozzle. This height serves as a safety margin and possibly for accommodating a high-level alarm.

   b. Large condensate pot (Fig. 17.1*c, d,* and *f*) only

   ■ Allow at least 12 in or 20 percent of total range of the level instrument (whichever is greater) between the bottom of the tubesheet and the lower nozzle.

   ■ Increase drum size to provide the required surge volume in the drum as per item 3 above.

   c. Small condensate pot (Fig. 17.1*e*) only

   ■ A height of 32 in is usually a satisfactory range for level control.

   ■ A drum about 4 ft high, with the midpoint about the same height as the bottom reboiler tubesheet, is often satisfactory (234).

   d. Level condensate pot (Fig. 17.1*f*) only (author's guideline)

   ■ Set the highest liquid level to match the expected liquid level in the reboiler under the most severe turndown conditions required. Keep in mind that these may occur under startup conditions, when the reboiler is clean and the heat transfer coefficient is high.

5. In steam reboilers, a small atmospheric vent should be provided on top of the condensate drum and always left cracked open (234) (except when steam chest pressure dips below atmospheric). This will

prevent noncondensable buildup, which may reduce reboiler efficiency; if CO_2 is one of the noncondensables, it can also cause corrosion. Similarly, when a heating medium other than steam is used, adequate condensate drum venting is required.

Loop seal. In some low-pressure steam reboilers the condensate pot is replaced by a loop seal (Fig. 17.1h). In this arrangement, increasing the flow to the reboiler raises the pressure in the reboiler shell, which in turn lowers the liquid level in the reboiler and exposes more tube area. The dynamics of this system are similar to that of the condensate outlet scheme. The height of liquid in the loop is typically 5 to 10 ft. The system can be troublesome when the reboiler heat load or the steam mains pressure tend to fluctuate, and it is usually best avoided.

17.1.4 Reboiling with sensible heat

Typical examples are reboiling with hot oil, hot gas, or by subcooling a refrigerant stream. Heat input is regulated by changing the heating medium flow through the reboiler, which in turn varies both the heat transfer coefficient and the ΔT across the reboiler. The reboiler response depends on whether the heat transfer coefficient variation or the ΔT variation is the dominant factor affecting heat input. Generally at high reboiler ΔT, the heat transfer coefficient effect dominates, while the ΔT effect becomes more dominant at low reboiler ΔT.

A change in heating fluid flow propagates rapidly through the reboiler, effecting an immediate, albeit nonlinear, change in the heat transfer coefficient of the heating fluid (which is usually the controlling coefficient). This results in a rapid but nonlinear response of reboiler heat input to a heating medium flow change.

The ΔT response to a heating medium flow change is slower and far more complex. Consider stepping up the heating medium flow. The flow change will propagate rapidly, and the heating medium outlet temperature will first rise, giving a higher ΔT and enhancing boilup. As boilup increases, more heat is extracted from the heating medium, and its outlet temperature falls, reducing reboiler ΔT and boilup. This in turn raises outlet temperature, and the process continues, oscillating back and forth until the reboiler reaches a new equilibrium. This may give rise to a slow, sluggish response. When ΔT effects dominate, the reboiler often needs detuning, as too rapid a response may generate unsteady boilup.

When ΔT effects dominate, the response can further be complicated when the heating medium is a circulating fluid. A change in reboiler heat duty may significantly affect the overall heat load on the circulating fluid system. Unless the system temperature is well-dampened or tightly controlled (this may not always be achievable), reboiler heat load changes may alter the reboiler inlet temperature. This will have a secondary effect on reboiler ΔT. If troublesome, this problem can be tackled by controller detuning or by implementing a Btu control system (below).

The above considerations generally apply regardless of the control scheme used. A discussion of the common control schemes used for reboiling with sensible heat follows.

Sensible-heat reboilers are most commonly controlled by the bypass scheme in Fig. 17.2a. This scheme is inexpensive and can almost always be implemented, but it suffers from sluggishness. An increase in reboiler duty closes the bypass valve, which initially raises the reboiler flow. This is accompanied by a rise in reboiler pressure drop, which routes some of the added flow back through the bypass. Eventually, the system will reach equilibrium, but only after some back-and-forth flow fluctuations. This flow sluggishness may compound the temperature sluggishness described earlier.

The above flow sluggishness can be eliminated using the three-way valve in Fig. 17.2b, as recommended by Shinskey (362). The three-way valve can be substituted by two separate but interlinked valves, one in the exchanger path and the other in the bypass so that one opens while the other closes. Alternatively (301, 332), the sluggishness can be prevented by controlling reboiler flow with a flow controller and the bypass flow with a differential pressure controller (Fig. 17.2c).

Unless the heating medium flow rate to the reboiler is constrained by requirements in the heating medium system, it can be varied directly (Fig. 17.2d), without the use of a bypass. Direct flow variation is preferred whenever possible, because the bypass is energy-wasteful due to mixing of the cold outlet fluid with the hot bypassing fluid. Further, the direct flow scheme is less expensive than the schemes in Fig. 17.2b and c and is not prone to the sluggishness of the scheme Fig. 17.1a.

A common problem with sensible heat reboilers is variation in heating medium inlet temperature. A heat input (Btu) controller (Fig. 7.2e) can alleviate this problem. Heat input control can be implemented either with a direct (Fig. 7.2e) or a bypass control system. The Btu controller compensates for changes in heating medium temperature. Heat input controllers are particularly useful in services where the temperature difference across the exchanger is small (ΔT effects dominate), and small variations in heating medium temperature significantly affect boilup.

The dynamic response of the Btu control may create instability. A step increase in inlet temperature will directly and linearly increase heat duty, but changes in flow and outlet temperature have a slower and more interactive effect as described earlier. Because of this problem, it has been recommended (362) to have the Btu controller cascade onto a flow controller rather than control the valve directly.

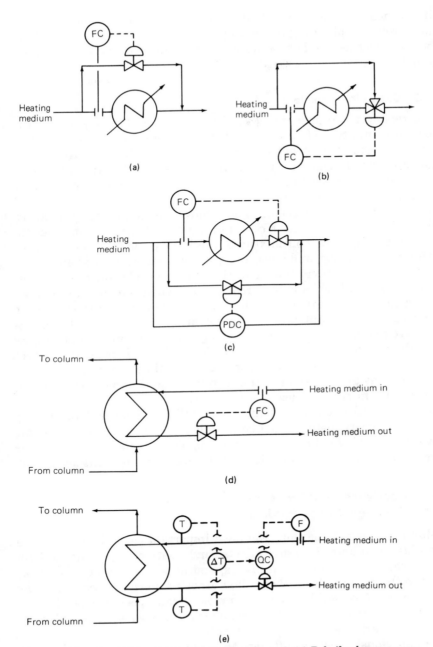

Figure 17.2 Reboiler sensible heating medium control. (a) Reboiler bypass, poor control; (b) reboiler bypass, good control; (c) reboiler bypass with PDC control; (d) direct flow control; (e) Btu control.

If the heating medium is fouling, the above control systems may lead to reboiler plugging or fouling during low-rate operation, when reboiler velocities are low. A system similar in principle to Fig. 17.7a can help boost velocities.

17.1.5   Direct-fired reboilers

Fired reboilers are principally used in refineries. Usually, these are fired with a mixture of gases vented from various units, supplemented by natural gas. A problem with direct-fired reboilers is variation of heat input due to changes in fuel gas composition. For this reason, it may be unsatisfactory to control the fuel flow rate to the furnace, and heat input control may be necessary.

Correction for composition can be achieved using the factor (heating value $\sqrt{\text{SG}}$), known as the *Wobbe index,* where SG is the gas specific gravity. For hydrocarbon mixtures, the Wobbe index varies linearly with specific gravity, and a specific gravity measurement alone is sufficient for composition correction. However, the relationship between gravity and the Wobbe index breaks down when components such as carbon monoxide, carbon dioxide, and nitrogen enter the system or when significant quantities of hydrogen are present. Thermal calorimeters are available for measuring fuel gas heating values, but in general, their response is too slow to be effectively used for heater control (362). In one typical case (370), the response time of the fuel gas calorimeter was greater than of the furnace itself; it was therefore useless for minimizing furnace disturbances.

A fired reboiler control system using the Wobbe index is shown in Fig. 17.3. The density meter should be installed directly in the fuel gas line, downstream of the knockout drum and any points where fuel gas streams are added, and where it would be unaffected by vibrations (370).

17.2   Condenser and Pressure Control—Single-Phase Product

Column pressure is perhaps the most important control variable in a distillation column. Pressure affects condensation, vaporization, temperatures, compositions, volatilities, and almost any process that occurs in the column. It has been the author's experience that a column will not achieve stable operation unless steady pressure can be sustained. Note that this does not preclude pressure variations, as long as these are performed slowly and steadily.

As previously described (Chap. 16), column pressure control is usually integrated with the condensation system. Pressure and condensation controls therefore need to be considered simultaneously. An ex-

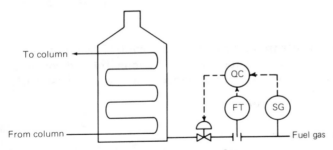

Figure 17.3 Fired reboiler heat input control.

cellent comprehensive treatment of this subject was presented in a paper by Chin (77). Other valuable reviews were presented by Buckley et al. (68), Boyd (58), Rademaker et al. (332), and Shinskey (362). The discussion below summarizes some of the highlights of these, and expands their treatment to include numerous other techniques and guidelines.

Some general guidelines for pressure and condensation control (77) are

1. Streams should enter the reflux drum at a velocity low enough to prevent disturbance to the liquid surface.

2. Vapor bypass piping and equalizing lines should be self-draining and should contain no low points where condensate can accumulate.

3. If the condensing temperature is high compared to ambient and the condensing range is narrow, condensation on the reflux drum walls may interfere with condensation or pressure control. Insulating the vapor space of the reflux drum should be considered.

4. Pressure-equalizing lines should be sized for negligible pressure drop. One designer (77) found that an equalizing line with at least 20 percent of the cross-sectional area of the column overhead vapor line is adequate.

Atmospheric columns. In atmospheric columns, pressure is controlled by having the column open to atmosphere, i.e., by variation of air flow in and vent gas out of the column through the column vent ("breathing"). If air ingress into the system is undesirable, an inerts purge is added at the vent, or the column is pressured to about 3 to 5 psig and operated as a pressure column.

When the column is open to the atmosphere, condensation rate can often be controlled by the level in the reflux drum or by temperature.

Techniques discussed in Sec. 17.3 are adaptable to columns open to atmosphere.

Control techniques. Pressure and condensation control techniques are classified into four categories: vapor flow variations, flooded condensers, coolant flow variations, and miscellaneous methods. These techniques are described below.

17.2.1 Vapor flow variations

This simple and direct method (Fig. 17.4a) is usually the best choice when the column has a vapor product (77). The controller directly manipulates vapor inventory and, therefore, column pressure. The equivalent method commonly used to control vacuum columns (Fig. 17.4b) has the pressure controller varying the quantity of spillback to the ejector suction. The spillback control method can also be applied to pressure columns where the vapor product is compressed.

With schemes 17.4a and b, accumulator level usually manipulates the rate of condensation. When the condenser is internal (Fig. 17.4c), condensation rate control is more difficult. If desired, a Btu controller (similar to Fig. 17.2e) can be used to maintain a constant condenser duty and reflux rate to the column.

Correct piping of the vapor product line is imperative to the success of the above control methods. In one troublesome case (203), pressure control using the Fig. 17.4a scheme was erratic because liquid accumulated in a low leg (Fig. 17.4d) backpressured the column and interfered with pressure control. The source of liquid was either liquid entrainment or atmospheric condensation or both. Elimination of the low leg solved the problem.

17.2.2 Flooded condensers

This method is most popular with total condensers generating liquid product. Some of the condenser surface is flooded with liquid at all times. The flow of condensate from the condenser is directly or indirectly manipulated to vary the flooded area. Little heat transfer takes place in the flooded regions, as heat is transferred by sensible heat exchange only (Sec. 15.10). To raise column pressure, the flow of condensate from the condenser is lessened. This increases the flooded area in the condenser and lowers the surface area exposed for vapor condensation. This in turn reduces condensation rate, thereby raising pressure.

Figure 17.5a illustrates a flooded condenser with the control valve located at the condenser outlet. The control valve required is small

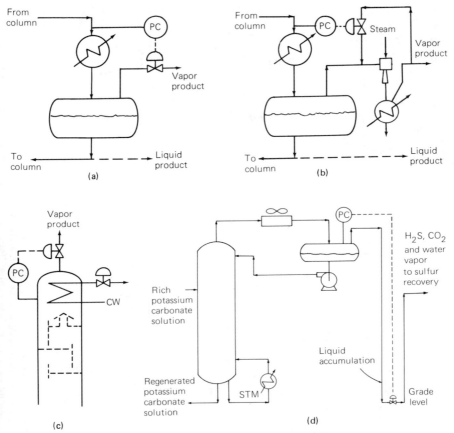

Figure 17.4 Pressure control by vapor rate variations. (a) Product rate variation, superatmospheric columns; (b) vapor spillback variation, vacuum columns; (c) product rate variation, with internal condenser; (d) product rate variation, poorly piped system. (*Part d from "Unusual Operating Histories of Gas Processing and Olefins Plant Columns," H. Z. Kister and T. C. Hower, Jr.,* Plant/Operations Progress, *vol. 6, no. 3, p. 153 (July 1987). Reproduced by permission of the American Institute of Chemical Engineers.*)

and should be located as close to the reflux drum as possible to maximize static head (77). This method is simple, linear, and maintains the same pressure in the column and the drum. It is therefore often favored (77).

The condenser outlet pipe may enter the drum either above or below the liquid level. Some designers (77) prefer entry above the liquid level so that drum level does not affect condenser level. Others (68) prefer entry below the liquid level so that cold liquid is introduced near the bottom of the drum. Introducing subcooled liquid above the liquid level may cause collapse of vapor when it meets the liquid and

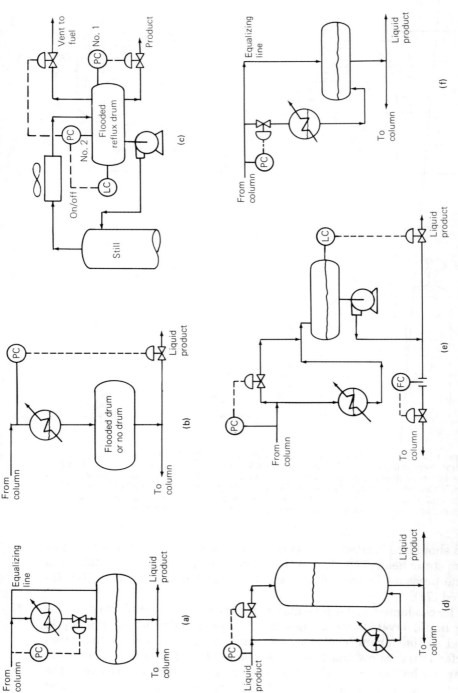

Figure 17.5 Pressure control by condenser flooding. (a) Control valve in condenser outlet; (b) flooded reflux drum; (c) flooded reflux drum with automatic noncondensables venting; (d) hot vapor bypass; (e) a poorly piped hot vapor bypass; (f) control valve in condenser inlet. (Part c from ..... "Operating Histories of Gas Processing and Olefins Plant Columns," H. Z. Kister and T.

result in pressure fluctuations. Extendingthe liquid pipe into the drum also has the advantage of providing a better seal to the condenser and avoiding vapor from blowing through the condenser as described Sec. 17.1.2, item 2.

This control method should not be used unless a pressure-equalizing line is included (77, 164). Without this line, pressure in the reflux accumulator will be unsteady. A smaller equalizing line is required when the subcooled liquid is introduced below the drum liquid level.

Figure 17.5b illustrates the flooded reflux drum method. Here the level control of the reflux drum is eliminated. The reflux drum runs full of liquid, and sometimes the reflux drum itself can be omitted. The pressure controller directly controls distillate flow. Due to the tight pressure control usually required, distillate flow controlled by this method is likely to fluctuate. These fluctuations may cause instability in downstream units. Distillate control by this method should be avoided (77, 301, 362) unless the product goes to storage.

This method is sometimes used to control reflux flow. This practice is generally not recommended. Here reflux flow rather than product flow is likely to fluctuate, and this can introduce instability into the column.

Accumulation of noncondensables in the drum may unflood the drum and interrupt the control action. If noncondensables accumulation is infrequent, manual venting from the top of the drum will be sufficient to restore satisfactory operation. If noncondensables accumulate frequently, or the column is run unattended, automatic venting is required.

Figure 17.5c illustrates an automatic vent system that worked well (203). A second pressure controller (PC No. 2), a level controller, and a control valve in the vent line are added. The set point of PC No. 2 is lower than that of the normal pressure controller (PC No. 1). When the drum is full, the level controller keeps PC No. 2 tripped off, and the vent valve is closed. Drum unflooding (due to noncondensable accumulation) is sensed by a drop in drum level. The lower level activates PC No. 2. Since the set point of PC No. 2 is lower than PC No. 1, it opens the vent valve. As the pressure falls, PC No. 1 closes, helping to build up the drum level. As soon as the drum refills, the level controller trips PC No. 2, and the vent valve closes.

Unless the product is subcooled and at a significantly higher pressure than storage, it is best to take the product to storage from downstream of the reflux pump (not as shown in Fig. 17.5c). If the product is taken directly from the drum (Fig. 17.5c), flashing may occur downstream of the control valve, or it may be difficult to get product into storage when storage pressure is high. Either may cause instability or back excessive liquid into the condenser, thereby reducing its capacity.

Figure 17.5d illustrates pressure control by a hot vapor bypass. The condenser is located below the liquid level in the drum (which may be horizontal or vertical). The condensate must be subcooled so that liquid surfaces in the drum are cooler than in the condenser. This causes a difference in vapor pressure large enough to transport condensate from the condenser into the drum. When column pressure rises, the pressure controller closes the valve. This reduces condensation at the drum surface and lowers the drum surface temperature, and, therefore, the drum vapor pressure. This enhances the vapor pressure difference between the condenser and the drum, which in turn forces more liquid out of the condenser and into the drum. This exposes additional condensing surface in the condenser and increases condensation rate.

The hot vapor bypass arrangement has the advantages of eliminating the need for a condenser support structure, easy access for maintenance, a small control valve, and fast response (423). These advantages can translate into handsome savings in steelwork, platforms, trolleys, and maintenance. These savings can be major in large installations, especially where a battery of condensers rather than a single exchanger is used.

The hot vapor bypass arrangement, however, has several disadvantages, and its operation can be troublesome (67, 68, 77, 258, 332, 362). Because of the liquid leg between the accumulator and the drum, this technique can suffer from interaction between the drum and the condenser liquid level (258, 301, 362) and from U-tube oscillations (258, 301). To minimize the interaction, the pressure controller should be tuned much tighter than the drum level controller (301, 362). There is an inverse response problem (67, 68); when pressure rises, it closes the bypass valve, which initially causes a further increase in pressure until the condenser level begins to fall. There is a reliance on subcooling in order to enable upward liquid movement and condensation of hot vapor against the drum surface (77). Operation may be troublesome if the drum liquid surface is agitated; a case of instability due to inadvertent agitation has been reported (77). Since the success of this technique is contingent on maintaining a small vapor pressure difference, the reflux drum vapor space needs to be insulated to minimize interference from rainstorms (77, 164). The above instabilities appear to be more pronounced with narrow boiling-range mixtures (77, 164), and at high pressures (164), where small temperature changes have a large effect on the split of overhead flow between the condenser and bypass. At the other extreme, the incidence of Rayleigh fractionation with wide-boiling mixtures (heavy components condensed out without mixing with the remaining mixture, see Sec. 15.10) can also interfere with this control system (423). The amount of subcooling and vapor bypass rates can only be determined empirically, and design is difficult. Simplified sizing procedures, however, are available (100, 164). Finally, leakage of vapor through the bypass valve at the closed position can substantially reduce condenser capacity as demonstrated in one reported case history (239). In summary, this method has some major advantages, but its operation can be troublesome.

Correct piping is mandatory for the success of the hot vapor bypass control method. Bypass vapor must enter the vapor space of the reflux drum (Fig. 17.5d). The bypass should be free of pockets where liquid can accumulate; any horizontal runs should drain into the reflux drum. If noncondensables are likely, vents are required on the condenser and drum. The condenser vent can be directed to the vapor space of the drum. Most important, liquid from the condenser must enter the reflux drum well below the liquid surface. The bottom of the drum (Fig. 17.5d) is the most suitable location. In one case (164) subcooled liquid entered the drum vapor space (presumably due to unflooding of the liquid inlet). The vapor space was 100°F hotter; rapid condensation sucked the liquid leg between the drum and condenser into the drum in seconds.

A poorly piped variation of this system (Fig. 17.5e) caused pressure fluctuations and inability to keep column pressure constant in one case (194). The author is familiar with two more troublesome cases with a similar piping arrangement, while a fourth similar case was reported by Hollander (164). With the Fig. 17.5e scheme, subcooled liquid mixes with dew point vapor. Collapse of vapor takes place at the point of mixing. The rate of vapor collapse varies with changes in subcooling, overhead temperature, and condensation rate. Variation of this collapse rate induces pressure fluctuations. The above problem (194) was completely eliminated by separating the liquid line from the vapor line, and extending the liquid line well below the liquid surface. The vapor line entered at the previous inlet.

Figure 17.5f shows a flooded condenser scheme similar to that of Fig. 17.5a, but with the control valve located at the condenser inlet. This method is inferior compared to Fig. 17.5a (77). It requires a larger control valve, is more difficult to understand, and it affects condensation at a lower temperature. The condenser outlet line must enter the reflux drum well below the liquid level. A pressure-equalizing line as in the method shown in Fig. 17.5a is also required.

In addition to the above limitations, this method is also prone to liquid hammering if the valve excessively closes. In one case (381), the valve closed fully under some startup conditions. Vapor downstream of the valve rapidly condensed, causing liquid to be rapidly drawn from the reflux drum, which in turn generated a liquid hammer that shook the whole unit. The problem was solved by changing the valve so that it would not fully close (381).

17.2.3 Coolant flow variations

Pressure can be controlled by adjusting the flow rate of coolant to the condenser. Figure 17.6a is an arrangement sometimes used with cooling water. The response may be slow or nonlinear in a manner analogous to a reboiler heated by a hot liquid stream (described earlier in Sec. 17.1.4). With cooling water, this arrangement can promote fouling, particularly during low-flow operation, when cooling water velocity is low and exit temperature is high. The author is familiar with many experiences of accelerated fouling in cooling water condensers using the Fig. 17.6a control scheme; water outlet temperatures were as high as 180 to 200°F. When distillate temperature is sufficiently high, it may even lead to boiling of water in the condenser (68; see Sec. 17.3), with the associated safety hazards. At the other extreme, very low velocities may also lead to freezing during winter. It is recommended (68, 77, 164) to keep water temperature below 120 to 130°F at the condenser outlet at all times, and to avoid cooling water velocities lower than 3 to 5 ft/s (58, 237).

This system can be made to work in several cases if designed carefully. The key point to consider is low-rate operation, and to ensure that under such conditions both velocity and cooling water outlet temperature are satisfactory. The following guidelines for using this technique have been recommended (68):

1. Provide an override control from the cooling water outlet temperature that will not allow it to rise excessively.

2. Provide a limiter, or even better, a flow override, that will prevent cooling water velocities from falling below the minimum recommended above.

3. If seasonal or daily cooling water temperature variations are sufficiently wide, use a small and a large valve in parallel (67, 68). Cooling water valves often have poor turndown and tend to open widely in summer and narrowly in winter, leading to flow instabilities. It is best to have these valves in a split-range arrangement so that the small one (which should be adequate for winter needs) opens first, then the large one.

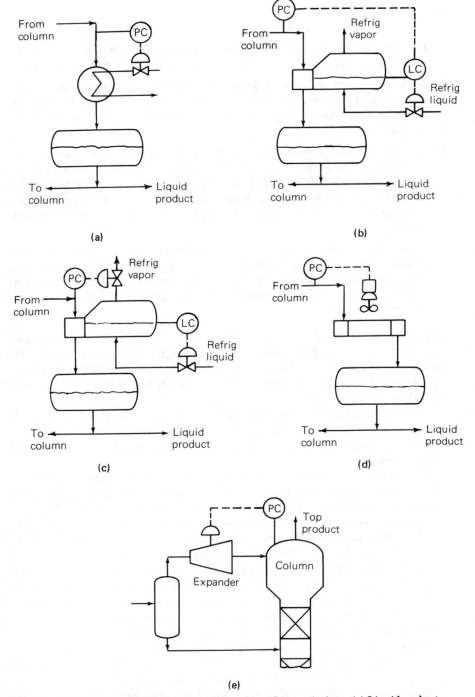

Figure 17.6 Pressure control by condensing medium flow variations. (a) Liquid coolant; (b) refrigeration coolant, level control; (c) refrigeration coolant, outlet flow control; (d) air coolant, fan motor speed or fan pitch variation; (e) turboexpander inlet guide vanes variation.

4. If keeping a high velocity is a problem, consider the system in Fig. 17.7a. This, however, may induce excessively high temperatures.

5. It is important to have at least some degree of subcooling, or the system will lack sensitivity. It has been recommended (67, 68) to have at least about 5 percent of the total heat load utilized for subcooling.

6. During low-rate operation, an inert gas can be injected into the process side of the condenser (173). This lowers the heat transfer coefficient. The controller will compensate by raising water flow into the condenser. The greater water flow will lessen the outlet temperature. Successful application of this technique has been reported (173).

7. Reversing condenser flows from countercurrent to cocurrent can lower condenser heat transfer. By similar action to item 6 above, this can induce lower condenser outlet temperatures and a greater water flow rate (173). However, this may permanently reduce condenser capacity.

Figure 17.6b is an arrangement often used with refrigerated condensers. The condenser is a kettle reboiler, boiling refrigerant liquid. To raise column pressure, the pressure controller sets the refrigerant liquid level at a lower set point. This exposes tube surface and reduces the flow of refrigerant to the exchanger.

The response of this arrangement is slow, as it has to work through varying the refrigerant liquid level, but it is usually satisfactory. Some designers (173) do not favor this arrangement, because of the slow response.

An alternative arrangement (173, 362) holds refrigerant level above the condensing tubes and controls the column pressure by manipulating a control valve in the refrigerant vapor line (Fig. 17.6c). This arrangement is prone to liquid carryover from the condenser, especially if the refrigerant side pressure suddenly drops. It also requires an additional control valve. The added control valve pressure drop may require exhausting the refrigerant vapor to a lower interstage pressure; in this case it will be energy-wasteful. Alternatively, if the vapor is exhausted to the same compressor interstage pressure, the added pressure drop will reduce condenser capacity. The control response of this technique has been reported to be excellent (173).

Figure 17.6d shows a coolant flow variation arrangement for air condensers. The controller varies fan speed or fan pitch to control pressure. This arrangement is energy-efficient and minimizes fan power consumption, but it also entails high maintenance and requires the use of a variable-pitch fan or a variable-speed motor. Fan pitch control also seems to feature a dead band or hysteresis (332, 362), which lessens the effectiveness of control actions. Control by a variable-speed motor is often preferred (362), but may be expensive.

Fan speed or fan pitch control generally works well with induced-draft fans, but is less satisfactory with forced-draft fans. In the latter case, precipitation tends to affect condensation to a far greater degree than variations in fan speed or pitch. Where precipitation severely interferes with control action, a roof some distance above the air cooler may be desirable.

A method more suitable for unroofed installations using forced-draft fans is controlling the pressure by manipulating the louvers. If the louvers are located above the condenser, they provide rain protection. With this control system, attention should be paid to the mechanical design of the motor-operated louvers (77). The louver control system also loses some of the energy efficiency advantage of the variable-pitch fan and variable-speed motor methods. In addition, louvers tend to be large, awkward, and easily damaged, and when a battery of condensers is used, also quite expensive (362).

If a battery of condensers is used, often only few of the condensers need to have louvers, fan pitch, or motor speed controls. The others are controlled by switching fans on or off when the modulating unit reaches the limits of its control range. The switching can be done automatically (362).

Figure 17.6e shows a turboexpander inlet guide vanes control, commonly used for controlling pressure of gas plant cryogenic demethanizers (149). Manipulating the guide vanes changes the expander speed and power and, therefore, the amount of condensation in the expander. This system usually gives rapid response and good control (149).

17.2.4 Miscellaneous pressure control methods

In this section, miscellaneous methods of pressure control are reviewed. Figure 17.7a shows a scheme where pressure is controlled by liquid recirculation. Either the inlet or the outlet coolant temperature can be controlled. Inlet temperature control is better if the coolant outlet temperature approaches the distillate temperature.

This scheme overcomes the low-velocity problem that cripples the Fig. 17.6a scheme and also greatly improves on its linearity and speed of response. Another advantage of this method is being able to recover heat at the highest possible level. In effect, it represents a "hot liquid belt" system and has a major energy-saving advantage when this heat can be utilized. An additional advantage of this method is that it max-

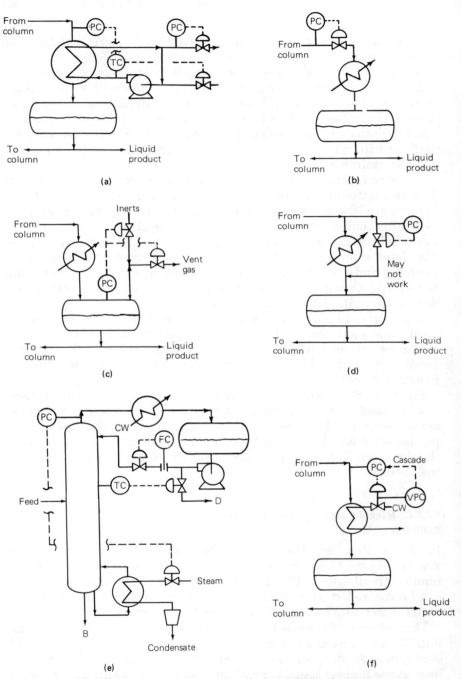

Figure 17.7 Miscellaneous pressure control methods. (a) Recirculation temperature control; (b) Column overhead control; (c) inerts pressure control; (d) drained condenser bypass; (e) pressure-reboil control; (f) floating pressure control, combined with coolant flow manipulation.

imizes the coolant temperature; this is important with high-freezing-point condensates.

A disadvantage of this system is the need to include a pump. In addition, its response is still somewhat slower than flooded condenser methods (77). If there is no advantage for recovering heat at a higher level with this system, it will consume more energy than other control systems. An analogous system can be used in air condensers by mixing recirculated air with fresh air. This method is often used in freezing climates (77).

Figure 17.7b shows a scheme where the pressure controller controls the column overhead stream. This system is wasteful of energy unless there is excessive temperature difference between the heating and condensing media. Even then it leads to a higher energy consumption in the column and at the reflux pump. The system requires that the reflux drum be open to a constant-pressure reservoir (e.g., atmosphere). The control valve required is large and may be impractical with large overhead lines (77). The system also requires a large condenser area and additional head at the reflux pump. Troublesome experiences with this system have been reported during winter and low-rate operation (164). The high pressure drops possible under these conditions resulted in sluggish control, loss of reflux pump suction, and even loss of control.

The advantages of the system are fast response and the ability to run the condenser and reflux drum at a lower pressure than the column. The latter may be a distinct advantage when the column is revamped to operate at higher pressure. Because of its fast response, at least one designer (258) feels that this is the best method when tight pressure control is required with total condensers. Others (e.g., 77, 164) do not favor this scheme because of its high costs and its potential problems during low-rate operation.

Figure 17.7c shows pressure control using inerts. As column pressure falls, and when the quantity of lights in the reflux drum is insufficient to permit continuous venting, an inert gas is admitted, usually on pressure control, to raise column pressure. Sometimes the inerts are admitted on flow control, and the drum is continuously vented on pressure control.

This method is costly and incurs a loss of inerts. If the inerts are recovered, it adds the cost of a recovery system. Sometimes, some of the heavy "inerts" may dissolve in the liquid and create problems downstream. The method has the advantage of fast response, but should generally be avoided because of its high cost.

Figure 17.7d shows a system which appeared to have worked in some instances (164, 332) but not in others (77, 164, 234). The author is also familiar with a mixed bag of experiences with this method. This method may appear similar to the hot condenser bypass, but here the condenser is drained and not flooded. It is hard to understand how this method works; one explanation (332) suggests that feeding some vapor below the condenser may restrict the downflow of condensate. This method is not recommended because it may not work (234, 77).

Figure 17.7e shows column pressure control by adjusting column boilup. This method is complex, but it has worked smoothly in some instances (234). Either a flooded or a nonflooded reflux drum can be used; in the latter case, reflux drum level can regulate the rate of condenstaion. Bottom flow is regulated by the bottom sump level. This method may be beneficial in some stripping columns receiving subcooled feeds, where feed temperature variations can affect column pressure to a larger extent than overhead condenser action.

One drawback of this system is that the distillate control valve may at times dump liquid out of the reflux drum faster than the reboiler can produce vapor to make up for this (234). Using a reflux drum low-level override which cuts back distillate or reflux flow was advocated (234) as a guard against draining the reflux drum. Another drawback of this system is a possible interaction between the pressure and temperature controllers.

Split range control. Often, a split-range pressure control is used. With this method, pressure may be normally controlled by venting a small vapor product stream. If the pressure falls and vent closure is insufficient to reinstate it, one of the other control methods automatically cuts in.

Floating pressure control. While tight pressure control is mandatory in distillation columns, it may be desirable to slowly manipulate the set point of the pressure controller. This is the essence of Shinskey's "floating pressure control" (362). The method minimizes column pressure without violating constraints such as maximum condenser or column capacity, or the ability to send column products to downstream units.

The prime purpose of floating pressure control is to reduce energy consumption. Generally, the lower the pressure, the easier the separation, and the smaller the reflux and boilup requirements. The method has the added advantage of reducing column relief frequency, because the pressure is kept at a greater margin below the relief pressure. Floating pressure is not always effective in saving energy; in many cases, the savings are marginal. Detailed discussion of the effectiveness and constraints of this method are available elsewhere (200, 362).

Shinskey (362) developed a system that enables automatic floating pressure control (Fig. 17.7*f*) for the common case where pressure reduction is limited by the available condenser capacity. The system features a valve position controller (VPC) that slowly manipulates the pressure controller set point. Suppose column pressure is lowered when the condenser control valve opens (e.g., Fig. 17.4*a, c;* 17.5*a, b;* 17.6*a.*) A valve opening greater than about 80 percent will indicate that the condenser operates beyond its desirable limit, and the position controller will slowly raise the pressure controller set point to keep the condenser "on control." On the other hand, a valve opening less than about 70 percent will signify some spare capacity is available at the condenser. The VPC will then slowly lower the pressure controller set point until the valve opening returns to within 70 to 80 percent. The VPC will thus minimize column operating pressure without allowing the pressure to go "off control." The VPC manipulation is tuned extremely loosely in order to prevent any interference with the minute-to-minute tight column pressure control.

The VPC can be readily incorporated in a conventional analogue control system. It is frequently also incorporated in advanced control systems.

17.3 Condenser and Pressure Control—Two-Phase Products

When it is desired to produce a liquid overhead product but total condensation is impractical because of the presence of inerts, the column produces an overheads inerts product stream in addition to the "normal" liquid product. The most suitable control method depends on the nature, quantity, and purity requirements of the products. The following situations can be distinguished.

1. *The quantity of inerts is small; losses of vaporized liquid in the gas are of little economic consequence:* This case coincides with venting small quantities of inerts from the condenser and the reflux drum. In this case, the control methods shown in Figs. 17.5*a–d* and *f;* 17.6; or 17.7*a, c* are suitable. The inerts as vented either manually or on flow control (except 17.7*c*).

2. *The quantity of inerts is significant or large; the inerts are either too light to significantly affect liquid product purity, or their nature and economics are such that it is desirable to maximize their content in the liquid:* In this case, it is almost always desirable to maximize liquid condensation in order to maximize recovery of liquid from the gas. A typical example is venting light gases such as methane from a column separating light organics. In this case, the control methods shown in Figs. 17.4*a–c* and 17.7*c* are suitable; the conden-

sation rate is usually set at a maximum, but can be varied manually.

3. *The vapor contains components that can condense out and are undesirable in the liquid in excessive quantities; vaporized liquid losses in the gas are of little economic consequence:* This case is identical to case 2 above, except that the condensation is not maximized but is set (usually manually) at a rate that will ensure that the undesirable components remain in the vapor. This case occurs when the product value of the gas is much the same as the liquid, or if the vaporized liquid is recovered from the gas in a downstream facility. As in case 2, the control systems in Figs. 17.4a–c and 17.7c are suitable.

4. *The vapor contains components that can condense out and are undesirable in the liquid in excessive quantities; vaporized liquid losses in the gas incur a significant economic penalty:* Here excessive condensation will render the liquid product off-spec on lights; insufficient condensation will cause too much liquid product to escape in the vapor stream, incurring an economic penalty. In this case, in addition to column pressure control, the rate of condensation must be controlled to obtain the desired vapor-liquid split. This case is perhaps the most common in the chemicals industry and is discussed in detail below.

Usually, the additional control for splitting the vapor and liquid products is the condensate temperature (or composition, if an analyzer is available). The temperature measurement point should be as close to the condenser as possible to avoid the dynamic lag associated with the reflux drum (which is commonly 5 to 10 minutes). It has been recommended (68) to locate the thermocouple in the liquid line just beneath the condenser and above the reflux drum. Figure 17.8a–d shows the common control schemes. It is frequently preferred to manipulate the condensation rate by varying the coolant rate.

If the coolant is cooling water, the schemes shown in Fig. 17.8a–d suffer from the problems described in Sec. 17.2.3 with reference to Fig. 17.6a. In one case (68), an atmospheric column controlled by a system similar to Fig. 17.8a (minus the pressure control, vent fully open to the atmosphere) experienced boiling of cooling water during low-rate operation. The volume of steam generated was too large to pass through the control valve, cooling water circulation ceased, and uncondensed product escaped to atmosphere. Fortunately, the column was quickly shut down before any damage was done. Techniques discussed earlier with reference to Fig. 17.6a also apply here. Similarly, the miscellaneous scheme shown in Fig. 17.7a can be effectively sub-

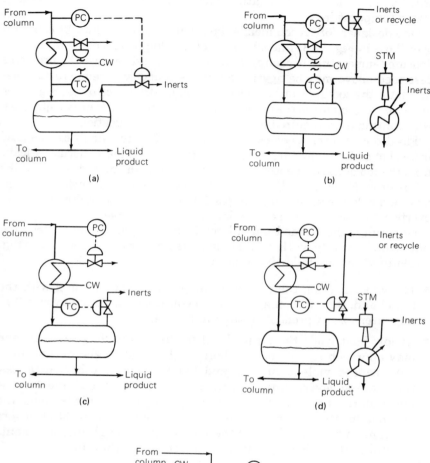

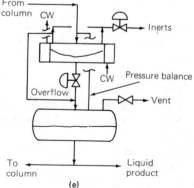

Figure 17.8 Condenser and pressure control, two-phase products. (*a*) PC on inerts stream, superatmospheric; (*b*) PC on inerts stream, vacuum; (*c*) PC on coolant, superatmospheric; (*d*) PC on coolant, vacuum; (*e*) flooded partial condenser arrangement.

stituted for coolant flow variation, but it may lead to excessive cooling water temperatures.

One designer (68) recommends the arrangements in Fig. 17.8c and d over those in Fig. 17.8a and b because of the sluggish response of the condensate temperature to cooling water changes, and because the problems described above may be aggravated if the condensate temperature manipulates the cooling water. However, a major drawback of the Fig. 17.8c and d arrangements is that they may not permit a sufficiently tight pressure control, as is required for most distillation applications.

Flooded condenser schemes shown in Fig. 17.5a to c and f can be used instead of coolant flow variations. In such cases, the inerts normally leave from the top of the condenser instead of the reflux drum (Fig. 17.8e). If the reflux drum is not flooded, a pressure balance line must be included; otherwise, a stable pressure will be impossible to keep in the reflux drum. An overflow line should also be included in this arrangement.

Generally, the flooded partial condenser scheme (Fig. 17.8e) compares unfavorably with the cooling water variation schemes (Fig. 17.8a to d) for the following reasons:

- The vapor leaving the condenser may not thoroughly mix with the liquid and, therefore, may not reach equilibrium. This may result in a greater loss of product in the inerts stream.

- At low heat loads the liquid level in the condenser runs high and may approach the top of the shell. High liquid levels are known to have caused violent surging and hammering; in other instances, they caused severe entrainment (68). Note that the liquid level is not uniform; because of pressure drop, the level is higher near the exit nozzles than near the inlet nozzle. In one incident (381), severe entrainment occurred under these conditions; rough measurements showed that the level gradient was feet rather than inches.

At the design stage, the high level problem is best avoided by providing adequate clearance between the top of the tubes and the shell. This, however, aggravates the product loss problem. During operation, the high level problem may be alleviated by injecting inerts into the condenser inlet. Injecting the inerts partially blankets the tubes and thereby lowers the liquid level. Monitoring the level in the condenser is recommended (68), and a high level override control can be hooked to the inerts injection valve.

Wild (426) described a horizontal steam reboiler case history which is analogous to the problem above. A high liquid level in the reboiler shell (Fig. 15.10f) caused liquid hammer. The remedy was to inject inerts when the level rose excessively.

Temperature and Composition Control

The function of composition control extends beyond assuring adequate product purity. The composition controller manipulates a stream such as reboil, reflux, or product. Unstable composition control will disturb this stream, and the disturbances will unsettle the column. Two primary methods are used for composition control: temperature control or analyzer control. Temperature control is cheaper, faster, and far more popular. Analyzers provide more direct and more accurate control of product composition and are favored where these qualities translate into profits. Both temperature and analyzer controls have serious limitations. Failure to recognize these limitations and plan for them will render composition control poor or unstable.

This chapter reviews the application of temperature and analyzer control in distillation columns, highlights the problem areas for each technique, examines the pitfalls of composition controls and the consequences of overlooking them, and provides guidelines for good composition control practices.

18.1 Temperature Control

Column temperature control is perhaps the most popular means of controlling product composition. The control temperature is used as a substitute to product composition analysis. A change in control temperature represents a corresponding variation in the concentration of key components in the product. For instance, a rise in top section control temperature represents a rise in the concentration of the heavy key component in the top product.

Temperature control is an easy and inexpensive means of controlling product composition. It uses a high-reliability, low-maintenance measuring element which suffers from little dynamic lag and down-time. These advantages make temperature control far more popular than analyzer control. The main drawbacks of temperature control are that the control temperature may not always correlate well with product composition or may not always be sufficiently sensitive to variations in product composition. The degree to which these drawbacks can be troublesome depends on

1. The location of the control temperature
2. Effect of concentration of nonkey components on the control temperature
3. Effect of pressure (and differential pressure) on the control temperature
4. The service, column design, control system, and sources and magnitude of disturbances

The first three factors, and some aspects of the fourth, are discussed below. Other aspects related to the fourth factor were discussed in Chap. 16.

18.2   Criteria for Locating the Control Temperature

Three major criteria are used for determining the location of the temperature control point (301, 332, 400): sensitivity, correlation with product(s), and dynamic response. Sensitivity is the most important; dynamic response is least important.

Sensitivity. Figure 18.1 shows typical temperature and composition profiles for binary distillation. Temperature is insensitive to composition below point A or above point B. Therefore, all trays above tray 8 and below tray 36 are unsuitable for temperature control from a sensitivity standpoint. Of the trays between trays 8 and 36, some are more sensitive to composition than others.

One case has been reported (237) where the control temperature was located at the reboiler outlet of a column producing a pure bottom product. At that point, the control temperature was insensitive to composition. This, together with the operator's natural reaction, promoted flooding. The problem was solved by switching the controller from automatic to manual.

In services where the difference between column top and column bottom temperature is small, it may be difficult to find any point where temperature is sufficiently sensitive to composition. Sometimes the

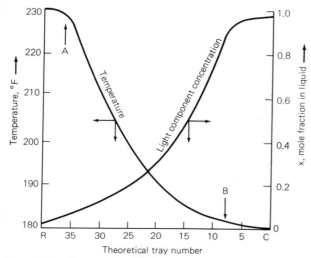

Figure 18.1 Temperature and composition profiles for a binary separation.

use of differential temperature control (Sec. 18.8) can help; at other times, analyzer control becomes necessary. Hatfield (151a, 151b) described a column separating a minimum-boiling toluene azeotrope from toluene. The temperature difference across the entire column was only 10°F, and the temperature throughout the column was insensitive to composition. The severe control problems experienced in that tower were at least party due to the insensitive temperature controller. Eliminating the temperature controller, among other changes, solved the problem.

In multicomponent separations, the sensitivity of temperature to the key components is important. Figure 18.2a and b shows composition and temperature profiles for a depropanizer separating propane (C_3) and lighter components from butane (C_4) and heavier components. The temperature is sensitive to the composition of the keys between tray 3 and tray 13. Below tray 3 and above tray 13 the temperature is more sensitive to the concentration of nonkeys than to the concentration of the keys. Trays 8 to 10 show some tendency toward retrograde distillation (recognized by the maximum in the C_4 concentration curve) and are best avoided. Moczek et al. (287) provide a detailed demonstration of the anomaly in temperature response in the retrograde distillation region. This leaves trays 3 to 7 and trays 11 to 13 as those suitable for temperature control.

If an intermediate component is present, the sensitivity of tray temperature to its concentration must also be taken into account (287,

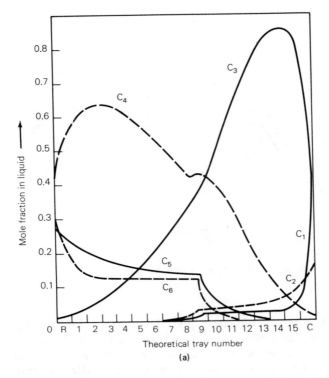

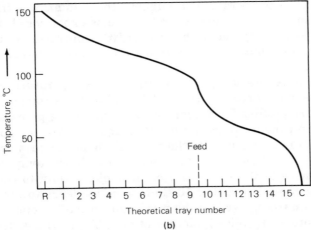

Figure 18.2 Depropanizer composition and temperature profiles. (*a*) Composition; (*b*) temperature. (*C. Judson King*, Separation Processes, 2d ed. Copyright © 1980 by McGraw-Hill, Inc. Reprinted by permission.)

332). The sensitivity of the control temperature to key components' concentration must also exceed its sensitivity to any expected pressure variations. This is most important under vacuum.

Correlation between product composition and tray composition. The closer the control temperature to the product, the better the correlation between the product composition and the tray composition. This consideration favors, in the binary example (Fig. 18.1), tray 8 for top product control and tray 36 for bottom product control. Similarly, trays 3 and 13 are favored by this consideration for the depropanizer example (Fig. 18.2). However, the correlation between product and tray composition may also be good for other trays.

It is important to realize that column temperature (or composition) is normally controlled in only one spot in the column (Chap. 16). If both product purities are equally important, the location sought should give the best possible correlation with both product compositions, not just with one. Rademaker et al. (332) proposed a steady-state computational procedure which surveys the combined effect of a disturbance and a corrective action on both product compositions. The corrective action varies with the control temperature location. Different control temperature locations are tried, and the location that generates the least undesirable overall variation to product purities is selected.

Dynamic response. The closer the control temperature to the manipulated stream, the faster the response of the temperature to a change in the manipulated stream. This is particularly true when the manipulated stream affects the liquid flow rate. Liquid moves down the column in a tray-by-tray manner, first increasing tray liquid holdup, and only then flowing over the weir to the tray below. Because of this lag, a controller located far below the reflux inlet will keep calling for more reflux for some time after a corrective action has been taken.

Another dynamic consideration is associated with the column feed. Generally, most disturbances enter the column at the feed. If the temperature controller is far from the feed, the disturbance may propagate a long way before the corrective action is taken. On the other hand, if the control tray is located near the feed, it will tend to take excessive corrective action and may be unstable. This consideration is most important when feed composition changes are frequent, and is detailed elsewhere (400).

Secondary criteria. Besides the above major criteria, Rademaker et al. (332) list several secondary criteria proposed in the literature. These include

1. *Linearity of response:* It is best to locate the control temperature where the steady-state responses to a positive and a negative change in the corrective action are least unequal. While this is not always achievable due to other considerations, locations where the response is highly unequal should be avoided (68). For instance, if the control temperature is too close to the bottom, an increase in boilup may only slightly raise the control temperature. On the other hand, an equal decrease in boilup can induce lights to reach the control point accompanied by a large drop in control temperature. When such a point is used for temperature control, controller tuning becomes extremely difficult (68).

2. *Maximum temperature or concentration gradient:* This point maximizes the sensitivity to product composition. While other considerations may override this, neither gradient should change sign in the neighborhood of the control tray (332). In Fig. 18.2a, the composition gradient of the key components changes signs at the maxima (stages 2, 3, 9, and 14). This consideration will eliminate trays 3, 4, and 13 from the list of remaining desirable control trays in the depropanizer example.

3. *Anticipatory action:* Locating the control temperature far from the top or bottom of the column can help the control system anticipate the disturbances traveling up or down the column. Note, however, that it may be just as bad to be too early as it is to be too late (332). This criterion is usually a minor consideration only (332).

4. *Minimizing composition changes inside the column:* This can improve the column dynamic response, but it is secondary to minimizing product composition changes.

18.3 Procedures for Finding the Best Temperature Control Location

Traditionally, the above analysis has been used for locating the control temperature. Accordingly, the preferred control trays will be tray 8 or 36 in the binary example above, and trays 5 or 12 in the multicomponent example. The final choice will move further away from the insensitive region to accommodate for uncertainty, so that trays 12 or 32 in the binary example and trays 6 or 11 in the multicomponent example will be selected.

Although the traditional procedure is often satisfactory as a first approximation, it fails to account for variations in pressure, nonkeys' concentrations, and column material balance. Unless the above anal ysis is repeated for several perturbations in these variables, it may yield a poorly located control temperature.

Tolliver and McCune (402, 403) proposed an improved procedure, based on the sensitivity of the column temperature profile to material balance variations. Accordingly, column temperature profiles are evaluated at different distillate-to-feed (D/F) ratios, while either reboiler duty or reflux flow is kept constant. The best temperature control point is where temperature variations are largest and most symmetrical (403). The analysis can be readily performed using steady-state computer simulation.

The D/F variations examined should coincide with those expected. They typically range from less than ± 0.1 percent for columns making high-purity splits to ± 5 percent for most common units (403). In most cases, it is sufficient to perform the calculation at the nominal D/F ratio and for one set of values above and below the base case (403).

Figure 18.3 shows examples of applying this procedure to benzene-toluene columns with different feed points and different feed compositions. Accordingly, trays 7, 10, and 5 or 10 are the best control trays in Fig. 18.3a, b, and c, respectively. Figure 18.4, based on the column in Fig. 18.3a, shows how a variation in control tray temperature affects product composition with a correctly located and an incorrectly located control tray. When the temperature variation is caused by a change of pressure or in the concentration of a nonkey component, it will produce a steady-state offset in product composition. A disturbance in the material or energy balance will cause a similar temperature variation until corrected by the control action; in this case, the offset will only be temporary. Figure 18.4 shows that the offset in either case is minimized when the control tray is selected in accordance with Tolliver and McCune's procedure (403). A dynamic analysis by these authors (403) indicated that the control tray thus selected tends to have the fastest, most linear dynamics.

Although Tolliver and McCune's procedure has a sound basis and was demonstrated to work well in some cases, the amount of experience reported with it is relatively small, particularly with multicomponent splits (e.g., Fig. 18.2). Contrary to the traditional criteria, this procedure can choose a stripping section control tray even when the distillate is more important, and vice versa (403). In this situation it is questionable whether that choice would actually be the best.

For these reasons, the author suggests caution with the use of Tolliver and McCune's procedure (403), at least until sufficient experience is gathered with its use. It is perhaps best to apply this method in conjunction with the traditional criteria, especially in

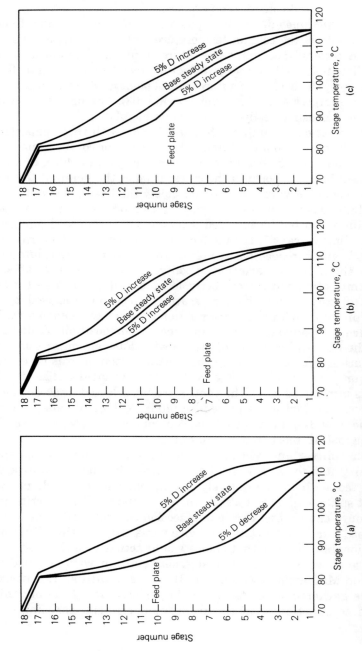

Figure 18.3 Application of Tolliver and McCune's procedure for finding the best temperature control location to benzene-toluene columns with different feedpoints and different feed compositions. (*a*) 65 mole % benzene feed to stage 10; (*b*) 15 mole % benzene feed to stage 7; (*c*) 35 mole % benzene feed to stage 9. (*Reprinted by permission. Copyright © Instrument Society of America 1980. From T. L. Tolliver and L. C. McCune, In Tech-September 1980.*)

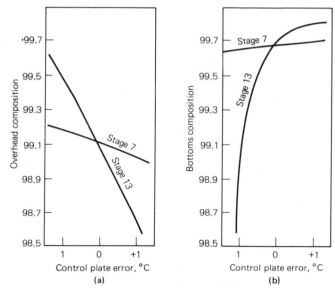

Figure 18.4 Effect of variation in control tray temperature on product composition with a correctly and an incorrectly located control tray. Same column as in Fig. 18.3a. (a) Overhead composition; (b) bottom composition. (*Reprinted by permission. Copyright © Instrument Society of America 1980. From T. L. Tolliver and L. C. McCune*, In Tech-September 1980.)

multicomponent distillation. This approach is discussed by Thurston (400), and has successfully been used by the author several times.

18.4   Temperature Control of Sharp Splits

Prior to the development of Tolliver and McCune's procedure (402, 403), Boyd (58, 59) applied an almost identical analysis to a benzene-toluene fractionator performing a sharp split. The results of this analysis are shown in Fig. 18.5. Figure 18.5a is analogous to Fig. 18.3; the only difference is that Boyd used product impurity as the parameter instead of D/F. Since the impurity level is related to the D/F ratio via the component balance equation, the two techniques are essentially identical.

Figure 18.5a illustrates an important consideration for sharp splits. A top section control tray (e.g., tray 15) will adequately control both top and bottom product compositions for plots 2 to 6. For plots 1 and 8 to 10, this control tray will retain the top product on-spec, but will permit wide variations of bottom product composition. Conversely, a control tray located in the bottom section (e.g., tray 30) will give good control of the bottom composition at all plots. It will also give good control

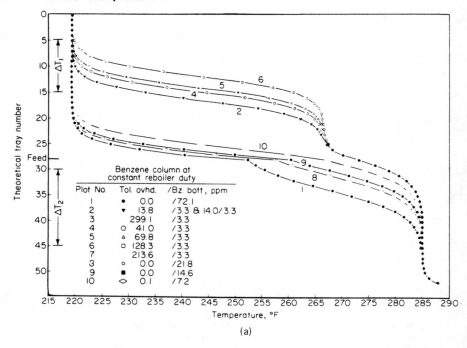

(a)

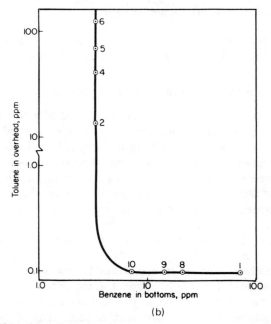

(b)

Figure 18.5 Relationships between temperature profile and product compositions in a benzene-toluene column performing a sharp split. (*a*) Temperature profile vs. tray number at various product purities. (*b*) Top purity vs. bottom purity at different temperature profiles. (*From "Fractionation Column Control," D. M. Boyd,* Chemical Engineering Progress, *vol 71, no. 6, p.55 (June 1975). Reproduced by permission of the American Institute of Chemical Engineers.*)

of the top composition for runs 1 and 8 to 10, but will permit wide variations of top product compositions for runs 2 to 6.

The reason for this behavior is illustrated in Fig. 18.5b. Because of the high product purity, the point of inflection is sharp. Plots 2 to 6 represent conditions where more of the heavy component enters the column than is withdrawn at the column bottom. The excess heavy component will ascend into the top section. The top section temperature will progressively climb as the heavies' migration intensifies. Due to the sharpness of the inflection point, there will be little lights in the bottom section once the "excess heavies" condition is established. The temperature of the bottom section will therefore remain constant throughout the excess heavies condition. The converse applies to plots 1 and 8 to 10; here less of the heavy component enters the column than is withdrawn at the bottom. The difference is made up by the light component. The quantity of lights migrating to the bottom will have little influence on top section temperatures, but will greatly lower the bottom section temperature.

In sharp splits such as that shown in Fig. 18.5, neither a top section temperature controller nor a bottom section temperature controller will be capable of adequately a controlling both product purities over the entire operating range. In most cases, one of the two products is selected as the more important, and the control tray is located in the section from which this product exits. The other product purity is allowed to vary. Alternatively, an average temperature control scheme can be used (58, 59, 68) and effectively overcome the problem. This is described in the next section.

The problem depicted above is not unique to systems similar to Boyd's. It is common when the top and bottom products have widely different boiling points while intermediate components are absent [e.g., separation of peanut butter from hydrogen, to use one author's extreme example (68)]. In such cases, separation is invariably sharp because of the wide boiling point difference. Average temperature control has successfully been implemented in such systems (68).

18.5 Average Temperature Control
Including double differential temperature control. For sharp splits and azeotropic distillation break points.

Sharp splits. Boyd (58, 59) developed a double differential temperature control scheme, which is essentially an average temperature control scheme that compensates for pressure and differential pressure variations. The scheme overcame the problem described in the previous section and was demonstrated to maintain tight control on both top and bottom purities. Luyben (262) proposed an average temperature control technique for sharp splits, and analytically demonstrated its superior performance compared to a single temperature control. Buckley et al. (68) report many successful industrial applications of controlling sharp splits by an average temperature calculated from four or five temperature sensors located around the normal temperature break.

Boyd's scheme (Fig. 18.6) utilizes a differential temperature measurement in the top section and a differential temperature measurement in the bottom section. The bottom section differential temperature is subtracted from the top section differential temperature (or vice versa), and the difference is used for temperature control.

It can be argued that the use of a double differential temperature is superfluous in many situations and that it would suffice to control by an average between a control temperature in the bottom section (e.g., tray 30) and the one in the top section (e.g., tray 15). Since the temperatures of trays 5 and 45 are practically unaltered in all of Boyd's plots (Fig. 18.5a), they can be treated as constant at a given pressure and differential pressure. Simple arithmetic will then show that con-

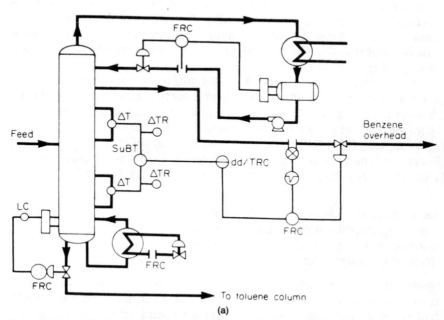

Figure 18.6 Boyd's double differential temperature control. (a) Control scheme. (*Part a from "Fractionation Column Control," D. M. Boyd,* Chemical Engineering Progress, *vol. 71, no. 6, p.55 (June 1975). Reproduced by permission of the American Institute of Chemical Engineers.*)

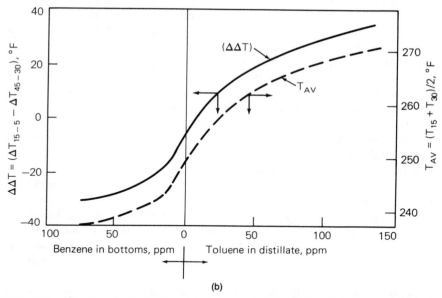

Figure 18.6 (*Continued*) (*b*) Relationship between double differential temperature and composition compared to relationship between average control temperature and composition (based on data in Fig. 18.5 *a*). (*Part b is based on Boyd's data from same reference as part a.*)

trolling the double differential temperature is equivalent to controlling the sum of tray 15 and tray 30 temperatures, or more conveniently, their average. Figure 18.6*b* shows that both alternatives give a similar variation with product composition.

In Boyd's example, both pressure and differential pressure variations had a significant effect on the control temperature (59). To compensate for these variations, it was necessary to use the double differential temperature (see Sec. 18.8). If pressure and differential pressure variations have little effect on the control temperatures, it would suffice to use the average between the tray 15 and tray 30 temperatures.

Boyd (58) emphasized that his method only applies for some high-purity separations, i.e., when product purities are in the parts per million range, but is not suitable when the purity is in the percent range. Bonilla (53) confirmed this, and demonstrated that this method was unsuitable in one case where the impurities were in the percent range.

Azeotropic distillation break points. In azeotropic distillation, a relatively sharp temperature break point often occurs. This break point is indicative of the transition from an aqueous environment to an organic environment; the entrainer usually persists in the column as far down as this break point. For control purposes, this break point is somewhat analogous to the break point of sharp splits. The main difference is that in sharp splits there is one point of inflection in the top section and one in the bottom section (Fig. 18.5a), while in azeotropic distillation only one point inflection may exist.

Bozenhardt (59a) reports an experience where a single tray temperature control was not sensitive to the break moving above or below it. This is analogous to the problem of controlling a sharp split from a single temperature point (Sec. 18.4). Using a two-temperature average below the feed gave a considerable improvement, but did not cure the problem. Increasing the number of averaged temperatures first to three, and then to five, gave little further improvement and ran into sensitivity and dynamic problems.

To solve the problem, Bozenhardt (59a) replaced the average temperature control by a "break point position control." This control acted to keep the break point between two desired trays. Temperatures at eight different bottom section trays were measured, and temperature differences calculated for each adjacent pair. The largest difference identified the break point interval. An integer index $(-3, -2, -1, 0, 1, 2, 3)$ was assigned to each interval, denoting its position relative to the desired break location. The controller acted to bring this index to zero. Consider temperature measurements on trays 9, 11, 13, 17, 21, 27, 33, and 39, with tray 39 being the uppermost, and a desired break point location between trays 17 and 21. If the largest temperature difference is between trays 17 and 21, the control signal will be zero and boilup rate (say) will remain unaltered. If the largest temperature difference moves to between trays 21 and 27, the control signal will become 1, and boilup will be reduced to lower the breakpoint. The control action will progressively intensify if the control signal moves to 2 and 3.

Bozenhardt (59a) experienced some cycling with his break point position control. He overcame this by introducing a mild and limited two-tray average temperature control within the desired interval.

18.6   Effect of Nonkeys on Temperature Control

When nonkey components enter the column, the light nonkeys rise to the top, while heavy nonkeys descend to the bottom. The concentrations of nonkeys undergo little stage-to-stage variation, except for the regions near the very top, near the very bottom, or near the feed (Fig. 18.2a). As long as the control temperature is kept out of these regions and the concentration of nonkeys in the column feed remains constant, temperature control will not be adversely affected by nonkeys.

Consider an increase in the content of heavy nonkeys in the feed.

Above the feed region, concentrations and temperatures will barely change, and a top section temperature controller will be unaffected. Below the feed region, the concentrations of heavy nonkeys will increase, and a bottom section control temperature will rise. The temperature controller will interpret the rise to signal a decline in the light key concentration, and will counteract it (e.g., by cutting reboil). This in turn will induce the light key to descend into the control location. Some of the light key will end in the bottom product.

In brief, an increase in heavy nonkeys' concentration will "fool" a bottom section temperature controller into letting more light keys out of the bottom, but will have little impact on a top section temperature controller. Conversely, an increase in light nonkeys in the feed will fool a top section temperature controller into letting more heavy non-keys out of the top, but will barely affect a bottom section temperature controller. In two different troublesome cases (239, 378), the top section temperature controller of an isobutane–normal butane splitter was frequently fooled into letting normal butane into the top product each time propane concentration in the column feed suddenly rose. In one of these (378), the problem was cured by using an analyzer/temperature control (see Sec. 18.3). The author is familiar with several similar experiences of temperature controller fooling.

The above fooling problem is most severe when the nonkey concentration in the section harboring the controller is high or changeable. A large boiling point difference between the relevant key and nonkey compared to the boiling point difference between the keys is also conducive to temperature controller fooling. For instance, if the isobutane–normal butane splitter above were operated at 60 psig, and the temperature were controlled where the mixture contained about 75 mole percent isobutane, a 1 mole percent change in key concentration will change the boiling point by 0.2°F. An identical temperature change would be effected by a 0.3 mole percent change in propane content at the same point.

Intermediate key components can also fool a temperature controller. These tend to build up in close to the feed, often in the neighborhood of the preferred temperature control location, and their accumulation rate and location are seldom predictable. Techniques for preventing intermediate component accumulation are discussed in Sec. 13.7.

Severe interference of nonkeys or intermediate keys may entirely preclude satisfactory temperature control without composition compensation. An excellent example has been described by Anderson and McMillan (13). Their column (Fig. 18.7) separated water and heavy ether from a mixed alcohol stream, the top product consisting of a volatile water-ether azeotrope and a volatile ether-alcohol azeotrope. Wa-

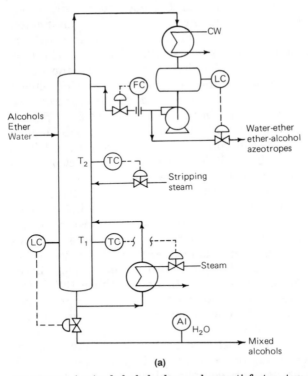

(a)

Figure 18.7   A mixed alcohol column where satisfactory temperature control could not be achieved. (*a*) Original control system; (*b*) modified control system. (*Based on J. S. Anderson and J. McMillan, I. Chem. E. Symp. Ser. 32, p. 6:7, London, 1969. Reprinted courtesy of the Institution of Chemical Engineers, UK.*)

ter was conducive to the formation of these azeotropes, and live steam was injected in the column to prevent drying out so that the azeotropes always formed. The control system (Fig. 18.7*a*) led to excessive losses of alcohol in the distillate. Their investigation revealed that near the bottom of the column, the temperature was insensitive to composition and therefore unsuitable for temperature control. Further up in the column, the temperature was more sensitive to the concentration of the heavy nonkeys than to the concentration of the keys. The problem was finally solved by replacing the temperature control by analyzer control (Fig. 18.7*b*). This lowered the alcohol losses by more than 80 percent. It also eliminated the need for steam injection, since sufficient water was always available in the feed to form the azeotropes, and the apparent water shortage was merely an outcome of the temperature control deficiency.

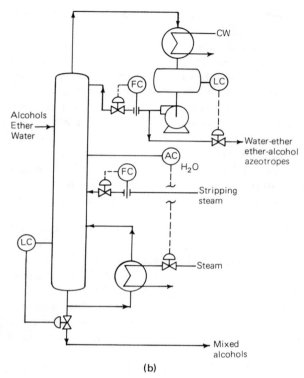

Alcohols
Ether →
Water

Water-ether
ether-alcohol
azeotropes

H_2O

Stripping
steam

Steam

Mixed
alcohols

(b)

Figure 18.7 (*Continued*)

18.7 Effects of Pressure on Temperature Control

Column pressure affects boiling points and, therefore, the control temperature. The temperature controller interprets a pressure change to signal a composition change, and is "fooled" by it. Consider a drop in column pressure. The control temperature will fall, and the temperature controller will counteract it (e.g., by raising boilup). This will induce heavies to ascend into the control location. Some of the heavies will end up in the distillate. Conversely, lights will be induced into the bottom product upon a pressure rise.

For adequate composition control, the control temperature must be relatively insensitive to the expected pressure changes. This condition prevails in most superatmospheric separations of components whose boiling points are not close. The tight pressure control practiced in most columns suppresses pressure variations, thereby assisting the temperature control. Under these circumstances, the effect of pressure on the control temperature is seldom troublesome.

In low-pressure (particularly vacuum) services, and in close separations, the control temperature is sensitive to pressure variations, and temperature control may become troublesome. As column pressure diminishes, its effect on the boiling point becomes more pronounced. In deep vacuum, even minor pressure changes can largely alter column temperatures. In close separations, temperature variation with composition is small, making even slight effects of pressure on temperature appear relatively large. For instance, in the isobutane–normal butane example given earlier, a 1 mole percent change in key component concentration will change the boiling point by 0.2°F; an identical temperature change would be affected by an 0.2-psi variation in pressure.

When the control temperature and the column control pressure are measured at two different locations, the control temperature may also be affected by changes in the differential pressure between the two locations. For instance, if column top pressure is controlled, and the control temperature is located near the bottom, a rise in differential pressure will increase the pressure at the temperature control location. The control temperature will rise, the controller will interpret the change as a rise in heavy key concentration, and will counteract it (e.g., by decreasing boilup).

Changes of differential pressure between the temperature control and pressure control locations are normally small compared to variations in column pressure. For this reason, the control temperature is seldom sensitive to differential pressure variations in other than low-pressure (particularly deep vacuum) services. Even then, the effects of differential pressure changes can be minimized by locating the temperature and pressure control measurement points close together.

The effect of pressure on the control temperature can be minimized by adequate selection of the temperature control location. Generally, all column temperatures have a similar sensitivity to pressure, but the sensitivity of temperature to composition varies widely from tray to tray. Therefore, locating the control temperature in a region highly sensitive to composition reduces its relative sensitivity to pressure changes (see Fig. 18.4).

When a well-located control temperature is excessively sensitive to pressure changes, pressure compensation is often added. The most common pressure compensation technique is differential temperature control (see Sec. 18.8). Alternatively, a control temperature can be linearly compensated for pressure changes.

The linear compensation technique calculates a compensated temperature T_c (°F) using the measured temperature, T (°F), the measured pressure, P (psia), and a compensation reference pressure, P_c

(psia). The relationship often used is (362)

$$T_c = T - a\,(P - P_c)$$

where a is a constant or an expression approximating the inverse slope of the vapor pressure curve (dT/dP, °F/psi) at the product composition. Unless a complex relationship is used for a, this technique has a low accuracy and is only correct for small variations (332). Detailed discussion and more accurate formulations suitable for computers are discussed elsewhere (68). A case history where pressure compensation by correlation contributed to performance improvement of temperature controllers has been reported (59a).

18.8 Differential Temperature Control

A differential temperature control (e.g., Fig. 18.8a) is in essence a pressure-compensated temperature control. The control tray temperature is measured in the usual manner. A second temperature is measured at a point where temperature is relatively insensitive to composition, such as near the bottom or the top of the column. The second temperature is subtracted from the first, giving a differential temperature measurement. This differential temperature is used for control. Since the second temperature varies little with composition, the differential temperature will reflect the composition variations measured by the first temperature. When column pressure changes, both temperatures change equally, but the temperature difference remains constant.

The success of the differential temperature control technique depends on finding a suitable second temperature. If none of the products is relatively pure, such a point may not be found (418), and this technique may be troublesome. Figure 18.8b shows the behavior of a differential temperature controller in the deisobutanizer (Fig. 18.8a) described by Webber (418). When the bottom product is relatively pure (to the left of the maximum in Fig. 18.8b), the controller functions normally. A fall in differential temperature signals depletion of lights in the bottom, the controller will reduce boilup, and both the differential temperature and the bottom composition will rise and return to their desired values. The same control action to the right of the maximum (Fig. 18.8b) will reduce the differential temperature, which in turn will further lower boilup, and so on, causing a "runaway" rise in the bottom lights content.

Finding a suitable second temperature is even more difficult when the product contains a significant or changeable amount of nonkeys. The concentrations of nonkeys tend to vary rapidly near the region where the product is relatively pure (e.g., stages 1, 2, 14, and 15 in Fig. 18.2a), causing the second temperature to change. Webber (418)

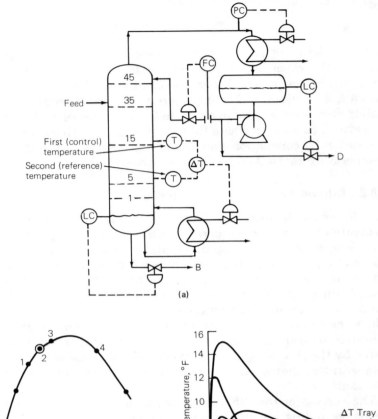

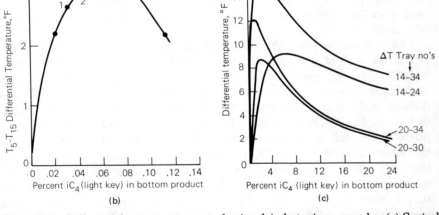

Figure 18.8 Differential temperature control using deisobutanizer examples. (a) Control system, Webber's case history (418); (b) ΔT vs. bottom impurity dependence, Webber's case history (418); (c) ΔT vs. bottom impurity behavior for different control temperature locations, Vermilion's case history (411). (Part b from W. O. Webber, Pet. Ref., May 1959.) Reproduced courtesy of Hydrocarbon Processing. (Part a based on W. O. Webber, above reference; part c based on W. L. Virmilion, Oil and Gas J., August 21, 1961. Reproduced courtesy of Oil and Gas Journal.)

could successfully avoid "bouncing" across the Fig. 18.8b maximum for as long as the bottom was free of nonkeys. When a second feed, containing heavy nonkeys, was later added to Webber's deisobutanizer, bouncing across the maximum made the differential temperature control inoperable (418). To minimize this problem, it is usually best to locate the second temperature a few trays below the top or above the bottom tray (e.g., Fig. 18.8a). In Webber's example (418), this was not enough to prevent the system from becoming unstable.

It should be noted that the above is only troublesome when none of the products is pure. Webber (418) recommended the differential temperature techniques only for "binary distillation with at least one product relatively pure" and spelled out more detailed guidelines in his article. The author has experienced several successful applications of this technique in essentially binary separations with pure products. Others (59, 301, 332) have reported unsteady operation of differential temperature when operating near the Fig. 18.8b maximum.

A good knowledge of the column temperature profile, coupled with a good choice of control trays, can sometimes make the differential temperature control technique work even in the presence of nonkeys and when neither product is pure. Vermilion (411) made this system work in another deisobutanizer, the feed to which contained a substantial fraction of both light and heavy nonkeys. The bottom sections contained 40 trays; Vermilion used tray 14 [from the bottom; this is similar to Webber (418)] as the first temperature control location, and tray 34 as the second. In that case, composition and temperature variations near the feed were relatively small (411). Figure 18.8c demonstrates the importance of correctly choosing the differential temperature measurement location.

In Vermilion's case, the light key content of the bottom product was about 5 percent, and operation was maintained to the right of the maximum (411); this is the reverse of Webber's case (418). During one period, at which the light key content was lowered to 2 percent, the same control system worked well after control valve action was reversed. Vermilion reported stable and accurate composition control with his scheme. "Bouncing across the maximum" only occurred at startups and in major upsets (411).

Pressure-drop changes between the two temperature measurement trays can induce a maximum similar to that of Figure 18.8b when the differential temperature is sensitive to pressure drop changes (59, 68). For instance, in the Fig. 18.8a system, a rise in differential temperature signals a rise of lights in the bottom and will increase boilup. The higher boilup will enhance the pressure drop between the trays, which will raise the differential temperature. If the differential pressure change affects the differential temperature to a larger degree than the composition change, the control action will be the same as that on the right-hand side of the Fig. 18.8*b* maximum. This is seldom troublesome in pressure distillation, it but may affect low-pressure (particularly deep vacuum) separations. If operation is to the right of the maximum [e.g., Vermilion's (411) scheme], a differential pressure change will produce an offset rather than bounce the differential temperature across the maximum, and will therefore be far less troublesome.

The differential temperature can be calculated between two trays in the same column section or between two trays in opposite column sections. The former technique has the two measurement trays closer together and is therefore less sensitive to differential pressure changes and to measurement lags. This makes it more suitable for low-pressure (particularly vacuum) towers. The former technique, however, has some difficulty coping with large feed composition changes (53, 418). In some separations of chemically similar components (53), locating the two trays in opposite sections has been shown to substantially improve the handling of feed composition fluctuations.

A differential temperature controller may also be sensitive to startup conditions or to abrupt feed changes. In one ethanol-water column equipped with a differential temperature controller across the top section, a low differential temperature was experienced (362) without any reflux at all or when the column was cold. The controller interpreted this as excessive product purity and cut reflux, which was not the desired control action. This behavior also destabilizes the control action following severe disturbances, as experienced by Shinskey (362) and others (411, 418).

Boyd (58, 59) applied a double differential temperature control to a column producing high-purity products. Although the main purpose of this system was to optimize the location of the control tray, it was also effective in compensating for both pressure and differential pressure variations. This system is described in detail in Sec. 18.5.

Luyben (261) analytically studied the application of a double differential temperature control to a deisobutanizer. His study indicated that this technique not only effectively compensates for pressure and differential pressure variation, but can also move the location of the maximum in Fig. 18.8*b* and *c* to a composition where it would not be troublesome. This, however, was achieved at the expense of having to control a very small differential temperature range (about 3°F). Others (53) also found the very small range to be a handicap with this technique.

18.9   Vapor Pressure Control

Vapor pressure control is normally used when satisfactory pressure compensation is difficult to achieve or when product vapor pressure is important. Successful applications of vapor pressure control in ethanol-water and acetic acid–water columns have been reported (361, 362).

A popular vapor pressure transmitter is Foxboro's differential vapor pressure (DVP) cell (Fig. 18.9). A bulb filled with a reference liquid is inserted in the column and is connected to one end of a differential pressure transmitter. The other end of the transmitter is connected directly to the column in the same elevation as the bulb. The reference liquid is selected so that it has the same vapor pressure as that on the tray, and is often a sample of the desired tray composition. The same vapor pressure in the bulb as on the tray signals a satisfactory tray composition. A rise in tray vapor pressure compared to the reference liquid signals an excessive presence of lights; a fall in tray vapor pressure compared to the reference liquid signals depletion of lights.

The vapor pressure controller has a far greater sensitivity to composition changes than a temperature controller (68, 332, 362), gives fast response, provides an accurate measurement in binary systems, and is relatively inexpensive. This technique is popular in services where pressure compensation of the control temperature is needed and is difficult to achieve, such as low-pressure (particularly vacuum) distillation and close separations. A typical example is ethanol-water columns (361). The technique is also useful where a good sensitivity of control temperature to composition and good correlation between product composition and control temperature are difficult to meet simultaneously (Sec. 18.2).

The DVP cell is generally unsuitable for controlling Reid vapor pressure (RVP) of petroleum products, because the RVP is measured at 100°F. This temperature is usually about 200°F lower than the column temperature, resulting in poor correlation between the tray va-

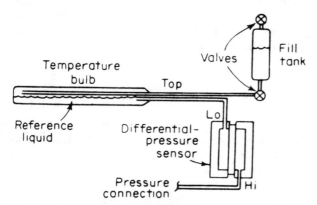

Figure 18.9 The Foxboro differential vapor pressure (DVP) cell (*From F. G. Shinskey*, Distillation Control, *second edition. Copyright © by McGraw-Hill, Inc. Reprinted by permission.*)

por pressure and the RVP. A different analyzer is usually used if the RVP is to be controlled (332). Although the instrument is primarily suitable for binary systems, it can also be used in some multicomponent separations, although with caution (68). Frequent calibration of the instrument against laboratory data, and adequate instrument location are important.

A DVP should not be inserted at the terminal point of a column, but a few trays away from it (362). This minimizes the impact of nonkey components (Fig. 18.2a), of subcooling, and of superheating. This location also improves the sensitivity of the measurement, since composition changes are greater at points further away from the products (Fig. 18.1a).

The reference liquid must be stable over a long period of time (68, 362). Any changes will affect its vapor pressure and, therefore, the measurement validity. Its vapor pressure–temperature dependence should be similar to that of the column liquid. The reference liquid is often a sample of column liquid, but this will be unsatisfactory if this sample reacts or degrades over a period of time. Filling the bulb is an art, and should avoid trapping air bubbles, as entrapped air will have a large effect on the vapor pressure. Leakage in or out of the bulb must be avoided.

Correct installation is essential and should take care of the following (362). The temperature bulb has a top and a bottom and must be installed correctly to ensure that the tube between the bulb and the cell is always filled with liquid. The connections to the column should be at the same elevation to avoid a difference in liquid head. The pressure connection should be short and direct to avoid trapping liquid.

The DVP responds to pressure changes much faster than to composition changes. This may cause erroneous readings and sometimes unsteady control (68). Snubbers are therefore often installed in the pressure impulse line (68). Finally, the DVP often has to be located well above the ground, making maintenance and troubleshooting inconvenient.

18.10 Analyzer Control

On-line composition is usually measured by gas chromatographs. Other analyzers include infrared and ultraviolet analyzers, mass spectrometers, boiling point analyzers, wet-chemical analyzers, flash point analyzers, and refractive index analyzers.

Analyzers have the advantage of directly measuring the product quality, but also have the drawbacks of high maintenance and slow dynamic response. They have a greater downtime than other instruments, and may be particularly troublesome when the stream analyzed is fouling or periodically contains impurities which interfere with the analyzer internals.

Sampling, sample conditioning, and transfer of a sample from the column to the analyzer are common weak spots in analyzer control systems, and frequently the cause of poor control. Plugging, two-phase sampling, reaction, and excessive transfer lags are common trouble-spots. A partial phase change must be avoided upstream of the analyzer, because fractionation will result, and the analyzer will not see the entire sample.

Analyzers are prone to large measurement lags which translate into response delays in a control system. The detrimental effect of these delays on analyzer control has been demonstrated (259, 309). The main sources of lags are process lags (e.g., the accumulator residence time if the sample is withdrawn downstream of the accumulator); *sample transfer lags* (the dead time in the line from the sampling point to the analyzer), and the *analyzer transfer lag* (the time it takes to transfer the components from the analyzer sample valves to the detector). Finally a chromatograph only performs an analysis at discrete time intervals, known as *sample intervals.* An additional lag occurs if the analyzer is shared among several streams. Shinskey (362) provides a detailed description of control lags when using gas chromatograph analyzers.

A typical overall lag is 10 to 20 minutes, and it is not unusual to find a 30-minute lag or higher. It has been recommended (258, 309, 332, 422) to restrict the sample transfer lag to less than 1 minute, and the sample interval to less than 2 minutes. In many situations, these targets are not readily achievable. Because of limitations associated with the electrical classification of the analyzer and the hazard rating of the area, and in order to ease maintenance, analyzers frequently need to be mounted in a protected shed, a fair distance from the column. The time it takes the samples to travel the distance can easily exceed 1 minute, even with high-velocity samples.

With analyzer controllers, it is essential to minimize the sample transfer lags. Vapor samples are preferred, since they can travel faster. A liquid sample is often vaporized upon withdrawal if sample lines are long. Heat tracing and insulation are usually required to keep the sample vaporized. To maintain high velocities, and at the same time avoid miniscule sample transfer lines that tend to break, the sample drawn is often much larger than the analyzer requires, with its unused portion returned to the process. This is considered an absolute must for unvaporized liquid samples (362).

Analyzer readings are often "smoothed" to prevent bumpy control or action resulting from a spurious jolt in the analyzer system. This smoothing imposes an additional delay to the analyzer control action.

Analyzers are expensive. Compared to a temperature controller, analyzers command higher capital, operating, and maintenance costs. Unless they give significantly better composition control, and the penalty for inferior control is high, the additional costs cannot be justified. This makes analyzer control mostly suitable either to large installations, where even small improvements in composition control enhance profitability, or to columns where a large differential exists between product values. Columns producing high-purity products, performing close separations, or having difficulty with temperature control are prime candidates for analyzer control (386).

The ability of analyzer control to improve product purity depends on the performance of the rest of the control system. In one case (259, 309), adding analyzer control to an unstable column experiencing frequent upsets in feed enthalpy did little to improve column control; in fact, the analyzer could be operated on automatic less than half the time. Once the instability was eliminated, the expected control improvement was achieved. In another case (378), an analyzer controller responded far worse to feed step changes than a temperature controller; it has been recommended (378) to maintain extremely stable feed flow and feed composition when using analyzer control. In a third case (309), however, an analyzer control system was demonstrated to tolerate a reasonable degree of feed fluctuations.

In summary, analyzer control has the advantage of directly controlling product purity, and in many cases, providing better composition control. However, analyzers tend to be troublesome; suffer from dynamic lags, high downtime, and low reliability; and are expensive. For these reasons, they are only used for composition control in applications where improved composition control is highly beneficial and the above limitations are not overly restrictive.

18.11   Sample Points for Top Product Analyzer Control

The sample point location affects the correlation between the composition measurement and the product composition. It also affects the measurement accuracy, speed of response, the nature of the sampling (sometimes also the analyzer), and the reliability of the system. The pros and cons of different sampling locations for column composition control are discussed below.

Recommendations for best sample location assume that the analyzer controls composition without cascading onto a temperature controller. In this case, the dynamics of the analyzer system is a prime composition control consideration. If the analyzer cascades onto the column temperature controller (Sec. 18.13), the dynamics of the analyzer system is a secondary factor. In this case, the considerations below still apply, but the best sample location shifts away from the location that gives best dynamic response to one which is least troublesome. The following sampling locations are used for controlling top product composition:

Downstream of the accumulator (usually at the discharge of the reflux pump). This is frequently the most convenient location. It has the advantages of directly sampling the product stream; minimizing the effects of vaporization, condensation, entrained vapor bubbles, or entrained liquid drops in the sample lines (the reflux or product are usually subcooled), frequently leading to the shortest sample lines; easiest maintenance and operation; and lowest cost. The main drawback of this location is the large time lag in the accumulator (Sec. 15.14). Because of this lag, the sample arriving at the analyzer is several minutes out of date. Other drawbacks are being prone to uneven mixing in the reflux drum and additional dead time in the sample line (because liquid rather than vapor is sampled). Sampling at this location often leads to slow and sluggish control because of these lags (332). These drawbacks are considered severe enough to offset the advantages of this location, and most authors recommend against sampling from this location (258, 301, 309, 332). A case where analyzer control at that point was unsatisfactory, and the analyzer needed to be reinstalled on the top tray has been reported (378), but in other cases (406), sampling at this point gave good control.

The overhead vapor line or top vapor space. Dynamically, this is the most desirable location. Measurement lags, both those between the column and the sampling point and those in the sample line (vapor sample) are minimized. However, this location also has several drawbacks.

Samples drawn at this location often suffer from nonreproducibility. Liquid droplets due to entrainment or atmospheric condensation may affect sample composition. The amount of liquid caught by the sample may vary, and this too affects the measurement. Large composition gradients may exist in the vapor phase above the top tray and persist in the overhead line. In one case (309), the heavy key concentration measured near the wall of the overhead line was half the concentration measured near the centerline, and the concentration gradient was unsteady. To mitigate the above problems it is best to sample from the top of a horizontal leg in the overhead line, and to provide a sample knockout pot for removing droplets, but this may not be sufficient to entirely eliminate the nonreproducibility problem.

Samples from the overhead line give an inferior correlation with product composition when the condenser is partial because of the additional separation stage at the reflux drum. This may not be troublesome if the noncondensables constitute only a small portion of the total product, but will become progressively more troublesome with larger noncondensable portions. Vacuum services, where inerts are always present, can be most sensitive to that problem.

Other drawbacks of this location include long sample lines; a need to keep the sample in the vapor phase (e.g., by using heat tracing); it is difficult to operate and maintain; and it is often expensive.

Despite all the above drawbacks, at least one designer (258) favors this location, even when the condenser is partial, because of its superior dynamic response. Others (309) recommend against this location. The author shares the latter view.

Liquid line from the condenser to the accumulator. This location is only feasible when the line is continually filled with condensate. This location gives a good dynamic response (unobstructed by the accumulator lag), and a representative sample (some vapor bubbles may be present, but because vapor density is much smaller than liquid, this has little effect on the analysis). Drawbacks of this location include an inaccurate correlation with product composition when the reflux drum is vented; fairly long sample lines; additional dead time in the sample line (because liquid and not vapor is sampled). This method evades the major drawbacks to the other two, and is frequently recommended (258, 301, 309, 332) whenever feasible. The author shares this view.

A top section tray. The main reason for using this location is to ease the requirements on the selectivity and the sensitivity of the analyzer. As one descends the column, the impurity level increases, and a less-sensitive or -selective analyzer can be employed.

This location gives good dynamic response but has several drawbacks. The correlation with product composition is poor, and often is reflux-dependent (406); sample lines are long; additional dead time in the sample line (if liquid is sampled); entrainment of liquid in vapor or vapor in liquid can be troublesome, and a sample knockout pot is usually required; it is difficult to operate and maintain; and it is expensive.

The control tray is best positioned close enough to the top to assure good correlation with the product composition, but at the same time far enough from the top to suit the analyzer sensitivity or selectivity requirements. About 5 to 15 trays from the top is typical in high-purity separations. The top tray should be avoided. Poor mixing on that tray may cause reflux to be sampled, and this location then becomes equivalent to downstream of the reflux drum. In one case (309), attempts to use the top tray failed for that reason. Because of these advantages, it is best to use this location only if analyzer sensitivity or selectivity is a major consideration. Successful experiences with this technique have been reported (378, 406, 422).

18.12   Sample Points for Bottom Product Analyzer Control

The general comments at the opening of the discussion on locating sample points for top product analyzer control also apply here and will not be repeated. The following sampling locations are used for controlling bottom product composition.

The bottom stream. This is usually the most convenient location, is analogous to sampling the top product downstream of the accumulator, and has similar advantages and disadvantages. Perhaps the main difference is that the bottom sump may impose a smaller time lag than the accumulator. The sample connection should be positioned at the top of a horizontal leg in the bottom product line in order to minimize solids entering the sampling system., As with its analogous counterpart, the drawbacks are considered severe enough to offset the advantages, and most authors recommend against sampling from this location (258, 309, 332). The author does not share this view. One experience where sampling from this location gave good control has been reported (406).

The bottom downcomer. Dynamically, this is the most desirable location, since it avoids the time lag in the bottom sump and gives faster response. Its major drawback is that the downcomer liquid may contain a large and variable quantity of entrained vapor, particularly in high-pressure, high-liquid-rate services, and this will affect the sample composition. Further, the quantity of vapor caught by the sample may vary. A sample knockout pot is helpful for easing this problem, but it may only provide a partial cure. A second major drawback is an inferior correlation with product composition, because the product is sampled upstream of the final separation stage (e.g., the reboiler). Other drawbacks include long sample lines and greater difficulty to maintain and operate. Despite these drawbacks, the fast dynamic response makes it a favorite, and it has been recommended by several authors (258, 309, 332). The author shares this view only in situations where is it unlikely to be troublesome.

The top product of the next column *(Fig. 18.10a)* . This is sometimes practiced in order to exclude oil, tars, dirt, or heavy components from the sample, analyzer, and sample conditioning system. This practice also gives a more direct measurement of product purity when the lights specification applies to the product leaving the next column. However, the time lags in this system can be enormous, making its dynamics disastrous. In one case (203), adding such an analyzer loop (Fig. 18.10*a*) made composition control worse than it was without the analyzer. The system response was tested by disconnecting the analyzer control and introducing a step change at the stripper temperature. The analyzer reading changed steadily over a 5-hour period following the change, and only then stabilized. The arrangement shown in Fig. 18.10*b* overcame that problem, giving good control while eliminating heavies from the sample. The little flashpot shown was fabricated from an old steam trap.

Unless the dynamics of the system are well known, and the lags in the system are acceptable, sampling from the top of the next column should be avoided. The sample flashpot technique (Fig. 18.10*b*) can be used to exclude the heavies from the sample before it enters the sample conditioning system.

On a bottom section tray. This is analogous to locating the top product sample point on a top section tray. The considerations are identical to that case. Successful experiences in this location have been reported (13, 422).

18.13 Analyzer/Temperature Control

Composition can be temperature-controlled, with an on-line analyzer adjusting the controller set point (Fig. 18.10). This approach merges the advantages of temperature and analyzer control while evading most of their drawbacks. While the principal control action is rapidly performed by the temperature controller, the analyzer slowly adjusts the temperature set point to prevent product purity from wandering off. In this capacity, a delayed analyzer response becomes tolerable, and its time lags become a secondary consideration. Further, should the analyzer become inoperative, the temperature controller will maintain automatic control of the column. The fast temperature controller action also renders this control method far less sensitive to upsets and step changes than an analyzer-only system. Analyzer/temperature control also permits placing the analyzer in a more convenient location (e.g., reflux pump discharge) without suffer-

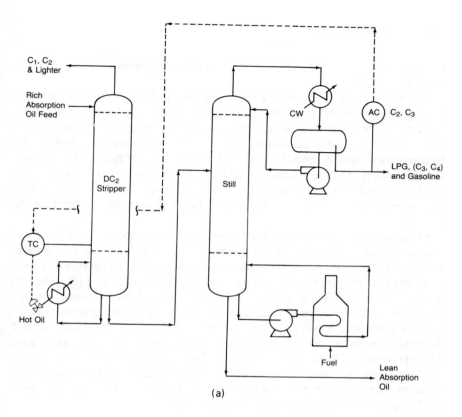

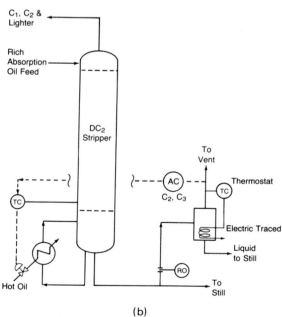

(b)

Figure 18.10 Analyzer located in the next column product line. (a) Original system that performed poorly; (b) modified system that worked well. (*From "Unusual Operating Histories of Gas Processing and Olefins Plant Columns," H. Z. Kister and T. C. Hower, Jr. Plant/Operation Progress, vol. 6, no. 3, p. 153 (July 1987). Reproduced by permission of the American Institute of Chemical Engineers.*)

ing from excessive dynamic lags, as has been demonstrated by field experience (378).

Fourroux et al. (122) found the same oscillation period when comparing the response of an analyzer/temperature control to that of an analyzer only control. This can be expected if the temperature controller is tuned for slow response, because the analyzer will adjust the temperature controller set point faster than the temperature controller can react. In the unusual case of slow temperature control, an analyzer/temperature control is therefore best avoided (406). The analyzer/temperature control relies on fast temperature control for its success.

Stanton and Bremer (378) demonstrated the superiority of the response of the analyzer/temperature control system over both a temperature control and a direct analyzer control in a 72-tray deisobutanizer. A temperature control alone produced a product purity offset due to variations in nonkeys; an analyzer-only control had a long lineout time and was sensitive to feed flow changes. An analyzer/temperature control gave a fast response and eliminated all these ill effects. Other favorable experiences with analyzer/temperature control have also been reported (25, 89, 203, 287, 379). Two references (25, 379) contain in-depth descriptions of tuning considerations and of other accessories that can improve system operation, particularly if the analyzer control is performed through a computer system.

Miscellaneous
Column Controls

While the major control considerations have been discussed in the previous three chapters, several loose ends remain. These loose ends pertain to controls that usually command less attention than others, such as reflux and level controls, and to those controls that are not applied in the majority of columns but are critical to some. The latter group includes side-stream drawoff controls, differential pressure control, and feed preheat control. Poor configuration of any of these loose ends can be just as troublesome as the poor practices described in relation to the major controls.

This chapter reviews considerations related to controlling these loose ends, critically examines the benefits and shortcomings of some common schemes used for their control, outlines pitfalls in their implementation, and supplies guidelines for avoiding these pitfalls.

19.1 Level Controls

Accumulator level. Accumulator level is usually controlled by manipulating liquid product flow, reflux flow, or condensation rate. The latter is often used when the column has a vapor product. It is normally unnecessary to tightly control accumulator level when the level manipulates product flow. In trains of columns, it is even undesirable (332), because the accumulator volume should be used to smooth product flow variations, and the only problem is to prevent the accumulator from overfilling or emptying out. However, when accumulator level controls reflux or condensation rate (schemes 16.4c and d) a tighter level control is desirable (332).

The accumulator level controller must be properly tuned, otherwise it may interact with composition or pressure control loops. Smoothest flow manipulation is accomplished using cascade control. Further discussion is available elsewhere (68, 301, 362).

Sump level.   Usually, the bottom level manipulates the bottom flow or boilup. If the bottom sump is unbaffled, it is commonly used as a liquid source for both the reboiler and bottom product. In this case, for certain reboiler types (e.g., thermosiphon), the level must be controlled within a narrow range in order to supply a constant head to the reboiler. Level fluctuations may lead to unsteady reboiler action and column upsets (Sec. 15.3). If a baffle is installed, tight level control is normally not required, particularly if bottom product goes to storage.

In many situations, a smooth bottom product flow is desired. Some means of achieving this have been discussed in guideline 11 in Sec. 16.6. For maximum smoothing, the bottom product should be flow-controlled, with the set point adjusted by a cascade level (or in scheme 16.4e, temperature) controller (68). Eliminating the bottom baffle can also reduce bottom flow fluctuations (258). A nonlinear level controller (below) can also help.

Nonlinear level controller.   When a level controls a product flow, disturbances from level fluctuations are transmitted to the downstream unit. The disturbances may be amplified if the pressure difference between the column and the downstream unit is small.

This situation can be improved by adding surge volume, but this remedy is expensive and often undesirable. A simpler cure is using a nonlinear level controller (301). The output of this controller is shown in Fig. 19.1. The nonlinear controller allows the level to fluctuate between points A and B without taking strong control action. This subdues flow fluctuations during most of the normal operating period.

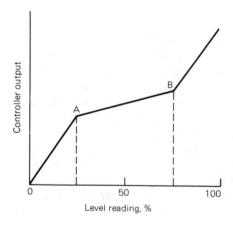

Figure 19.1 A nonlinear level controller designed to reduce manipulated stream flow fluctuations.

When the level goes outside range A-B, strong action is required to prevent the sump from emptying out or overfilling, and the higher gain is used.

19.2 Reflux Control

Internal reflux control. It was stated earlier (Sec. 16.6, guideline 7) that a major strength of scheme 16.4d is its ability to minimize the impact of disturbances in the cooling medium on the column. A similar capability can be incorporated into the alternative schemes (16.4a–c, e) by adding simple instrumentation to set up an internal reflux controller (Fig. 19.2). This controller automatically adjusts the reflux rate for changes in reflux subcooling.

Upon entry into a column, a subcooled reflux quickly heats up to its boiling point by condensing column vapor. The liquid downflow (or "internal reflux") in the top section therefore equals the external reflux plus that condensate. Frequently, the condensate makes up about 10 to 30 percent of the internal reflux flow, and it is not uncommon to find situations where it exceeds 40 percent of the internal reflux. Subcooling variations alter the quantity of condensate formed and are therefore reflected as changes in internal reflux. Inside the column, only internal reflux matters, and this will fluctuate even when the external reflux stream is flow-controlled.

In scheme 16.4d, the action of the level controller (which controls the reflux) eliminates these fluctuations. The internal reflux controller (IRC) achieves the same function by computing the column internal reflux and controlling it at a desired value. A simple, approximate correlation often used is (68)

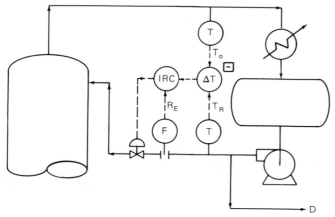

Figure 19.2 Internal reflux control.

$$R_I = R_E[1 + k(T_O - T_R)]$$

where R_E and R_I are the external and internal reflux, respectively, in consistent weight units; T_O and T_R are the overhead and subcooled reflux temperatures, respectively, in consistent units; and k is a constant, roughly equal to the liquid heat capacity divided by the latent heat (a more accurate value is obtained from a computer simulation). R_E, T_O, and T_R are measured (Fig. 19.2), and R_I is computed from the above equation. This computation can be readily done by conventional instrumentation; details are described elsewhere (68).

Consider a sudden rise in subcooling. T_R will drop and raise R_I. The IRC will close the valve to bring R_I to its set value. External reflux will be reduced, but the column will see no change, because the IRC will maintain the internal reflux unaltered. Similarly, an IRC can be incorporated in scheme 16.4b (where temperature controls reflux) by cascading the temperature to the IRC set point.

Experience of major control improvements due to IRC addition to many air-condensed columns has been reported (259). The benefits are not unique to air-condensed columns and can also be accomplished with other condensing media (e.g., cooling water).

Gravity reflux control cycling. Figure 19.3 shows an application of scheme 16.4d to a gravity reflux system. Level in the condenser liquid compartment (which is equivalent to the reflux drum) is controlled by

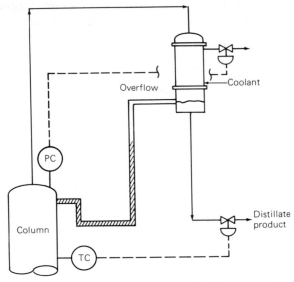

Figure 19.3 Application of scheme 16.4d to a gravity flow reflux system.

the overflow reflux line. The flow in the column overhead line is set by the pressure difference between the top of the column and the bottom of the condenser, and by the hydraulic resistance in this route. This system can experience severe oscillations and instability (68).

Consider a rise in liquid subcooling. Initially, this will lower the vapor pressure of the condensed liquid and, therefore, the pressure in the liquid compartment. The pressure difference between the top of the column and the liquid compartment will rise. This will enhance overhead vapor flow while retarding reflux flow.

The added overhead vapor flow from the column will condense and column pressure will fall. The pressure controller will cut back cooling. At the same time, the reflux deficiency will warm up the column. The temperature controller will cut back distillate flow.

The extra condensate and the liquid diverted by distillate cutback will now overflow into the reflux route. The greater condensing duty and the coolant cutback will lower subcooling and raise the pressure in the liquid compartment. This in turn will increase the driving force for reflux flow. The additional reflux, as well as that accumulated in the overflow line while the pressure drop was low, will now siphon into the column. The column will cool. At the same time, due to the higher pressure in the liquid compartment, the overhead vapor rate will decline. Column pressure will rise, and the pressure controller will increase cooling. Subcooling will increase, and the cycle will repeat.

Schemes 16.4a, b, and e do not suffer from the above instability. This is particularly true for schemes 16.4a and e, where reflux flow is maintained steady by a flow controller. In the Fig. 19.3 arrangement, these schemes can be configured with either the distillate or the reflux on overflow. If reflux is an overflow, the reflux flow controller manipulates a valve in the distillate line (e.g., Fig. 19.4a).

It has been recommended (68) to avoid scheme 16.4d in gravity reflux systems whenever possible. If it is still desired to use that scheme, there are numerous corrective actions that can subdue the above cycling. Buckley et al. (68) list several, including minimizing subcooling, using a high head and high hydraulic resistance in the reflux line, and minimizing pressure drop in the vapor line. Finally, the scheme in Fig. 19.4b was found effective in suppressing cycling and was therefore recommended (68). Additional details on the Fig. 19.4 schemes follow.

Sutro weir dividers.   A device often used for separating gravity reflux and distillate in continuous distillation systems is a Sutro weir divider. There are several variations, two of which (68) are shown in Fig. 19.4. It is suitable for small-scale systems and can eliminate the need for flow metering. Unlike the reflux splitter (below), it decouples

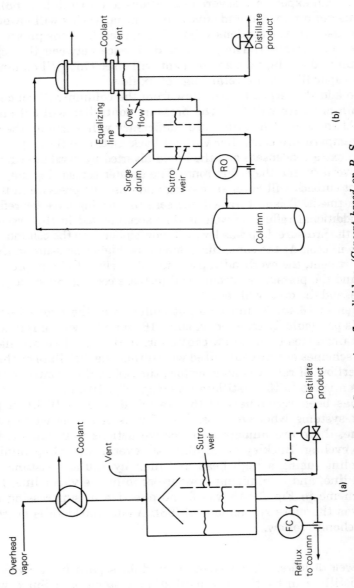

Figure 19.4 Common examples of Sutro weir reflux dividers. (*Concept based on P. S. Buckley, W. L. Luyben, and J. P. Shunta,"Design of Distillation Column Control Systems,"* Instrument Society of America, 1985.)

(a)

(b)

the reflux and distillate flows and is easy to incorporate into a MB control scheme. Buckley et al. (68) recommends the Fig. 19.4*b* arrangement when scheme 16.4*d* needs to be used (e.g., where metering reflux is undesirable). The surge drum dampens reflux variations due to "reflux cycles"; Buckley's experience was that arrangement 19.4*b* can reduce the cycling amplitude by a factor of 10 or more (68).

Reflux splitters. A reflux splitter (Fig. 19.5a) is an on/off solenoid-operated device actuated by a timer. The timer setting corresponds to a fixed reflux–distillate flow ratio. This setting is manipulated manually. The solenoid operates a slide gate that diverts all the liquid either to the column or to the product route. The solenoid operates at a high frequency so that neither reflux nor distillate flow is interrupted.

Reflux splitters are extensively used in batch distillation (44, 106), but they have also been used in small continuous distillation columns. They have a distinct cost advantage in small columns. Even more important, flow measurement in small reflux lines is difficult (106) because of the small orifice bore required. The problem is most acute in corrosive and fouling applications, where the orifice bore may change during service. Reflux splitters offer a low-cost alternative for controlling small columns.

A reflux splitter can be installed with an external gravity-reflux system (Fig. 19.5*b*), an internal condenser system (Fig. 19.5c), and a pumped reflux system (Fig. 19.5*d*). Venting near the top of the reflux splitter is important if vacuum creation is undesirable (130) or when noncondensables may be present. Figure 19.5a shows a vented reflux splitter construction (130).

A reflux splitter is difficult to incorporate in continuous distillation MB control schemes because it couples variations of the material balance stream (i.e., the distillate) with reflux rate changes. This is especially troublesome when reflux ratio is high, and a rise in distillate rate will effect a severalfold increase in reflux rate. The author experienced one case of poor continuous distillation control which was attributed to the above shortcoming. The author recommends restricting application of reflux splitters to batch distillation, where they work well. In continuous distillation, Sutro weir dividers are far more suitable.

19.3 Sidestream Drawoff Controls

In many multicomponent distillations, a single column yields three or more product streams (Fig. 19.6). Besides the top and bottom products, each additional product is usually withdrawn as a side stream from an intermediate location in the column. As a general rule of thumb, a side product from the rectifying section is taken as a liquid, in order to

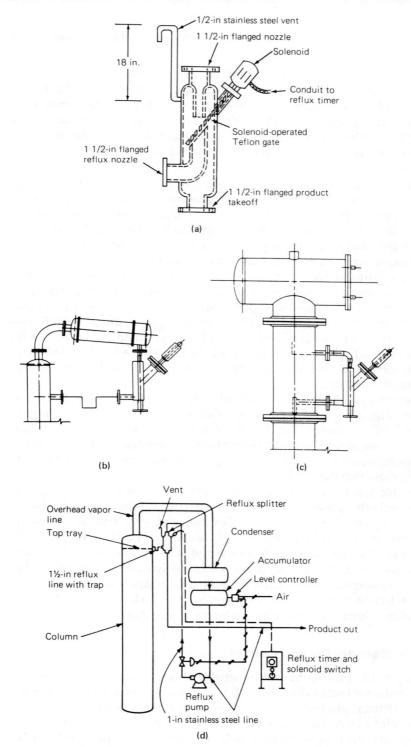

Figure 19.5 Reflux splitter arrangements. (a) Vented construction, atmospheric column; (b) with gravity reflux; (c) with internal reflux; (d) with pumped reflux, atmospheric column. (Parts a and d from Joseph F. Galluzzo, excerpted by special permission from Chemical Engineering, February 13, 1967; copyright © by McGraw-Hill, Inc., New York, N Y 10020; parts b and c reprinted courtesy of Nutter Engineering.)

minimize its lights contents, while a side product from the stripping section is taken as a vapor, in order to minimize its heavies contents (64, 68). A side-stream stripper or rectifier can be used to complete side-stream purification; side-stream strippers are extensively employed in refinery fractionators. In most applications (with the exception of refinery fractionators), the number of side products does not exceed one.

Side products complicate column control. Each side product adds two control variables (side product lights and heavies concentrations), but only one manipulated stream (side product flow) to the column (Sec. 16.2). Only one of these two compositions can be controlled by the additional manipulated stream; the other is either allowed to vary, or "rides" on the nearest end-product composition. Alternatively, both lights and heavies content of the side product can be controlled, at the expense of letting the composition of one of the end products vary. Side drawoffs also escalate the potential for interaction among control loops.

For these reasons, it is often difficult to devise a satisfactory MB control system for the column. Frequently, one or more of the products is flow-controlled (or kept at a fixed ratio to the feed or to another product), and the flow (or ratio) is manually adjusted by the operator to achieve the desired purity.

Control scheme selection for side-drawoff columns is governed by the relative quantity and economic value of each product stream. Generalization is difficult and will not be attempted here. Some common considerations are described below; further discussion is available elsewhere (64, 68, 260, 263, 287, 332, 362). Luyben's articles (260, 263) are particularly illuminating, and were extensively used as a basis for the following discussion.

19.3.1 Two prominent products

In many side-drawoff columns, the side draw is very small compared to the other product streams. This is a common situation when the side stream serves to remove an intermediate impurity (Sec. 13.7). In many other side-drawoff columns, distillate is withdrawn as a side product a few trays below the top of the column, leaving an upper "pasteurizing" section for separating light ends as a small vent stream. Common examples are an ethylene plant C_2 splitter, where small quantities of methane are "pasteurized" out of the ethylene product, and an alcohol still, where "heads" are pasteurized out of the alcohol product. The reverse situation has the bottom product drawn as a vapor side product a few trays above the bottom, leaving a small heavy end stream to exit from the bottom.

In all these situations, the two prominent products are MB-controlled using one of the normal schemes (Fig. 16.4), as if the small stream does not exist. There is little incentive to tightly control the composition of the small stream, and it is often assigned a non-MB control. The small stream may be withdrawn on flow control, flow-to-feed-ratio control, or flow-to-main-product-ratio control (Fig. 19.6a–c). A generous flow or ratio setting is usually fixed as a means of positively preventing impurity accumulation. If this leads to excessive product losses, the small flow can be manipulated by a temperature (or composition) controller in the pasteurizing section. For simplicity, Fig. 19.6 shows the small stream to be on flow control, but the discussion below also applies when the small steam is temperature- or ratio-controlled as described above.

When the small stream is the side product (Fig. 19.6a), any of the Fig. 16.4 MB control schemes can be adopted with no additional limitation. On the other hand, the configurations in Fig. 19.6b and c may be less compatible with some of the Fig. 16.4 schemes than with others. This is discussed below.

19.3.2   MB Control with the Fig. 19.6b Configuration

The Fig. 19.6b configuration is fully compatible only with scheme 16.4d. With schemes 16.4a, b, and e, distillate flow is manipulated to control accumulator level. When combined with configuration 19.6b, the level control loop of schemes 16.4 a, b, or e becomes "nested" in the temperature loop, i.e., the accumulator level controller can only control level if the temperature control loop is working. Consider a drop in accumulator level. The level controller will cut back the sidedraw flow. This will have no direct effect on the accumulator level, and the level will continue to drop. Cutting the distillate draw rate will increase the internal reflux to the column and cool it down. This will eventually be picked by the temperature controller, which will either increase boilup (schemes 16.4a and e) or cut reflux (scheme 16.4b). Accumulator level will have to wait for this last action to occur before it starts replenishing. Since the changes involved can be slow, particularly in large columns, level control may be poor, and long-period cycles may develop. Considerable surge capacity at the reflux accumulator will be required to accommodate the slow control.

A related limitation impairing the compatibility of configuration 19.6b with schemes 16.4a, b, and e is the strong interaction between the accumulator level and column temperature control. Consider a rise in column temperature. The temperature controller will decrease boilup (schemes 16.4a and e) or increase reflux (scheme 16.4b) to com-

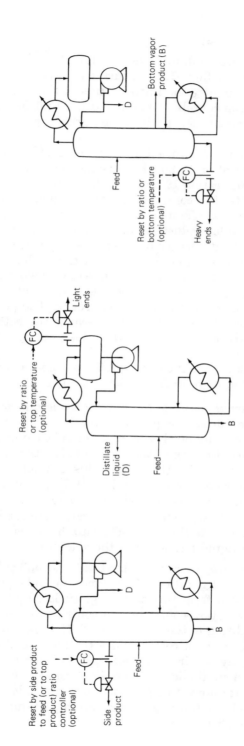

Figure 19.6 Sidedraw control configurations, two prominent products. (a) Small side product stream; (b) small light ends stream; (c) small heavy ends stream.

pensate. In either case, accumulator level will fall. The level controller will reduce side product flow. This in turn will raise the internal reflux rate, which will amplify the initial action of the temperature controller, but after a considerable time lag. If the control temperature stabilized in the meantime, it will be disturbed by the delayed amplification.

One case of a high-purity separation has been reported (287) where a combination of configuration 19.6b with scheme 16.4d was preferred to a combination of configuration 19.6b with scheme 16.4b. The former was believed to give better MB control of the side product, presumably due to the above limitations. When most of the reflux to the top of the column is drawn as the side product, the internal reflux to the section below the side draw is the difference between two large streams. Any change in one of the streams will effect a big fractional change in the internal reflux. Liquid traffic in the column section below the side draw (with any of the Fig. 16.4 control schemes) will swing widely; in some cases, drying up of trays or packing may result. The problem is most pronounced with schemes 16.4a and e, because the bottom section control temperature is more sensitive to feed changes than to changes in the small reflux rate.

Two techniques can alleviate the above problems and render configuration 19.6b compatible with schemes 16.4a, b, and e.

1. *Controlling the internal reflux to the section below the side draw:* Subtracting the measured side-product flow from the measured reflux flow (the latter may need correction for subcooling; see Sec. 19.2) gives the internal reflux to the section below the side draw. An internal reflux controller (IRC) uses this computed internal reflux to manipulate side-product flow (Fig. 19.7a). A limitation of this technique is that the internal reflux is calculated as a small difference between two large numbers, and can therefore be in error. The error escalates as the internal reflux becomes a smaller fraction of the total liquid traffic above the side draw.

2. *Using a total drawoff tray* (Sec. 4.9): This total drawoff replaces the partial drawoff in configuration 19.6b, and the side product is drawn on level control from this tray (Fig. 19.6d). Reflux to the tower section below is flow-controlled. In some cases, an external drum is used in addition to the total drawoff. The main limitation of this technique is its high cost. Besides the additional controls, this configuration may require additional column height in order to accommodate sufficient liquid head for flow control. Ample pipe length to diameter ratio must be provided to permit adequate flow measurement; with large flows, the horizontal pipe required can be quite long and difficult to support.

Both techniques eliminate the problem of internal reflux swings by maintaining a controlled liquid flow to the section below the side

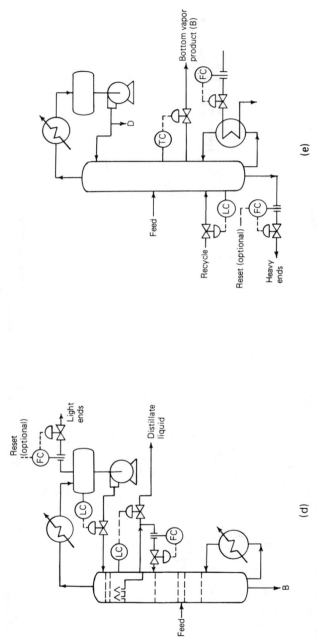

Figure 19.6 (*Continued*) Sidedraw control configurations, two prominent products. (*d*) Small light ends stream, greater control flexibility; (*e*) small heavy ends stream utilizing an external recycle to prevent control fluctuations.

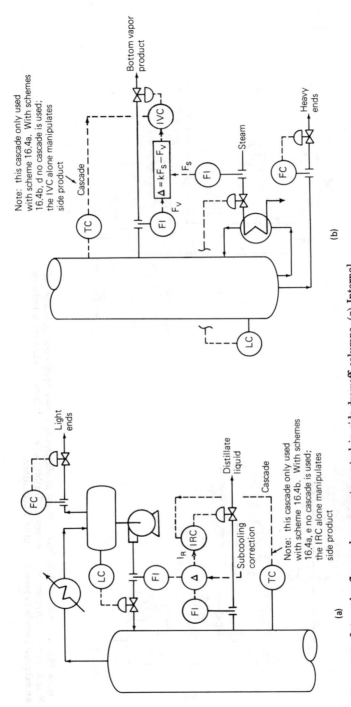

Figure 19.7 Internal reflux and vapor rate control in side-drawoff columns. (*a*) Internal reflux control, liquid side product; (*b*) internal vapor control, vapor side product.

Bottom vapor product

Note: this cascade only used with scheme 16.4a. With schemes 16.4b, d no cascade is used; the IVC alone manipulates side product

Cascade

IVC

$\Delta = kF_s - F_V$

F_V

F_s

FI

FI

Steam

TC

FC

Heavy ends

LC

(b)

Light ends

FC

Distillate liquid

Cascade

IRC

I_R

Subcooling correction

LC

FI

FI

Δ

TC

Note: this cascade only used with scheme 16.4b. With schemes 16.4a, e no cascade is used; the IRC alone manipulates side product

(a)

draw. Both techniques also eliminate the level nesting and level-temperature interaction problems that cripple the compatibility of configuration 19.6b with schemes 16.4a, b, and e. With schemes 16.4a and e, the liquid flow to the section below the side draw is on flow (or internal reflux) control; with scheme 16.4b, a temperature control is cascaded onto the flow (or internal reflux) controller. Both techniques preclude the use of scheme 16.4d, since they essentially put the side product on accumulator level control.

In summary, it is best to use configuration 19.6b with the desired MB control scheme as per Sec. 16.6. If the desired control scheme is 16.4a, b, or e, internal reflux control (Fig. 19.7a) should be incorporated. This internal reflux control cannot be used with scheme 16.4d.

When the internal reflux in the section below the side draw is low compared to the side product flow, scheme 16.4d should be avoided. The IRC in Fig. 19.7a may work well, it but will become increasingly troublesome as the internal reflux becomes a smaller fraction of the total liquid traffic above the side draw. In the latter case, it may pay to go to configuration 19.6d.

19.3.3   MB Control with the Fig. 19.6c Configuration

The Figure 19.6c configuration is fully compatible only with scheme 16.4e. With schemes 16.4a,b, and d, bottom product is manipulated to control bottom level. When combined with configuration 19.6c, the level control loop of schemes 16.4a, b, or d becomes "nested" in the temperature loop, i.e., the bottom level controller can only control level if the temperature control loop is working. This is analogous to the compatibility problem of configuration 19.6b with schemes 16.4a, b, and e. Also analogous to the previous compatibility problem is an interaction between the temperature controller and the bottom level controller when configuration 19.6c is joined with schemes 16.4a,b, or d.

In contrast to the above analogy, the nesting and interaction problems have only a slight adverse effect on the compatibility between scheme 16.4a and configuration 19.6c. This combination often works well as long as the temperature remains on control (260). This is because level control is achieved by manipulating column vapor, which rapidly responds to changes. Consider a drop in bottom level in a column using scheme 16.4a jointly with configuration 19.6c. The bottom level controller will reduce the vapor side product rate. More vapor will ascend up the column, and the control temperature will rise almost instantaneously. The temperature controller will lower boilup. Assuming the reboiler responds quickly (a good assumption if the reboiler is controlled by a vapor inlet scheme, see Sec. 17.1.2), the bottom level will begin replenishing soon after the initial fall in level.

Note that the above only applies to scheme 16.4a with a fast-responding reboiler. With schemes 16.4b and d, as well as scheme 16.4a with a slow-responding reboiler (e.g., a reboiler using the condensate outlet valve, Sec. 17.1.2), the responses involve liquid manipulation steps and are inherently slower. In these cases, the nesting and interaction problems will be more troublesome.

When vapor flow up the column is small compared to the side draw rate, configuration 19.6c (with any of the Fig. 16.4 control schemes) can induce large swings in column vapor traffic above the side draw, and a severe interference with the pressure controller. This is analogous to the problem described with configuration 19.6b. The vapor swings are most pronounced with schemes 16.4b and d, because the top section temperature is more sensitive to the feed than to the changes in the small vapor flow.

Two techniques can alleviate the above problems and render configuration 19.6c compatible with schemes 16.4a, b, and d.

1. *Controlling the internal vapor flow to the section above the side draw:* Reboiler heat duty is measured and divided by the latent heat of the boiling mixture; the measured side product flow is subtracted from the quotient to give the internal vapor rate in the section above the side draw. In a steam (or condensing vapor) reboiler, the internal vapor rate is computed as a constant times the measured steam rate less the measured side product flow, with the constant equal to the ratio of the latent heat of steam to that of the boiling mixture. An internal vapor controller (IVC) uses this computed internal vapor to manipulate product flow (Fig. 19.7b). A limitation of this technique is that internal vapor is computed as a small difference between two large numbers and can therefore be in error. The error escalates as the internal vapor rate becomes a smaller fraction of the total vapor traffic below the side draw.

   The internal vapor rate controller can be substituted by a differential pressure controller measuring pressure drop in the section above the side draw (260, 263, 362). This option, however, is less satisfactory (260, 263) because of the shortcomings inherent in pressure-drop controllers (Section 19.4).

2. *Utilizing an external recycle stream (260, 263; Fig. 19.6e):* This stream is often the condensed bottom product. This configuration was found simple and effective (260, 263) with scheme 16.4a (as shown in Fig. 19.6e). In principle, it can also be used with scheme 16.4b and d, but no field experience has been reported with these schemes. A major drawback of this scheme is that it wastes column capacity and utilities.

In an analogous manner to the techniques described in conjunction with configuration 19.6b, the above techniques eliminate the vapor swings, level nesting, and level-temperature interaction problems. With scheme 16.4a, a temperature control manipulates the internal vapor flow above the side draw (directly in configuration 19.6e, or by cascading onto the IVC in Fig. 19.7b). With schemes 16.4b and d, this internal vapor is flow-controlled (in that case, the Fig. 19.6e temperature controller is replaced by a flow controller). Both techniques preclude the use of scheme 16.4e.

In summary, it is best to use configuration 19.6c with the desired MB control scheme as per Sec. 16.6. If the desired control scheme is 16.4b or d, IVC should be incorporated; if it is 16.4a, IVC is optional but advantageous. This IVC cannot be used with scheme 16.4e.

When the internal reflux in the section above the side draw is low compared to the side product flow, scheme 16.4e should be avoided. The internal vapor control in Fig. 19.7b may work well, but it will become increasingly troublesome as the internal vapor becomes a smaller fraction of the total liquid traffic below the side draw.

Subject to availability of an external recycle stream, as well as spare capacity in the column section below the side draw, configuration 19.6c can also be used. However, this scheme is energy-wasteful and will usually accomplish little more than the IVC.

19.3.4 One prominent product

Many side-drawoff columns are used for purification of a crude product containing small amounts of light and heavy boilers. Typically, the purified product stream constitutes 90 to 95 percent of the feed. Solvent purification and refrigerant purification are two typical examples. The prime consideration is keeping the product on-spec. A secondary consideration is maximizing product recovery, but since the top and bottom streams are small, and often recycled, this may be a low-priority consideration.

Figure 19.8a shows one control system frequently used for such services. The upper and lower temperature controllers maintain product lights and heavies content at the desired levels by manipulating the respective product streams. The side product is drawn using an IRC (in case of a vapor product, an IVC) system, similar to that described in Fig. 19.7. The bottom level manipulates boilup, and the accumulator level manipulates reflux. The bottom level and temperature controllers are sometimes interchanged, so that the level controls the bottom flow and that the lower temperature controls boilup. Similarly, the top pressure and temperature controllers are interchanged when this can tighten pressure control (see Sec. 17.3; similar to the arrangements discussed in Fig. 17.8). A system similar to that in Fig. 19.8a has been proposed by Shinskey (362).

The Figure 19.8a system (including the variations described above)

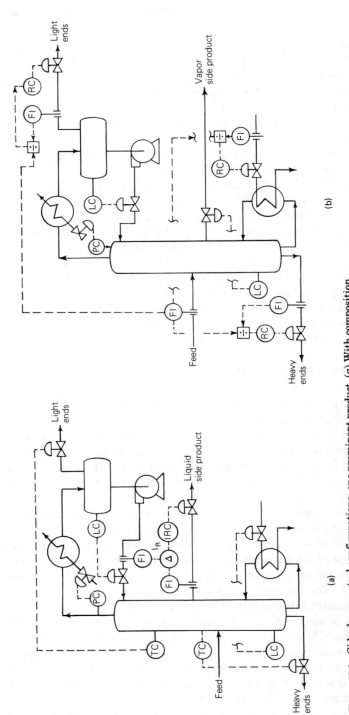

Figure 19.8 Side-draw control configurations, one prominent product. (*a*) With composition control and internal reflux control to section below side draw; (*b*) simple ratio control configuration.

offers efficient product recovery and energy usage, but its control action can be poor. The temperature controllers manipulate small streams, so that control may be extremely slow and sluggish. Because of the small amount of low and high boilers, a suitable control temperature location may be hard to find. If analyzers need to substitute the temperature controller, the method will become expensive. Finally, some interaction may occur between the two temperature controllers. In summary, this method is efficient, but may give poor control.

Figure 19.8b shows a simpler system (64, 68) that ratios the products and boilup rate to the feed and draws the side product on bottom level control. The system contains no composition control and relies on setting the ratio controls so that the product is overpurified, at the expense of a lower recovery. The bottom level loop is nested and relies both on the accumulator level controller and on the column pressure controller. When the internal vapor rate is low in relation to the vapor side product, it may be unstable, and significant overrefluxing and overreboiling may be needed to circumvent this problem. Generous sizing of the column base has been recommended (68) to accommodate some of the above shortcomings.

The Fig. 19.8b system can be modified to become an improved version of the Fig. 19.8a system by cascading the composition controls onto the product ratio controls (68). If needed, an internal vapor flow control can be used to control the side-draw rate, and the bottom level controller cascaded onto the steam ratio control.

In addition to the above methods, any of the control schemes described in Sec. 19.3.1 can also be used. This is simply accomplished by designating one of the product streams as the "small product" and drawing it on flow control. This is accomplished at a small economic penalty of lower product recovery compared to the Fig. 19.8a system.

19.3.5 Three prominent products

In some side-drawoff columns, three or more products are prominent. Typical examples are an ethanol-water still (Fig. 19.9) and some alkylation depropanizers and deisobutanizers. The control philosophy is to pick the two most important product purities and have these compositions controlled. Other product purities are allowed to vary. Any additional nonprominent streams are drawn on flow or ratio control.

The ethanol still in Fig. 19.9 (362) generates five products. Of these, the vents and heads products are treated as nonprominent products. The vents are drawn on pressure control (condensation rate is maintained "free", i.e., uncontrolled). The heads product flow is manipulated by controlling its ratio to the ethanol product. Excessive alcohol

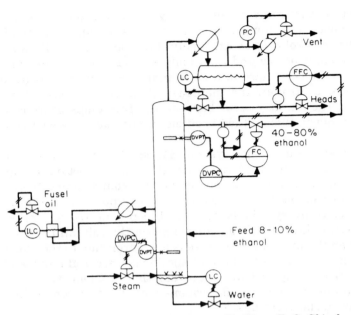

Figure 19.9 Control of an ethanol-water still. (*From F. G. Shinskey* Distillation Control, *second edition. Copyright © by McGraw-Hill, Inc. Reprinted by permission.*)

losses in this stream can be tolerated because these are recovered downstream.

The three remaining streams, ethanol, water, and fusel oil, are treated as prominent products. The prime product specs are alcohol in water and water in alcohol; these are composition-controlled (in this case, using vapor pressure controllers, see Sec. 18.9). The flywheel in this system is the fusel oil stream, and its ethanol content is allowed to vary somewhat. This bears a much lower economic penalty than allowing alcohol to escape in the water or permitting water to dilute the alcohol. In this specific case, the fusel oil is cooled and decanted, and the water phase returned to the column. The column in Fig. 19.9 can be viewed as two merged columns—the top section controlled using scheme 16.4*d*, the bottom using scheme 16.4*a*.

Doukas and Luyben (95) proposed the idea of dynamically altering the location of the side drawoff in order to gain an extra degree of freedom. Figure 19.10 shows a simplified flowsheet of their idea. The side stream is drawn from one or two trays at a time. Top and bottom compositions are controlled in the usual manner. One additional composition (e.g., lights in the side stream) is controlled by varying the relative contribution of each of the two active draw trays to the side product, and/or by changing the active draw trays. For instance, when

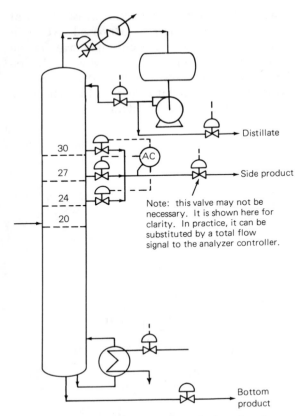

Figure 19.10 Dynamic variation of side draw location.

trays 24 and 30 are active, lights in the side draw can be reduced by gagging the tray 30 valve while opening the tray 24 valve wider. Alternatively, the tray 30 valve can be closed and the tray 27 valve opened. The scheme, which can be implemented using conventional analog hardware (95), looks good on paper, but no field experiences have yet been reported.

The composition control loops can be interactive. Crude oil distillation and other refinery fractionators are examples where these loops are very interactive (362). In such services, undecoupled composition (or temperature) controls are usually unsatisfactory, and product streams are drawn on flow control.

19.3.6 Side-stream strippers control

Side-stream strippers are common in refinery fractionators. The control system depends on whether the stripper feed is withdrawn from the main fractionator via a total drawoff or a partial drawoff. Total drawoffs (normally utilizing chimney trays) are used when a pumparound stream is taken out from the same chimney tray as the side draw. The pumparound is usually removed on flow control. The preferred side-draw controls (234) are shown in Fig. 19.11*a*. Sometimes the chimney tray overflow arrangement shown in Fig. 19.11*a* is substituted by a bona fide chimney tray level controller.

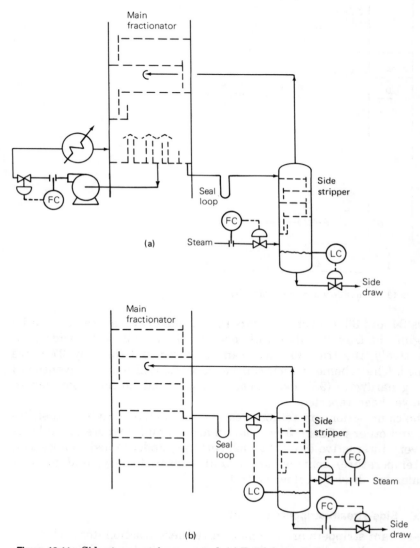

Figure 19.11   Side-stream stripper control. (*a*) Total drawoff; (*b*) partial drawoff. (*Concept based on N. Lieberman,* Chemical Engineering, *September 12, 1977.*)

The method shown in Fig. 19.11a requires a large surge volume at the bottom of the side stripper. As the pumparound rate is frequently much greater than the side-draw rate, the feed rate to the stripper may swing. The large surge volume is required in order to handle these swings without allowing liquid level at the stripper base to rise above the reboiler return nozzle, because this can dislodge trays (Sec. 13.2).

Partial drawoffs (commonly utilizing downcomer trapouts) are used when a side draw does not share a pumparound drawoff. Figure 19.11b shows the preferred controls (234). Operator action is required to ensure that the correct distillate quantity is drawn and to prevent drying of trays below the drawoff. The dryout problem can often be mitigated by drawing the side product from the bottom seal pan (i.e., just above a chimney tray). In both arrangements (Fig. 19.11a and b), note the seal loop in the line from the main fractionator to the stripper. This loop prevents vapor backflow at low liquid rates (Sec. 5.1).

19.4 Differential Pressure Control

When the "free" stream in an MB control system (Sec. 16.2) is the boilup rate, it is sometimes manipulated by the differential pressure across a column section or across the entire column (Fig. 19.12a). Column pressure drop is primarily a measure of column vapor load, although it is also influenced by the liquid load. Therefore, controlling differential pressure generally maintains a uniform vapor load in the column.

Differential pressure control automatically safeguards against flooding, and is therefore attractive when operation nears the upper capacity limit. It also automatically sustains high column loadings when the feed rate plunges; this is a distinct advantage when the column contains low-turndown internals. Differential pressure control is particularly beneficial when a column receives a fluctuating feed and the feed flow is relatively large relative to the column internal flows. The controller response is fast and suffers little dynamic lag because it works by varying boilup and vapor rates.

A past practice promoted differential pressure control in small, oversized columns as a means of simplifying instrumentation and control (68). These columns were overrefluxed and overreboiled by setting a relatively high differential pressure, so that the products were far purer than the specifications required. This permitted doing away with composition and temperature controllers and with MB control schemes. Nowadays, this practice is becoming unpopular due to its energy wastefulness and tight column designs (68).

Differential pressure control is popular in batch stills (44). Here

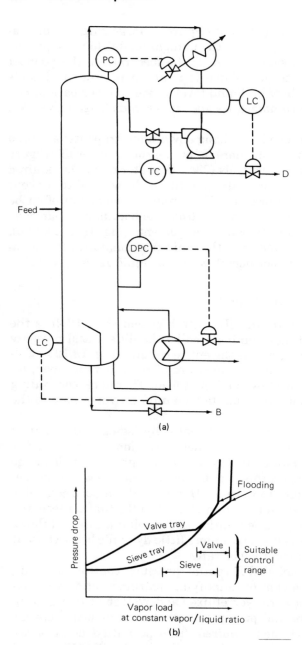

Feed

(a)

(b)

Figure 19.12 Differential pressure control. (a) Typical scheme; (b) typical pressure drop–load diagram, depicting suitable control range.

the vaporization rate gradually drops off with time (i.e., as the mixture becomes heavier); this prolongs separation time. A differential pressure controller sustains a uniform vaporization rate, thereby speeding the separation.

Differential pressure control is often troublesome (below) and is therefore generally unpopular. The author recommends avoiding this method (except as an override control) whenever possible. The override control can accomplish the prime objective (automatic protection against flooding) while sidestepping most of the shortcomings. Adding surge capacity upstream of the column is a far superior way of smoothing feed fluctuations than differential pressure control. The following drawbacks render continuous (as distinct from override) differential pressure control undesirable:

1. Differential pressure measurement is generally one of the least reliable measurements in a column (Sec. 5.4). Low pressure drops (e.g., packed towers) and density variations (e.g., vacuum towers) make a reliable measurement particularly difficult to achieve.

2. The suitable control range of differential pressure controllers may be narrow, especially with valve trays (Fig. 19.12b). In the valve-throttling range of a valve tray, and below the weep point of a sieve tray, pressure drop may be insensitive to vapor loads. Differential pressure control may therefore be difficult to apply under turned-down conditions. Packed towers are not prone to this limitation (44), because pressure drop tends to be sensitive to vapor load over the entire operating range (Fig. 14.3).

   The turndown problem can be overcome by cascading the differential pressure controller onto a heating medium flow controller. Under turned-down conditions, the differential pressure control is taken off cascade, and the heating medium is flow-controlled. The Fig. 19.12a scheme, for example, will then revert to scheme 16.4b.

3. Unless the cascade arrangement in item 2 above is used, the differential pressure control method will lead to overrefluxing and overreboiling at low rates and will therefore be energy-wasteful.

4. Differential pressure control may temporarily adversely affect product purity. Consider a rise in liquid feed to the column in Fig. 19.12a. Pressure drop in the lower section will rise, and the differential pressure controller will lower steam rate. This will raise the concentration of lights in the bottom product. Eventually, the problem be rectified by the temperature controller, which will reduce reflux, but the bottom product may temporarily go off-spec.

5. Differential pressure control is a non-fail-safe control. Consider a reflux failure (e.g., reflux pump trips) in Fig. 19.12a. Pressure drop will fall, and the controller will call for more boilup. The use of a differential pressure control may thereby also enhance the column relief requirements (Sec. 9.7).

6. A differential pressure controller located in one column section (e.g., the bottom in Fig. 19.12a) will not protect the column against flooding in another column section (the top in Fig. 19.12a). A differential pressure measurement across the entire column may not be sufficiently sensitive to identify flooding until it has propagated through a few trays.

19.5   Feed Preheat Control

Feed preheating (or precooling) is usually practiced for heat recovery or to attain the desired vapor and liquid traffic above and below the feed. If preheating supplies only a small fraction ($<$ 10 to 20 percent) of the column heat input, or the preheater duty is nonfluctuating, preheat control may be unnecessary. The preheater may be then run at full capacity, or be regulated manually by adjusting its bypasses. This is seldom satisfactory when the preheater supplies a major portion of the column heat input or uses a fluctuating heat source. In one feed-bottom interchanger (362), lack of control triggered a periodic cycle in bottom liquid level. The bottom flow was drawn from a weir compartment of a kettle reboiler and therefore tended to fluctuate (Sec. 15.7). A rise in bottom level increased bottom flow, which in turn increased preheat. This reduced column liquid downflow, causing a fall in bottom level several minutes later. This in turn reduced bottom flow and preheat, and increased column liquid downflow. This reraised bottom level, and the cycle was then repeated.

The objective of the preheat control system is to supply the column with a feed of consistent specific enthalpy (enthalpy per unit mass). With a single-phase feed, this translates into a constant feed temperature; with a partially vaporized feed, this translates into a constant fractional vaporization. Maximizing feed temperature (if desired) is usually performed manually, by an advanced control system, or by a valve position controller similar to that used in floating pressure control (Sec. 17.2.4).

The feed enthalpy is normally inferred from a temperature measurement of the feed leaving the preheater, and preheat is manipulated to control this temperature. This is satisfactory when the feed is a single-phase fluid, and often also with partially vaporized wide-boiling mixtures at superatmospheric pressures, but not with partially vaporized narrow-boiling mixtures. In the latter case, fractional vaporization (and feed enthalpy) varies considerably over a narrow temperature range, and minor temperature changes can induce large shifts in fractional vaporization. If the feed rate is of the same order as the column internal vapor and liquid traffic, these shifts will cause large swings in column vapor and liquid traffic, and possibly lead to premature flooding. In one case (239), these swings caused column product to go off-spec; in another (259) they rendered the control system inoperable.

With partially vaporized feeds under vacuum, feed temperature varies largely with pressure as well as fractional vaporization and will not provide a reliable measure of feed enthalpy. The consequences of preheater outlet temperature control will be similar to those described above.

The above swings can be eliminated by controlling the preheater heat duty or, better still, the preheater heat duty per unit feed flow (362), instead of the preheater outlet temperature. For steam (or condensing vapor) preheaters, the duty per unit feed flow equals the ratio of the measured steam (or vapor) flow to the measured feed flow times a constant, the constant being the steam latent heat. For a sensible-heated preheater, the above ratio is multiplied by the measured hot-side temperature difference, and the constant is the average hot-fluid heat capacity. For two or more feed preheaters, it is best to compute their total heat duty on-line and ratio it to the feed (68, 259). The computation can be readily performed using conventional analog instrumentation. Similar techniques cured the above-cited swing problems (239, 259).

When column differential pressure is controlled (see previous section), the preheat can be manipulated to control top section differential pressure, while boilup is manipulated to control bottom section differential pressure (332, 362). This system has the advantage of preventing feed disturbances from interfering with differential pressure control, but is prone to the shortcomings described earlier.

The manipulative arrangements used for preheater control are similar to those used for reboiler control (Sec. 17.1). Alternatively, some feed can be bypassed around the preheater, and the bypass flow manipulated using arrangement similar to Fig. 17.2a and b. With sensible-heating media, manipulating a feed bypass gives faster and more linear response compared to manipulation of a heating-side bypass (362), and should therefore be preferred. Unless a fluctuating heat source is used, and/or the preheater supplies a large fraction of the column heat duty, preheater control is far less critical than reboiler control, and less sophisticated control arrangements are normally appropriate.

19.6 Disturbance Amplification in a Column Train

Feed disturbances. Disturbances in the feed or operation of a column are often transmitted to downstream columns (e.g., Fig. 19.13a). The disturbance can be either amplified or dampened, with or without a significant time lag. If large enough, it can upset downstream columns. Generally, the smaller the feed rate compared to the internal vapor and liquid traffic, the less sensitive the column to feed disturbances. Superfractionators (high reflux ratios) seldom suffer from feed disturbances, while low-reflux-ratio columns suffer most.

Factors that commonly cause disturbances to be amplified are

1. Insufficient surge volume between columns. This usually compels level controller tuning for a fast response (in the bottom sump or reflux drum), causing flow fluctuations.

2. A high ratio of the feed flow rate to the flow rate of the product which is passed to the next column. This is especially true for small bottom streams passed on to the next column. Any feed or even reboiler variations often cause significant fluctuations in bottom flow.

3. A preferential baffle in the bottom compartment (Sec. 4.5) increases the tendency of a bottom sump level and, therefore, bottoms flow, to fluctuate.

4. Vapor feed from one column to the next. Vapor product is usually on pressure control, with no surge. When such a feed is used, disturbances should be expected and accommodated in downstream equipment.

5. Packed columns have lower pressure drop and lower liquid inventory than tray columns and, therefore, a smaller dampening effect on fluctuations.

Some factors that can mitigate disturbance amplification in a column train are

1. Feed surge (below).

2. Level controllers for a stream flowing into the next column should act as slowly as possible (without causing the level to rise or fall beyond its desired range).

3. Feed forward control loops can often be used to dampen fluctuations in a column train, particularly if the disturbances occur stepwise rather than as a continuous fluctuation.

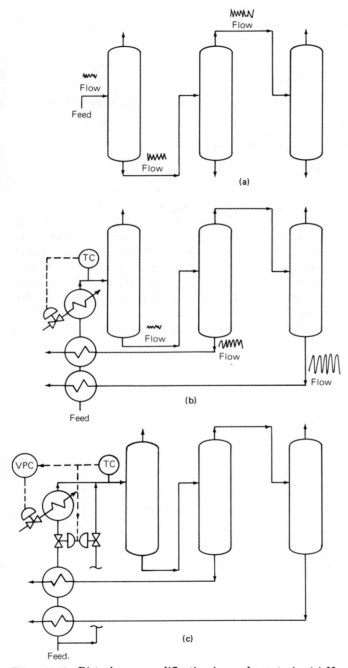

Figure 19.13 Disturbance amplification in a column train. (*a*) No heat interaction; (*b*) with heat interaction, poor control; (*c*) with heat interaction, good control. (*Parts b and c based on Dale E. Lupfer,* Proc. 53rd Ann. Convention of the Gas Processors Assoc., March 25-27, 1974. *Reproduced courtesy of the Gas Processors Association.*)

Feed surge.   Buckley et al. (68) recommend installing a separate feed surge upstream of a column in a column train when

- Top or bottom product impurity specification is less than 0.1 mole or weight percent.
- Vapor from the column furnishes reboil heat to another column(s).
- Column has a side draw that is a major fraction of the feed.
- Column operates close to its maximum capacity.
- Column train has primitive controls.

The list is not intended to be comprehensive. Some guidelines for sizing this intermediate surge have also been presented (68).

Intermediate surge can be provided at a lower temperature. This is often practiced when the intermediate material is thermally unstable. To minimize heat losses, it is usually desirable to devise a control strategy that preferentially feeds the hot material into the next column (e.g., Fig. 19.14). However, such a strategy will be self-defeating if implemented at the expense of excessive flow fluctuations to the downstream column. In Fig. 19.14, column 1's bottom flow controller must be set low enough so that the column level controller always stays on control.

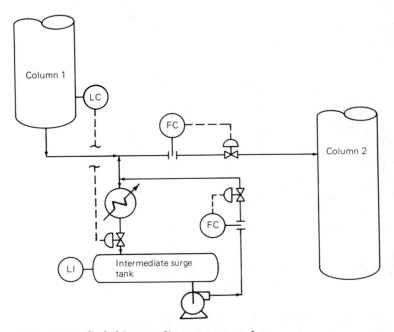

Figure 19.14   Cooled intermediate surge control.

Heat recycle. When heat is recycled in a column (e.g. feed-bottom interchanger) or in a column train (e.g., Fig. 19.13b), so are disturbances. These may amplify disturbances to column feed, causing severe oscillations. It is essential in such cases to ensure that feed preheat control be effective and insensitive to fluctuations. The scheme in Fig. 19.13b should be avoided unless the duty of the auxiliary exchanger is large compared to that of the feed-bottom interchangers. When the duty of the auxiliary exchanger is relatively small, this scheme oscillated, while the scheme in Figure 19.13c worked well (258). A slow valve position controller (VPC) was included in the latter scheme to minimize feed flow through the bypass. The VPC received the temperature controller output signal (which was proportional to the valve position) as the valve position measurement. The VPC set point was fixed such that the bypass opening was kept to the minimum required for control. Usually control was accomplished by the feed and bypass valves alone. If the bypass valve approached the closed position, or opened too far, the VPC would lower or raise heat input to the auxiliary exchanger, respectively, so that the temperature controller always stayed on control.

Control of heat-recovery systems, in which the overhead vapor from one column is used as reboil for other columns, has been discussed by Buckley et al. (63, 68).

20

Columns That Did Not Work: Case Histories

"If anything can go wrong, it will."
Murphy's Law

Case histories of columns that did not work are the heart of this book. Analysis of these case histories formed the basis for contents of the book, the topics discussed, and the depth of coverage of each topic. The lessons learned were formulated into guidelines, and the guidelines can be found in each chapter.

What are these case histories? The following sixteen tables contain about 300 case histories extracted from the literature of columns that did not work. Each case was abstracted, and the major lessons learned highlighted. Can these case histories happen to us? If you believe Murphy's law, you have this question answered.

20.1 Criteria for Columns That Did Not Work

Compilation of the case histories in this chapter required setting criteria as to what constitutes a "column that does not work." The criteria used by the author are subjective and can be criticized for being somewhat arbitrary. In choosing these criteria, the author was solely guided by his interpretation of the title of this chapter.

The author's criteria were

- Columns that performed well and whose performance was improved (e.g., capacity raised by replacing trays by packing; control system worked well, but was further improved by implementing computer control) were excluded. These, in the author's opinion are not "columns that did not work"; they worked well, but were further improved.

- Incidents of corrosion and fouling were only included if a feature unique to the column design, operation, or control contributed to their occurrence. For instance, tales of accelerated corrosion due to liquid maldistribution, or poor removal of water from a hydrocarbon column, constitute "columns that did not work." On the other hand, an incident where the wrong corrosion inhibitor was applied or a pH controller malfunctioned is related to handling the chemicals rather than to the column; it would have occurred in any other piece of equipment processing the chemicals; it therefore does not constitute a "column that did not work."

- Only specific incidents were included. For instance, a statement such as "Leakage from chimney trays in refinery, vacuum columns can be reduced by seal-welding tray sections" does not constitute a case history. On the other hand, a statement such as "One vacuum column experienced severe chimney tray leakage at low-rate operation. Seal-welding tray sections reduced leakage to acceptable levels" does.

- Only incidents previously reported in the published literature are included. New incidents reported in this book have been excluded. The author based the contents of the book on the following case history list (Sec 1.1), and aimed at avoiding a bias toward his own experiences.

TABLE 20.1 Problems Caused by Appearance of Unexpected Components, Reactions, or VLE pitfalls

Case no.	Ref.	Type of plant	Type of column	Brief description	Some morals
101	98	Solvent recovery	Isopropyl acetate recovery column	Poor decanter phase separation resulted from slow hydrolysis of isopropyl acetate to isopropanol, which was soluble in both organic and aqueous phases	Slow reactions may occur in chemical systems, especially when material is recycled.
102	278	Acetylene	Solvent water stripper	Solvent losses were far greater than design. Unsuccessful extrapolation of VLE data was one of the causes. Increasing number of trays and raising reflux helped reduce losses.	Caution is required when extrapolating VLE data.
103	180	Butadiene	Butadiene refining column	A detonation which demolished the column and caused widespread damage occurred due to vinylacetylene doubling in concentration and detonating in the absence of air. The increase in concentration resulted from leakage of butadiene (light key) during total reflux operation.	A very thorough analysis of the accident is presented. Recommendations include avoiding total reflux operation in such services; keeping vinyl acetylene concentrations low and continuously monitored in this type of column; and others.
104	131	First-of-a-kind process	Vacuum refining column	Lower trays in column plugged and later buckled due to excessive pressure drop. Plugging was caused by polymerization of the product. Product polymerization required the presence both of reactor catalyst and air. Traces of the former were carried over; air entered due to a substantial leak. Repairing the air leak cured the problem.	Air leakage and/or catalyst carryover into a column can induce an undesirable reaction.

611

TABLE 20.1 Problems Caused by Appearance of Unexpected Components, Reactions, or VLE pitfalls (Continued)

Case no.	Ref.	Type of plant	Type of column	Brief description	Some morals
105	98		Aqueous feed/solvent column	Removal of the volatile solvent from an aqueous feed solution caused dissolved solids to precipitate in the packing below the feed, plugging the column. The solution was replacing the packing by sieve trays.	
106	275 vol 1 case 363	Vitamin A intermediate	1′ Pentol still (high vacuum)	A small amount of caustic, left over in the piping from a shutdown cleaning operation, was drawn into the still. The mixture and still exploded, causing fatalities, injuries, and an extensive damage.	Do not use a chemical wash unless the chemical can enter the column during operation without causing a hazard.
107	96		Propargyl bromide	An explosion occurred when this material was distilled under pressure. Tests showed that although stable at atmospheric distillation, it detonates when distilled under pressure.	Pressure (and/or temperature) can have a strong effect on a reaction during distillation.
108	96	Nitro chemicals	Solvent recovery steam distillation	Nitrocellulose precipitated out of the solvent and exploded.	
109	338	Carbon tetrachloride	Trichloroethylene (TCE) and carbon tetrachloride (CTC) separation	The concentration of TCE in CTC was higher than expected. A total reflux test showed that separation near the column bottom was worse than expected. Either VLE nonideality or decomposition of chlorinated ethanes at the reboiler temperature was the culprit.	Total reflux tests can give valuable insight into poor performance. Watch out when extrapolating VLE data.

110	Chemicals	Absorption of HF from HCL gas by wash with aqueous HCL	HF absorption was poor. HF escaping in the column overhead destroyed the downstream glass plant. The cause was that most of the "HF" in the feed was in the form of carbonyl fluoride. This component was sparingly soluble in water, but hydrolyzed slowly to HF.	Carefully examine the chemistry of the components before choosing a separation process.
111	Refinery HF alkylation	Depropanizer/HF stripper	Depropanizer overhead went to an HF stripper. Stripper bottoms was the propane product, while stripper overhead was recycled to the depropanizer overhead. When ethane entered the depropanizer due to an upstream unit upset, it entrapped in the overhead system and could not get out. Depropanizer pressure climbed and excessive venting was needed.	The stripper bottom temperature was reduced temporarily to allow ethane into the propane product. Within two shifts the trapped ethane was cleared.
112	Nitro chemicals	O-nitrotoluene recovery vacuum column (batch)	The residues were held at 150°F and air admitted. A previously unknown exotherm set in, causing an explosion.	Recommendations were to modify the plant and process.
113	Nitro chemicals	Nitrotoluene distillation	An alternative still was used for the first time for this product. Blockages in the condenser led to excessive pressure drop and overheating in the still. This led to a runaway decomposition and an explosion.	Ensure adequate temperature and pressure drop monitoring in this type of plant. Take special precautions when heating unstable compounds.

TABLE 20.1 Problems Caused by Appearance of Unexpected Components, Reactions, or VLE pitfalls (*Continued*)

Case no.	Ref.	Type of plant	Type of column	Brief description	Some morals
114	14a	Offshore gas processing	Amine absorber/ regenerator	H_2S in the absorber lean gas was well above specification. This was caused by aldehyde inadvertently backing from a storage tank into the amine charge. The aldehyde reacted more strongly with H_2S than the amine, and H_2S could not be properly stripped in the regenerator.	A new solvent charge solved the problem. Piping was modified to prevent recurrence.
115	209a	Ethylene oxide (EO)		Polymerization of EO in a distillation column caused overheating which caused a decomposition reaction. The polymerization may have been catalyzed by iron carried over from an upstream column. Several similar incidents are said to have occurred.	Avoid contamination of ethylene oxide (209a).
116	209a	Cumene oxidation		An explosion occurred in the base of a distillation column containing 65% cumene hydroperoxide in cumene. An interruption in bottom takeoff may have caused overheating or air leakage into the column may have caused an explosion. There were fatalities and the column blew 600 ft into the air, then fell on other equipment.	Use low-inventory separation equipment, rather than distillation, for concentrating cumene hydroperoxide (209a).

TABLE 20.2 Unique Features of Multicomponent Fractionation

Case no.	Ref.	Type of plant	Type of column	Brief description	Some morals
201	203	Gas processing	Lean oil still	Column did not achieve required separation because of insufficient reflux induced by an undersized reflux orifice plate. Problem was difficult to diagnose because of unusual behavior of top temperature.	In some multicomponent distillations, normal top and bottom temperature can be observed even if adequate separation is not achieved.
202	203	Gas processing	Lean oil still	An added preheater which performed better than design caused column to "hiccup" and empty itself out every few hours either from the top or bottom.	Oversized preheaters can cause fractionation difficulties. Bypasses around preheaters can eliminate this problem.
203	268			Separation efficiency of a component appeared extremely poor, while that of other components was OK. This was caused by analyzer error, and was discovered by calculating a component balance.	Suspect unexpected lab analyses, and check by material and component balances.
204	239	Refinery	Debutanizer	The column feed was rich (72%) in butane. A few degrees extra preheat caused a large increase in feed vaporization, accompanied by a large drop in stripping vapor rate. This increased butane in bottom product. The problem was solved by controlling the flow of steam to the preheater.	Excess preheating can cause fractionation difficulties. When most of the feed boils over a narrow temperature range, heat input to the preheater needs to be controlled.
205	237	Refinery	Vacuum tower	Top tower temperature was reduced by 130°F by cooling top pumparound. This caused an unexpected rise in column pressure. The pressure rise was caused by the flash equilibrium behavior of the precondenser. When temperature was reduced, less liquid was condensed in the precondenser, and less lights were absorbed.	The absorption effect may be important in some systems.

TABLE 20.3 Problems Related to Column Size

Case no.	Ref.	Type of plant	Type of column	Brief description	Some morals
301	203	Olefins	Depropanizer	Instability and fluctuations occurred because of the dumping loads in a section of operation below the column just below the feed.	Ensure adequate hydraulics above and below alternate feed points.
302	278	Acetylene	Solvent water stripper	Some causes of excessive solvent losses were an incorrect efficiency estimate and excessive entrainment. Performance was improved by adding trays and mist elimination pads under top section trays, and raising reflux.	Mist elimination pads have been useful for eliminating entrainment.
303	85	Refining or petrochemical	Steam distillation splitter of an olefinic naphtha-type hydrocarbon feed	Bottom section was oversized and operated at low rates. Separation was poor because of valve tray weeping. Increasing loadings solved it.	
304	85	Refinery	Crude residuum vacuum distillation column	Tray efficiency was lower than expected. Column contained valve trays and operated at low liquid loads and with wide variations in vapor loads.	Beware of overpredicting efficiency at very low liquid loads.
305	55	Whiskey distillery		Excessive entrainment was experienced with tunnel trough trays at close (18-in) spacing. The problem was resolved by installing a 2-in-thick demisters directly on top of the troughs.	A useful technique for minimizing entrainment at close tray spacings.

306	239	Refinery	Catalytic cracker fractionator	The design vapor rate in the slurry section of a catalytic cracker fractionator did not allow for vaporization which occurs when a bottom feed with 300°F superheat contacts column liquid. Column therefore prematurely flooded. Problem solved by injecting subcooled quench liquid to desuperheat the feed. At a later stage, subcooled quench was replaced by a lighter liquid that vaporized, and premature flooding reoccurred.	Account for vaporization due to superheat in column loading calculation. When attempting to reduce column loads by quenching a superheated feed, use a sufficiently subcooled liquid.
307	338		Desorption of methanol, acetone, and ammonia from water, using air	Column failed to achieve design separation. Design efficiency was predicted from air humidification and oxygen stripping studies in a single-plate laboratory column. Wall and downpipe mass transfer enhanced efficiency in the lab column. This led to optimistic efficiency predictions.	Caution is required when predicting efficiency by adding lab-measured mass transfer resistances.
308	19	Pharmaceuticals	IPA-Water Azeotropic distillation using benzene and IPE entrainer	The 2-ft ID column containing metal mesh packing achieved 12-in HETP. This was considerably higher than the design HETP. Column HETP was scaled up from small-diameter columns that had good and frequent liquid distribution and redistribution.	Pay attention to differences in distribution equipment when scaling up packed tower efficiency.

TABLE 20.4 Problems Related to Tray or Downcomer Layout or Tray Type

Case no.	Ref.	Type of plant	Type of column	Brief description	Some morals
401	85	Refinery	Crude residuum vacuum distillation column valve trays	Weeping at low liquid loads in a pumparound section did not permit circulating liquid. Replacing valves equipped with turned-down nibs (to prevent sticking) with valves that seat flush with the floor solved the problem.	Avoid trays equipped with turned-down nibs when liquid leakage is critical.
402	296	Refinery	Vacuum tower valve trays	A portion of the valves on three pumparound trays was removed and their opening blanked. Ends of distribution pans and draw pans were seal-welded. Leakage was reduced, separation and heat recovery were improved.	Reducing valve density and seal-welding of pans can reduce leakage.
403	232	Refinery	Main fractionating tower valve trays	This column was prone to pressure surges because of accidental introduction of water. Valve trays needed replacing approximately once per year. Cast iron bubble-cap trays were used in a very similar unit and could weather such surges.	
404	239	Refinery	Fluid catalytic cracker stripper valve trays	Vapor load near the column top was three times lower than near the column bottom, causing turndown problems. Successful resolution was achieved by different modifications to top 40% of trays. Modifications included (a) blanking half the valves, (b) using leak-resistant valve units, each containing a lightweight plate located below the normal disk; (c) retraying with bubble-cap trays.	A number of useful techniques are successful for overcoming valve tray turndown problems.

405	237	Refinery	Reboiled absorber	Single-pass trays were replaced by two-pass trays in a capacity revamp. Higher capacity was achieved, but reducing flow path from 36 to 18 in lowered efficiency to an extent that original trays had to be reinstalled.	Beware of efficiency loss when increasing number of tray passes.
406	237	Refinery	Deethanizer valve trays	A dramatic increase in efficiency resulted from retraying column with leak-resistant valve units, each containing a lightweight plate located below the normal disk.	As for 404.
407	108	Chemicals	Oxidation reaction effluent absorbers, 14 ft ID sieve trays	Heavy flow-induced vibrations occurred in two towers in a train of four while operating at low rates. The vibrations caused cracking of beam to support welds and of some trays. Increasing gas flow rates overcame the problem.	As for 408.
408	62	Chemicals	Five columns pressure and vacuum 5- to 25-ft ID sieve and valve trays	Flow-induced vibrations at gas rates close to the weep point caused damage to trays, support beams, and beam to column supports, at times within hours of operation at the damaging vapor rates. In one case total internal collapse resulted; in another, shell cracking occurred. Successful cures included reducing fractional hole area, stiffening support beams, and avoiding low operating rates.	Effective techniques have been proposed for preventing flow-induced vibrations. A predictive method has also been developed.

TABLE 20.4 Problems Related to Tray or Downcomer Layout or Tray Type (*Continued*)

Case no.	Ref.	Type of plant	Type of column	Brief description	Some morals
409	209	Olefins	High-pressure condensate stripper	Column flooded prematurely, presumably due to foaming. Capacity was increased by 50% by replacing original sieve trays by valve trays that had half the weir height, and 50% more downcomer top area. Extensive performance data are presented.	Pay attention to downcomer performance in systems with high foaming tendencies.
410	239a	Refinery HF alkylation	Iso stripper valve trays	Upon turndown, products went off-spec due to column efficiency loss. The efficiency loss was caused by weeping of the non-leak-resistant valve caps. Problem was overcome by a large reduction in column pressure.	Pressure reduction can help boost vapor velocity and counter weeping.
411	237	Pharmaceuticals	Methanol dehydrator sieve trays	Column flooded prematurely in the upper part of the bottom section due to insufficient fractional hole area on the sieve tray. This was presumably caused by the trays being sized to accommodate the vapor loads at the tower base. Since water has a higher latent heat than methanol, the vapor load near the feed was higher.	Ensure that trays are sized for the maximum vapor and liquid loads anticipated in the tower.
412	14a	Offshore gas processing	Amine absorber valve trays	At 25% of the design throughput, the lean gas H_2S content was above specification. Blinding 60% of the valves solved the problem.	Reducing valve density can improve efficiency at low rates.

| 413 | 150a | Refinery | Replacement of carbon steel valve trays by stainless steel sieve trays reduced frequency of required tray cleaning to remove scale. | |
| 414 | 150b | Chemicals | 10-ft ID, containing 72 trays below feed | Comuns flooded prematurely, with bottleneck just below feed. Bottleneck was identified by gamma scanning. Two trays below the feed were modified by reworking some unusual design features and lowering outlet weir. Column capacity did not change, but gamma scan showed improvement on modified trays. Modifying the next 10 trays below the feed improved capacity by 15%. Modifying the next 17 trays down gave another 15% capacity enhancement. Each step was guided by gamma scans. | Beware of unusual design features on trays. Gamma scans are effective for detecting tower bottlenecks. |

TABLE 20.5 Problems Related to Packing Size and Type

Case no.	Ref.	Type of plant	Type of column	Brief description	Some morals
501	203	Olefins	Process water stripper	Unsuitable packings caused capacity restriction. Carbon steel packing corroded, ceramic packing chipped. Using undersized stainless steel packing did not solve the problem.	Ensure adequate choice of packing material and size.
502	203	Gas processing	Hot pot absorber	Plastic packing melted upon startup because of reaction and possibly also hot spots.	Carefully examine suitability of plastic for this service.
503	349	Ammonia	Hot pot absorber	On many occassions, plastic packing melted upon solution circulation (power) failure. Absorber feed was cooled by the regenerator reboiler. This cooling was interrupted when circulation ceased, causing hot feed to enter the column.	Either avoid plastic packing in ammonia hot pot absorbers or instrument absorbers to avoid hot feed entrance upon power failure.
504	237	Refinery	Refinery fractionators	Grid-type packing showed high capacity and excellent resistance to plugging and coking in two fractionator wash sections. Coked sections, however, could not be cleaned at turnaround and needed replacement.	
505	76	Multicomponent separation, two-feed column		Following replacement of trays by structured packing, bottom product could not meet specifications. HETP below the feed was half that expected. Raising reflux ratio did not improve separation. Problem was solved by preheating the highly subcooled feed to its bubble point.	Highly subcooled feed can cause a premature column bottleneck.
506	219			High temperature during startup caused aluminum packing to lose strength and become compressed. This incurred excessive pressure drop.	Consider abnormal operation when using aluminum packing.

TABLE 20.6 Foaming

Case no.	Ref.	Type of plant	Type of column	Brief description	Some morals
601	203	Gas processing	Hot pot absorber (packed)	Foaming occurred because a corrosion inhibitor injected into the gas well ended up in the column. Laboratory tests did not identify the problem. Tests under actual operating conditions did. Antifoam injection solved it.	Tests for foaming are best carried out under actual plant operating conditions.
602	85		Selective absorption of a light gas constituent from a gas stream using a nonvolatile solvent	Foaming occurred, but batchwise addition of antifoam did not improve performance. This led to the conclusion that no foaming occurred. Lab tests at plant conditions finally indicated foaming. Continuous antifoam injection solved the problem.	Batchwise antifoam addition may be ineffective.
603	85		Absorber-stripper, removing CO_2 from H_2-rich gas	Too much antifoam was added to overcome a foaming problem. This interfered with tray froth action, causing poor absorption.	Too much antifoam can be as detrimental as too little.
604	85	Refinery	Atmospheric crude distillation	Serious foaming occurred when processing one type of crude. In this case, the residue was retained in the system at elevated temperatures for a relatively long time.	Foaming tendency is sensitive to the nature of feedstock and to operating conditions.

TABLE 20.6 Foaming (*Continued*)

Case no.	Ref.	Type of plant	Type of column	Brief description	Some morals
605	239	Refinery	Visbreaker fractionator	At high visbreaker conversions, the column bottom would foam and carry over into the distillate. Injecting 10 ppm of silicone defoamer into the vapor space above the fractionator bottom solved problem.	As for 604.
606	33	Chemicals	Aldehyde column	Premature flooding occurred due to foaming in downcomers from just below the feed up. This was detected by gamma-ray scans. Adding antifoam improved performance but was undesirable in the process. Downcomer enlargement in the trouble area solved the problem.	Gramma rays are effective in detecting downcomer foaming. Enlarging downcomers and adding antifoam are effective in overcoming a foam problem.
607	299	Refinery	FCC unit secondary absorber	At least 15 instances of foaming occurred when light cycle oil (LCO) was used as the lean oil. Increasing tray spacing, increasing downcomer area, reducing pressure drop, increasing temperature, and injecting antifoam were only partially successful remedies. Replacing LCO by naphtha effectively cured problems. In other cases, use of antifoam, injecting naphtha into LCO, and raising pressure were effective remedies.	
608	298	Refinery	Preflash tower	Several columns experienced foaming problems. Problems successfully solved by antifoam injection.	

624

609	408 407	Sulpholane extraction process	Extractive stripper	Foaming occurred near the top of the column. Bubbling air through a mixture sample failed to detect foaming, but Oldershaw and pilot column tests indicated foaming. Injecting antifoam or a small quantity of kerosene effectively suppressed foaming.	Foaming can sometimes be suppressed by adding a small quantity of heavy material.
610	407	Butadiene (acetonitrile solvent)	Extractive distillation pilot column, simulating plant column by using pentane/isoprene and acetonitrile	Foaming occurred and was promoted by increased water content. Oldershaw column tests were in good agreement with pilot column results. Foaming was suppressed by antifoam injection.	Variations in water content may affect foaming.
611	135	Gas processing	Natural gasoline absorber	Foaming occurred in this absorber which used crude oil as solvent. Foaming and antifoam effectiveness were successfully tested in a foam test apparatus. Problem was solved by intermittent antifoam injection.	As per 613
612	50	Separation of light mono and diolefins in a reboiled absorber using DMF (dimethyl formamide)		Foaming occurred, but bubbling air through solvent samples failed to detect it. Initial antifoam addition was unsuccessful because of poor dispersal. Effective dispersal of antifoam solved the problem. The investigation preceding the cure showed that foaming is temperature-sensitive. Foaming was not observed in a similar, but oversized column.	A thorough investigation and analysis of a foaming problem is described including symptoms and tests. Foam testing under other than actual plant conditions can mislead. Foaming can be sensitive to temperature and downcomer size. Effective antifoam dispersal is important.

TABLE 20.6 Foaming (Continued)

Case no.	Ref.	Type of plant	Type of column	Brief description	Some morals
613	26	Gas processing	Amine regenerator	A foaming problem in this packed column was cured by batchwise antifoam injection. Previously, when bed height was low because of packing migration through the supports, foaming was not observed.	Batchwise antifoam injection is sometimes effective.
614	373	Gas processing	Gas treating	Foaming was caused by a corrosion inhibitor used in the boiler feed water. Steam condensate used for solvent makeup contained the inhibitor.	Beware of external corrosion inhibitors entering a foaming system.
615	182 132	Heavy water using GS process	"Cold" tower, H_2S-water contactor	Foaming occurred when the suspended solids content of the feed water was high. Antifoam injection suppressed the foaming. Some tray designs coped better with foaming than others. Pressure-drop measurements and gamma scans were useful for diagnosing foam flood.	Suspended solids promote foaming in this service. Gamma rays are effective in diagnosing foaming problems.
616	71	Refinery	Oil absorber	Foaming occurred, primarily near the top of the column. Removing the top two trays in order to provide more disengagement space did not solve the problem. Gamma scans identified the foaming, and antifoam injection cured it.	Gamma scans are useful for diagnosing foaming problems.

617	Gas processing	MEA absorber	A foaming problem was treated with limited success by antifoam injection, frequent activated-carbon regeneration, and reduced MEA circulation. Problem was finally solved by using finer filter elements and adding a filter to the lean amine circuit.	Improving filtration can effectively cure some foaming problems.
618	Chemicals		Small amounts of sodium chloride impurity precipitated out of solution and caused foaming. Problem solved by antifoam injection.	Particulates in liquid can cause foaming.

TABLE 20.7 Troublesome Tray Column Internals

Case no.	Ref.	Type of plant	Type of column	Brief description	Some morals
701	12	Refinery		Excessive liquid carryover from the top of the tower was experienced at less than design rates. Problem believed to be caused by an excessively tall outlet weir on the bottom seal pan and by a submerged reboiler return nozzle.	Ensure sufficient distance from bottom tray seal pan weir to bottom tray. Avoid submerging the reboiler return nozzle.
702	12	Refinery	Presumably a fractionator	A restrictive design of a transition tray converting single-pass flow to two-pass flow caused premature flooding.	Avoid restrictive transition tray designs.
703	12	Refinery	Absorber	Choking of an outlet line from a downcomer trapout limited liquid flow, which in turn reduced absorber capacity. Increasing the height of the trapout pan did not help. Degassing the drawoff liquid in a separate enlarged pan solved the problem.	Either avoid vapor in liquid outlets or design for it.
704	12	Refinery		Vapor choking of a long line from column to reboiler caused premature tower flooding. To solve the problem, the draw pan was converted to a degassing pan and the line was sloped and vented.	As for 703.
705	391	Gas processing	Crude oil stabilizer	Degassed liquid on a draw tray caused excessive backup of aerated liquid in the downcomer. The downcomer was submerged below the liquid level on the draw tray. The backup caused premature column flooding.	Ensure adequate height and allow for differences in aeration when using submerged downcomers.

628

706	57	Refinery	Unstable operation and premature flooding were experienced at 90% of design rates. Either block valves in a liquid drawoff line to a side reboiler or insufficient downcomer area were the culprit.	Gamma scans are useful in diagnosing problems.
707	57	Refinery	Premature flooding resulted from absence of downpipes on a chimney tray, which forced liquid flow down vapor risers. Problem fixed by installing an external downpipe.	Gamma scans are useful in diagnosing problems. Liquid downflow through vapor risers should be avoided.
708	255	Stripper column	A carpenter's sawhorse was found in the bottom of the vessel. Following its removal, the bottom pump experienced loss of suction at low levels due to vortexing.	Vortex breakers should be routinely installed at liquid outlets.
709	255	Small-diameter steam stripper	Level float in a column bottom sump "bounced" and finally broke due to impingement of entering steam. Problem fixed by installing a shielding baffle over the level connection.	Avoid impingement of bottom feeds (or reboiler return inlets) on instrument connections.
710	98	Chemicals	Vapor from a forced-circulation reboiler caused vibration. This resulted in loosening of tray fasteners and tray failure. Increasing nozzle size and stiffening the support beams solved the problem.	Do not undersize reboiler return and bottom feed nozzles.

TABLE 20.7 Troublesome Tray Column Internals (*Continued*)

Case no.	Ref.	Type of plant	Type of column	Brief description	Some morals
711	85	Refinery	Reboiled deethanizing absorber	An undersized liquid drawoff line from a drawoff box caused vaporization in the line. This backed up liquid in the downcomer. The liquid overflowed into the section below. Poor serparation resulted.	Do not undersize liquid outlet lines.
712	375	Refinery		A column flooded prematurely after being switched to a new service. Flooding was caused by an excessive pressure drop in a kettle reboiler circuit backing up liquid to above the reboiler return nozzle. Elevating vapor inlet above liquid level solved the problem. Lack of level indication made diagnosing the problem difficult.	Reboiler vapor should enter above liquid level. Level indication should be provided in columns. Radioactive scans are useful for liquid level detection at column bottoms.
713	296	Refinery	Vacuum column	Leakage from gasketed draw trays increased in service until no level could be maintained on the trays. Different gasketing materials and putty did not solve the problem. Seal welding significantly reduced leakage.	Gasketing may be ineffective in preventing leakage in such services.
714	152	Refinery	Catalytic cracker gasoline debutanizer	Following a retray, the column did not make design separation, and operation was erratic. Column feed was 80°F hotter than tray liquid, and entered the tray a short distance upstream of the downcomer. Insufficient mixing caused vaporization in the downcomer. Problem was solved by feeding into a new chimney tray.	Do not enter hot feeds close to the tray outlet downcomer.

715	4	Refinery	Column consisted of two sections, separated by an upward bulging internal head, which served as a draw pan. The liquid outlet was 3 in above the lowest point; water accumulated below that. When hot oil later filled the pan, a pressure surge occurred and damaged trays.	Ensure adequate drainage of trapout pans.
716	231	Refinery	A valve tray used in trapout service excessively leaked. The leakage was eliminated by seal-welding tray sections, and by welding a strip onto the periphery of the tray, a few inches from the support ring.	A couple of successful techniques for minimizing trapout tray leakage.
717	231	Refinery	A leaking valve tray in trapout service was replaced by an all-welded chimney tray, which was seal-welded to the tray ring. Leakage was eliminated.	An all-welded chimney tray can eliminate trapout tray leakage.
718	237	Refinery	Upon retray, the new trays were rotated 90° to their original orientation. Internal piping was required from the new intermediate product draw sump to its draw nozzle and level gage. A flange on the internal draw line leaked, and starved the line of liquid. Poor separation resulted. The internal level gage lines plugged.	Avoid flanged internal piping whenever possible if leakage can be troublesome. Avoid internal instrument lines.

Combination tower (row 715)
Midsize crude fractionator (row 717)
Wax fractionator (row 718)

TABLE 20.7 Troublesome Tray Column Internals (*Continued*)

Case no.	Ref.	Type of plant	Type of column	Brief description	Some morals
719	237	Refinery	Lube oil prefractionator	The amount of side product that could be withdrawn was restricted. The restriction was caused by an over-flowing draw pan. The pan overflowed because the side product control valve was located on the horizontal pipe between the draw nozzle and the first elbow turning down.	Locate control valves downstream of long vertical runs of outlet pipes.
720	237	Refinery		Plant test data on one pumparound section showed that the efficiency of the top tray was halved when its liquid distributor was removed.	Use liquid distributors for feeds into large columns.
721	237	Refinery	Asphaltic crude tower	Column efficiency severely dropped following a capacity revamp. Problem was caused by the bottom downcomer being converted into a drawoff box with no overflow, coupled with an undersized side-stripper overhead line. Excessive pressure drop in this overhead line backed liquid into the drawoff box, and this liquid flooded the column.	Ensure draw trays are equipped with overflows. Beware of undersized lines.

722	237	Refinery	Depropanizer	A once-through thermosiphon reboiler was starved of liquid because the overflow weir of the trapout pan feeding the reboiler was level with a seal pan weir. Problem was solved by raising trapout pan weir by 6 in.	Trapout pan weir should be higher than seal pan weir when feeding a once-through thermosiphon reboiler.
723	6	Refinery		Bubble caps located beneath an internal cold reflux pipe suffered chloride-compound corrosion.	Condensation of corrosive components can occur at cold internals in the tower.
724	237	Refinery	Lube oil vacuum tower	Lube oil was drawn as a side cut from a trapout tray (total drawoff). When all tower baffle trays were replaced by valve trays, lube oil rate declined by 12% due to leakage at the trapout tray. To restore the original draw rate, the trapout tray was replaced first by a bubble-cap tray, then by a valve tray with venturi openings; finally the outlet nozzles were expanded. Each of these steps progressively further lowered the lube oil rate. Installation of a seal-welded chimney trapout tray solved the problem and achieved a lube oil rate 19% above original.	A leakproof chimney tray should be used for total drawoffs.

TABLE 20.7 Troublesome Tray Column Internals (*Continued*)

Case no.	Ref.	Type of plant	Type of column	Brief description	Some morals
725	150a	Chemicals		Column and its piping, operating at 30 percent of design rates, were shaken by water hammer at feed sparger. Due to oversized discharge orifice area, feed liquid probably ran out of upstream orifices, sucking vapor in via downstream orifices. This vapor collapsed onto the subcooled liquid. Hammer was eliminated by orienting sparger orifices upward to keep it full of liquid at low rates. A deflection bar was added above the orifices to prevent impingement onto the tray above.	Consider low-rate operation when designing spargers.

TABLE 20.8 Troublesome Packed Column Internals

Case no.	Ref.	Type of plant	Type of column	Brief description	Some morals
801	268			Capacity was restricted by the packing supports. The supports were 6-in-wide channels with the wide axis horizontal. If these were vertical, they would have been better structurally and provided more open area.	Provide sufficient open area at the packing supports.
802	410, 215	Ammonia	Hot pot absorber	Severe localized corrosion of tower shell occurred due to impingement of liquid accumulated in a gas inlet distributor. Distributor was modified to eliminate accumulation. Corrosion at another spot was caused by gas issuing from the modified distributor impinging on tower wall. Poor wetting of the tower wall and gas maldistribution caused other localized corrosion incidents. Using stainless steel shingles to protect wall areas was effective in checking corrosion, but later led to stress corrosion.	Pay attention to inlet gas distributors. Maldistribution can lead to corrosion in corrosive services.
803	239	Refinery	FCC fractionator, slurry oil pumparound section	Cool slurry (700°F) did not evenly contact the entering hot (980°F) reactor vapor. Spreads of up to 90°F were observed at the same elevation in a 14-ft tower. A grid packing with a spray distributor was used.	In some sections, overheating of slurry may occur.

TABLE 20.8 Troublesome Packed Column Internals (*Continued*)

Case no.	Ref.	Type of plant	Type of column	Brief description	Some morals
804	237	Refinery	Crude tower	Pieces of packings squeezed through a grid bar support and got stuck in a product pump suction screen. A mesh screen was installed over the grid bars to prevent packing migration. The mesh reduced open area and caused premature flooding.	Avoid restrictive packing supports, or stack larger-size packings on top of the support.
805	237	Refinery	Crude tower	A pressure surge caused breakage of clamps holding together a bed limiter. Sections of the limiter were dislodged, and packings were damaged and carried over. This resulted in poor separation.	Strengthen sections of bed limiters in services prone to pressure surges. Avoid pressure surges in random-packed columns.
806	349	Ammonia	Hot pot absorber	A maldistribution problem caused a CO_2 slip six times greater than design. A water test of the distributors at shutdown detected the problem. Distributor modification solved it.	Water tests are useful for checking distributor performance.
807	237	Refinery	Debutanizer	Two debutanizers operated in parallel and in identical service. One had an HETP of 39 in, the other 72 in. The only difference was that the latter did not have redistributors.	Redistribution is often essential for achieving good packed-column efficiency.

808	346	Olefins	Water wash tower 13.5-ft ID	Ladder-pipe distributor plugged after 3 days in service. Less than 1 lb of solids was sufficient to plug 80% of perforations. Problem eliminated by installing Y-strainers and enhancing distributor perforation diameters. Good packing performance was achieved even though tray support rings were not removed.	Ensure a filter upstream of perforated distributors. Avoid small perforations.
809	26	Gas Processing	Amine regenerator	Pieces of polypropylene saddles deformed at temperatures of about 250°F and passed through the support screen. Pieces were found in downstream equipment and blocked booster pump suction. When the still was opened, only 1 ft of the original 20 ft was found. Despite the loss, the amine was adequately regenerated. Repacking with ceramic saddles solved problem.	Watch out for plastic packing migrating through a support plate in hot services.
810	76	Aromatic hydrocarbon binary separation in a pilot column		Bubble point feed entered the column via a pipe distributor with underside perforations. When the feed became partially vaporized, lights moved into the bottom section, almost contaminating bottom product. Installing a chimney tray below the distributor eliminated the problem.	Distributors that work well with liquid feeds may not work well with flashing feeds.

TABLE 20.8 Troublesome Packed Column Internals (*Continued*)

Case no.	Ref.	Type of plant	Type of column	Brief description	Some morals
811	304	Petrochemicals	Xylene fractionator	Reflux entered via a ladder-pipe distributor, and the center bed was irrigated by an orifice distributor, both of standard construction. These were replaced by a lateral arm and orifice deck high-performance distributors. HETP in the beds was lowered by 20 to 40%.	As for 812.
812	289	Miscellaneous	Six columns: debutanizer; xylene tower; ethylene oxide absorber; Selexol towers, 3- to 14-ft ID	In each case, a standard liquid distributor was replaced by a high-performance distributor. In all cases, substantial improvements in separation efficiency resulted.	Well-designed high-performance distributors can substantially improve column efficiency.
813	75	Hydrogen peroxide	4-ft-diameter column	The column performed poorly after its ceramic random packings were replaced by structured packing. Replacing the vapor distributor by an improved, higher-pressure-drop type solved the problem.	Pay attention to vapor distribution.
814	75	Specialty chemicals	6.5-ft-diameter column, fouling service	The column, which used conventional pan-type liquid distributors with large orifices, could not achieve design separation efficiency. Problem was solved by replacing all distributors with a proprietary two-stage high-performance distributor.	Orifice distributors can be troublesome in fouling services.

815	75	Gas Processing arctic service	Triethylene glycol (TEG) dehydrator	Following replacement of trays with structured packing, the product failed to meet specs and turndown was poor. Glycol rate had to be doubled to achieve dehydration. The most likely cause was gas maldistribution induced by an inlet vapor velocity head of 56 in of water. The column achieved design performance after new structured packing as well as new vapor and liquid distributors were installed.	As for 813.
816	141	Petrochemicals	Styrene-ethylbenzene splitter	The products did not meet design specs. Replacing the liquid distributor by a high-performance distributor solved the problem.	As for 812.
817	183		CO_2 absorption from gas, three columns, 15-ft-diameter, 50-ft-tall beds	Severe gas maldistribution was measured and caused uneven velocity and pressure profiles throughout the entire bed. The maldistribution was initiated at the bottom feed inlet. Efficiency in the column experiencing maldistribution was roughly half that measured in another similar column that had a specially designed gas inlet sparger.	As for 813.
818	219	Test column	C_6-C_7 separation	Replacing a notched-trough distributor by a drip pan distributor effected a 30-40% reduction in HETP.	As for 812.

TABLE 20.8 Troublesome Packed Column Internals (*Continued*)

Case no.	Ref.	Type of plant	Type of column	Brief description	Some morals
819	237	Refinery FCC gas plant	C_3–C_4 splitter	Trays were replaced by structured packing in the top section. A chimney tray installed beneath the packing had undersized downpipes that were not liquid-sealed. This led to chimney tray liquid overflowing the risers, causing maldistribution and possible local flooding. This led to a drop in efficiency as production rates were raised. Problem was solved by eliminating the offending details.	Beware of undersizing chimney tray downpipes. Ensure chimney tray downpipes are liquid-sealed.
820	150a	Chemicals	Acetic acid scrubber from off-gas; 2 ft ID	An odor and excessive acetic acid emissions problem resulted from poor scrubbing. Scrubbing was poor because an open vapor riser was right under the liquid feed point to the distributor. The incoming liquid went into the riser, bypassing the distributor. Removing and blanking the riser solved the problem.	Pay attention to feeding liquid onto distributors.
821	150a	Chemicals batch processing unit	A single-bed column	The column did not achieve a relatively simple separation. Separation was poor because of a past modification of drilling 4-in holes in the distributor floor. This caused liquid to bypass the distribution orifices. Problem was solved by equipping holes with risers.	

822	Ammonia	Selexol absorber and regenerator	Unit did not meet design specifications. Problem solved by replacing standard distributors and redistributors in the towers with high-performance distributors.	As for 812
823	Refinery	Crude vacuum tower	Plugging of several spray nozzles caused packing fouling. Plugging resulted from relying on filter systems that were too distant from the column and using carbon steel piping downstream of filters.	Rust particles from piping can plug distributors. Locate filters close to column and ensure absence of rust particles downstream
824	Refinery	FCC main fractionator	Disk and donut trays just above the feed and trays below the HCO draw were replaced by grid. Erratic temperatures and poor heat transfer resulted, caused by vapor maldistribution. A V-shaped wedge baffle was installed directly at the vapor inlet, but did not help. Baffle and grid coked after 10 months operation, causing a capacity bottleneck.	Vapor maldistribution can be detrimental to grid performance. Avoid baffling vapor inlets in this service.
825	Refinery	FCC main fractionator	Vapor inlet had a baffle about midway in the vapor inlet zone. Coke grew on the baffle, starting on the back of the baffle up through the top of the grid packing above the feed, and needed to be dynamited out	Avoid inlet baffles in this service

TABLE 20.9 Installation Mishaps

Case no.	Ref.	Type of plant	Type of column	Brief description	Some morals
901	203	Olefins	Process water stripper	A bottom downcomer installed backward caused a restriction between the downcomer bottom and the seal pan wall. Cyclic flooding resulted.	Ensure adequate inspection even when internal is hard to get at.
902	203	Olefins	Water-wash tower	A welding rag left in the column reflux line found its way to and partially blocked the reflux distributor to two-pass trays. Excessive entrainment resulted.	Inspect for debris at points where lines leading to the column were modified.
903	55	Whiskey distillery	A whiskey still	Significant entrainment and weeping were experienced as a result of leaking bubble-cap trays at very low liquid rates. Gasketing solved problem.	Bubble-cap trays may not prevent leakage unless properly gasketed.
904	12	Refinery	Stripping tower	Bottom seal pan inadvertently blocked off during a revamp in which a reboiler was replaced by steam injection. Column operated, but flooded prematurely.	Closely inspect modification areas in the tower following revamps.
905	255			A hatchway was left unbolted in a preferential baffle separating column bottom drawoff and reboiler compartments. This caused poor reboiler performance.	Ensure adequate inspection of column bottom section.
906	85		Extractive distillation	Bolts at the flanges of a sectionalized drawoff pan feeding a once-through thermosiphon reboiler were left handtight. Flange leakage caused excessive reboiler outlet temperature.	Ensure adequate tightening of bolts on draw pan joints.

No.	Service	Equipment	Description	Recommendation
907	295		Drying column	Tray perforations varied from one column section to another. During construction, tray sections were mixed up. Resolving problem was costly.	Inspecting engineers should be alert to inspection pitfalls ahead of column assembly.
908	295			Trays were specified to be 316SS; but four of the column trays installed, and many nuts and bolts were 304SS. These would have failed in this service.	Ensure adequate material inspection.
909	364	LPG		During commissioning, reboiler pump strainers were broken due to blockage by debris. Pieces of strainer casings damaged the pumps. Strainer casings with extra support bars avoided recurrence.	It is best to keep debris out of outlet lines. Attention is required to the mechanical strength of strainer casings.
910	34	Soda ash recovery unit	Ammonia still, 10-ft diameter	Ceramic packings were dumped through a chute installed at the manhole. This caused pieces of packing to stratify in layers on an inclined plane ("hill" formation), as well as breakage. This resulted in poor liquid distribution, low efficiency, and low capacity.	Avoid hill formation while packing a column. Plan to avoid breakage while dumping ceramic packings.
911	102		Light hydrocarbon column	Premature flooding occurred on the third tray from the bottom because of variations on this tray that caused vapor leakage into the downcomer. Gamma scans identified troublespot.	Abnormalities on only one tray can bottleneck an entire column.
912	238	Gas processing	Debutanizer	The carcass of a dead rat lodged in the kettle reboiler inlet nozzle, and backed up liquid into the tower. When level reached the reboiler return nozzle, the column flooded prematurely.	At shutdown, keep manholes closed when no one is in the column.

TABLE 20.9 Installation Mishaps (*Continued*)

Case no.	Ref.	Type of plant	Type of column	Brief description	Some morals
913	192	Caustic absorber		An internal pipe on a bottoms drawoff side nozzle was bent upward instead of downward. This caused vapor rather than liquid to escape out of the bottom.	As for 905.
914	219	Test column		Screening chips from breakage of ceramic saddles during shipment was troublesome. It was necessary to resort to picking out chips by hand over 144 ft^3 of packing.	
915	336		Propylene-propane splitter	The column was being retrayed. Existing tray panels would not fit through manholes and had to be hot-cut first. This added 4 days to the retray. Column sway during a violent storm caused a further delay. Despite the difficulties, the retray was completed within the time available.	Ensure column internals are manhole-removable. This paper contains some excellent data on timing and labor requirements for retraying field work.
916	150a	Refinery	Naphtha splitter	Column flooded prematurely after valve trays were replaced by sieve trays. Flooding was caused by large pieces of scale and debris restricting rectifying section downcomer clearances. Design clearances were 1 in; installed were ⅝ to ¾ in. Clearance problem was due to scale left on tray support rings when new panels were installed; this raised the panels.	Properly inspect downcomer clearances following installation.

TABLE 20.10 Water In Otherwise Dry Columns

Case no.	Ref.	Type of plant	Type of column	Brief description	Some morals
1001	203	Gas processing	Deethanizer	Small quantities of water accumulated in a refluxed deethanizer and caused column to hiccup, or empty itself out either from top or bottom every few hours. Cured by replacing reflux by oil absorption.	Refrigerated columns may suffer from water accumulation. Using absorption oil instead of refluxing column can solve problem.
1002	296	Refinery	Atmospheric crude column	Weep holes in the bottom seal pan plugged and trapped water. The water vaporized at startup, causing a pressure surge that lifted trays off their supports. Problem was solved by installing downpipes (extending below bottom liquid level) to drain the pan.	Ensure weep holes operate properly.
1003	296	Refinery		Undrained water in stripping steam line entered tower upon startup and caused a pressure surge which dislodged trays.	A valve right at the column flange with a blowdown drain just upstream can eliminate this problem.
1004	232	Refinery	Debutanizer	Column internals and reboiler tubes severely corroded after the water drawoff control valve on the reflux drum boot plugged. Manual draining was too inconsistent to prevent water (saturated with H_2S) refluxing into the tower. Continuous water flushing with an external water source prevented recurrence.	Avoid small-port control valves in this service. Continuous water flushing of water draw line can prevent blockage.

TABLE 20.10 Water In Otherwise Dry Columns (*Continued*)

Case no.	Ref.	Type of plant	Type of column	Brief description	Some morals
1005	4	Refinery	Vacuum tower	All trays were bumped by a pressure surge at startup. A block valve was opened to establish flow of hot circulating oil to a pump. A pocket of water, trapped between the block valve and a second block valve at the pump suction flashed on contact with the hot oil, resulting in a pressure surge.	Ensure adequate drainage. Even small quantities of water can cause major damage in vacuum towers. Carefully analyze any design changes.
1006	4	Refinery	Combination tower	Nearly all trays were damaged by a pressure surge. A low point was formed in a long horizontal line from a coke drum to the tower. Condensate collected at the low point. When the coke drum was heated, the low point was lifted, dumping water into the tower and causing a pressure surge.	Assure positive drainage of lines connecting two vessels to avoid water trapping.
1007	4, 5	Refinery	Catalytic cracker fractionator	A pressure surge severely damaged trays, support beams, and internal piping. At startup water was trapped above a block valve in the vertical (downflow) fractionator feed line. A drain just above the block valve was plugged. When the block valve was opened, the water was dumped into hot oil, causing a pressure surge.	Ensure complete draining of feed lines before opening. Avoid block valves in vertical lines where water can be trapped.

1008	4	Refinery	Combination tower	A pressure surge severely damaged trays at startup. Steam bleeds were used to keep the upper part of the tower free of air. Some condensed steam drained into hot oil that was introduced near the bottom, resulting in a pressure surge.	Use fuel gas rather than steam to keep column free of air. Start with warm rather than hot oil.
1009	3	Refinery	Vacuum tower	A pressure surge severely damaged trays at shutdown. The surge occurred when hot oil leaked into the tower and contacted condensate from steaming out the column feed furnace.	Prevent leakage of hot oil into the column while steaming out is in progress.
1010	233	Refinery	Alkylation depropanizer	Column feed contained strongly acidic components, which dissolved in small quantities of water and caused a severe and recurring corrosion failure problem. The rate of corrosion failure was greatly reduced by adopting an effective dehydration procedure at startup.	An effective dehydration procedure was developed. Acid-free butane was totally refluxed while drains were intermittently opened until all water was removed.
1011	7, 23	Refinery	Vacuum tower	Pressure surges occurred upon feed introduction and caused tray damage in several cases. The startup procedure did not use oil circulation to flush out water. The surges resulted from pockets of water remaining in draw pans and pumparound circuits.	Oil circulation is essential for flushing out water prior to hot feed introduction in this service.
1012	7, 23	Refinery	Crude tower	Damage to trays occurred at shutdown. The damage was caused by exposing column internals to cold water and air prematurely. Steam cooling of the tower was not adequate prior to air and water introduction.	Ensure adequate column cooling before introducing air or water at shutdown.

TABLE 20.10 Water In Otherwise Dry Columns (*Continued*)

Case no.	Ref.	Type of plant	Type of column	Brief description	Some morals
1013	239	Refinery	Combination tower	Tower trays were repeatedly upset due to pressure surges resulting from water accumulating in the tower during short unit outages. The source of water was condensation of purge steam used under the column relief valves to prevent their inlets from plugging.	Beware of relief valve steam purges.
1014	358	Organic chemicals	High-molecular-weight (106+) separation, large-diameter vacuum column	Severe tray damage occurred due to a pressure surge. The pressure surge occurred when water reached the reboiler. The source of water was a cooling water leak in the condenser. Water found its way down because of a liquid distribution problem.	
1015	7	Refinery	Vacuum tower	A side-stream accumulator was lifted off its foundation by a pressure surge at startup. During hot oil recirculation, back flush of the spare pump was being attempted. The pump discharge valve was cracked open before the suction valve was closed. A pocket of water in the pump or its piping was sucked into the hot oil, causing the surge.	Ensure adequate drainage of spare pumps before connecting to a hot oil system.
1016	7	Refinery	Vacuum tower	The majority of trays and a number of tray supports were damaged by a pressure surge. The column was being restarted following an interruption; 280°F oil was circulated through the tower and heater, and the column was under full vacuum. Source of water was condensed steam that was accumulated in one heater pass.	Several recommendations were made, including (a) avoid starting up under full vacuum; (b) pump the tower out whenever temperature falls below 300°F during an interruption; (c) monitor coil outlet temperatures; (d) take steps to prevent water accumulation in heater coils.

TABLE 20.11 Startup/Shutdown Difficulties

Case no.	Ref.	Type of plant	Type of column	Brief description	Some morals
1101	203, 196	Olefins	Low-temperature fractionator	Column could not be started up because vapor would not allow liquid to descend into downcomer. This caused excessive entrainment.	A method for diagnosing and solving sealing problems was developed. Raising pressure can help solve a sealing problem.
1102	239	Refinery gas plant	Debutanizer	Excessive tube leakage in a reboiler caused reboil, reflux, and column instability, and inability to run at design pressure.	Sometimes an extremely complex-looking problem has a simple cause.
1103	364	LPG	Lean oil stripper	Column was pressured up through a connection in the overhead system while liquid circulated through its valve trays. The gas could not travel downward, causing mechanical damage to top 12 trays. This later resulted in premature flooding.	Always pressure columns from the bottom up, especially when column contains valve trays.
1104	295			Water used in prestartup wash was heavy in solids, and laid a thick mud deposit on trays and exchangers.	Check source of wash water prior to wash.
1105	440	Chemicals	Column stripping inorganic chemicals from aqueous stream	Polypropylene packings melted because of evaporation of a water seal which desuperheated stripping steam issuing from a submerged bottom feed distributor. The incident occurred during a brief maintenance shutdown. Resolidified plastic which oozed through the packing support later caused pump damage.	

TABLE 20.11 Startup/Shutdown Difficulties (Continued)

Case no.	Ref.	Type of plant	Type of column	Brief description	Some morals
1106	210		A large distillation column, made in two halves in series	The two halves were connected by a large vapor line containing a bellows. Steaming the line during shutdown caused one end of the bellows to rise excessively above the other.	Steaming should be avoided unless the column and auxiliaries are mechanically designed for it.
1107	210		Column distilling flammable material	A reflux line was fixed rigidly to brackets welded to the shell of the column. At startup, differential expansion of the hot column and the cold line tore one of the brackets off the column, causing a leak of flammable vapor.	Check supports of auxiliary lines before commissioning. Inspect effect of thermal expansion on supports as the column heats up.
1108	210	Chemicals		Ammonia flowed from storage backward through a leaking valve into the reflux drum of a column that was shut down. From there it flowed into the column and out of an open end in the bottom line.	The possibility of reverse flow should be taken into account when preparing the blinding schedule.
1109	210	Chemicals		Toxic gas leaked from a blowdown header back into a shut-down column through a closed valve, and killed an operator who was draining the column.	As for 1108.
1110	275 vol 2 case 838		C_4 hydrocarbons	A reboiler outlet isolation valve was blinded on the reboiler side at shutdown. Liquid remained trapped in the valve bonnet. After the reboiler was cleaned it was water-washed, then drained. To facilitate draining, the blind was removed and the reboiler valve opened. The trapped hydrocarbons were released and exploded in the column, killing a worker and lifting trays.	Always blind on the column side when people work in the column. Watch out for gas release from rubbery deposits. Open valves during purging and flushing to remove undesirable materials.

650

1111	Chemicals	Stripper column	An infrequently used intermediate drawoff line was plugged just above an isolation valve, which was at grade. When the isolation valve was being removed for maintenance, the plug suddenly cleared, spraying the worker with water and sludge.	Check for obstructions when purging a column prior to shutdown. Isolate infrequently used drawoff lines at the column to avoid dead legs.
	275 vol 2 case 924			
1112	Chemicals	Solvent recovery still	At shutdown the still was being steamed. After steaming, cold water was applied to cool the column. The sudden cooling caused a partial vacuum, and the still imploded. The open vent did not have sufficient capacity to relieve the vacuum.	Either design the vessel for vacuum (e.g. mechanically, or by providing an adequate vacuum breaker or vent) or avoid rapid cooling after steaming.
	275, vol. 2, case 1088			
1113	Ammonia	Hot pot regenerator	Several startup/shutdown accidents in hot pot regenerators have been described. In all cases, the system was being water-flushed, with the absorber under pressure of insoluble gas. The water absorbed small amounts of the gas (natural gas, hydrogen, nitrogen) in the absorber, and released it in the regenerator or its piping. When the gas was combustible, explosions occurred once hot work was performed inside the regenerator or on its vent line. When the gas was nitrogen, a suffocating atmosphere resulted inside the regenerator.	Blind column connections and maintain good ventilation during hot work. Watch out for absorption of gases in wash water. The absorbed gases can be released at undesirable locations. These gases can be trapped in the system even after the wash.
	276			
1114	Refinery		Propane leaked into the steam system via a steam purge connection. This resulted in propane issuing from a fire-suppressant steam purge nozzle.	Ensure proper blinding.
	237			

TABLE 20.11 Startup/Shutdown Difficulties (*Continued*)

Case no.	Ref.	Type of plant	Type of column	Brief description	Some morals
1115	237	Refinery		High-point vent on a distillation column was left open, causing product loss for an entire week.	Ensure adequate closure of all high-point vents following commissioning.
1116	7	Refinery	Crude tower	Crude tower gases were released to the atmosphere and detonated. This followed unblinding and valve removal from a line which contained the gases. Blind removal followed a breakdown in communication.	Blind removal should only be authorized by a written permit.
1117	7	Refinery	Catalytic cracking unit columns	A supplier error caused the unit to be purged with a gas containing 93% oxygen. Several explosions and fires resulted.	Always test purge gas before use.
1118	7	Refinery	Caustic scrubbing system	Hydrogen sulfide was liberated to atmosphere when a caustic scrubbing system was acid-washed.	Consider reaction of wash chemicals with deposits in the column. Adequately plan route of disposal and/or venting of reaction products.
1119	7	Refinery	Crude and vacuum tower	Caustic backflowed into column steam lines. This resulted in caustic being sprayed over a wide area via an atmospheric steam vent.	Pay special attention to blinding before performing a chemical wash.
1120	148	Ammonia	Benfield hot pot absorber/regenerator	Frequent plugging occurred at the lean solution pump suction strainer. Problem persisted despite frequent strainer cleaning. The plugging was caused by particulate matter, including rust, which remained in the system after washing. Unit operated well following a rewash.	It pays to check for particulates at the completion of a wash.

652

1121	Gas processing	Nitrogen rejection unit (NRU) LP stripper	At startup, turboexpander lube oil leaked into the stripper feed and plugged the column. The NRU was shut down, and a hydrocarbon solvent was used to clean the system.	
1122	Chemicals	Vacuum column containing unsaturated fatty acid	A shut-down column contained flammable gas. Work was performed on a ground-level exchanger in the column product line. The product line was purged by inert gas pumped backward into the column, and out through the top of the column. Air managed to get into the column, and an explosion which damaged trays occurred.	Ensure proper blinding.
1123	Chemicals	Vacuum column containing unsaturated fatty acid	Immediately after shutdown, the bottom manhole was opened while column contained 400°F liquid. Air was sucked in due to condensing vapor. A violent combustion resulted, causing injury and widespread tray damage.	Drain liquid and adequately cool column before introducing air.
1124	Gas processing	Glycol dehydrator bubble-cap trays	Downcomer liquid seal was lost on bubble-cap trays. Seal was reestablished by blocking in the gas flow to the tower and continuing glycol circulation for 30 min before reestablishing gas flow.	

TABLE 20.11 Startup/Shutdown Difficulties (*Continued*)

Case no.	Ref.	Type of plant	Type of column	Brief description	Some morals
1125	7	Refinery	Alkylation unit fractionator	An explosion and fire were caused by a butane release during a startup following an interruption. After bolts in the reboiler pump suction screen housing were loosened, heavy deposits trapped in the plugged suction isolation valve broke loose, releasing 150-psi butane to atmosphere.	Ensure proper blinding. Watch out when cracking flanges in plugging services.
1126	7	Refinery	Coker sponge absorber	The light distillate product pump, which supplied lean oil to the absorber, lost suction at startup. Gas from the absorber backed through the lean oil line and traveled into the hydrotreater charge pump, causing it to gas up and lose suction. This resulted in hydrotreater catalyst damage.	Beware of reverse flow.
1127	18	Ammonia recovery	Ammonia absorber	Gas overpressure lifted three storage tanks off their plinths. One split at the base, releasing ammonia liquor. The tanks received liquid feed from the bottom of a high-pressure absorber. The rundown line branched off a pumparound circuit at an elevated position. The pump failed during a startup, and gas from the column backflowed through the upper leg of the pumparound into the tank's rundown line.	Beware of reverse flow. Rundown lines should branch off near the bottom of a pumparound circuit.

1128	Chemicals	18 incidents in various columns, 2.5- to 12-ft diameters	In each incident, one or many trays were damaged. Over half the incidents were caused by bottom liquid level rising above the reboiler return nozzle. Other prime culprits were local vacuum in a column and poor installation.	Beware of excessive bottom level. The paper contains an invaluable list of techniques for preventing this.
1129	Chemicals	Packed rectifier separating low boilers from much higher boilers.	Following operation with near-design boilup, column would flood whenever feed was shut off. Feed entered the column base subcooled; 40% of the boilup was condensed while preheating the feed below the packed section. When feed was shut off, the heat sink was eliminated, causing this boilup to enter the packing and flood it. Problem was solved by installing an override controller that reduced steam flow in response to excessive pressure drop.	Pay attention to the effects of subcooling on capacity under startup/shutdown conditions.
1130	Chemicals	Vacuum tower	A sudden loss of vacuum from the top of the tower caused vapor backflow through the column. This exerted a downward force on the trays, which in turn caused the trays to collapse.	

TABLE 20.12 Operational Difficulties

Case no.	Ref.	Type of plant	Type of column	Brief description	Some morals
1201	203	Olefins	Debutanizer	An end-of-run capacity limitation caused an off-spec product. A feed stream was bypassed around the column. This overcame the problem, with a surprising beneficial side effect.	Bypassing a column feed can unload the column and have suprisingly beneficial side effects.
1202	12	Refinery	Stabilizer reboiled by a fired heater	Premature tower flooding occurred when level in the bottom of the column rose above reboiler return nozzle and backpressured a uniquely designed bottom surge drum. Level rose either because vapor was present in liquid line to heater or liquid was entrained in drum vapor. Provisions which lowered surge drum level solved the problem.	Either avoid vapor in liquid lines or design for it. Always monitor tower bottom level.
1203	296	Refinery	Vacuum tower	An automatic steam cutout on high level was installed in the stripper section of the tower. It saved the tower many times from tray damage.	
1204	231	Refinery	Fractionators	In one case, it was thought that a drawoff tray was leaking, but the problem turned out to be a control valve stuck half open. In another case, pumparound performance was improved simply by opening two discharge valves that were pinched back for an unknown reason.	Do not overlook the obvious.

1205	98		A feed tank was emptied for the first time, stirring up settled solids from the bottom. The solids plugged a column.	Avoid emptying out feed tanks while they are feeding a column.
1206	210		A temperature controller at the column base went out of order. Seven hours later, flammable liquid spilled out of the reflux drum. Several abnormal instrument readings were overlooked during this period.	Do not ignore abnormal instrument readings.
1207	275	Anthraquinone	Still separating hydrogen peroxide from organics	A feed tank was switched. At the switch, the feed filter appeared to block. The liquid level dropped at the column vaporizer, resulting in hydrogen peroxide concentration and an explosion.	Columns should be designed so that feed failure is not hazardous.
1208	96	Insecticide	Hexane recovery from residue	Excessive concentration of residue caused an explosion.	Avoid excessive concentration of unstable chemicals in column bottom.
1209	239	Refinery	Fluid catalytic cracker absorber-stripper using a recontact drum	Excessive condensation in the drum during winter repeatedly caused buildup of ethane, until the stripper flooded. Keeping drum warm by cutting cooling to the condenser prevented recurrence.	
1210	239	Refinery	Catalytic naphtha reformer absorber	At low rates, lean solvent was subcooled and absorbed light components (ethane). This increased the internal circulation rates of lights and wasted energy. Preheating the solvent with waste heat solved the problem.	An excessive subcooling problem can be prevented by preheating the solvent.

TABLE 20.12 Operational Difficulties (*Continued*)

Case no.	Ref.	Type of plant	Type of column	Brief description	Some morals
1211	239	Refinery	FCC unit Propylene-propane splitter	A small amount of KOH solution from an upstream dryer was carried over into the column. Once in the column, KOH precipitated, and plugged sieve decks equipped with ³⁄₁₆-in holes. Deposits were removed by acid wash.	Small perforations plug
1212	238	Gas processing	Glycol dehydration	Poor dehydration was caused by a leaking feed-effluent exchanger that leaked cold, wet glycol into the heated dry glycol flowing to the top of the contactor.	
1213	352	Chemicals	Large distillation column	A trapped component periodically built up in the upper section of the column. When it built up, the control temperature rose and increased reflux, eventually causing premature flooding. Gamma scans diagnosed the problem. Taking a purge stream from this section solved the problem.	Gamma scans are effective in diagnosing the nature of a flooding problem. Side draws are effective in preventing trapping of an intermediate component.
1214	238	Gas processing	Depropanizer	Flooding occurred in the rectifying section even though measured pressure drop was low. Pressure drop was low because only the top tray was flooded. Flooding was caused by top downcomer plugging with corrosion products.	Flooding may occur even when column pressure drop is low.
1215	71	Chemicals	Batch distillation still	Leakage of steam from the reboiler contaminated overhead product with water. A radiotracer technique diagnosed the source of leak and measured the rate of leakage.	Radiotracer techniques are useful for diagnosing exchanger leakage problems.

1216	71	Phenol	A fractionation column	A salt precipitated out in the lower section of the column. The deposits restricted vapor upflow and caused poor fractionation. Gamma scans identified the problem, tray design changes solved it.	Gamma scans are useful for diagnosing plugging problems.
1217	7	Refinery	Vacuum tower	Most of the trays were torn off their supports following a pressure surge. The surge was caused by vaporization of a slug of flushing oil. The slug entered the tower from a spare bottoms pump that was inadequately isolated.	Ensure proper isolation of spare pumps in services prone to pressure surges.
1218	125	Ammonia	MDEA absorber	The main pump delivering semilean amine from the desorber to the absorber was shut down for maintenance and the spare pump was started. The spare pump delivered one-third of the normal flow, causing poor absorption. This could not be tolerated downstream, so the plant was shut down. Shortly after the spare pump was switched off, hydrogen escaped through the seals of the main pump and fired. The incident was caused by backflow through the main pump, induced by failure of the main pump nonreturn valve.	Beware of reverse flow. Nonreturn valves cannot be relied on to prevent reverse flow. The article contains a comprehensive description of the incidents and measures to avoid recurrence.
1219	19	Pharmaceuticals solvent recovery	IPA-water azeotropic distillation using benzene and IPE entrainers	Small amounts of methanol and acetone in the IPA reduced separation efficiency. Problem was solved by removing all high volatiles before starting the azeotropic distillation.	Impurities can interfere with azeotroping.

TABLE 20.12 Operational Difficulties (*Continued*)

Case no.	Ref.	Type of plant	Type of column	Brief description	Some morals
1220	19	Pharmaceuticals solvent recovery	IPA-water azeotropic distillation using benzene and IPE entrainers	Cold entrainer reflux reduced column capacity. Problem was solved by reheating the reflux.	In azeotropic distillation, subcooled reflux can lower column capacity.
1221	237	Pharmaceuticals	Methanol dehydrator	Excessive caustic injected into feed was entrained into the methanol-rich top section and precipitated there due to water vaporization. The deposits plugged the 3/16-in sieve tray holes. This induced premature flooding. Problem solved by on-line water wash, effected by raising boilup and cutting reflux for a few hours, thus inducing water up the column. Longer-term solution was cutting caustic injection.	A useful on-line washing technique was devised. Small perforations plug
1222	237	Pharmaceuticals	Methanol stripper 3-ft-ID tower with 16 trays in bottom section, random packing in top section.	Column ran flooded with off-spec product for several months because bottom sump liquid level was above the reboiler return inlet. During this period, the packing support also collapsed, presumably due to pressure surges. The flooding was eliminated by lowering the liquid level below the reboiler return inlet.	As for 1224.
1223	429a	Gas processing	Amine regenerator	The large stripping heat supplied was insufficient to adequately strip H$_2$S from a rich amine solution due to excessive amine circulation rate. An eightfold cut in circulation permitted an eightfold cut in stripping heat simultaneous with a major improvement in H$_2$S stripping.	Good plant testing is invaluable for defining and overcoming plant problems.

660

1224	150α	Chemicals		A temporary loss of bottoms pump caused base level to rise above the reboiler return nozzle. This caused bottom trays to collapse. Reflux was raised to meet purity with fewer trays, resulting in flooding.	Avoid base level rise above the reboiler return nozzle.
1225	150α	Natural gas	Amine regenerator valve trays	Premature flooding occurred after several months in service. Iron and other metal carbonates formed deposits 1 in thick on some of the top trays. The solids originated in the natural gas stream. Problem was solved by cleaning, and recurrence prevented by annual acid wash-out.	
1226	150α	Chemicals	Water scrubber	Fungus growth caused orifice pan distributor to plug and overflow.	
1227	75α	Heavy water	Heavy water distillation	Seal oil leaking into the feed from pumps covered column packing with a thin hydrophobic film. This reduced wettability, which in turn cut efficiency in half. A detergent wash was only partially effective for oil removal. Problem solved by injecting a low concentration of nonvolatile surfactant during several days of operation.	In situ scouring can help eliminate an oil problem without shutdown. Solvent selection is critical and may require extensive off-line experimentation.

TABLE 20.13 Reboilers That Did Not Work

Case no.	Ref.	Type of plant	Type of column	Brief description	Some morals
1301	203	Olefins	Demethanizer	Condensate seal was lost on a vertical thermosiphon reboiler causing loss of heat transfer.	Losing reboiler condensate seal can lower reboiler heat transfer rate.
1302	203	Gas processing	Demethanizer	Erratic vertical thermosiphon reboiler action resulted from a piece of masking tape stuck in a reboiler flange.	Do not use masking tape as flange covers.
1303	134			Heat transfer from a vertical thermosiphon reboiler declined with time. Cause was accumulation of residue in liquid line to reboiler.	Draw off residue from base of reboiler, not from base of column, at an adequate rate.
1304	134			Kettle reboiler supplied insufficient heat. Reason was excessive pressure drop in the vapor line from the reboiler causing low liquid level in the reboiler shell.	Ensure vapor lines are adequately sized. Level indicators on kettle reboilers are helpful.
1305	28		Recycle reactants separator (C_6 range)	Vertical thermosiphon reboiler performed poorly because of inerts accumulation on its condensate side. Venting was inadequate.	Ensure adequate venting on reboiler condensate side.
1306	232	Refinery		Inability to vent accumulated CO_2 from the steam (tube) side of a horizontal reboiler caused corrosion and tube leakage near the floating head. Problem was solved by extending an upper tube to make up a vent tube from the floating head to a vent valve located at the channel head.	A novel technique was developed to solve problem.

1307	Air separation	Air separation column reboiler	Poor venting of noncondensables from the condensing side of the reboiler interrupted thermosiphon action. This in turn reduced the effectiveness of removal of hydrocarbon impurities from the reboiler liquid. Hydrocarbon accumulation caused an explosion.	Ensure adequate venting on reboiler condensate side. Avoid buildup of hazardous impurities in reboiler.
1308	Refinery	Depropanizer	A newly installed preheater had several tube leaks. Column feed backflowed into the steam supply, and from there into the steam side of the reboiler. The volatile feed gas-blanketed the reboiler, causing erratic behavior in the column.	What may appear to be a reboiler tube leak can be caused by a leak elsewhere.
1309	Refinery	Toluene column	Restricting liquid circulation through a thermosiphon reboiler almost tripled heat transfer coefficient. The high rate of circulation evidently interfered with nucleation.	Watch out for excessive circulation rates.
1310	Refinery	Depropanizer	High pressure drop in the reboiler caused liquid to back up above the reboiler return nozzle and flood the column. The problem was caused by introducing kettle reboiler feed at a point from which liquid could not easily spread, failure to allow for head over the kettle overflow baffle, and low liquid driving head from the column.	
1311	Refinery	Depropanizer	A once-through thermosiphon reboiler could not be started up because tray weeping at startup starved the reboiler of liquid. A valved dump line connecting reboiler liquid and bottom sump was added and solved problem.	A dump (or startup) line connecting column bottom sump and reboiler liquid line should not be overlooked for once-through reboilers.

TABLE 20.13 Reboilers That Did Not Work *(Continued)*

Case no.	Ref.	Type of plant	Type of column	Brief description	Some morals
1312	237	Refinery	Depropanizer	Slug flow in an oversized outlet line from a once-through thermosiphon reboiler caused fluctuations in column pressure and bottom level.	Beware of oversized reboiler outlet lines.
1313	134	Refinery		The performance of a forced circulation reboiler was poor, and was much the same whether power to the pump was on or off. Problem was caused by NPSH required exceeding NPSH available.	Ensure pump system compatibility in forced-circulation reboiler systems.
1314	87	Refinery	200–240°F hydrocarbon separation	Column product failed to meet specification because of puffing in a once-through horizontal thermosiphon reboiler. The puffing caused some liquid to bypass the trapout pan. The puffing was caused by vapor binding at the distribution baffles. Drilling vent holes in the baffles improved operation.	Ensure distribution baffles in horizontal thermosiphon reboilers are adequately vented.
1315	290	Chemicals		Poor heat transfer occurred in a vertical thermosiphon reboiler heating 440°F column bottom (expected outlet temperature was 550°F) by liquid Dowtherm which entered at 725°F. Film boiling caused the problem. Reversing the Dowtherm flow from cocurrent to countercurrent solved the problem.	Reversing flow direction can overcome film boiling problems.
1316	209a	Cumene oxidation	Cumene hydroperoxide (CHP) concentration	Reboiler used to concentrate CHP exploded. As a result of a late change in the design, the low level alarm was set too low.	Watch out for low liquid levels when concentrating unstable substances.

TABLE 20.14 Condensers That Did Not Work

Case no.	Ref.	Type of plant	Type of column	Brief description	Some morals
1401	134	Hydrocarbon gases condensation		Low heat transfer in a horizontal, two-pass, in-tube condenser was caused by an undersized condensate line using gravity flow.	Ensure adequate condensate removal.
1402	134			Capacity of a horizontal, in-shell condenser was well below design. Inlet vapor entered in the middle; ends were inert-blanketed. Vents solved problem.	Ensure adequate venting.
1403	134			A horizontal in-tube condenser with axial inlets and outlets did not achieve design capacity. Axial outlet did not permit condensate drainage.	Ensure adequate condensate removal.
1404	381			Vapor entering a vertical downflow in-shell condenser contained a high-molecular-weight condensable material and a low-molecular-weight inert. Poor condensation was caused by channeling that caused inert blanketing.	Use sealing strips; ensure adequate condenser pressure drop.
1405	381			Horizontal in-shell condenser with vapor upflow and condensate downflow did not reach design capacity. This was caused by liquid entrainment at excessive vapor velocities.	Avoid excessive velocities in vapor upflow condensers.

665

TABLE 20.14 Condensers That Did Not Work (*Continued*)

Case no.	Ref.	Type of plant	Type of column	Brief description	Some morals
1406	239	Refinery	Debutanizer	The ability to condense the overhead product was lost because of vapor blanketing in the condenser shell. Venting solved the problem. A newly installed, nitrogen-purged instrument caused the problem.	Ensure adequate venting when inerts are likely to be present.
1407	237	Refinery		A new set of condensers was added in parallel to an existing set in order to increase condensation capacity. Instead of increasing, condensation capacity decreased. Vapor maldistribution was the cause.	Total condensers are best added in series. If added in parallel, beware of maldistribution.
1408	70	Refinery	Crude tower	Column pressure and product gas rate sharply fluctuated during low-rate operation in this and other units. The condenser was a partial condenser located at ground level, with an elevated reflux drum. Problem was caused by slug flow in the riser from the condenser to the drum.	Size risers to avoid slug flow even during low-rate operations.
1409	254	Chemicals		Vacuum jets overloaded with uncondensed vapor. A downflow in-tube condenser was used with large baffle windows. Reducing baffle windows solved the problem.	

666

1410			Absorption plant stripper	Rayleigh fractionation occurred with a wide-boiling mixture. Some vapor, which left the condenser uncondensed, mixed with condensate in the condensate outlet pipe, causing a sudden 10°F temperature rise due to vapor condensation. The system still worked, but not much leeway was left. Situation could have been remedied by venting or by injecting liquid near the back of the condenser.	
1411	7	Refinery	Crude distillation	Overhead to crude exchangers were equipped with impingement plates on both sides for 180° rotation. The lower impingement plates in three exchangers collapsed, blocking their outlets. A sudden pressure rise followed and lifted several atmospheric relief valves. Recurrence was prevented by removing the lower impingement plates.	
1412	177a	Refinery		Sour hydrocarbons from column condense in the shell of water-cooled submerged condenser. Tube bundle severely corroded due to concentration of acidic components near outlet, and needed frequent replacement. Adding a vent line from condenser to reflux drum and modifying controls to supply a small purge stream to condenser feed doubled tube life.	A valuable technique for alleviating accumulation of undesirable gaseous components.

TABLE 20.15 Control Systems That Did Not Work

Case no.	Ref.	Type of plant	Type of column	Brief description	Some morals
1501	203	Gas processing	Lean oil still	Inerts accumulation in flooded reflux drum caused unflooding of the drum and poor control. Manual venting could not solve problem because plant was not continuously attended.	A simple automatic venting system was developed and solved the problem.
1502	203	Gas processing	Deethanizer	Controlling bottom impurity by an analyzer located in the next column overhead did not work because of excessive dynamic lags.	A simple sampling system was developed to obtain an adequate sample from the deethanizer bottom stream.
1503	203	Gas processing	Hot pot regenerator	A low leg in the column overhead line filled with liquid and backpressured the column, causing control instability.	Avoid low legs in column overhead lines.
1504	13		A mixed alcohol-ether column	Excessive alcohol losses occurred because a temperature control point sensitive to the key products and at the same time insensitive to other components could not be found. The column separated volatile azeotropes from mixed alcohols. Analyzer control solved the problem.	In some multicomponent separations, adequate temperature control cannot be achieved. Analyzer control is then required.
1505	68	Chemicals		Controlling reflux drum temperature by throttling cooling water to condenser caused boiling of cooling water when control valve closed. This resulted in atmospheric product release.	Pay attention to low-rate operation of control systems throttling cooling water to condensers.

1506	381		Condensation in a horizontal in-shell partial condenser with liquid outlet at the bottom and vapor outlets at the top was controlled by varying liquid level in the condenser. Excessive entrainment was caused by condenser pressure drop building a large hydraulic gradient.	Avoid high levels and high pressure drops in such condensers.
1507	381		Closure of a control valve in column overhead line to an air condenser caused rapid condensation downstream of the valve and a severe liquid hammer. Control valve was modified so that it would not shut.	
1508	268	Heavy ends removal column	Column was unable to meet bottom design purity even at higher than design reflux rates. Top purity was on spec. Problem was caused by control system setting too low a top product rate. The remainder of the light component was forced to leave out of the bottom.	Ensure proper material balance control.
1509	210	Chlorine	Radioactive bromine, used as a tracer in a brine stream which was electrolyzed to make chlorine, ended up in the column. It concentrated at the base, and interfered with the action of the nucleonic level controller, eventually flooding the column.	Nucleonic level devices are affected by radioactive materials.
1510	210		Failure of a column bottom level controller caused gas to enter the product storage tank and rupture it.	Pay attention to column bottom level indication.

TABLE 20.15 Control Systems That Did Not Work *(Continued)*

Case no.	Ref.	Type of plant	Type of column	Brief description	Some morals
1511	97	Chemicals	Recovery of epichlorohydrin from tars	The thermowell used for controlling heat input into the column was located at the reboiler outlet. It fouled up. The operators tried to control heat input by watching the column top temperature. This was unsuccessful; the reboiler overheated, resulting in an explosion.	In heat-sensitive services, provide an alternative temperature indication above (but near) the bottom. Maintain thermowells near the column bottom clean.
1512	97	Chemicals	Column vaporizing cumene from cumene hydroperoxide	A "duplicate" column was installed. In the duplication process, the reboiler was deepened. The setting of the low-level alarm did not take the deepening into account. This resulted in the reboiler exploding.	Be alert to any hardware differences when duplicating instrumentation from an "identical" or "similar" unit.
1513	97	Hydrogen peroxide	Hydrogen peroxide–water still	A low level signal at the reboiler served as a safety device to prevent low liquid level. The level float, which was located in the reboiler boiling liquid, failed to detect a low level condition. This caused an explosion.	Locate level devices on a bridle and not in the boiling liquid.
1514	275	Organic chemicals	Vacuum distillation of a toluene cut batch distillation	Foaming occurred at the kettle. The level indicator therefore failed to detect the low level condition. Because of the low level, the still temperature indicator showed a vapor temperature which was lower than the liquid temperature. The low temperature increased the heat input. An exothermic reaction took place, causing eruption of residue.	As for 1513 and 1511. Watch out for temperature indicators located below the bottom tray.

1515	Refinery	239	Combination tower	Bottom liquid level rose above the vapor inlet nozzle because of a faulty level controller. The submergence backpressured the coke drum upstream. When the operator noticed this, he quickly lowered the bottom level. This caused foamover (a "champagne bottle" effect) in the coke drum.	Avoid liquid level rising above the bottom feed nozzle. Avoid excessively rapid draining of column liquid.
1516	Refinery	239		Rust layer formed on the inside of the channel head of a reboiler to the level where the steam condensate normally ran. This indicated that 20% of the heat transfer area was waterlogged and ineffective.	Beware of controlling reboilers by a valve directly on the condensate drain line, and the steam supply pressure is low.
1517	Refinery	239	Large debutanizer	Column was limited by overhead condensing capacity during summer. A new condenser was purchased, but never used because just before its installation it was discovered that the control valve in the condenser vapor bypass (process side) leaked. Blocking in the bypass increased condenser capacity by 50%.	Never overlook the obvious.
1518	Refinery	239	Isobutane–n-butane splitter	A sudden rise in propane (light nonkey) occurred. The top temperature controller counteracted the falling temperature by increasing n-butane (heavy key) in the top product.	Temperature controllers can be fooled by sudden changes in nonkey concentration.

TABLE 20.15 Control Systems That Did Not Work (Continued)

Case no.	Ref.	Type of plant	Type of column	Brief description	Some morals
1519	239	Refinery		The heat input control valve was located in the 30-psig steam line to the reboiler. The condensate was at 20-psig. When the valve was throtled, condensate would back up into the reboiler and waterlog tubes. The column would call for more heat and the valve would reopen until the condensate drained. It would then throttle, and the cycle was repeated.	Problem was overcome by relocating the valve to the condensate line.
1520	237	Refinery	High-pressure absorber using gasoline to absorb propane from vent to fuel gas	Loss of bottom level indication resulted in column flooding. Gasoline spilled over to the top knockout drum, thence to the fuel system, and ended spilling out of burners, causing several heater fires.	Ensure adequate level indication.
1521	237	Refinery	Depropanizer	Reflux drum level indicator, level gage, and level alarm were connected to the same taps. The upper tap plugged, and all became erratic. This caused liquid to flow into the flare because of excessive level. Later, the reflux pump blew a seal because of cavitation resulting from low level.	Level indicator and level gage should not share the same tapping.
1522	77	Refinery	Column producing narrow-boiling-range distillate	Column pressure was controlled using a hot vapor bypass scheme (partially flooded condenser). Severe pressure and reflux drum level upsets occurred whenever the reflux drum surface was inadvertently agitated.	Avoid reflux drum agitation when using this control scheme.

672

1523	77	Column equipped with a total condenser	Column pressure was controlled by a valve located in the condenser bypass. The condenser was located above the reflux drum and drained freely (no liquid held in the condenser). This method did not work.	Beware of the condenser bypass control method when the condenser is not partially flooded.
1524	164		Condenser controlled using a hot vapor bypass. Subcooled liquid leaving the condenser was mixed with the hot bypass vapor prior to entering the reflux drum. Severe shock condensation occurred. Problem was solved by entering the vapor and liquid separately into the drum.	With this scheme, vapor and liquid should enter the reflux drum separately. Liquid should enter below the liquid surface.
1525	164		Condenser controlled using a hot vapor bypass (partially flooded condenser). Subcooled liquid entered the reflux drum vapor space (presumably due to unflooding the liquid inlet), and contacted drum vapor that was 100°F hotter. The rapid condensation sucked the liquid leg between the condenser and drum in seconds.	Ensure the point of liquid introduction into the reflux drum is always below the liquid surface when this control scheme is used.
1526	238	Gas Processing Glycol regenerator	The regenerator (kettle reboiler) draw compartment level exceeded the overflow baffle. The high level was undetected because of oil accumulation in the compartment. The level gage was fooled by the low density of the oil.	Check whether any oil can be skimmed before trusting level indication in aqueous systems.

TABLE 20.15 Control Systems That Did Not Work (*Continued*)

Case no.	Ref.	Type of plant	Type of column	Brief description	Some morals
1527	238	Gas processing	Amine absorber	A hydrocarbon phase settled above the amine in the bottom sump. Due to the lower density of this phase, the level indicator read low. Liquid level rose above the vapor inlet nozzle, while level indication was still normal. This prematurely flooded the tower. The flooding flushed the hydrocarbon layer overhead. Once the layer was flushed, the flooding stopped by itself.	Ensure adequate skimming to avoid settling of oil above an aqueous phase.
1528	239	Refinery	Isobutane–normal butane splitter	An advanced feed-forward control system was installed. Each time a power outlet was used in the control room, the feed-forward system was affected just like running an electric appliance interferes with TV reception. This caused erratic reflux and reboil behavior.	
1529	239	Refinery	Debutanizer	An advanced feed-forward control system caused steam flow, reflux, pressure, and temperature to drop. Switching the steam flow to manual resuscitated the column. Problem was caused by a plugged tap of a steam flow meter. The malfunctioning meter misled the controller.	

1530	66, 65	Chemicals	12-ft-diameter, 100-tray column with valve trays	Control of bottom level by manipulating boilup was unstable due to inverse response. Stepping up reboiler steam displaced tray liquid into the column base so that bottom level rose instead of falling. Problem was solved by manipulating reflux flow to control bottom level.	Column inverse response can be troublesome when boilup is manipulated by bottom level. An extensive analysis of inverse response, including predictive equations, is presented.
1531	52	Chemicals	Bubble-cap column	Column acted like a stripper; reflux-to-distillate ratio was 0.43. When reflux flow was on accumulator level control, a small change in heat input led to large changes in reflux flow. Reflux flow at times fell below the minimum required for tray wetting.	Avoid controlling reflux by accumulator level in "stripping" types of columns operated at low reflux ratios.
1532	52	Chemicals	Same as 1531	Control of bottom level by manipulating boilup was sluggish and unsatisfactory due to reboiler inverse response. Increasing reboiler heat input backed up liquid into the column base so that bottom level rose instead of falling. Problem was solved by controlling bottom level by manipulating bottom flow and bottom composition by manipulating boilup.	Inverse response of the reboiler may be troublesome when boilup is manipulated by bottom level.
1533	300	Refinery	FCC debutanizer	The column was highly sensitive to changes in ambient conditions. Reflux was controlled by a tray temperature and distillate was on accumulator level control. Interchanging these controls desensitized the column and improved its stability.	The modified system is better suited for minimizing upsets due to disturbances in the column overhead system.

TABLE 20.15 Control Systems That Did Not Work (*Continued*)

Case no.	Ref.	Type of plant	Type of column	Brief description	Some morals
1534	269	Ethylbenzene Unit	Ethylbenzene-xylene splitter	Reflux-to-distillate ratio was 70:1. Boilup was controlled by bottom composition, distillate by accumulator level, and reflux by distillate composition. Control was unstable and extremely slow. System was modified to control reflux by accumulator level, boilup by steam flow, and distillate by set flow adjusted manually for distillate composition. This was better but still slow. Tight product quality was achieved when manual setting of distillate flow was replaced by an on-off control.	An extensive analysis of the system and its dynamics is presented.
1535	418	Refinery	Alkylation unit deisobutanizer, several columns	Boilup was controlled by bottom section ΔT. This worked well when the bottom contained no components heavier than C_4. Occasionally, the control became unstable when lights content in the bottom was high. At a later stage, a feed containing heavy nonkeys was added to the column, and the ΔT control became unstable. Changing to straight temperature control improved stability.	ΔT control may be unsatisfactory when product impurities are relatively high, especially if the impurities are nonkey components. A thorough analysis of the behavior is presented.
1536	59	Aromatic hydro-carbons	Benzene column	Double differential temperature control gave stable control of top and bottom product purities. Both product purities were in the parts per million range. A conventional temperature control was unable to accomplish this.	An effective technique was developed for composition control of high-purity separations. A thorough analysis is presented.

1537	259	Natural gas liquids	Debutanizer	The column feed was preheated by the bottoms then by a steam preheater. Preheater steam was controlled by the feed temperature downstream. The feed enthalpy fluctuated with fluctuations in column bottom flow. This interfered with the column product analyzer control. Problem was cured by a feed enthalpy controller which regulated preheater steam flow.	Analyzer control may fall short of achieving its objectives if column is unstable. A feed enthalpy control may be needed if heat input to the feed fluctuates.
1538	378		Deisobutanizer	Top section temperature controller responded well to feed changes but produced an offset when lights were present. Installing IR analyzer control with a sampling point at accumulator outlet gave poor control. Relocating the sampling point to a tray near the column top was better, but control was destabilized by rapid feed disturbances. A cascade system using a chromatograph sampling at the accumulator outlet to adjust the set point of the temperature controller gave good response, eliminated the offset, and handled feed disturbances well.	An analyzer-temperature cascade can give better control than temperature or direct analyzer control. Sampling at the accumulator outlet can be troublesome with direct analyzer control. Temperature control can be troublesome in the presence of nonkeys. Direct analyzer control can be troublesome in the presence of feed disturbances.
1539	362	Alcohol	Ethanol-water column	Reflux was controlled by a top section differential temperature controller. A low differential temperature signaled excess reflux, but also occurred without any reflux at all or when the column was cold.	Beware of differential temperature control problems during startups or severe disturbances.

677

TABLE 20.15 Control Systems That Did Not Work *(Continued)*

Case no.	Ref.	Type of plant	Type of column	Brief description	Some morals
1540	309		Depropanizer	Analyzer control was troublesome when sample point was located in the overhead vapor line. Isobutane concentration was twice at the center of the line than near the wall, and the concentration gradient was unsteady.	Nonreproducibility of samples drawn from the column overhead line can render direct analyzer control troublesome.
1541	309		Debutanizer	Column feed was heated by a feed-bottom interchanger, then by a steam preheater. Feed temperature was controlled by adjusting preheater steam. Boilup was manipulated by a tray analyzer and bottom flow by the base level. A disturbance in steam pressure at times rendered the feed temperature control loop inoperative, leading to analyzer control cycling. Interchanging the level and analyzer controls eliminated the problem.	Beware of disturbance amplification via a feed-bottom interchanger. Some control schemes handle such disturbances better than others.
1542	411	Refinery	Alkylation unit deisobutanizer	Boilup was controlled by the bottom section ΔT controller. By a careful choice of the ΔT measurement locations, and operation to the right of the maximum in the curve of ΔT versus bottom composition, the system was made to work even in the presence of significant fraction of nonkeys.	There are cases where ΔT control can be made to work even when a substantial fraction of nonkeys is present in the feed (compare 1535). A thorough analysis is presented.

1543	406	Refinery	Naphtha splitter and deisobutanizer	Column temperature controls were unable to prevent periodic off-spec product resulting from feed fluctuations. Replacing temperature controls by analyzer controls eliminated problem and gave smooth, tight composition control.	Analyzer control can give superior performance to temperature control.
1544	71	Olefins	Stripper	The base level controller failed at startup, and liquid level in the column rose to fill half the column. This caused excessive heavies in the top product, possibly due to liquid carryover. The problem was diagnosed using gamma scans. Cutting feed rate provided short-term solution. Using a gamma-ray absorption level indicator provided a longer-term solution.	
1545	7	Refinery	Alkylation depropanizer	Both the level indicator and level controller failed on the overhead receiver, which separated liquid HF from liquid hydrocarbons. HF overflowed into the hydrocarbon product route, which included a bed of solid KOH. Violent reaction between KOH and HF overpressured the vessel, causing multiple explosions and rupture of the vessel.	As for 1520.

TABLE 20.15 Control Systems That Did Not Work (*Continued*)

Case no.	Ref.	Type of plant	Type of column	Brief description	Some morals
1546	362			Column bottoms, drawn from the weir compartment of a kettle reboiler, preheated column feed. Preheat was not controlled. A rising bottom level increased bottom flow and feed preheat. The greater preheat reduced column downflow, bottom level, and bottom flow. This in turn reduced preheat and raised bottom level. A cycle developed.	Controlling preheat could have avoided the problem.
1547	151*a,b*	Pharmaceuticals	Column fractionating a minimum boiling toluene azeotrope from toluene	Distillate and bottoms were controlled by accumulator and sump levels, respectively, feed and reflux on flow control and boilup was temperature-controlled. Tower pall rings were replaced by higher-capacity rings (bottom) and wire-mesh structured packing (top) to increase capacity and reduce reflux. The column was sensitive to ambient disturbances (e.g., rainstorms). The reflux reductions escalated this sensitivity to an extent that annulled the revamp benefits. The temperature control was ineffective due to its narrow range of variation. Problems were solved by controlling boilup on sump level and bottom product on flow control.	Indirect MB control can be troublesome where ambient variations are a prime source of disturbances. A non-MB control can perform well where fluctuations in feed rate and composition are minor. Column temperature control may be ineffective when its range of variation is narrow.

1548	315a	Gas plant	Stripper	At a capacity-boosting revamp a preheater was added to supplement boilup requirements. The preheater heat input control valve was in the 100-psig steam line to the preheater. At turndown to near the pre-revamp rates, the preheater heat duty fell. The inlet valve closed, dropping the preheater pressure below the condensate header, and the preheater would stop operation.	Problem overcome by shifting heat load from the reboiler to the preheater, thus keeping the preheater above its turndown limit.
1549	237	Pharmaceuticals	Methanol stripper (stripping methanol from water)	Control temperature was located at the reboiler outlet, where the mixture was almost pure water and temperature was insensitive to composition. This, combined with the operator's natural reaction, promoted flooding.	Manual operation of the boilup rate was more satisfactory than this temperature control.
1550	59a		Azeotropic alcohol/ether/water/heavies column	Heavies, alcohol/ether/water azeotrope, and dry alcohol product were the bottoms, distillate, and bottom section side-draw products, respectively. Azeotrope was split into an aqueous purge and an organic distillate/reflux stream in top decanter. Column experienced erratic operation and excessive alcohol losses and steam consumption. Problems were mitigated by changing boilup control from a single temperature to a two-temperature average near the water/organic break point, adding a second temperature control on bottom tray to replace bottom flow control, and changing pneumatics to DCS to improve tuning and responses.	Average temperature control can be advantageous near sharp composition break points. In columns with four products, two composition controls are often better than one.

TABLE 20.15 Control Systems That Did Not Work (*Continued*)

Case no.	Ref.	Type of plant	Type of column	Brief description	Some morals
1551	59a		Azeotropic alcohol/ether/ water/heavies column (same column as in 1550)	Above problems were completely eliminated by further changes including (*a*) implementing break-point position control. This technique subtracted adjacent temperature readings, identified the largest difference as the break-point interval, and used its position as the control signal; (*b*) adding analyzer control on the aqueous/organic decanter split; (*c*) adding pressure compensation to temperature controls; (*d*) adding feed-forward controls.	Break-point position control can be beneficial for azeotropic distillation. In columns with four products, three composition controls are often better than two. Pressure compensation can improve temperature control.

TABLE 20.16 Safety Valves, Relief, and Venting

Case no.	Ref.	Type of plant	Type of column	Brief description	Some morals
1601	414		10 × 100 ft distillation column	Column relief requirement was based on failure of the steam flow controller. The relief requirement was more than halved by adding a restriction orifice in the steam supply line.	A method for reducing relief discharge requirements.
1602	414		15 × 100 ft column processing high-boiling hydrocarbons	Column relief requirement was based on the loss of cooling to the condenser and full steam on the reboiler. Adding additional (redundant) controls to shut off steam to the reboiler on high temperature or pressure effected a severalfold reduction in relief requirement.	A method for reducing relief discharge requirements.
1603	414		8-ft-diameter column	Top section of column was destroyed when cooling water valve to condenser was inadvertently shut. Steam supply was controlled by column dP; the loss of cooling caused the controller to open. Two failures thus occurred simultaneously; relief capacity was designed only for one.	Carefully examine behavior of control system when determining relief discharge requirements.
1604	414			Column top head was torn loose and tray remnants blown out as a result of inadequate relief capacity. The reason for inadequate relief capacity was identical to 1603.	As for 1603.

683

TABLE 20.16 Safety Valves, Relief, and Venting (*Continued*)

Case no.	Ref.	Type of plant	Type of column	Brief description	Some morals
1605	131			A runaway reaction occurred in a column and could only be stopped by water flooding. The flooding blew the relief valve, which was not yet connected to the quench/flare system, causing an atmospheric discharge of noxious fumes.	Column startup should not proceed unless its relief devices are properly connected to the vent header.
1606	96	Chemicals	Chloro-nitro compound distillation	The vacuum system failed. This permitted a temperature rise to the self-accelerating decomposition level. An explosion resulted.	Column protective system should cater to uncontrolled temperature rise when materials are thermally unstable.
1607	239	Refinery	Depropanizer	Column pressure reached 450 psig. The inlet line to the relief valve (set at 300 psig) was plugged by corrosion products. Both the pressure controller and high-pressure alarm came off the same transmitter and gave no indication of high pressure. Problem was only discovered when the feed pump could not maintain flow to tower.	In fouling services, prevent plugging of inlet line to relief valve (e.g, using a rupture disk). A sensing element for an alarm or trip must be separate from the sensing element used for control.
1608	239	Petrochemicals	Ethylbenzene fractionator	Column was reboiled by a fired heater. Heater fuel was controlled by a tray temperature, and there was a high-temperature trip at the process side heater outlet. When the circulation pump briefly failed, the column cooled, and the controller increased heating rate. The trip failed to function. When circulation was reestablished, an extremely high vaporization rate resulted and produced a pressure surge that dislodged trays.	Where trips are critical, install a high-reliability trip system. Test it regularly. Carefully examine the behavior of the control system when determining trip and relief requirements.

1609	239	Refinery	Isobutane–n-butane splitter	Total condenser was close to maximum capacity, and no adequate venting was available. Each time propane in the column overhead would rise, the relief valve would lift.	Either provide additional condensing capacity or adequate venting facilities when light nonkeys are likely to occur in a totally condensed system.
1610	7	Refinery	Catalytic cracking fractionator	Massive carryover of liquid from the reflux drum destroyed internals of the overhead compressor and damaged its turbine driver. The incident followed a fire at the product pump, which made it and its spare inoperable. The accumulator level rose past the compressor trip level, but the trip failed to activate.	Trips cannot always be counted on. Liquid entry to compressors should always be avoided.
1611	16a	Ethylene		Failure of level controller on column caused cold liquid to pass out of relief valve and into the carbon steel flare header. This overchilled and cracked the header. A vapor cloud formed and ignited, causing several fatalities.	Adequate liquid level monitoring is of prime importance in columns.
1612	209a	Ethylene oxide (EO)		Five separate incidents have been described in which external fires caused overheating, which in turn led to decomposition reactions and explosions in EO distillation columns or their auxiliaries. At least one involved a fatality; in some, the column was destroyed.	Pay attention to equipment layout and fireproof insulation.

References

1. Albright, M. A., "Packed Tower Distributors Tested," *Hydrocarbon Proc.* 63(9), 1984, p. 173.
2. AIChE Equipment Testing Procedures Committee, *AIChE Equipment Testing Procedure—Tray Distillation Columns,* 2d ed., 1987.
3. American Oil Company (AMOCO), *Hazard of Steam,* 2d ed., Chicago, Ill., 1984.
4. American Oil Company (AMOCO), *Hazard of Water,* 6th ed., Chicago, Ill., 1984.
5. American Oil Company (AMOCO), "Safe Ups and Downs," 3d ed., Chicago, Ill., 1984.
6. American Petroleum Institute, *Guide for Inspection of Refinery Equipment,* chap. VI, "Pressure Vessels," 4th ed., API, Washington, D.C., Dec. 1982.
7. American Petroleum Institute, "Safety Digest of Lessons Learnt," Publication 758, Sections 2–4, 1979–1981.
8. American Petroleum Institute, "Guidelines for Confined Space Work in the Petroleum Industry," API Publication 2217, June, 1984.
9. American Petroleum Institute, *Recommended Practice for the Design and Installation of Pressure-Relieving Systems in Refineries,* API RP 520, Part I—*Design,* 4th ed., 1976; Part II—*Installation,* 2d ed. (reaffirmed), 1973, API, Washington, D.C.
10. American Petroleum Institute, *Guide for Pressure-Relieving and Depressuring Systems,* API RP 521, 2d ed., Sept., 1982, API, Washington, D.C.
11. American Society of Mechanical Engineers, "Unfired Pressure Vessels," Section VIII, *ASME Boiler and Pressure Vessel Code,* ASME, New York, 1974.
12. Andersen, A. E., and J. C. Jubin, "Case Histories of the Distillation Practitioner," *Chem. Eng. Prog.* 60(10), 1964, p. 60.
13. Anderson, J. S., and J. McMillan, "Problems in the Control of Distillation Columns," *I. Chem. E. Symp. Ser. 32,* 1969, p. 6:7.
14. Anderson, R. H., G. Garrett, and M. Van Winkle, "Efficiency Comparison of Valve and Sieve Trays in Distillation Columns," *Ind. Eng. Chem. Proc. Des. Dev. 15*(1), 1976, p. 96.
14a. Anon., "Absorber Changes Solve Offshore High -H_2S Problems," *Oil & Gas J.* May 23, 1988, p. 40.
15. Anon., "Delayed Coking Is Topic of Experience Exchange," *Oil & Gas J.* July 27, 1987, p. 51.
16. Anon., "GPA Cryogenics Panel Discussion Members Explore Cryogenic Plant Start-up," *Oil & Gas J.* July 18, 1977, p. 60.
16a. Anon., "Miscellaneous Case Histories," in C. H. Vervalin (ed) *Fire Protection Manual,* vol. 2, Gulf Publishing, Houston, Texas, 1981, p. 29.
17. Anon., "New Detector Spots Coke, Foam Level in Drums", *Oil & Gas J.* July 19, 1982, p. 183.
18. Anon., "Over Pressuring of a Storage Tank", *Loss Prevention Bulletin* no. 75, June, 1987, p. 19.
19. APV DH-682, *Distillation Handbook,* 2d ed., Chicago, Ill.
20. Armer, A., "Efficient Condensate Removal," *Hydrocarbon Proc.* 67(1), 1988, p. 81.

21. Bain, J. L., and M. Van Winkle, "A Study of Entrainment, Perforated Plate Column—Air-Water System," *AIChE J.* 7(3), 363, 1961.
22. Ballard, D., "Cut Energy, Chemical and Corrosion Costs in Amine Units," *Energy Prog.* 6(2), 1986, p. 112.
23. Ballmar, R. W., "Towers Are Touchy," *API Proc., Section III—Refining, 40,* 1960, p. 279.
24. Barber, A. D., and E. F. Wijn, "Foaming in Crude Distillation Units," *I. Chem. E. Symp. Ser. 56* 1979, p. 3.1/15.
25. Bartman, R. V., "Dual Composition Control in a C_3/C_4 Splitter," *Chem. Eng. Prog.* 77(9), 1981, p. 58.
26. Baumer, J. A., "DEA Treats High-Volume Fractionation Plant Feed," *Oil & Gas J.* March 15, 1982, p. 63.
27. Beaverstock, M. C., and P Harriott, "Experimental Closed Loop Control of a Distillation Column," *Ind. Eng. Chem. Proc. Des. Dev.* 12(4), 1973, p. 401.
28. Bell, K. J., "Coping with an Improperly Vented Condenser," *Chem. Eng. Prog.* 79(7), 1983, p. 54.
29. Bennett, A. W., G. F. Hewitt, H. A. Kearsey, R. K. F. Keeys, and D. J. Pulling, "Studies of Burnout in Boiling Heat Transfer," *Trans. Inst. Chem. Eng. (London) 45,* 1967, p. T319.
30. Berman, H. L., "Fired Heaters," *Chem. Eng.,* June 19, 1978, p. 99; July 31, 1978, p. 87; Aug. 14, 1978, p. 129; and Sept. 11, 1978, p. 165.
31. Bernard, J. D. T., and R. W. H. Sargent, "The Hydrodynamic Performance of a Sieve-Plate Distillation Column," *Trans. Inst. Chem. Eng. (London) 44,* 1966, p. T314.
32. Bertram, C. G., "Sizing and Specifying Level-Controlled Condensate Pots," *Hydrocarbon Proc.* 60(8), 151, 1981.
33. Betts, B. W., and H. N. Rose, "Radioactive Scanning of Distillation Columns," Joint Symposium on Distillation, The University of Sydney/The University of NSW (Australia), 1974.
34. Biales, G. A., "How Not to Pack a Packed Column," *Chem. Eng. Prog.* 60(10), 1964, p. 71.
35. Biddulph, M. W., "How Stable Are Split-Flow Distillation Columns?," *Proc. Eng.* Feb., 1987, p. 61.
36. Bikerman, J. J., *Foams,* Springer Verlag, New York, 1973.
37. Billet, R., "Development and Progress in the Design and Performance of Valve Trays", *Br. Chem. Eng.* 14(4), 1969, p. 489.
38. Billet, R., *Distillation Engineering,* Chemical Publishing Company, New York, 1979.
39. Billet, R., *Energieeinsparung bei Thermischen Stofftrennverfahren,* Hüthig Verlag, Heidelberg, 1983.
40. Billet, R., "Packed Column Analysis and Design," *Proceedings of the 1st Glitsch Packed Column Workshop,* Glitsch, Inc., 1987.
41. Billet, R., *Verdampfung und Ihre Technischen Anwendungen,* Verlag Chemie, Weinheim, 1981.
42. Billet, R., S. Conrad, and C. M. Grubb, "Some Aspects of the Choice of Distillation Equipment," *I. Chem. E. Symp. Ser. 32,* London, 1969, p. 5:111.
43. Billet, R., and J. Máckowiak, "How to Use the Absorption Data for Design and Scale-Up of Packed Columns," EFCE Working Party on Distillation, Absorption, and Extraction Meeting in Helsinki, June, 1982.
44. Block, B., "Control of Batch Distillations," *Chem. Eng.* Jan. 16, 1967, p. 147.
45. Bluhm, W. C., "Protective Facilities for Refinery Process Units," *API Proc.,* Section III—Refining, 39, 1959, p. 507.
46. Bolles, W. L., "Estimating Valve Tray Performance," *Chem. Eng. Prog.* 72(9), 1976, p. 43.
47. Bolles, W. L., "Multipass Flow Distribution and Mass Transfer Efficiency for Distillation Plates," *AIChE J.* 22(1), 1976, p. 153.
48. Bolles, W. L., "Optimum Bubble-Cap Tray Design," *Petr. Proc.* Feb. 1956, p. 65; March, 1956, p. 82; April, 1956, p. 72; May, 1956, p. 109.
49. Bolles, W. L. (Monsanto Co.), Private communication.
50. Bolles, W. L., "The Solution of a Foam Problem," *Chem. Eng. Prog.* 63(9), 1967, p. 48.

51. Bolles, W. L., and J. R. Fair, "Performance and Design of Packed Distillation Columns," *I. Chem. E. Symp. Ser. 56,* 1979, p. 3.3/35.
52. Bojnowski, J. J., R. M. Groghan, Jr., and R. M. Hoffman, "Direct and Indirect Material Balance Control," *Chem. Eng. Prog. 72*(9), 1976, p. 54.
53. Bonilla, J. A., "Better Control for C_3/C_4 Split," *Hydrocarbon Proc. 55*(11), 1976, p. 240.
54. Bonnell, W. S., and J. A. Burns, "Startup-Shutdown Procedures for a Large Crude Oil Distillation Unit," *API Proc.,* Section III—*Refining, 40,* 1960, p. 285.
55. Bosworth, C. M., "Alcohol Rectification," *Chem. Eng. Prog. 61*(9), 1965, p. 82.
56. Bouck, D. S. (BP Oil), Private communication, 1986.
57. Bouck, D. S., and C. J. Erickson, "Gamma Scans—A Look into Troubled Towers," Paper presented at the AIChE Annual Meeting, Miami Beach, Fl., November, 1986.
58. Boyd, D. M., Jr., "Continuous-Distillation Column Control," in Schweitzer, P.A. (ed.) *Handbook of Separation Techniques for Chemical Engineers",* McGraw-Hill, New York, 1979, p. 1-179.
59. Boyd, D. M., "Fractionation Column Control," *Chem. Eng. Prog. 71*(6), 1975, p. 55.
59a. Bozenhardt, H. F., "Modern Control Tricks Solve Distillation Problems," *Hydrocarbon Proc. 67*(6), 1988, p. 47.
60. Bradford, M., and D. G. Durrett, "Avoiding Common Mistakes in Sizing Distillation Safety Valves," *Chem. Eng.* July 9, 1984, p. 78.
61. Branan, C., *The Fractionator Analysis Pocket Handbook,* Gulf Publishing, Houston, Texas, 1978.
62. Brierly, R. J. P., P. J. M. Whyman, and J. B. Erskine, "Flow Induced Vibration of Distillation and Absorption Column Trays", *I. Chem. E. Symp. Ser. 56,* 1979, p. 2.4/45.
63. Buckley, P. S., "Control of Heat-Integrated Distillation Columns", in T.F. Edgar (ed.) *Chemical Process Control 2: Proceedings of the Engineering Foundation Conference,* The American Institute of Chemical Engineers, New York, 1982, p. 347.
64. Buckley, P. S., "Controls for Sidestream Drawoff Columns," *Chem. Eng. Prog. 65*(5), 1969, p. 45.
65. Buckley, P. S., R. K. Cox, and D. L. Rollins, "Inverse Response in a Distillation Column," *Chem. Eng. Prog. 71*(6), 1975, p. 83.
66. Buckley, P. S., R. K. Cox, and D. L. Rollins, "Inverse Response in Distillation Columns," Paper presented at the AIChE Annual Meeting, Houston, Texas, March 16–20, 1975.
67. Buckley, P. S., "Material Balance Control in Distillation Columns," Paper presented at the AIChE Workshop on Industrial Process Control, Tampa, Fl. Nov. 11–13, 1974.
68. Buckley, P. S., W. L. Luyben, and J. P. Shunta, *Design of Distillation Column Control Systems,* Instrument Society of America, Research Triangle Park, NC, 1985.
69. Butterworth, D., "Film Condensation of Pure Vapor," in Armstrong, R.C. and associates (ed.), *Heat Exchanger Design Handbook,* vol. 2, rev.1, sec. 2.6.2, Hemisphere Publishing, 1985.
70. Cady, P. D., "How to Stop Slug Flow in Condenser Outlet Piping," *Hydrocarbon Proc. Pet. Ref. 42*(9), 1963, p. 192.
71. Charlton, J. S. (ed.), *Radioisotope Techniques for Problem Solving in Industrial Process Plants,* Gulf Publishing, Houston, Texas, 1986.
72. Charlton, J. S., and M. Polarski, "Radioisotope Techniques Solve CPI Problems," *Chem. Eng.,* January 24, 1983, p. 125; and February 21, 1983, p. 93.
73. Chase, J. D., "Sieve Tray Design," *Chem. Eng.* July 31, 1967, p. 105; Aug. 28, 1967, p. 139.
74. Chen, G. K., "Packed Column Internals," *Chem. Eng.* March 5, 1984, p. 40.
75. Chen, G. K., "Troubleshooting Distribution Problems in Packed Columns," *Chem. Eng. (Suppl.) (London)* Sept. 1987, p. 10.
75a. Chen, G. K., and K. T. Chuang, "Recent Developments in Distillation," *Hydrocarbon Proc.,* 68(2), 1989, p. 37.
76. Chen, G. K., T. L. Holmes, and J. H. Shieh, "Effects of Subcooled or Flashing Feed on Packed Column Performance," *I. Chem. E. Symp. Ser. 94,* 1985. p. 185.

77. Chin, T. G., "Guide to Distillation Pressure Control Methods," *Hydrocarbon Proc. 58*(10), 1979, p. 145.
78. Chisholm, D., "Fogging," in Armstrong, R.C. and associates (eds.) *Heat Exchanger Design Handbook*, vol. 2, rev.1, sec. 2.6.7, Hemisphere Publishing, 1985.
79. Chiu, C., "Apply Depressuring Analysis to Cryogenic Plant Safety," *Hydrocarbon Proc. 61*(11), 1982, p. 255.
80. Clay, H. A., T. Hutson, Jr., and L. D. Kleiss, "Effect of Load and Pressure on Performance of a Commercial Bubble-Tray Fractionating Column," *Chem. Eng. Prog. 50*(10), 1954, p. 517.
81. Collier, J. G., "Boiling within Vertical Tubes," in Armstrong, R.C., and associates (eds.), *Heat Exchanger Design Handbook*, vol. 2, rev.1, Sec. 2.7.3, Hemisphere Publishing, 1985.
82. Collins, G. K., "Horizontal Thermosiphon Reboiler Design", *Chem. Eng.*, July 19, 1976, p. 149.
83. Colwell, C. J., "Clear Liquid Height and Froth Density on Sieve Trays," *Ind. Eng. Chem. Proc. Des. 20,* 1981, p. 298.
84. Curry, R. N., *Fundamentals of Natural Gas Conditioning*, PennWell, Tulsa, Oklahoma, 1981.
85. Custer, R. S., "Case Histories of Distillation Columns," *Chem. Eng. Prog. 61*(9), 1965, p. 89.
86. Davies, J. A., "Bubble Trays—Design and Layout," *Pet. Ref. 29*(8), 1950, p. 93; *29*(9), 1950, p. 121.
87. Davies, J. A., "Trouble with Transients," *Chem. Eng. Prog. 61*(9), 1965, p. 74
88. Davies, J. A., and K. F. Gordon, "What to Consider in Your Tray Design", *Petro/Chem. Eng.* Oct., 1961, p. 230; Nov., 1961, p. 250; Dec., 1961, p. 228.
89. Deshpande, P. B., *Distillation Dynamics and Control,* Instrument Society of America, Research Triangle Park, North Carolina, 1985.
90. Detman, R. F., "How Weir Location Affects Sieve Tray Pressure Drop," *Hydrocarbon Proc. 42*(8), 1963, p. 147.
91. Dickinson, W. S., "Flooding in a Two-Section Fractionator," *Chem. Eng. Prog. 60*(10), 1964, p. 73.
92. Diehl, J. E., and C. R. Koppany, *Chem. Eng. Prog. Symp. Ser. 92*(65), 1969, p. 77.
93. Doig, I. D., "Incorporating Flexibility into the Design of Fractional Distillation Plant," *Aust. Chem. Eng.* May, 1971, p. 3.
94. Dolan, M. J., "Packed Column Internals," *Chem. Eng.* May 14, 1984, p. 5, Chen, G. K., ibid., p. 5.
95. Doukas, N., and W. L. Luyben, "Control of Sidestream Column Separating Ternary Mixtures," *Instrum. Technol. 25*(6), 1978, p. 43.
96. Doyle, W. H., "Industrial Explosions and Insurance," *Loss Prevention 3,* 1969, p. 11.
97. Doyle, W. H., "Instrument-Connected Losses in the CPI," *Instrum. Technol. 19*(10), 1972, p. 38.
98. Drew, J. W., "Distillation Column Startup," *Chem. Eng.*, Nov. 14, 1983, p. 221.
98a. Driskell, L. R., "Piping of Pressure Relieving Devices," in C.H. Vervalin (ed.) *Fire Protection Manual*, vol.1, 3d ed., Gulf Publishing, Houston, Texas 1985, p. 274.
99. Dubeau, Y., "Temperature Indicators Locate Distillation Column Liquid Level," *Chem. Eng.* Jan. 7, 1985. p. 103.
100. Durand, A. A., "Sizing Hot-Vapors Bypass Valve," *Chem. Eng.* Aug. 25, 1980, p. 111.
101. Dylag, M., and L. Maszek, *Int. Chem. Eng. 27*(2), 1987, p. 358.
102. Eagle, R. S., "Trouble-shooting with Gamma Radiation," *Chem. Eng. Prog. 60*(10), 1964, p. 69.
103. Eckert, J. S., "Design of Packed Columns," in P.A. Schweitzer (ed.) *Handbook of Separation Techniques for Chemical Engineers,* McGraw-Hill, New York 1979, p. 1-221.
104. Eckert, J. S., "Design Techniques for Sizing Packed Towers," *Chem. Eng. Prog. 57*(9), 1961, p. 54.
105. Eckert, J. S., "Problems of a Packed Column," *Chem. Eng. Prog. 61*(9), 1965, p. 89.
106. Ellerbe, R. W., "Batch Distillation," in P.A. Schweitzer (ed.) *Handbook of Separation Techniques for Chemical Engineers,* McGraw-Hill, New York, 1979, p. 1-147.

107. Ellingsen, W. R., "Diagnosing and Preventing Tray Damage in Distillation Columns," DYCORD 86, IFAC Proceedings of International Symposium on Dynamics and Control of Chemical Reactors and Distillation Columns, Bournemouth, U.K., Dec. 8–10, 1986.
108. Erskine, J. B., and W. Waddington, "Investigation into the Vibration Damage to Large Diameter Sieve Tray Absorber Towers," Paper No. 211, International Symposium on Vibration Problems in Industry, UKAEA and NPL, Keswick, England, 1973.
109. Ewach, J. (Koch Engineering Co.), Private communication, 1988.
110. Fadel, T. M., "The Safe Way to Install Restriction Orifices," *Chem.* Eng., April 13, 1987.
111. Fadel, T. M., "Selecting Packed-Column Auxiliaries," *Chem. Eng.* January 23, 1984, p. 71.
112. Fair, J. R., "How to Predict Sieve Tray Entrainment and Flooding", *Petro/Chem. Eng. 33*(10), 1961, p. 45.
113. Fair, J. R., "Vaporizer and Reboiler Design," *Chem. Eng.* July 8, 1963, p. 119; Aug. 5, 1963, p. 101.
114. Fair, J. R., "What You Need to Design Thermosiphon Reboilers," *Pet. Ref. 39*(2), 1960, p. 105.
115. Fair, J. R., and A. Klip, "Thermal Design of Horizontal Reboilers," *Chem. Eng. Prog. 79*(3), 1983, p. 86.
116. Fane, A. G., and H. Sawistowski, "Plate Efficiencies in the Foam and Spray Regimes of Sieve-Plate Distillation," *I. Chem. E. Symp. Ser. 32,* 1969, p. 1:8.
117. Fasesan, S. O., "Weeping from Distillation/Absorption Trays," *Ind. Eng. Chem. Proc. Des. Dev. 24,* 1985, p. 1073.
118. Fell, C. J. D., and R. M. Wood, "Industrial Distillation—An Old Art but a New Science," *Proc. Chem. Eng.* Oct., 1977, p. 31.
119. Finch, R. N., and M. Van Winkle, "A Statistical Correlation of the Efficiency of Perforated Trays," *Ind. Eng. Chem. Proc. Des. Dev. 3*(2), 1964, p. 106.
120. Fisch, E., "Winterizing Process Plants," *Chem. Eng.*, Aug. 20, 1984, p. 128
121. Formisano, F. A., "Method Quickly Troubleshoots Packed-Column Problems," *Chem. Eng.* Nov. 3, 1969, p. 108.
122. Fourroux, M. M., F. W. Karasek, and R. E. Wightman, "High-Speed Chromatography in Closed-Loop Fractionator Control," *ISA J. 7*(5), 1960, p. 76.
123. Frank, O., "Shortcuts for Distillation Design," *Chem. Eng.*, March 14, 1977, p. 110.
124. Frank, O., and R. D. Prickett, "Designing Vertical Thermosyphon Reboilers," *Chem. Eng.*, Sept. 3, 1973, p. 107.
125. Fromm, D., and W. Rall, "Fire at Semi-Lean Pump by Reverse Motion," *Plant/Operations Prog 6*(3), 1987, p. 162.
126. Fulham, M. J., and V. G. Hulbert, "Gamma Scanning of Large Towers," *Chem. Eng. Prog. 71*(6), 1975, p. 73.
127. Fulks, B. D., "Planning and Organizing for Less Troublesome Plant Startups," *Chem. Eng.* Sept. 6, 1982, p. 96.
128. Gale, J. A., "Costs of Distillation Internals," *I. Chem. E. Symp. Ser. 61,* 1981, p. 57.
129. Gallier, P. W., and L. C. McCune, "Simple Internal Reflux Control," *Chem. Eng. Prog. 70*(9), 1974, p. 71.
130. Galluzzo, J. F., "Installing a Reflux Splitter under Positive Pump Pressure," *Chem. Eng.*, Feb. 13, 1967, p. 180.
131. Gans, M., S. A. Kiorpes, and F. A. Fitzgerald, "Plant Startup—Step by Step," *Chem. Eng.* Oct. 3, 1983, p. 74.
132. Garvin, R. G., and E. R. Norton, "Sieve Tray Performance under GS Process Conditions," *Chem. Eng. Prog. 64*(3), 1968, p. 99.
133. Gibson, G. J., "Efficient Test Runs," *Chem. Eng.* May 11, 1987, p. 75.
134. Gilmour, C. H., "Troubleshooting Heat Exchanger Design," *Chem. Eng.* June 19, 1967, p. 221.
135. Glausser, W. E., "Foaming in a Natural Gasoline Absorber," *Chem. Eng. Prog. 60*(10), 1964, p. 67.
136. Glitsch Field Services, Inc., "Rapid Identification and Correction of Column Operating Problems," Bulletin 343, Dallas, Texas, 1984.
137. Glitsch, H. C., "Mechanical Specifications of Trays," *Pet. Ref. 39*(8), 1960, p. 91.

692

138. Glitsch, Inc., "Ballast Tray Design Manual," 3d ed., Bulletin No. 4900, Dallas, Texas, 1974.
139. Glitsch, Inc., "Glitsch Ballast Trays," Bulletin No. 159/160 (revised), Dallas, Texas, 1983.
140. Glitsch, Inc., "The Glitsch Minute Manway," Dallas, Texas.
141. Glitsch, Inc., "High Performance Distributors," *The Glitsch Column,* No. 262A.
142. Glitsch, Inc., "Tower Packings and Internals," 3d ed., Bulletin No. 217, Dallas, Texas, 1983.
143. Glitsch, Inc., "17 Critical Questions and Answers about Trays, Column Internals and Accessories," Bulletin no. 674, Dallas, Texas, 1985.
144. Glitsch, Inc., "44 Frequently-Asked Questions and Answers about Trays and Packing," Bulletin No. 681R1, Dallas, Texas, 1983.
145. Golden, S. W. (Glitsch), Private communication, August, 1987.
146. Golden, S. W., and M. J. Binkley, "Crude Tower Modification Stabilizes Operations", *Oil & Gas J.* July 30, 1984, p. 197.
147. Graf, K., "Correlations for Design, Evaluation of Packed Vacuum Towers", *Oil & Gas J.* May 20, 1985, p. 60.
148. Grover, B. S., and E. S. Holmes, "The Benfield Process for High Efficiency and Reliability in Ammonia Plant Acid Gas Removal—Four Case Studies," in *Nitrogen 1986,* The British Sulphur Corp. Ltd., Amsterdam, April 20–23, 1986. p. 101,
149. Guffey, C. G., and W. A. Heenan, "Process Control of Turboexpander Plants," *Hydrocarbon Proc. 63*(5), 1984, p. 71.
150. Haas, J. R. (UOP), Private communication, 1987.
150a. Harrison, M.E., and J.J. France, "Distillation Column Troubleshooting," *Chemical Engineering,* March, 1989, p. 116; April, 1989, p. 121; May, 1989, p. 126; and June, 1989, p. 139.
150b. Harrison, M. E., "Gamma Scan Evaluation Techniques and Applications for Distillation Column Debottlenecking," presented at the AIChE Meeting, Houston, Texas, April 2–6, 1989.
151. Haselden, G. G., "Scope for Improving Fractionation Equipment," *Chem. Engr. (London) 299/300,* 1975, p. 439.
151a. Hatfield, J. A., "High Efficiency Tower Packings and Responsive Control Schemes," Chem. Proc. Sept., 1988, p. 130
151b. Hatfield, J. A. (Merck & Company), Private communication, Oct. 1988.
152. Hausch, D. C., "How Flooding Can Affect Tower Operation," *Chem. Eng. Prog. 60*(10), 1964, p. 55.
153. Hayes, A. H., and R. M. Melaven, "Safe Ups and Downs for Refinery Units," *API Proc., Section III—Refining, 40,* 1960, p. 270.
154. Head, J., and J. Rumley, "Production Design for Floating Platforms," *Chem Eng. (London)* July, 1987, p. 17.
155. Helzner, A. E., "Operating Performance of Steam-Heated Reboilers," *Chem. Eng.* Feb. 14, 1977, p. 73.
156. Hepp, P. S., "Internal Column Reboilers—Liquid Level Measurement," *Chem. Eng. Prog. 59*(2), 1963, p. 66.
156a. Hernandez, R. J., and T. L. Hunrdeman, "Solvent Unit Cleans Synthesis Gas," *Chem. Eng.,* Feb., 1989, p. 154.
157. Hesselink, W. H., and A. Van Huuksloot, "Foaming of Amine Solutions," *I. Chem. E. Symp. Ser. 94,* 1985, p. 193.
158. Hills, P. D., "Designing Piping for Gravity Flow," *Chem. Eng.* Sept. 5, 1983, p. 111.
159. Ho, G. E., R. L. Muller, and R. G. H. Prince, "Characterisation of Two-Phase Flow Patterns in Plate Columns," *I. Chem. E. Symp. Ser. 32,* 1969, p. 2:10.
160. Hoek, P. J., "Large and Small Scale Liquid Maldistribution in a Packed Column," Ph.D. thesis, University of Delft, The Netherlands, 1983.
161. Hoek, P. J., and F. J. Zuiderweg, "The Influence of Channelling on the Mass Transfer Performance of Packed Columns," Paper presented at the AIChE Annual Meeting, Chicago, Nov., 1985.
162. Hoerner, B. K., F. G. Wiessner, and E. A. Berger, "Effect of Irregular Motion on Absorption/Distillation Processes," *Chem. Eng. Prog. 78*(11), 1982, p. 47.

163. Hofhuis, P. A. M., and F. J. Zuiderweg, "Sieve Plates: Dispersion Density and Flow Regimes," *I. Chem. E. Symp. Ser. 56,* 1979, p. 2.2/1.
164. Hollander, L. "Pressure Control of Light-Ends Fractionators," *ISA J. 4*(5), 1957, p. 185.
165. Holmes, T. L., and G. K. Chen, "Design and Selection of Spray/Mist Elimination Equipment," *Chem. Eng.,* Oct. 15, 1984, p. 82.
166. Horner, G., "How to Select Internals for Packed Columns," *Proc. Eng.* May, 1985, p. 79.
167. Horner, G., "Selecting Suitable Materials for Tower Packings and Internals", *Chem. Eng. (London)* Nov., 1984, p. 22.
168. Horner, G. V., "Tips for Installing Structured Packings and Internals", *Chem. Eng. (Suppl.) (London)* Sept. 1987, p. 8.
169. Howard, W. B., "Hazards with Flammable Mixtures," *Chem. Eng. Prog. 66*(9), 1970, p. 59.
170. Hower, T. C., Jr. (C. F. Braun, Inc.), Private communication, 1987.
171. Hsieh, C. L., and K. J. McNulty, "Weeping Performance of Sieve Trays and Valve Trays," Paper presented at the AIChE Annual Meeting, Miami Beach, Fl., Nov., 1986.
172. Huang, C. J., and J. R. Hodson, "Perforated Trays Designed This Way," *Pet. Ref. 37*(2), 1958, p. 104.
173. Hughart, C. L., and K. W. Kominek, "Designing Distillation Units for Controllability," *Instrum. Technol. 24*(5), 1977, p. 71.
174. Hughmark, G. A., "Designing Thermosiphon Reboilers", *Chem. Eng. Prog. 60*(7), 1964, p. 59; and *65*(7), 1969, p. 67.
175. Hughmark, G. A., and H. E. O'Connell, "Design of Perforated Plate Fractionating Towers," *Chem. Eng. Prog. 53*(3), 1957, p. 127.
176. Hunt, C. d'A, D. N. Hanson, and C. R. Wilke, "Capacity Factors in the Performance of Perforated Plate Columns," *AIChE J. 1*(4), 1955, p. 441
177. Interess, E., "Practical Limitations on Tray Design," *Chem. Eng.* Nov. 15, 1971, p. 167.
177a. Irhayem, A. Y. N., "Purging Prevents Condenser Corrosion," *Chem. Eng.,* August 15, 1988, P. 178.
178. Jacobs, J. K., "Reboiler Selection Simplified," *Hydrocarbon Proc. Pet. Ref. 40*(7), 1961, p. 189.
179. Jamison, R. H., "Internal Design Techniques," *Chem. Eng. Prog. 65*(3), 1969, p. 46,
180. Jarvis, H. C., "Butadiene Explosion at Texas City," *Loss Prevention 5,* 1971, p. 57; R.H. Freeman and M.P. McCready, ibid., p. 61; R.G. Keister, B.I. Pesetsky, and S.W. Clark, ibid., p. 67.
181. Johnson, D. L., and Y. Yukawa, "Vertical Thermosiphon Reboilers," *Chem. Eng. Prog. 75*(7), 1979, p. 47.
182. Jones, D. W., and J. B. Jones, "Tray Performance Evaluation," *Chem. Eng. Prog. 71*(6), 1975, p. 65.
182a. Junique, J. C., "Flush or Blow Lines Adequately," *Hydrocarbon Proc. 67*(7), 1988, p. 55.
183. Kabakov, M. I., and A. M. Rozen, "Hydrodynamic Inhomogeneities in Large-Diameter Packed Columns and Ways to Eliminate Them," *Khim. Prom.,* no. 8, 1984, p. 496. *The Soviet Chemical Industry* 16(8), 1984, p. 1059.
183a. Kaiser, V., and A. Devos, "Review of Liquid Distributor Design for Packed Columns," presented at the AIChE Meeting, Houston, Texas, April 2–6, 1989.
184. Kalbassi, M. A., M. M. Dribika, M. W. Biddulph, S. Kler, and J. T. Lavin, "Tray Efficiencies in the Absence of Stagnant Zones," *I. Chem. E. Symp. Ser. 104,* 1987, p. A511.
185. Keller, C. L., "Future Requirements for Unrestricted Entry into Confined Spaces", *Plant/Operations Prog. 6*(3), 1987, p. 142.
186. Kelley, R. E, T. W. Pickel, and G. W. Wilson, "How to Test Fractionators," *Pet. Ref. 34*(1), 1955, p. 110; and *34*(2), 1955, p. 159.
187. Kern, D. Q., *Process Heat Transfer,* McGraw-Hill, New York, 1950.
188. Kern, R., "How to Design Overhead Condensing Systems," *Chem. Eng.* Sept. 15, 1975, p. 129.

189. Kern, R., "How to Design Piping for Reboiler Systems," *Chem. Eng.* Aug. 4, 1975, p. 107.
190. Kern, R., "Layout Arrangements for Distillation Columns," *Chem. Eng.* Aug. 15, 1977, p. 153.
191. Khamdi, A. M., A. I. Skoblo, and Yu K. Molokanov, "Some Questions Concerning the Hydraulic Gradients of Plate Columns," *Khim Tekhnol Topliv Masel 8(2)*, 1963, p. 31.
192. Kister, H. Z., "Column Internals," *Chem. Eng.* May 19, 1980, p. 138; July 28, 1980, p. 79; Sept. 8, 1980, p. 119; Nov. 17, 1980, p. 283; Dec. 29, 1980, p. 55; Feb. 9, 1981, p. 107; April 6, 1981, p. 97.
193. Kister, H. Z., *Distillation Design,* to be published, McGraw-Hill, New York, 1991.
194. Kister, H. Z., "Practical Distillation Technology," Notes for continuing education seminar sponsored by *Chemical Engineering.*
195. Kister, H. Z., "Pressure Variations in Fractional Distillation," M.E. Thesis, University of NSW, Australia, 1976.
196. Kister, H. Z., "When Tower Startup Has Problems," *Hydrocarbon Proc. 58(2),* 1979, p. 89.
197. Kister, H. Z., and J. D. Doig, "Computational Analysis of the Effect of Pressure on Distillation Column Feed Capacity," *Trans. Inst. Chem. Engrs. (London) 57,* 1979, p. 43.
198. Kister, H. Z., and I. D. Doig, "Distillation Pressure Ups Thruput," *Hydrocarbon Proc. 56(7),* 1977, p. 132.
199. Kister, H. Z., and I. D. Doig, "Guidelines for the Variation of Distillation Column Feed Capacity with Pressure", *PACE* April, 1978, p. 23.
200. Kister, H. Z., and I. D. Doig "When Would Floating Pressure Strategy Save Energy?," *Chem. Eng. Prog. 77(9),* 1981, p. 55.
201. Kister, H. Z., and J. R. Haas, "Entrainment from Sieve Trays in the Froth Regime," *Ind. Eng. Chem. Res.* 27, 1988, p. 2331.
201a. Kister, H. Z. and J. R. Haas, "Predicting Entrainment Flooding on Sieve and Valve Trays," paper presented at the AIChE Meeting, Houston, Texas, April 2–6, 1989.
202. Kister, H. Z., and J. Haas, "Sieve Tray Entrainment Prediction in the Spray Regime," *I. Chem. E. Symp. Ser. 104,* 1987, p. A483.
203. Kister, H. Z., and T. C. Hower, Jr., "Unusual Case Histories of Gas Processing and Olefins Plant Columns," *Plant/Operations Prog. 6(3),* 1987, p. 151.
204. Kister, H. Z., W. V. Pinczewski, and C. J. D. Fell, "Entrainment from Sieve Trays Operating in the Spray Regime," *Ind. Eng. Chem. Proc. Des. Dev. 20(3),* 1981, p. 528.
205. Kister, H. Z., W. V. Pinczewski, and C. J. D. Fell, "The Influence of Operating Parameters on Entrainment from Sieve Trays," Paper presented in the 90th National AIChE Meeting, Houston, Texas, April, 1981.
206. Kistler, R. S., and A. E. Kassem, "Stepwise Rating of Condensers," *Chem. Eng. Prog. 77(7),* 1981, p. 55.
206a. Kitterman. L. (Glitsch, Inc.), private communication, December 1988.
206b. Kitterman, L., "Things I Have Seen," unpublished paper, April 1988.
207. Kitterman, L., "Tower Internals and Accessories," Paper presented at Congresso Brasileiro de Petroquimica, Rio de Janeiro, Nov. 8–12, 1976.
208. Kitterman, L., and M. Ross, "Tray Guides to Avoid Tower Problems," *Hydrocarbon Proc. 46(5),* 1967, p. 216.
209. Kler, S. C., R. J. P. Brierley, and M. C. G. Del Cerro, "Downcomer Performance at High Pressure, High Liquid Load Analysis of Industrial Data on Sieve and Valve Trays," *I. Chem. E. Symp. Ser. 104,* 1987, p. B391.
209a. Kletz, T. A., "Fires and Explosions of Hydrocarbon Oxidation Plants," *Plant/Operation Prog., 7(4),* 1988, p. 226.
210. Kletz, T. A., *What Went Wrong?,* 2d ed., Gulf Publishing, Houston, Texas, 1988.
211. Koch Engineering Company, Inc., "Design Manual—Flexitray, Bulletin 960-1, Wichita, Kansas, 1982.
212. Koch Engineering Co., Inc., *Packed Column Internals,* Bulletin No. K1-4, no date, and Bulletin KI-5, 1987.
213. Koch Engineering Co., "Quick Opening Manway," Bulletin QO MW-1, 1987.

214. Koch Engineering Co., Knight Division, "Knight Tower Packings," Bulletin TP-108A, no date.
215. Kolff, S. W., "Corrosion of a CO_2 Absorber Tower," Plant/Operations Prog. 5(2), 1986, p. 65.
216. Kouloheris, A. P., "Foam: Friend and Foe," *Chem. Eng.* Oct. 26, 1987, p. 88.
217. Kouri, R. J., and J. J. Sohlo, "Liquid and Gas Flow Patterns in Random and Structured Packings," *I. Chem. E. Symp. Ser. 104,* 1987, p. B193.
218. Kreis, H., and M. Raab, "Industrial Application of Sieve Trays with Hole Diameters from 1 to 25 mm with and without Downcomers," *I. Chem. E. Symp. Ser. 56,* 1979, p. 3.2/63.
219. Kunesh, J. G., "Practical Tips on Tower Packing," *Chem. Eng.* December 7, 1987, p. 101.
220. Kunesh, J. G., "Research Finding on New Tower Packing Material," Presentation to the Los Angeles Section of the AIChE, April, 1985.
221. Kunesh, J. G., L. Lahm, and T. Yanagi, "Commercial Scale Experiments That Provide Insight on Packed Tower Distributors," *Ind. Eng. Chem. Res. 26(9),* 1987, p. 1845.
222. Kunesh, J. G., L. L. Lahm, and T. Yanagi, "Controlled Maldistribution Studies on Random Packing at a Commercial Scale," *I. Chem. E. Symp. Ser. 104,* 1987, p. A233.
223. Kupferberg, A., and G. J. Jameson, "Pressure Fluctuations in a Bubbling System with Special Reference to Sieve Plates," *Trans. Inst. Chem. Eng. (London) 48,* 1970, p. T140.
224. Lahm, L., Jr., and T. Yanagi, "Liquid Distribution in a 4 Foot Diameter Packed Bed," Paper presented at the AIChE Annual Meeting, Los Angeles, Calif., Nov., 1982.
224a. Lee, A. T., and L. Kitterman, "Liquid Distributor for Packed Tower," U.S. Patent No. 4,729,857, March 8, 1988.
225. Lees, F. P., *Loss Prevention in the Process Industries,* vols. 1 and 2, Butterworths, London, 1980.
226. Leibson, I., R. E. Kelley, and L. A. Bullington, "How to Design Perforated Trays," *Pet. Ref. 36(2),* 1957, p. 127.
227. Lemieux, E. J., and L. J. Scotti, "Perforated Tray Performance," *Chem. Eng. Prog. 65(3),* 1969, p. 52.
228. Lenz, A. T., "Viscosity and Surface Tension Effects on V-Notch Weir Coefficients", *Trans. Am. Soc. Civ. Eng. 108,* 1943, p. 759.
229. Leslie, V. J., and D. Ferguson, "Radioisotope Techniques for Solving Ammonia Plant Problems," *Plant/Operations Prog. 4(3),* 1985, p. 144.
230. Leung, L. S., B. E. T. Hutton, and D. J. Nicklin, "A Second Mode of Operating Packed Columns and Wetted-Wall Columns," *Ind. Eng. Chem. Fund. 14(1),* 1975, p. 63.
231. Lieberman, N. P., "Common Crude Unit Problems, Remedies," *Oil & Gas J.* Aug. 11, 1980, p. 115.
232. Lieberman, N. P., "Design Processes for Reduced Maintenance," *Hydrocarbon Proc. 58(1),* 1979, p. 89.
233. Lieberman, N. P., "Drying Light End Towers Is Critical for Preventing Problems," *Oil & Gas J.* Feb. 16, 1981, p. 100.
234. Lieberman, N. P., "Instrumenting a Plant to Run Smoothly," *Chem. Eng.* Sept. 12, 1977, p. 140.
235. Lieberman, N. P., "Packing Expands Low-Pressure Fractionators," *Hydrocarbon Proc. 63(4),* 1984, p. 143.
236. Lieberman, N. P. (Process Improvement Engineering), Private communication, Aug., 1986.
237. Lieberman, N. P., *Process Design for Reliable Operation,* 2d ed., Gulf Publishing, Houston, Texas, 1988.
238. Lieberman, N. P., *Troubleshooting Natural Gas Processing,* PennWell Publishing, Tulsa, Oklahoma, 1987.

696

239. Lieberman, N. P., *Troubleshooting Process Operations,* 2d ed., PennWell Publishing, Tulsa, OK, 1985.
239a. Lieberman, N. P., and G. Liolios, "HF Alky Unit Operations Improved by On-Site Troubleshooting to Boost Capacity, Profit," *Oil & Gas J.* June 20, 1988, p. 66.
240. Lim, C. T., K. E. Porter, and M. J. Lockett, "The Effect of Liquid Channelling on Two-Pass Distillation Plate Efficiency," *Trans. Inst. Chem. Engs. (London) 52,* 1974, p. 193.
241. Lin, Y. N., "Wax Problems in Natural Gas Plants," *Hydrocarbon Proc. 48(2),* 1969, p. 89.
242. Lo Pinto, L. "Fog Formation in Low-Temperature Condensers," *Chem. Eng.* May 17, 1982, p. 111
243. Lockett, M. J., *Distillation Tray Fundamentals,* Cambridge University Press, Cambridge, England, 1986.
244. Lockett, M. J., "The Froth to Spray Transition on Sieve Trays," *Trans. Inst. Chem. Engs. (London), 59,* 1981, p. 26.
245. Lockett, M. J., and S. Banik, "Weeping from Sieve Trays," *Ind. Eng. Chem. Proc. Des. Dev. 25,* 1986, p. 561.
246. Lockett, M. J., and A. A. W. Gharani, "Downcomer Hydraulics at High Liquid Flow Rates," *I. Chem. E. Symp. Ser. 56,* London, 1979, p. 2.3/43.
247. Lockett, M. J., R. D. Kirkpatrick, and M. S. Uddin, "Froth Regime Point Efficiency for Gas-Film Controlled Mass Transfer on a Two-Dimensional Sieve Tray," *Trans. Inst. Chem. Eng. (London) 57,* 1979, p. 25.
248. Lockhart, F. J., "Drainage Time of Bubble-Cap Columns", *Pet. Ref. 35(11),* 1956, p. 165.
249. Lockhart, F. J., and C. W. Leggett, "New Fractionating Tray Designs," in *Advances in Petroleum Chemistry and Refining,* vol. 1, Interscience Publishers, New York, 1958, p. 277.
250. Lockwood, D. C., and W. E. Glausser, "Are Level Trays Worth Their Cost?," *Pet. Ref. 38(9)* 1959, p. 281.
251. Looney, S. K., B. C. Price, and C. A. Wilson, "Integrated Nitrogen Rejection Facility Produces Fuel and Recovers NGL's," *Energy Prog., 4(4),* 1984, p. 214.
252. Lopez-Bonillo, F., M. Nolla, and F. Castells, "Tray Efficiency: The Influence of Plate Operating Regime," *I. Chem. E. Symp. Ser. 104,* 1987, p. B461.
253. Lord, R. C., P. E. Minton, and R. P. Slusser "Design Parameters for Condensers and Reboilers," *Chem. Eng.* March 23, 1970, p. 127.
254. Lord, R. C., P. E. Minton, and R. P. Slusser, "Guide to Trouble-Free Heat Exchangers," *Chem. Eng.* June 1, 1970, p. 153.
255. Love, F. S., "Troubleshooting Distillation Problems," *Chem. Eng. Prog. 71(6),* 1975, p. 61.
256. Lowry, J. A., "Evaluate Reboiler Fouling," *Chem. Eng.* Feb. 13, 1978, p. 103.
257. Ludwig, E. E., *Applied Process Design for Chemical and Petrochemical Plants,* 2d ed., vol. 2, Gulf Publishing, Houston, 1979.
258. Lupfer, D. E., "Distillation Column Control for Utility Economy," *Proceedings of the 53rd Annual Convention, Gas Processors Association,* March 25–27, Denver, Colorado, 1974, p. 159.
259. Lupfer, D. E., and M. W. Oglesby, "Automatic Control of Distillation Columns," *Ind. Eng. Chem. 53(12),* 1961, p. 963.
260. Luyben, W. L. "Control of Columns with Side Stream Draw-Off," in J.T. Ward (ed.), *Instrumentation in the Chemical and Petrochemical Industries,* vol. 3, Plenum Press, New York, 1967.
261. Luyben, W. L., "Feedback Control of Distillation Columns by Double Differential Temperature Control," *Ind. Eng. Chem. Fund. 8(4),* 1969, p. 739.
262. Luyben, W. L., "Profile Position Control of Distillation Columns with Sharp Temperature Profiles," *AIChE J. 18,* 1972, p. 238.
263. Luyben, W. L., "10 Schemes to Control Distillation Columns with Sidestream Drawoffs," *ISA J. 13(7),* 1966, p. 37.
264. McConnell, J. A., and W. W. Smuck, "Gamma Backscatter Technique for Level and Density Detection," *Chem. Eng. Prog. 63(8),* 1967, p. 79.
265. McCune, L. C., and P. W. Gallier, "Digital Simulation: A Tool for Analysis and Design of Distillation Columns," *ISA Trans. 12(3),* 1973, p. 193.

266. McDonald, A. C., "Winterizing Plant Instrumentation," *Instrum. Technol.*, Nov., 1977, p. 45.

267. McKee, H. R. "Thermosiphon Reboilers—A Review," *Ind. Eng. Chem.* 62(12), 1970, p. 76.

268. McLaren, D. B., and J. C. Upchurch, "Guide to Trouble-Free Distillation," *Chem. Eng.* June 1, 1970, p. 139.

269. McNeill, G. A., and J. D. Sacks, "High Performance Column Control," *Chem. Eng. Prog.* 65(3), 1969, p. 33.

270. McNulty, K. J., J. P. Monat, and O. V. Hansen, "Performance of Commercial Chevron Mist Eliminators," *Chem. Eng. Prog.* 83(5), 1987, p. 48.

271. Maddox, R. N., *Process Engineer's Absorption Pocket Handbook,* Gulf Publishing, Houston, 1985.

272. Maddox, R. N. and J. H. Erbar, *Gas Conditioning and Processing,* vol. 3; *Advanced Techniques and Applications,* Campbell Petroleum Series, 1980.

273. Mahiout, S., and A. Vogelpohl, "Mass Transfer of Highly Viscous Media on Sieve Trays," *I. Chem. E. Symp. Ser. 104,* 1987, p. A495.

274. Manifould, D., "Distillation Tray Maintenance," in *1977 NPRA Refinery and Petrochemical Plant Maintenance Conference,* Petroleum Publishing co., Tulsa, Oklahoma, p. 114.

275. Manufacturing Chemists' Association Inc., *"Case Histories of Accidents in the Chemical Industry,"* Washington, D. C., vol. 1, 1962; vol. 2, 1966; vol. 3, 1970.

276. Markham, R. S., and R. W. Honse, "Carbon Dioxide Stripper Explosion," *Ammonia Plant Safety 20,* 1978, p. 131.

277. Martin, C. L., J. L. Bravo, and J. R. Fair, "Performance of Structured Packings in Distillation Service—Experimental and Modeling Results," Paper presented at the AIChE National Meeting, New Orleans, La., March 6–10, 1988.

278. Martin, H. W., "Scale-up Problems in a Solvent-Water Fractionator," *Chem. Eng. Prog.* 60(10), 1964, p. 50.

279. Mason, G. S., "Good Design Simplifies Reboiler Maintenance," *Hydrocarbon Proc.* 62(1), 1983, p. 82.

280. Mathur, J., "Performance of Steam Heat-Exchangers", *Chem. Eng.* Sept. 3, 1973, p. 101.

281. Mayfield, F. D., W. L. Church, A. C. Green, D. C. Lee, and R. W. Rasmussen, "Perforated Plate Distillation Columns," *Ind. Eng. Chem.* 44(9), 1952, p. 2238.

282. Mayinger, F. "Boiling—The Stabilizer and Destabilizer of Safe Operation," *Int. Chem. Eng.* 26(3), 1986, p. 373.

283. Meier, W., and M. Huber, "Measurements of the Number of Theoretical Plates in Packed Columns with Artificial Maldistribution," *I. Chem. E. Symp. Ser. 32,* 1969, p. 4:31.

284. Meier, W., and M. Huber, "Methode zum Messen der Maldistribution in Füllkörperkolonnen," *Chem. Ing. Tech. 39,* 1967, p. 797.

285. Meier, W., R. Hunkeler, and D. Stöcker, "Performance of New Regular Tower Packing Mellapak," *I. Chem. E. Symp. Ser. 56,* 1979, p. 3.3/1.

286. Miller, J. E., "Include Tech Service Engineers in Turnaround Inspections," *Hydrocarbon Proc.* 66(5), 1987, p. 53.

287. Moczek, J. S., R. E. Otto, and T. J. Williams, "Control of a Distillation Column for Producing High-Purity Overheads and Bottoms Streams," *Ind. Eng. Chem. Proc. Des. Dev.* 2(4), 1963, p. 288.

288. Molyneux, F., *Chemical Plant Design,* Part I, Butterworths, London, 1963.

289. Moore, F., and F. Rukovena, "Liquid and Gas Distribution in Commercial Packed Towers," Paper presented at the 36th Canadian Chemical Engineering Conference, October 5–8, 1986; same paper published in *Chemical Plants and Processing* (European edition), August, 1987, p. 11.

290. Moore, J. A., comments in Rubin, F. L. (chairman), and P. Minton "Heat Exchangers That Did Not Work," Panel Discussion, AIChE Meeting, Anaheim, Calif., June 6–10, 1984.

291. Mueller, A. C., "Condensers," in Bell, K. J. and associates (eds.), *Heat Exchanger Design Handbook,* vol. 3, rev. 1, sec. 3.4, Hemisphere Publishing, 1985.

292. Muir, L. A., and C. L. Briens, "Low Pressure Drop Gas Distributors for Packed Distillation Columns", *Can. J. Chem. Eng. 64*, 1986, p. 1027.
293. Mukerji, A., "How to Size Relief Valves," *Chem. Eng.* June 2, 1980, p. 79.
294. Mulraney, D. M. (C. F. Braun Inc.), Private communication, 1983.
295. Murray, R. M., and J. E. Wright, "Trouble-Free Startup of Distillation Columns," *Chem. Eng. Prog. 63*(12), 1967, p. 40.
296. NPRA Panel Discussion, "Refiners Respond to Distillation Queries," *Oil & Gas J.* July 28, 1980, p. 189.
297. NPRA, Q & A Session on Refining and Petrochemical Technology, 1978, p. 15.
298. NPRA, Q & A Session on Refining and Petrochemical Technology, 1981, p. 21.
299. NPRA, Q & A Session on Refining and Petrochemical Technology, 1983, p. 52.
299a. NPRA, Q & A Session on Refining and Petrochemical Technology, 1988, pp. 17, 59, 60.
300. Nisenfeld, A. E., "Reflux or Distillate—Which to Control?," *Chem. Eng.* Oct. 6, 1969, p. 169.
301. Nisenfeld, A. E., and R. C. Seemann, *Distillation Columns,* Instrument Society of America, Research Triangle Park, North Carolina, 1981.
302. The Norton Company, "Norton Ceramic Intalox Saddles," Bulletin CI-78, Akron, Ohio, 1973.
303. Norton Company, "Design Information for Packed Towers," Bulletin DC-11, Akron, Ohio, 1977.
304. The Norton Company, "Intalox High Performance Separation Systems," Bulletin IHP-1 UK, Akron, Ohio, 1987.
305. Norton Company, "Packed Tower Internals," Bulletin TA-8OR, 1974.
306. Nutter Engineering Co., Brochure N-23-R, Tulsa, Oklahoma, 1980.
307. Nutter Engineering, *Float Valve Tray Design Manual,* Tulsa, Oklahoma, 1976.
308. Nutter Engineering Co., "Package Trays for Pipe-Size Process Vessels," Brochure N-12, Tulsa, Oklahoma.
309. Oglesby, M. W., and J. W. Hobbs, "Chromatograph Analyzers for Distillation Control", *Oil & Gas J.* Jan. 10, 1966, p. 80.
310. Owen, R. G., and W. C. Lee, "Some Recent Developments in Condensation Theory," *Chem. Eng. Res. Des. 61,* 1983, p. 335.
311. Palen, J. W., "Shell-and-Tube Reboilers," in Bell, K. J., and associates (eds.) *Heat Exchanger Design Handbook,* vol. 3, rev.1, Sec. 3.6, Hemisphere Publishing, 1985.
312. Palen, J. W., C. C. Shih, and J. Taborek, "Mist Flow in Thermosiphon Reboilers," *Chem. Eng. Prog. 78*(7), 1982, p. 59.
313. Palen, J. W., C. C. Shih, A. Yarden, and J. Taborek, "Performance Limitations in a Large-Scale Thermosiphon Reboiler," *Proceedings of the 5th International Heat Transfer Conference,* Tokyo, Japan 1974, p. 204.
314. Palen, J. W., and W. M. Small, "A New Way to Design Kettle and Internal Reboilers," *Hydrocarbon Proc. 43*(11), 1964, p. 199.
315. Palluzi, R. P., "Testing for Leaks in Pilot Plants," *Chem. Eng.* Nov. 9, 1987, p. 81.
315a. Pathak, V. K., and I. S. Rattan, "Turndown Limit Sets Heater Control," *Chem. Eng.,* July 18, 1988, p. 103.
316. Patterson, F. M. "Vortexing Can Be Prevented in Process Vessels and Tanks," *Oil & Gas J.* Aug. 4, 1969, p. 118.
317. Patton, B. A., and B. L. Pritchard, Jr., "How to Specify Sieve Trays," *Pet. Ref. 39*(8), 1960, p. 95.
318. Pauley, C. R., and B. A. Perlmutter, "Texas Plant Solves Foam Problems with Modified MEA System," *Oil & Gas J.* Feb. 29, 1988, p. 67.
318a. Perry, D. "Liquid Distribution for Optimum Packing Performance," Presented at the AIChE Meeting, Houston, Texas, April 2–6, 1989.
319. Perry, R. H. (ed.), *Chemical Engineers' Handbook,* 6th ed., McGraw-Hill, New York. 1984.
320. The Pfaudler Company, "Column Packings and Design Data," Bulletin SB-14-600-2, 1983.
321. Pinczewski, W. V., and C. J. D. Fell, "New Considerations in the Design and Operation of High-Capacity Sieve Trays," *Chem. Eng. (London),* Jan. 1977, p. 45.

322. Piqueur, H., and L. Verhoeye, "Research on Valve Trays—Hydraulic Performance in the Air-Water System," *Can. J. Chem. Eng. 54* June, 1976, p. 177.
323. Pluss, R. C., and P. Bomio, "Design Aspects of Packed Columns Subjected to Wave Induced Motions," *I. Chem. E. Symp. Ser. 104*, 1987, p. A259.
324. Porter, K. E., "Liquid Flow in Packed Columns," *Trans. Instn. Chem. Eng. (London), 46*, 1968, p. T69; also, Porter, K.E., V.D. Barnett, and J.J. Templeman, ibid. p. T74; and Porter, K.E., and J.J. Templeman, ibid, p. T86.
325. Porter, K. E., K. A. O'Donnell, and A. A. Zaytoun, "Gas Maldistribution in Shallow Large Diameter Packed Beds," *I. Chem. E. Symp. Ser. 73*, 1982, p. L28.
326. Prado, M., and J. R. Fair, "A Fundamental Model for the Prediction of Sieve Tray Efficiency," *I. Chem. E. Symp. Ser. 104*, 1987, p. A529.
327. Prahl, W. H. "Pressure Drop in Packed Columns," *Chem. Eng.*, Aug. 11, 1969, p. 89; and Nov. 2, 1970, p. 109.
328. Pratt, C. F., and S. Y. Hobbs, "Quick Kill of Foams on Fractionator Trays," *Chem. Eng.* Jan. 19, 1976, p. 134,
329. Priestman, G. H., and D. J. Brown "The Mechanism of Pressure Pulsations in Sieve-Tray Columns," *Trans. Inst. Chem. Eng. (London) 59*, 1981, p. 279.
330. Priestman, G. H., and D. J. Brown, "Pressure Pulsations and Weeping at Elevated Pressures in a Small Sieve Tray Column," *I. Chem. E. Symp. Ser. 104*, 1987, p. B407.
331. Priestman, G. H., and D. J. Brown, Private communication, 1987.
332. Rademaker, O., J. E. Rijnsdorp, and A. Maarleveld, *Dynamics and Control of Continuous Distillation Units,* Elsevier, Amsterdam, 1975.
333. Raper, J. A., W. V. Pinczewski, and C. J. D. Fell, "Liquid Passage on Sieve Trays Operating in the Spray Regime," *Chem. Eng. Res. Des. 62*, 1984, p. 111.
334. Reay, D. W. (BP Oil), Private communication, 1988.
335. Reay, D. W., "Vacuum Distillation of Heavy Residues—Meeting Changing Refinery Requirements," *I. Chem. E. Symp. Ser. 73*, 1982, p. D51.
336. Resetarits, M. R., J. Agnello, M. J. Lockett, and H. L. Kirkpatrick, "Retraying Increases C_3 Splitter Column Capacity," *Oil & Gas J.*, June 6, 1988, p. 54.
337. Richert, J. P., A. J. Bagdasarian, and C.A. Shargay, "Stress Corrosion Cracking of Carbon Steel in Amine Systems," Paper 187, NACE "Corrosion 1987" Meeting, San Francisco, Calif., March 9–13, 1987, also *Oil & Gas J.*, June 5, 1989, p. 45.
338. Rose, L. M., *Distillation Design in Practice,* Elsevier, Amsterdam, 1985.
339. Ross, S. "Mechanisms of Foam Stabilization and Antifoaming Action," *Chem. Eng. Prog. 63*(9), 1967, p. 41.
340. Ross, S. and G. Nishioka, "Foaminess of Binary and Ternary Solutions," *J. Phys. Chem. 79* (15), 1975, p. 1561.
341. Ross, T. K., and B. Haqjoo, "The Effect of Redistribution and Wetting on the Performance of Packed Distillation Columns," in Sawistowski, H. (ed.) *Final Report by the ABCM/BCPMA Distillation Panel,* Chem. Ind. Assoc., London, 1964, p. 217; also, discussion of the paper.
342. Roy, P., and A. C. Mercer, "The Use of Structured Packing in a Crude Oil Atmospheric Distillation Column," *I. Chem. E. Symp. Ser. 104*, 1987, p. A103.
343. Roza, M., R. Hunkeler, O. J. Berven, and S. Ide, "Mellapak in Refineries and in the Petrochemical Industry," *I. Chem. E. Symp. Ser. 104*, 1987, p. B165.
344. Rukovena, F. (Norton), Private communication, 1987.
345. Ryans, J. L., and D. L. Roper, *Process Vacuum System Design and Operation,* McGraw-Hill, New York, 1986.
346. Sauter, J. R., and W. E. Younts III, "Tower Packings Cut Olefin-Plant Energy Needs," *Oil & Gas J.,* Sept. 1, 1986, p. 45.
347. Sawistowski, H., "Hydrodynamisches Verhalten und Stoffübergang an Siebboden im Sprudel- und Sprühbereich," *Chem. Ing. Tech. 50*(10), 1978, p. 743
348. Schreiner, H., Plate Efficiency in Distillation Columns with Demisters, *I. Chem. E. Symp. Ser. 73*, 1982, p. L23.
349. Seidel, R. O., "Experience in the Operation of Activated Hot Potassium Carbonate Acid Gas Removal Plants (U.S.)," Seminar on Raising Productivity in Fertilizer Plants, Baghdad, Iraq, March 23–25, 1978.
350. Senecal, V. E., "Fluid Distribution in Process Equipment," *Ind. Eng. Chem. 49*(6), 1957, p. 993.

351. Sengupta, M., and F. Y. Staats, "Relief Valve Load Calculations," *API Pro 57*, 1978, p. 37.
352. Severance, W. A. N., "Advances in Radiation Scanning of Distillation Columns," *Chem. Eng. Prog. 77*(9), 1981, p. 38.
353. Severance, W. A. N., "Differential Radiation Scanning Improves the Visibility of Liquid Distribution," *Chem. Eng. Prog. 81*(4), 1985, p. 48.
354. Sewell, A., "Practical Aspects of Distillation Column Design," *Chem. Eng. (London)* no. 299/30, 1975, p. 442.
355. Sewell, A. (BP Chemicals), Private communication, 1987.
356. Shah, G. C., "Guidelines Can Help Improve Distillation Operations," *Oil & Gas J.* Sept. 25, 1978, p. 102.
357. Shah, G. C., "Troubleshooting Distillation Columns," *Chem. Eng.*, July 31, 1978, p. 70.
358. Shah, G. C. "Troubleshooting Reboiler Systems," *Chem. Eng. Prog. 75*(7), 1979, p. 53.
359. Shaw, R. J., J. A. Sykes, and R. W. Ormsby, "Plant Test Manual," *Chem. Eng.*, Aug. 11, 1980, p. 126.
360. Shellene, K. R., C. V. Sternling, D. M. Church, and N. H. Snyder, "Experimental Study of a Vertical Thermosiphon Reboiler," *Chem. Eng. Prog. Symp. Ser. 64*(82), 1968, p. 102
361. Shinskey, F. G., "Controlling Distillation Processes for Fuel-Grade Alcohol," *InTech 28*(12), 1981, p. 47.
362. Shinskey, F. G., *Distillation Control for Productivity and Energy Conservation*, 2d ed., McGraw-Hill, New York, 1984.
363. Shinskey, F. G., "The Material Balance Concept in Distillation Control," *Oil & Gas J.* July 14, 1969, p. 76.
364. Shtayieh, S., C. A. Durr, J. C. McMillan, and C. Collins, "Successful Operation of a Large LPG Plant", *Oil & Gas J.* March 1, 1982, p. 79.
365. Sigales, B., "How to Design Reflux Drums," *Chem. Eng.* March 3, 1975, p. 157; and Sept. 29, 1975, p. 87.
366. Silvey, F. C., and G. J. Keller, "Performance of Three Sizes of Ceramic Raschig Rings in a 4-ft Diam Column," *I. Chem. E. Symp. Ser. 32*, 1969, p. 4:18.
367. Silvey, F. C., and G. J. Keller, "Testing on a Commercial Scale," *Chem. Eng. Prog. 62*(1), 1966, p. 68.
368. Simmons, C. V., Jr. "Avoiding Excessive Glycol Costs in Operation of Gas Dehydrators," *Oil & Gas J.* Sept. 21, 1981, p. 121.
369. Simon, H., and S. J. Thomson, "Relief System Optimization," *Loss Prevention, 6*, 1972, p. 74.
370. Smallwood, R., "Designing Plant Models for Improved Control," *Oil & Gas J.* Dec. 8, 1980, p. 73.
371. Smith, B. D., *Design of Equilibrium Stage Processes*, McGraw-Hill, New York, 1963.
372. Smith, J. V., "Improving Performance of Vertical Thermosiphon Reboilers," *Chem. Eng. Prog. 70*(7), 1974, p. 68.
373. Smith, R. F., "Curing Foam Problems in Gas Processing", *Oil & Gas J.* July 30, 1979, p. 186,
374. Smith, V. C., J. C. Upchurch, and D.W. Weiler, "Advantages of Small Hole Sieve Trays in a Water Scrubber," *Chem. Eng. Prog. 77*(9), 1981, p. 48.
375. Snow, A. I., and W. S. Dickinson, "Analysis of Tower Flooding," *Chem. Eng. Prog. 60*(10), 1964, p. 64.
376. Solari, R. B., E. Saez, I. D'Apollo, and A. Bellet, "Velocity Distribution and Liquid Flow Patterns on Industrial Sieve Trays," Paper presented at the AIChE 90th National Meeting, Houston, Texas, April 5–10, 1981.
377. Standiford, F. C., "Effect of Non-Condensibles on Condenser Design and Heat Transfer," *Chem. Eng. Prog. 75*(7), 1979, p. 59.
378. Stanton, B. D., and A. Bremer, "Controlling Composition of Column Product," *Control Eng. 9*(7), 1962, p. 104.
379. Stanton, B. D., and M. A. Sterling, "A Better Way to Control Composition," *Hydrocarbon Proc. 58*(11), 1979, p. 275.

380. Steinmeyer, D. E., "Fog Formation in Partial Condensers," *Chem. Eng. Prog.* *68*(7), 1972, p. 64.
381. Steinmeyer, D. E., and A. C. Mueller, "Why Condensers Don't Operate as They Are Supposed To," panel discussion, *Chem. Eng. Prog. 70*(7), 1974, p. 78.
382. Stichlmair, J., *Grundlagen der Dimensionierung des Gas/Flussigkeit-Kontaktapparates Bodenkolonne,* Verlag Chemie, Weinheim, 1978.
383. Stichlmair, J., and A. Stemmer, "Influence of Maldistribution on Mass Transfer in Packed Columns," *I. Chem. E. Symp. Ser. 104,* 1987, p. B213.
384. Stichlmair, J., and S. Ulbrich, "Liquid Channelling on Trays and Its Effect on Plate Efficiency," *I. Chem. E. Symp. Ser. 104,* 1987, p. A555.
385. Stickkelman, R. M., and J. A. Wesselingh, "Liquid and Gas Flow Patterns in Packed Columns," *I. Chem. E. Symp. Ser. 104,* 1987, p. B155.
386. Strigle, R. F., Jr., *Random Packings and Packed Towers,* Gulf Publishing, Houston, Texas, 1987.
387. Strigle, R. F., Jr., and M. J. Dolan, "A New Approach to Packed Distillation Column Design," Paper presented at the AIChE Annual Meeting, Anaheim, Calif., June, 1982.
388. Strigle, R. F., Jr., and K. Fukuyo, "Cut C_4 Recovery Costs," *Hydrocarbon Proc.* *65*(6), 1986, p. 47.
389. Sulzer Brothers, Ltd., "Sulzer Separation Columns for Distillation and Absorption—Packings, Columns, Plants," Winterthur, Switzerland.
390. Svensson, H. V., "Distillation Problems on a Floating Plant," *Chem. Eng. Prog.* *78*(11), 1982, p. 43.
391. Talley, D. L., "Startup of a Sour Gas Plant," *Hydrocarbon Proc. 55*(4), 1976, p. 92.
392. Tammami, B., "Avoid Partial Condenser Flooding," *Hydrocarbon Proc. 66*(8), 1987, p. 43.
393. Tammami, B., "Simplifying Reboiler Entrainment Calculations," *Oil & Gas J.* July 15, 1985, p. 134.
394. Tarbutton, A. J., Jr., "Cryogenic Demethanizer Startup," *Proceedings of the 56th Annual Convention of the Gas Processors Association,* Dallas, March 21–23, 1977, p. 95.
395. Taylor, I. "Pump Bypasses Now More Important," *Chem. Eng.,* May 11, 1987, p. 53.
396. Thomas, W. J., "Sieve-Plate and Downcomer Studies in a Frothing System," in Sawistowski, H. (ed.) *Final Report by the ABCM/BCPMA Distillation Panel,* Chemical Industry Association, London, 1964. p. 181.
397. Thomas, W. J., and M. Campbell, "Hydraulic Studies in a Sieve Plate Downcomer System," *Trans. Inst. Chem. Eng. (London) 45,* 1967, p. T53.
398. Thomas, W. J., and A. N. Shah, "Downcomer Studies in a Frothing System," *Trans. Inst. Chem. Eng. (London) 42,* 1964, p. T71.
399. Thrift, G. C., "How to Specify Valve Trays," *Pet. Ref. 39*(8), 1960, p. 93.
400. Thurston, C. W., "Computer-Aided Design of Distillation Column Controls", *Hydrocarbon Proc. 60*(7), 1981, p. 125; and *60*(8), 1981, p. 135.
401. Todd, W. G., and M. Van Winkle, "Fractionation Efficiency—Eighteen Inch Valve Tray Column," *Ind. Eng. Chem. Proc. Des. Dev.11*(4), 1972, p. 578.
402. Tolliver, T. L., and L. C. McCune, "Distillation Control Design Based on Steady State Simulation," *ISA Trans. 17*(3), 1978, p. 8.
403. Tolliver, T. L., and L. C. McCune, "Finding the Optimum Temperature Control Trays for Distillation Columns," *InTech 27*(9), 1980, p. 75.
404. Treybal, R. E., *Mass Transfer Operations,* 3d ed., McGraw-Hill, New York, 1980.
405. Trowbridge, T. D., "Permit Space Standards Proposed by OSHA," *Chem. Eng. Prog. 83*(6), 1987, p. 68.
406. Tyler, C. M., "Process Analyzers for Control," *Chem. Eng. Prog. 58*(9), 1962, p. 51.
407. Van der Meer, D., "Foam Stabilisation in Small and Large Bubble Columns," VDI Berichte Nr. 182, 1972, p. 99.
408. Van der Meer, D., F. J. Zuiderweg, and H. J. Scheffer, "Foam Suppression in Extract Purification and Recovery Trains," *Proceedings of the International Solvent Extraction Conference* (ISEC 71), Soc. Chem. Ind. London, 1971, p. 350.

409. Van Winkle, M., *Distillation,* McGraw-Hill, New York, 1967.
410. Verduijn, W. D., "Corrosion of a CO_2-Absorber Tower Wall," *Plant/Operations Prog. 2*(3), 1983, p. 153.
411. Vermilion, W. L., "Precise Control of Alky-Unit Deisobutanizers," *Oil & Gas J.* Aug. 21, 1961, p. 98.
412. Voss, C. H., P. F. Baughn, and J. Pennington, "Stop Tray Leaks: Reduce Costs," *Hydrocarbon Proc. 60*(9), 1981, p. 131.
413. Walas, S. M., "Rules of Thumb for Selecting and Designing Equipment," *Chem. Eng.,* March 16, 1987, p. 75.
413a. Waliullah, S., "Do-It-Yourself Vortex Breakers," *Chem. Eng.,* May 9, 1988, p. 108.
414. Walker, J. J., "Sizing Relief Areas for Distillation Columns," *Chem. Eng. Prog. 66*(9), 1970, p. 38.
415. Wallsgrove, C. S., and J. C. Butler, "Process Plant Start-up," Continuing Education Seminar, The Center for Professional Advancement, East Brunswick, N.J.
416. Wareing, T. H., "Entry into Confined Spaces," in W. Handley (ed.) *Industrial Safety Handbook,* McGraw-Hill, London, 1969, p. 61.
417. Watkins, R. N., "Sizing Separators and Accumulators," *Hydrocarbon Proc. 46*(11), 1967, p. 253.
418. Webber, W. O., "Control by Temperature Difference?," *Pet. Ref. 38*(5), 1959, p. 187.
419. Weiss, S., and J. Langer, "Mass Transfer on Valve Trays with Modifications of the Structure of Dispersions," *Inst. Chem. Emg. Symp. Ser.* no. 56, 1979, p. 2.3/1.
420. Wheeler, D. E., "Design Criteria for Chimney Trays," *Hydrocarbon Proc. 47*(7), 1968, p. 119.
421. Wheeler, G. W., "The Don't Do's for Gas Plant Design and Construction," *Proceedings of the 63d GPA Annual Convention,* New Orleans, La., March 19–21, 1984, p. 23.
422. Wherry, T. C., and D. E. Berger "What's New in Fractionator Control," *Pet. Ref. 37*(5), 1958, p. 219.
423. Whistler, A. M., "Locate Condensers at Ground Level," *Pet. Ref. 33*(3), 1954, p. 173.
424. White, R. L., "On-Line Troubleshooting of Chemical Plants," *Chem. Eng. Prog. 83*(5) 1987, p. 33.
425. Wijn, E. F., "Pulsation of the Two-Phase Layer on Trays," *I. Chem. E. Symp. Ser. 73,* 1982, p. D79.
426. Wild, N. H., "Noncondensable Gas Eliminates Hammering in Heat Exchanger," *Chem. Eng.* April 21, 1969, p. 132; also Taborek, J., and N.H. Wild, discussion, *Chem. Eng.* Oct. 6, 1969, p. 7.
427. Wilson, K. B., "Nitrogen Use in EOR Requires Attention to Potential Hazards," *Oil & Gas J.* Oct. 18, 1982, p. 105.
428. Windebank, C. S. (chairman), Discussion of Papers Presented in Session 4, *I. Chem. E. Symp. Ser. 32,* 1969, p. 4:64,
429. Yanagi, T., and M. Sakata, "Performance of a Commercial Scale 14% Hole Area Sieve Tray," *Ind. Eng. Chem. Proc. Des. Dev. 21,* 1982, p. 712.
429a. Yarborough, L., L. E. Petty, and R. H. Wilson, "Using Performance Data to Improve Plant Operations," *Proc. 59th Annual Convention of the Gas Processors Associations,* Houston, March 17–19, 1980. p. 86.
429b. Yeoman, N., and the FRI Design Practices Committee, "FRI Standard Tray Data Sheet," presented at the AIChE Meeting, Houston, Texas, April 2–6, 1989.
430. Yilmaz, S. B., "Horizontal Shellside Thermosiphon Reboilers," *Chem. Eng. Prog. 83*(11), 1987, p. 64.
431. York, O. H., "Performance of Wire-Mesh Demisters," *Chem. Eng. Prog. 50*(8), 1954, p. 421.
432. York, O. H., and E. W. Poppele, "Two Stage Mist Eliminators for Sulfuric Acid Plants," *Chem. Eng. Prog. 66*(11), 1970, p. 67.
433. York, O. H., and E. W. Poppele, "Wire Mesh Mist Eliminators," *Chem. Eng. Prog. 59*(6), 1963, p. 45.
434. Yoshinaga, K., Y. Izawa, and T. Kotani, "Effects of Barge Motion on Absorption

Column," Paper presented in the 90th National Meeting of the AIChE, Houston, Texas, April 8, 1981.

435. Yuan, H. C., and L. Spiegel, "Theoretical and Experimental Investigation of the Influence of Maldistribution on the Number of Theoretical Stages in Packed Columns at Partial Reflux," *Proc. 2nd World Congress of Chemical Engineering,* Montreal, Canada, Oct., 1981, p. 274.

436. Zanetti, R., "Boosting Tower Performance by More than a Trickle," *Chem. Eng.* May 27, 1985, p. 22.

437. Zenz, F. A., "Calculate Capacities of Perforated Plates," *Pet. Ref. 33*(2), 1954, p. 99.

438. Zenz, F. A., "Design of Gas Absorption Towers," in P. A. Schweitzer (ed.) *Handbook of Separation Techniques for Chemical Engineers,* McGraw-Hill, New York, 1979, p. 3–49.

439. Zenz, F. A., "Designing Gas Absorption Towers," *Chem. Eng.* Nov. 13, 1972, p. 120.

440. Zilka, M. I., "The Chemical Engineer as a Super Sleuth," *Chem. Eng.* May 17, 1982, p. 121.

441. Zuiderweg, F. J., "Distillation—Science and Business," *Chem. Eng. (London)* 297, 1973, p. 404.

442. Zuiderweg, F. J., and P. J. Hoek, "The Effect of Small Scale Liquid Maldistribution on the Separating Efficiency of Random Packings," *I. Chem. E. Symp. Ser. 104,* 1987, p. B247.

443. Zuiderweg, F. J., P. J. Hoek, and L. Lahm, Jr., "The Effect of Liquid Distribution and Redistribution on the Separating Efficiency of Packed Columns," *I. Chem. E. Symp. Ser. 104,* 1987, p. A217.

444. Zuiderweg, F. J., H. Verburg, and F. A. H. Gilissen, "Comparison of Fractionating Devices," in P.A. Rettenburg (ed), *Proceedings of the International Symposium on Distillation,* Inst. Chem. Eng., London, 1960, p. 201.

Useful Conversions

To convert from	To	Multiply by
	Metric	
Atmospheres	Kilopascals	101.325
Barrels (oil)	Cubic meters	0.15899
Bars	Kilopascals	100.000
Btu	Joules	1055.1
Btu per hour	Watts	0.29307
Btu per pound	Joules per kilogram	2326
Btu per square foot per hour	Joules per square meter per second	3.1546
Centipoises	Newton second per square meter	0.001
Cubic feet	Cubic meters	0.028317
Degrees Fahrenheit	Degrees Celsius	Subtract 32, then divide by 1.8
Dynes per centimeter	Newton per meter	0.001
Feet	Meters	0.3048
Feet per second	Meters per second	0.3048
F-factor (English units, $ft/s\sqrt{lb/ft^3}$)	F-factor (metric units, $m/s\sqrt{kg/m^3}$)	1.2199
Gallons (US)	Cubic meter	0.003785
Gallons (US) per minute	Cubic meter per second	0.00006308
Gallons (US) per minute per inch	Cubic meter per meter per second	0.002484
Horsepower	Kilowatts	0.7457
Inches	Millimeters	25.4
Inches of water (at 60°F)	Millimeters of mercury at 0°C	0.5358
	Kilopascals	0.24884
Pounds	Kilograms	0.45359

To convert from	To	Multiply by
	Metric	
Pounds force	Newton	4.4482
Pounds per cubic feet	Kilograms per cubic meter	16.018
Pounds per square inch	Kilopascals	6.8948
Square feet	Square meters	0.09290
	English and nonstandard	
Bars	Pounds per square inch	14.504
Barrels (oil)	Gallons (US)	42
Btu	Kilocalories	0.252
Cubic feet	Gallons (US)	7.481
Cubic feet per second	Gallons (US) per minute	448.8
Feet	Inches	12
Feet per second	Gallons per minute per square foot (gpm/ft^2)	448.8
Gallons (US)	Gallons (Imperial)	0.8327
Inches of water (60°F)	Pound per square inch	27.706
(ρV^2)*	Velocity head	
	-inches of water	0.002990
	-pounds per square inch	0.0001079

*ρ = Density, lb/ft^3; V = Velocity, ft/s.

Revised Equations for Constructing Startup Stability Diagrams

Equations for constructing startup stability diagrams were presented in Kister, H.Z., *Hydrocarbon Proc.* Feb., 1979, p. 89. Below are revisions and updates to the set of equations in the above reference. The equation number and the symbols match those in the original reference.

1. Revision to Eq. (3)

$$P_T = K_1 + \frac{K_2}{\rho_v \rho_L} \left(\frac{V_T}{3600 A_h} \right)^2 \tag{3}$$

2. Revision to Eq. (4)

$$D_e = \frac{4(\text{Cross-sectional area})}{\text{Wetted perimeter}} \tag{4}$$

The cross-sectional area of the downcomer entrance, A_e, in square feet, is given by

$$A_e = \frac{l_D h_{\text{cl}}}{144} \tag{4a}$$

The wetted perimeter, P_e, in feet, at the downcomer entrance is given by

$$P_e = \frac{2(l_D + h_{\text{cl}})}{12} \tag{4b}$$

The equivalent diameter is obtained from Eqs. (4), (4a), and (4b)

$$D_{eq} = \frac{4 l_D h_{\text{cl}}/144}{2(l_D + h_{\text{cl}})/12} = \frac{l_D h_{\text{cl}}}{6(l_D + h_{\text{cl}})} \tag{4c}$$

When column diameter is greater than 4 ft, $l_D \gg h_{cl}$, and therefore

$$D_{eq} \approx \frac{h_{cl}}{6} \tag{4d}$$

Which gives

$$A_E = \frac{\pi}{4}\left(\frac{h_{cl}}{6}\right)^2 = \frac{\pi h_{cl}^2}{144} \tag{4e}$$

3. Revision to Eqs. (10), (11), and (12)

$$(V_D)_{cr} = 3600a_D F_1 F_2 (\rho_v \sigma)^{0.5} \tag{10}$$

$$F_1 = (960\,D_D/\sigma)^{0.4} \qquad \text{if } 960\,D_D/\sigma < 1 \tag{11}$$

$$\text{or } F_1 = 1 \qquad \text{if } 960 D_D/\sigma < 1 \tag{12}$$

For most columns greater than 4 ft in diameter, $960 D_D/\sigma > 1$, and Eq. (10) simplifies to

$$(V_D)_{cr} = 3600 a_D F_2 (\rho_v \sigma)^{0.5} \tag{10a}$$

Index

Boldface numbers indicate pages where major discussion takes place

About the Author

Henry Z. Kister is a recognized specialist with a strong background in all phases of distillation, including operation, startup, troubleshooting, design, control, research, and teaching. Currently leading distillation and absorption work at CF Braun, Inc., he is also their representative to the Fractionation Research Inc. (FRI) Technical Advisory Committee and serves on FRI's Technical and Design Practices Committees. Previously, he spent several years as a startup superintendent at ICI Australia. Henry Kister received his B.S. and M.S. degrees from the University of NSW (Australia). He is author of over 30 published technical papers and two patents.

About the Author